THE ORGANIZATION AND EXPRESSION OF THE MITOCHONDRIAL GENOME

DEVELOPMENTS IN GENETICS
Volume 2

Symbol design on cover by Gio Pomodoro

THE ORGANIZATION AND EXPRESSION OF THE MITOCHONDRIAL GENOME

Proceedings of the 12th International Bari Conference on the Organization and Expression of the Mitochondrial Genome held in Martina Franca, Italy, 23-28 June, 1980

Editors
A.M. KROON
and
C. SACCONE

1980

ELSEVIER/NORTH-HOLLAND BIOMEDICAL PRESS
AMSTERDAM · NEW YORK · OXFORD

ISBN for this volume: 0-444-80276-2
ISBN for the series: 0-444-80160-X

Published by:
Elsevier/North-Holland Biomedical Press
335 Jan van Galenstraat, P.O. Box 211
Amsterdam, The Netherlands

Sole distributors for the USA and Canada:
Elsevier North Holland Inc.
52 Vanderbilt Avenue
New York, N.Y. 10017

Library of Congress Cataloging in Publication Data

International Bari Conference on the Organization and Expression of the Mitochondrial Genome, 12th, Martina Franca, Italy, 1980.
The organization and expression of the mitochondrial genome.

(Developments in genetics ; v. 2)
Bibliography: p.
Includes indexes.
1. Mitochondria--Congresses. 2. Extrachromosomal DNA--Congresses. I. Kroon, A. M. II. Saccone, C. III. Title. IV. Series.
QH603.M5I54 1980 574.87'342 80-20398
ISBN 0-444-80276-2 (Elsevier North Holland)

Printed in The Netherlands

INTRODUCTION

E. QUAGLIARIELLO
Istituto di Chimica Biologica, Università di Bari, Italy

It is not only a great honour, but also a great pleasure for me to introduce this book, holding the proceedings of the International Conference on mitochondrial biogenesis, held in Martina Franca from June 23 to 28, 1980. It was the fifteenth in the series of yearly meetings on aspects of mitochondrial metabolism, function and biogenesis held in Italy and the twelfth held in Bari or its surroundings. It was the fourth conference within this series concentrating on aspects of mitochondrial biogenesis. For younger scientists this probably does not mean too much but the eldest ones, some of them present at the first Bari Conference in 1965, will certainly be able to share my sentiments. In 1965, a Symposium on the Regulation of Metabolic Processes in Mitochondria was organized in Bari by E.C. Slater, J. Tager, S. Papa and myself and the proceedings were published by Elsevier, BBA Library, volume 7. At that time the study of the mitochondrion was in its infancy, and mitochondrial biogenesis considered a curiosity. Only two years later, however, the same organizers thought that it would be very interesting to hold a Round Table Discussion dedicated solely to the problem of the origin of mitochondria. In only two years the progress made in this field was so large as to justify this choice. A book on the "Biochemical Aspects of the Biogenesis of Mitochondria" was produced, containing not only the papers but also the discussions typed simultaneously. After 1967 it became quite clear that mitochondrial biogenesis represented an independent field and in some laboratories, such as my laboratory in Bari and that in Amsterdam, or Münich, just to quote some of them, the groups working on mitochondrial biogenesis split off from those concentrating on other metabolic or biophysical aspects of mitochondrial function.

The "Mitochondrial Biogenesis" after its baptism rapidly began to reach old age. At that time we felt it was necessary to put in appropriate hands the organization of the future Bari Conferences on mitochondrial biogenesis.

In 1973 A.M. Kroon and C. Saccone organized the International Conference on "The Biogenesis of Mitochondria" in Rosa Marina, and in 1976 that on "The genetic function of mitochondrial DNA" in Riva dei Tessali. The meeting in Martina Franca was their third collaborative effort.

As an observer from outside, but still near to some aspects of this important field, such as that of the transport of proteins across the mitochondrial membranes, I wish to make some short and probably expected comments. Few fields like that of mitochondrial biogenesis have been born by putting together the results of so many disciplines such as Genetics, Biochemistry, Molecular Biology, Electron Microscopy, etc. The boundaries between them, at the beginning well-defined, are now practically erased.

Almost everybody speaks the same language and follows the same philosophy. This, in my opinion, is because mitochondrial biogenesis has, as few other fields, followed the most important progress of modern biology so closely. It has become at once one of the most modern fields of modern biology. The echos of the new discoveries about the genetic organization of mitochondria reach even the ears of the non-experts, giving rise to surprise and admiration. For this reason too I was most happy to witness the opening of the conference. Let me be proud to have been one of the first who has believed in and has supported this field.

It is not by chance that the organizers and the publisher, Elsevier-North Holland Biomedical Press, have chosen again the sign of Giò Pomodoro as the symbol of this conference. A few words of explanation. The symbol represents the mitochondrial genome. The colours, white, red and black are that of the alchemy and of the hermetic iconography. The shape is that of Castel del Monte, the castle of Frederic II which is in Puglia. How opportune that an artist has linked with this symbol our region "Puglia" with the mitochondrial genome!

The Conference of which these are the proceedings was sponsored mainly by the Consiglio Nazionale delle Ricerche (CNR) and also by the University of Bari.

CONTENTS

© 1980 Elsevier/North-Holland Biomedical Press
The Organization and Expression of the Mitochondrial Genome
A.M. Kroon and C. Saccone, editors

THE ORGANIZATION AND EXPRESSION OF THE MITOCHONDRIAL GENOME: INTRODUCTORY REMARKS AND SCOPE

C. SACCONE[1] and A.M. KROON[2]
[1]Istituto di Chimica Biologica, University of Bari and
[2]Laboratory of Physiological Chemistry, University of Groningen, The Netherlands

In the field of mitochondrial biogenesis the boundaries between different disciplines have been gradually dissolved in the course of time. The different topics are tending to fuse together. The present book contains the proceedings of a conference which had as the title: "The Organization and Expression of the Mitochondrial Genome". It was difficult to group the various papers into different sessions. Quite a lot of papers could find an appropriate position in more than one; sometimes in all the sessions. This clearly reflects, in our opinion, how the borders between the various disciplines have disappeared due to the emphasis put on the molecular aspects of biological processes. The proceedings[1] of the Bari Conference, organized in 1976 and entitled "The genetic function of mitochondrial DNA" were dominated by the various aspects of the "physical maps" of the mitochondrial genomes. The good correspondence between genetic and physical maps was one of the best achievements, which certainly encouraged people having different backgrounds to collaborate in common projects. In this book the main leit-motiv will be the base sequence analysis. The majority of problems has been tackled by determining directly the nucleotide sequences which, in turn, revealed a number of unexpected phenomena not only important for the specific field of mitochondrial biogenesis but for Molecular Biology as a whole. The non-universality of the genetic code is one of the best examples. However, sequence analysis is not solving all problems. We still believe that in the field of mitochondrial biogenesis it is important to pursue the study of many other aspects such as the purification of important mitochondrial enzymes, e.g. the DNA and RNA polymerases, or the purification of factors involved in protein synthesis. Assembly of the mitochondrion as well as the nucleocytoplasmic mitochondrial interrelationships are, of course, among the other exciting aspects to consider. Furthermore also the role of the mitochondrial genetic system during development and differentiation requires careful experimental approach. In agreement with these considerations the book is divided into four parts: Mitochondrial gene organization; Mitochondrial gene characterization; Mitochondrial replication, transcription and translation and Developmental and regulatory aspects of mitochondrial biogenesis.

Mitochondrial Gene Organization and Characterization.

The mitochondrial genomes of various organisms, including the kinetoplast DNA of trypanosomes, have been intensively studied during the last five years. The mitochondrial genome of yeast remains one of the favoured model systems in which the genetic approach is still of great importance. In yeast the order of genes is now well known and more and more light is shed on the structure of various genes. It is becoming clear that some genes may have a mosaic structure and some not. An intervening sequence in the large ribosomal gene may be either present or absent. The gene model including AT spacer regions followed by GC clusters, hypothesized several years ago[2], appears to be completely valid. The recent discovery[3] that yeast mosaic genes can code for a maturase protein, an enzyme supposed to be involved in the processing (maturation) of mitochondrial transcripts, is a most exciting observation, which can throw light upon the mechanism of splicing. Detailed information on this point can be found in several contributions in this book. Petite mutants continue to give useful information and still represent a very potent tool in the study of yeast mitochondriogenesis. The petite mitochondrial (mt) DNAs are now used as probes in other systems as well.

The genetic organization of the moulds *Neurospora crassa* and *Aspergillus niger* is also well established as far as the ribosomal and transfer RNA genes are concerned. Also here intervening sequences have been observed. Whether some of the genes for the mitochondrially coded polypeptides are mosaic also in these organisms remains to be investigated. Also the order of genes is not yet known for these organisms.

It has been reported that for human mitochondrial DNA the base sequence is known for 98%[4]. It may be expected that this will be of great help to solve many of the open questions still existing. Especially the establishment of the genetic function of the potential protein-products of so far unidentified reading frames, seems interesting. By the availability of the complete sequence the upper limit of the genetic function of mtDNA is set, the lower or actual limit has to await the characterization of all transcription and translation products. Comparison of the sequences and the latter products from different animals, from plants and from lower eukaryotes may be a firm basis for evolutionary studies. It may perhaps help answering the question whether mitochondria have developed within a cell or that they have evolved from endosymbionts.

Mitochondrial replication, transcription and translation.

The study of mitochondrial transcription is now actively undertaken in many laboratories and for various organisms. Many questions are still open. The role of the two strands of mtDNA during transcription, linked to the presence of double stranded mtRNA often reported in the literature, certainly needs elucidation. It also seems worthwhile to go back to the mechanism of action of the mtRNA polymerase, the characterization of which dates back to the 1973 Bari Conference held in Rosa Marina [5].

Agreement has been reached about the identification of mtDNA polymerase as DNA polymerase γ , at least for the vertebrate cells. The knowledge of these mitochondrial enzymes and of functionally related enzymes such as mitochondrial topisomerases or gyrases is far from complete.

Regarding mitochondrial translation, the possibility that not only a number of mitochondrial proteins synthesized in the cytoplasm and then transported, but also some mitochondrial proteins inside the organelles can be synthesized as precursors, is intriguing. This raises, of course, the problem of the role and fate of the precursors inside the mitochondrion and also in a more general context.

The role and the nature of cytoplasmic factors responsible for the activation of mitochondrial protein synthesis is also of great interest but grossly unknown.

Developmental and Regulatory aspects of Mitochondrial Biogenesis.

Several interesting problems will be treated in this respect in the fourth part of the book. The role of protein transport in the assembly and organization of mitochondria is a very fashionable topic in which great progress has been made recently. The apparent disagreements that existed a year ago seem to be solved. Transport of proteins through the mitochondrial membranes is not one standard process accompanied by proteolytic activity, but may follow various strategies.

The study of further control mechanisms in the interplay of various cell compartments during the process of mitochondrial biogenesis, is now approached from various angles. It may be expected that the combined results obtained with many different organisms, each having its special advantage for tackling part of the problem, may lead to insight into what is now still one of the most fascinating secrets of cell biology.

REFERENCES

1. Saccone, C. and Kroon, A.M. (Eds), 1976. The Genetic Function of Mitochondrial DNA, North-Holland Publ. Co., Amsterdam, Oxford, New York.
2. Bernardi, G. and Timasheff, S.N. (1970) J. Mol. Biol., 48, 43-52
3. Claisse, M.L., Slonimski, P.P., Johnson J. and Mahler H.R. (1980) Molec. Gen. Genet., 77, 375-387
4. Barrell, B.G., Anderson, S., Bankier, A.T., de Bruijn, M.H.L., Chen, E., Coulson, A.R., Drouin, J., Eperon, I.C., Nierlich, D.P., Roe, B., Sanger, F., Schreier P.H., Smith, A.J.H., Staden, R. and Young, J.G. (1980) Hoppe Seyler's Z. Physiol. Chem., 361, 493
5. Kroon, A.M. and Saccone, C. (Eds),1974. The Biogenesis of Mitochondria, Academic Press Inc., New York.

MITOCHONDRIAL GENE ORGANIZATION

The Organization and Expression of the Mitochondrial Genome
A.M. Kroon and C. Saccone, editors

THE KINETOPLAST DNA OF Trypanosoma brucei: STRUCTURE, EVOLUTION, TRANSCRIPTION, MUTANTS

P.BORST, J.H.J.HOEIJMAKERS*, A.C.C.FRASCH**, A.SNIJDERS, J.W.G.JANSSEN AND F.FASE-FOWLER

Section for Medical Enzymology and Molecular Biology, Laboratory of Biochemistry, University of Amsterdam, Jan Swammerdam Institute, P.O.Box 60.000, 1005 GA Amsterdam (The Netherlands)

Introduction

The position of kinetoplast DNA (kDNA) afficionados in Conferences such as these is usually a marginal one. True, the real workers on mitochondrial biogenesis are willing to admit that kDNA is a mtDNA and even that it was the first mtDNA to be discovered[1]. But there doubts set in. Is this immodest network of more than 10^4 catenated circles (see Fig. 1) really the mtDNA of trypanosomes? Is there no other DNA in the single, large mitochondrion that would more resemble the mtDNA found in all other mitochondria? Why is mitochondrial biosynthesis in these unicellular flagellates not inhibited by chloramphenicol[2] as it should in decent mitochondria? In short, is it appropriate to include people who work on kDNA at all in a serious Conference on mitochondrial biogenesis?

In view of such suspicions we are doubly honoured that the first paper in this Symposium is devoted to kDNA and to show our appreciation we shall try and demonstrate that kDNA is indeed a bona fide mtDNA, that there is no reason to continue the search for other mtDNAs than kDNA in trypanosomes and that there are even many striking parallels between mitochondrial biogenesis in trypanosomes and in that prima donna of biogenetic studies, the yeast Saccharomyces cerevisiae.

---

*Present address: Laboratory of Medical Microbiology, University of Amsterdam, Mauritskade 57, 1092 AD Amsterdam (The Netherlands).
**Permanent address: Instituto de Química Biológica, Facultad de Medicina, Universidad de Buenos Aires, Paraguay 2155 (5° piso), Buenos Aires (Argentina).

Abbreviations: kDNA, kinetoplast DNA; kb, kilo base pair(s); DBM-paper, diazobenzyloxymethyl-cellulose paper; mRNA, messenger RNA; rRNA, ribosomal RNA.

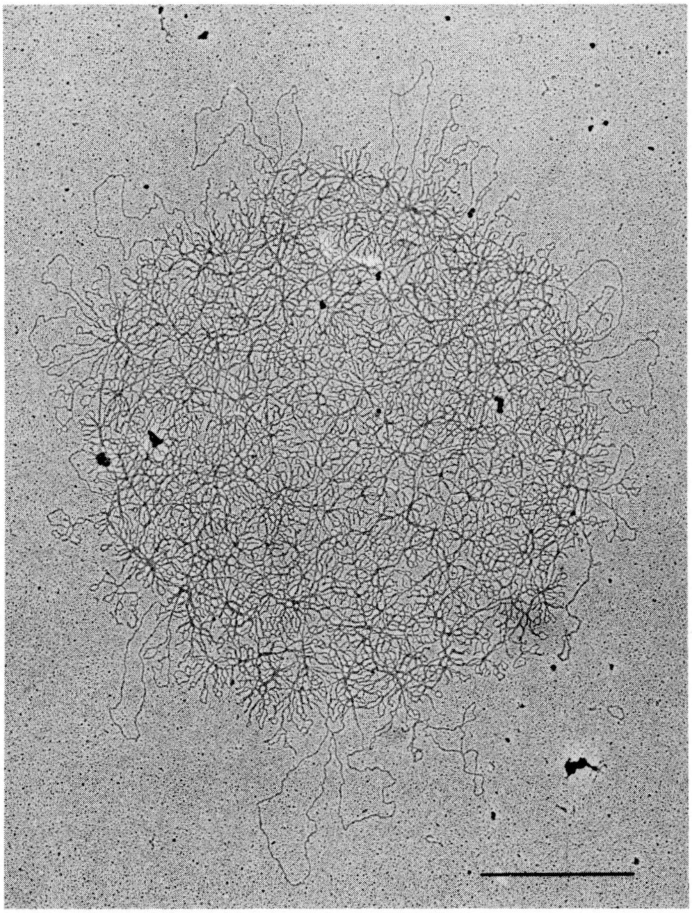

Fig. 1. Electron micrograph of a kDNA network from T. brucei, strain 31, spread by the micro-diffusion technique of Lang and Mitani (ref. 23). The network consists of 1 kb mini-circles and contains 20 kb maxi-circles, which extend from the edge. The bar is 1 µm. From ref. 3.

The maxi-circles of T. brucei kDNA

For the study of kDNA Trypanosoma brucei has a number of advantages[3,4]: the networks are among the smallest of all Kinetoplastidae and easy to isolate intact; the maxi-circle of 6 µm (about 20 kb) is the smallest known thusfar and there are many different stocks and well-defined related species available for the study of sequence evolution. In the context of this Conference, however, the most interesting property is the ability of T. brucei to completely repress mitochondrial biogenesis. In the bloodstream of the vertebrate host it lives by glycolysis alone, like anaerobic yeast, and only in the tse-tse fly (or in culture) a fully-developed mitochondrion is essential for survival. Hence, mutants containing defective kDNA can be maintained by syringe passage in the lab and they even occur in nature (e.g. Trypanosoma evansi and Trypanosoma equiperdum) where transmission by direct blood-blood contacts or flies replaces the syringe. Such mutants are fully analogous to the cytoplasmic petite mutants in yeast[5].

The maxi-circles that can be isolated from T. brucei kDNA networks have an unusually low density in CsCl (1.682 g/cm³), which is even lower than the 1.684 g/cm³ of yeast mtDNA[6]. Another unusual property of these maxi-circles is shown in the linearized map of restriction enzyme cleavage sites presented in Fig. 2. There is a 5 kb stretch in this 20.5 kb molecule, between the HhaI site L4 and the encircled TaqI site, which is virtually devoid of restriction enzyme cleavage sites[7]. Moreover, this 5 kb segment varies over 1.5 kb in size in the 9 strains analysed,

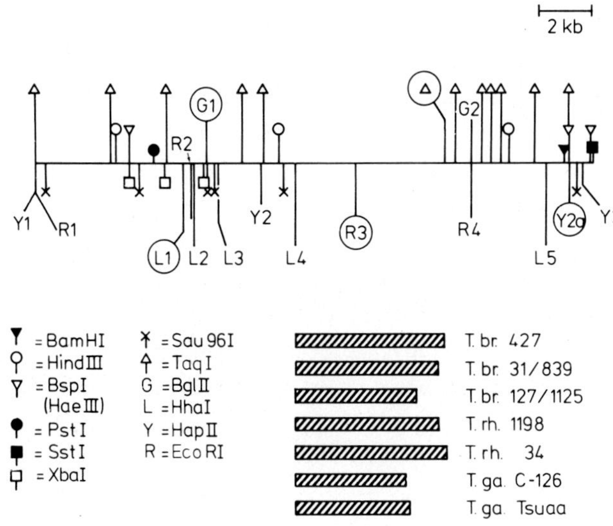

Fig. 2. The linearized physical map of the maxi-circle from T. brucei kDNA. The length is 20.5 kb. Polymorphic sites are circled. The hatched bars represent the size of the 'variable region'. The TaqI map is incomplete and there are several additional fragments in the span R4-R1. T. br, T.b. brucei; T. rh, T.b. rhodesiense; T. ga, T.b. gambiense.

even though the remainder of the sequence is highly conserved, as shown by the fact that we have found only 5 out of 44 restriction sites polymorphic in the maxi-circle of these strains. These observations taken together with the high AT content of this DNA suggest that the maxi-circle contains an AT-rich segment analogous to the AT-rich segment of *Drosophila* mtDNA. The fact that this segment has resisted cloning as recombinant DNA in *Escherichia coli*[8] fits this interpretation.

We note in passing that we have been unable to find major differences between the maxi-circles from *T. brucei*, *Trypanosoma rhodesiense* and *Trypanosoma gambiense*[9]. The latter are infective to man and cause sleeping sickness; the former is not. This confirms suspicions from isoenzyme analyses that these are not separate (sub-)species, but variants of a single species: *T. brucei*.

The mini-circles of T. brucei

The mini-circles of *T. brucei* are very heterogeneous in sequence and early DNA-DNA renaturation experiments suggested that they consist of more than 200 sequence classes with little or no sequences in common[10]. This picture has turned out to be an over-simplification in two respects:

1. Different mini-circle classes have a segment of about 20-25% of the mini-circle sequence in common. This follows from more detailed, renaturation studies[11,12], the finding that all mini-circles cloned as recombinant DNA in *E. coli* cross-hybridize[8,11] and from electron micrographs of briefly renatured mini-circles which contain the partly duplex molecules shown in Fig. 3 (see also ref. 12). The average size ($\pm$ standard deviation) of the duplex part in these molecules is 0.07 $\pm$ 0.017 μm and the duplex part may be located anywhere relative to the ends, even though the circles in this experiment were all cut with HindIII. The common segment in different circle classes is not identical, however. This follows from our finding that treatment of briefly renatured circles with S_1 nuclease does not yield a sharp and prominent band of 1/4 molecules in gels and from the recent sequence analysis of two cloned mini-circles in which the common segment contains frequent mis-matches[13]. This common segment could be involved in the replication of mini-circles.

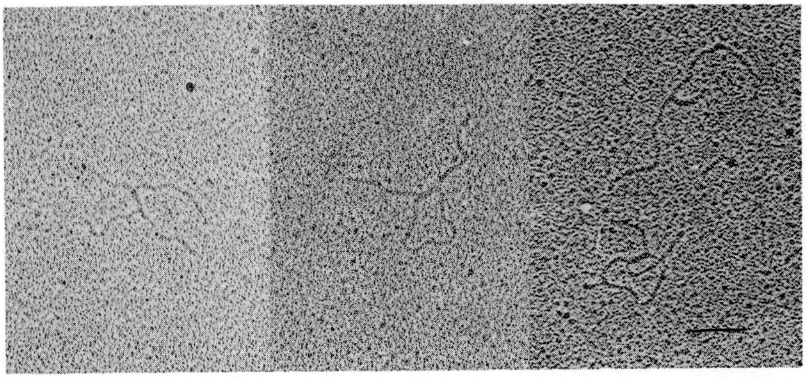

Fig. 3. Electron micrograph of briefly renatured cleaved mini-circles of T. brucei. kDNA from T. brucei, strain 427, was cleaved with endonuclease HindIII. The DNA was denatured, renatured and spread as described (ref. 20). The bar is 0.1 µm.

2. When T. brucei mini-circles are cut with restriction endonucleases that cleave a substantial fraction of the mini-circles more than once, the sub-mini-circle fragments obtained show a strikingly non-random size distribution as judged by agarose gel electrophoresis[4] (unpublished results). This suggests that the 75% of the mini-circles outside the common segment must have a low degree of homology as well, even though this has not been found in the DNA-DNA renaturation experiments.

These results give no clue as to mini-circle function and this remains an enigma.

Transcription of kDNA

To study transcription of kDNA we have followed two routes:

1. We have hybridized labelled total cellular RNA with restriction endonuclease digests of purified kDNA. If care is taken to remove all DNA from the RNA preparations, we only find hybridization with maxi-circle fragments, whereas mini-circles do not hybridize at all even though they are present at much higher concentration in kDNA than the maxi-circles.

2. We have hybridized labelled kDNA segments with RNA blots obtained by electrophoresing total cellular RNA through agarose followed by transfer of the RNA onto diazobenzyloxymethyl-cellulose

paper (DBM-paper). Since it is difficult to obtain kDNA completely free of nuclear DNA and impossible to get maxi-circles free of mini-circles, all these experiments were done with DNA fragments cloned as recombinant DNA in E. coli. Two types of DNA probes were used: two EcoRI fragments cloned in bacteriophage lambda.gt.WES (fragments RR2 and RR3, see Fig. 4 and ref. 8); and DNA complementary to two maxi-circle transcripts, cloned[14] in plasmid pBR322 (Tck-1 and Tck-3, see Fig. 4). Unfortunately, we have been unable to clone the largest EcoRI fragment and our maxi-circle probes, therefore, only cover half of the maxi-circle.

The results of the experiments on maxi-circle transcripts are summarized in Fig. 4 and will be reported in detail elsewhere. The main conclusions are the following:

a) The two most prominent RNA species detected in DNA and RNA blots are the 9S and 12S RNAs, first discovered by Simpson[15] as the main RNAs of high molecular weight in mitochondrial fractions of another representative of the Kinetoplastidae, Leishmania tarentolae. We return to these RNAs below.

b) At high RNA input all maxi-circle segments of 2 kb or more hybridize with the added RNA, with the possible exception of the 'variable region'. This has only been tested in large fragments which include neighbouring regions and we cannot rule out that it is not transcribed at all.

c) In addition to the 9S and 12S RNAs, we have identified six minor transcripts on RNA blots (see Fig. 4). All these minor RNAs (but not the 9S and 12S RNAs) are enriched in RNA that is retained on oligo(dT) cellulose and presumably they are mitochondrial messenger RNAs (mRNAs) or mRNA precursors, containing a poly(A) tail, as also observed in the minor RNAs of animal mitochondria.

d) One of the complementary DNA clones, Tck-3, hybridizes with a region twice as long as the complementary DNA itself and may, therefore, be derived from a split gene. Unfortunately, we have not been able to detect the transcript corresponding to this clone yet and this must be present in very low concentration.

e) None of the transcripts seen in RNA blots exceeds 1100 nucleotides. This is puzzling, because this would be too short to code for the apocytochrome b and subunit I of cytochrome c oxidase of yeast and human mitochondria. The same blots show

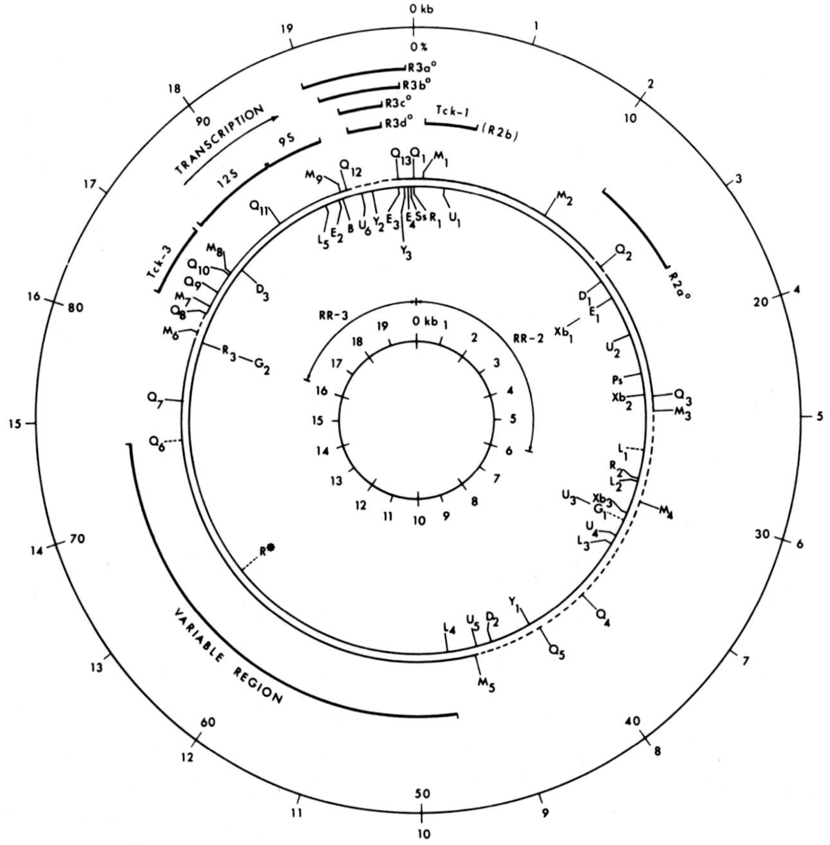

Fig. 4. A partial transcription map of the maxi-circle of T. brucei, strain 427. The location of recognition sites for most of the enzymes is taken from refs 6 and 7; the position for the MboI sites is unpublished. ---, Maxi-circle segments with many not-localized sites for the enzyme that borders the region. The location for the RNA species, indicated with °, is not further determined within RR2 or RR3 and placed arbitrarily. ..., Polymorphic sites (see refs 7 and 9). *, Polymorphic site absent in T. brucei strain 427 (ref. 7). B, BamHI; D, HindIII; E, HaeIII; G, BglII; L, HhaI; M, MboI; Ps, PstI; Q, TaqI; R, EcoRI; Ss, SstI; U, Sau-96I; Xb, XbaI; Y, HapII.

relatively intact cytoplasmic mRNAs - larger than 2000 nucleotides - so degradation seems unlikely. Longer RNAs could still be transcribed, however, from the EcoRI fragment 1, which has not been cloned.

The map in Fig. 4 only shows the transcripts that have been mapped with some precision, but from the DNA blotting experiments we know that several more could be present. We conclude that the transcripts found are of sufficient diversity to specify the usual set of mitochondrial gene products, with the possible exception mentioned.

The 9S and 12S RNAs are ribosomal RNAs

We have previously proposed[3,4] that the 9S and 12S RNAs are the ribosomal RNAs (rRNAs) of trypanosome mitochondria and this proposal is now supported by the following arguments:

1. The 9S and 12S RNAs are the only major high-molecular-weight RNAs ever found in trypanosome mitochondria.

2. They are present in about 1:1 stoicheiometry in our experiments (but contrast ref. 15).

3. They are not retained on oligo(dT) cellulose.

4. Their size and sequence is highly conserved among the kinetoplastid flagellates studied. We find no difference in apparent size between the RNAs from *Trypanosoma* and *Crithidia*. Cheng and Simpson have found no differences between the RNAs from *Leishmania* and *Phytomonas*[16]. In Northern gels we also find ready cross-hybridization between the DNA from *Trypanosoma* and 9S and 12S RNAs from *Crithidia*.

5. The RNAs hybridize with a single segment of the maxi-circle and this segment is approximately the same size as the RNAs, both in *T. brucei* (Fig. 4) and in *Crithidia*[17]. There is, therefore, one gene for each of these RNAs per maxi-circle.

6. We have provided Mr I.C.Eperon (MRC Laboratory of Molecular Biology, Cambridge, UK) with the 1.8 kb HindIII x BamHI fragment to which both RNAs hybridize and he has determined the nucleotide sequence of this segment. The sequence shows low but highly significant homology with the rRNA genes of human mtDNA and *E. coli*.

These results establish that the 9S and 12S RNAs are in fact rRNAs and that the maxi-circle codes for at least one function found on mtDNA in other organisms. The report by Simpson and Lasky[18] that 9S and 12S RNAs act as mRNAs in a wheat-germ system can be attributed to the presence of mitochondrial poly(A)$^{+}$ RNAs that co-migrate with 9S and 12S RNAs in the gel (cf. Fig. 4).

It is surprising that nature manages to make functional ribosomes with a sum total of 590 + 1080 = 1670 nucleotides worth of RNA. It is possible that this is done by using two or three of these RNAs per subunit. We find it more likely, however, that this ribosome will turn out to have a very high protein/RNA ratio. This trend to minimize the size of the rRNAs is already visible in the vertebrate mitochondrial ribosome, with its high number of relatively large proteins and we expect that it is taken to its extreme consequence in trypanosomes. The resulting ribosome could have very unusual properties and this might explain why we and others have been unable yet to isolate this putative ribosome and why mitochondrial protein synthesis in *Crithidia* appears to be totally unaffected by chloramphenicol, spiramycin or ethidium[2].

The 9S and 12S RNAs are transcribed from the same strand and probably away from the variable region. In *Drosophila* this is also observed, but the small rRNA lies 5' of the large rRNA[19]. We find the reverse for *T. brucei*.

Control of gene expression in mitochondria

We have recently compared the maxi-circle transcripts from cultured *T. brucei* (fully-developed mitochondrion) with those from bloodstream trypanosomes (mitochondrial biogenesis repressed). At equal RNA inputs the relative amounts of mtRNA are about 5-10 fold higher in culture than in bloodstream trypanosomes. Besides this quantitative difference we find no gross qualitative differences, however. All transcripts are present at approximately equal amounts in culture and bloodstream trypanosomes.

We interpret this as a dual control of mitochondrial biogenesis: the strong overall decrease in the level of transcripts should result in an overall decrease in the rate of mitochondrial protein synthesis. There must be an additional control, however, at the level of mitochondrial protein synthesis or membrane assembly, because no trace of holo-cytochrome *b* or cytochrome *c* oxidase is detectable in bloodstream trypanosomes, whereas the

TABLE I

THE STATE OF kDNA IN TRYPANOZOON SUB-SPECIES THAT HAVE LOST THE ABILITY TO MAKE FUNCTIONAL MITOCHONDRIA (I^- TRYPANOSOMES)

Organism	Strain	Oligo-sens. ATPase	DAPI stain kDNA	Organized networks present	Maxi-circles (µm)	Mini-circles (µm)	Mini-circle sequence heterogen.	Ref
T. brucei	427[a)]	+	+	+	6	+	+	6,7
	31	–	+	+	6	+	+	4,7
	LUMP 127[b)]	–	+	+	6	+	+	7
T. equiperdum	ATCC 30019	?	+	+	4	+	±	20
T. evansi	AMB3[c)]	–	+	+	–	+	–	4
	ILRAD B-32	?	?	+	–	+	–	4
	SAK	?	±	–	?	?	?	21
T. equiperdum	ATCC 30023	?	–	–	–	–		20

a) I^+ control. b) Clone 7. c) Referred to as Zwart$^+$ in ref. 4.

mitochondrial subunits of the ATPase complex are presumably synthesized, because the ATPase is normally sensitive to oligomycin[5].

Trypanosome 'petite' mutants

In the bloodstream forms of *T. brucei* the synthesis of the mitochondrial respiratory chain is repressed and ATP is only generated by the conversion of glucose into pyruvate. It is not surprising, therefore, that mutants that have lost the ability to make functional mitochondria will survive in the vertebrate host. These I^- mutants (insect minus) are of two types: (i) K^+ mutants which appear to have normal kDNA as judged by light microscopy of stained trypanosomes, and (ii) K^- or dyskinetoplastic mutants which have lost the normal kDNA networks. Our information on some of these mutants is summarized in Table I.

The first group of mutants is represented here by *T. brucei* strains 31 and LUMP 127. These strains contain kDNA that is normal by all criteria applied, including a detailed restriction enzyme analysis of maxi-circles[7]. Nevertheless, attempts to grow these trypanosomes in culture or in tse-tse flies (only strain 31) have failed and because the ATPase of these trypanosomes is insensitive to oligomycin, we are convinced that they have irreversibly lost the ability to make functional mitochondria. Analogous *T. brucei* mutants have been obtained by Mr S.L.Hajduk (personal communication). This suggests that maxi-circles can be rather stable even if rendered non-functional by mutations.

The second group of mutants has an altered maxi-circle. Our only representative of this group is *T. equiperdum* ATCC 30019 in which the maxi-circle differs from that of *T. brucei* by the deletion of one continuous segment of 7.5 kb (ref. 20). The third group is formed by the *T. evansi* strains AMB3 and ILRAD B-32, which completely lack maxi-circles. The fourth group consists of *T. evansi* SAK, which lacks a kDNA network but still contains DNA in the mitochondrion as shown by patchy DAPI staining[21]; whether these are mini-circle associations is not known. Finally, the *T. equiperdum* ATCC 30023 has lost all trace of kDNA and this is probably the case for most other dyskinetoplastic strains as well. These strains still contain a DNA that bands at the density of kDNA in CsCl gradients and this was originally thought to be an altered kDNA. We have recently found[22], however, that this is a nuclear satellite DNA present in all strains - dyskinetoplastic or normal.

An intriguing finding is that some I^- strains have no sequence heterogeneity in their mini-circles. This is not related in a simple fashion to the loss of maxi-circles or the ability to make functional mitochondria (see Table I). Possible explanations will be considered elsewhere.

In general, the results with the I^- mutants parallel those with the ρ^- mutants of yeast with one exception: Whereas ρ^- petites contain repetitions of wild-type segments, this has not been found for I^- mutants. However, this is hardly surprising since the maxi-circle represents only 10-15% of total kDNA.

Outlook

From the limited data available we infer that the mitochondrial genetic system in trypanosomes may be one of the most unusual in nature. Further analysis of the maxi-circle and its transcripts may, therefore, provide us with more surprises than mini-rRNAs and DNA networks.

Acknowledgements

We are indebted to Mr S.L.Hajduk (Department of Zoology, The University, Glasgow, UK) for his help in the isolation of several DNAs; to Dr J.Davison (Research Unit of Molecular Biology, International Institute of Cellular and Molecular Pathology, Brussels, Belgium) for providing cloned DNA and to Miss J.Van den Burg for help in some experiments. This work was supported in part by a grant to PB from the Foundation for Fundamental Biological Research (BION), which is subsidized by The Netherlands Organization for the Advancement of Pure Research (ZWO). Initial experiments were supported by a grant to PB from NATO (grant Nr. 559). ACCF was supported by a WHO Research Training Grant funded by the UNDP World Bank WHO Special Programme for Research and Training in Tropical Diseases (Nr. M8/181/4/F.78).

References

1 Bresslau, E. and Scremin, L. (1924) Arch.Protistenk. 48, 509.
2 Kleisen, C.M. and Borst, P. (1975) Biochim.Biophys.Acta 390, 78.
3 Borst, P. and Hoeijmakers, J.H.J. (1979) Plasmid 2, 20.
4 Borst, P. and Hoeijmakers, J.H.J. (1979) in: Extrachromosomal DNA: ICN-UCLA Symposia on Molecular and Cellular Biology (Cummings, D.J., Borst, P., Dawid, I.B., Weissman, S.M. and Fox, C.F., Eds), Vol. 15, Academic Press, New York, pp. 515.
5 Opperdoes, F.R., Borst, P. and De Rijke, D. (1976) Comp.Biochem. Physiol. 55B, 25.

6 Borst, P. and Fase-Fowler, F. (1979) Biochim.Biophys.Acta 565, 1.
7 Borst, P., Fase-Fowler, F., Hoeijmakers, J.H.J. and Frasch, A.C.C. (1980) Biochim.Biophys.Acta, submitted.
8 Brunel, F., Davison, J., Merchez, M., Borst, P. and Weijers, P.J. (1980) in: DNA - Recombination Interactions and Repair (Zadrazil, S. and Sponar, J., Eds), Pergamon, Oxford, pp. 45-54.
9 Borst, P., Fase-Fowler, F. and Gibson, W.C. (1980) Mol.Biochem. Parasitol., submitted.
10 Steinert, M., Van Assel, S., Borst, P. and Newton, B.A. (1976) in: The Genetic Function of Mitochondrial DNA (Saccone, C. and Kroon, A.M., Eds), North-Holland, Amsterdam, pp. 71.
11 Donelson, J.E., Majiwa, P.A.O. and Williams, R.O. (1979) Plasmid 2, 572.
12 Steinert, M. and Van Assel, S. (1980) Plasmid 3, 7.
13 Chen, K.K. and Donelson, J.E. (1980) Proc.Natl.Acad.Sci.U.S. 77, in press.
14 Hoeijmakers, J.H.J., Borst, P., Van den Burg, Weissmann, C. and Cross, G.A.M. (1980) Gene 8, 391.
15 Simpson, L. and Simpson, A.M. (1978) Cell 14, 169.
16 Cheng, D. and Simpson, L. (1978) Plasmid 1, 297.
17 Hoeijmakers, J.H.J. and Borst, P. (1978) Biochim.Biophys.Acta 521, 407.
18 Simpson, L. and Lasky, L. (1975) J.Cell Biol. 67, 402A.
19 Wolstenholme, D.R., Goddard, J.M. and Fauron, C.M.-R. (1979) in: Extrachromosomal DNA: ICN-UCLA Symposia on Molecular and Cellular Biology (Cummings, D.J., Borst, P., Dawid, I.B., Weissman, S.M. and Fox, C.F., Eds), Vol. 15, Academic Press, New York, p. 409.
20 Frasch, A.C.C., Hajduk, S.L., Hoeijmakers, J.H.J., Borst, P., Brunel, F. and Davison, J. (1980) Biochim.Biophys.Acta 607, 397.
21 Vickerman, K. (1977) Protozool. 111, 57.
22 Borst, P., Fase-Fowler, F., Frasch, A.C.C., Hoeijmakers, J.H.J. and Weijers, P.J. (1980) Mol.Biochem.Parasitol. 1, in press.
23 Lang, D. and Mitani, M. (1970) Biopolymers 9, 373.

The Organization and Expression of the Mitochondrial Genome
A.M. Kroon and C. Saccone, editors

THE PETITE MUTATION : EXCISION SEQUENCES, REPLICATION ORIGINS AND SUPPRESSIVITY.

Giorgio Bernardi, Giuseppe Baldacci, Gregorio Bernardi, Godeleine Faugeron-Fonty, Claire Gaillard, Regina Goursot, Alain Huyard, Marguerite Mangin, Renzo Marotta, and Miklos de Zamaroczy.
Laboratoire de Génétique Moléculaire, Institut de Recherche en Biologie Moléculaire, 2 Place Jussieu, F-75005 Paris, France.

Introduction

Last year we reported (1) restriction mapping and DNA: DNA hybridization experiments on the mitochondrial genomes of spontaneous petite mutants and of the corresponding parental wild-type cells of *Saccharomyces cerevisiae*. The main conclusions arrived at were : a) that the petite genomes are made of tandem repetitions of the DNA segment originally excised from the wild-type genome; b) that the excision mechanism is extremely conservative, in that the excised segment does not show any sign of the rearrangements found in ethidium-induced petites (2); this conclusion implies that every repeat unit of the petite genome contains a replication origin and, since petite genomes can arise from many regions of the wild-type genome, that the latter contains several origins of replication; c) that excision most frequently takes place in GC clusters; alternatively, it occurs elsewhere, most probably in AT spacers.These conclusions seem to apply to the vast majority, but not to the totality of spontaneous petite mutants. Fig. 1 shows the restriction map of the mitochondrial genomes of some of the spontaneous petites used in our work and the localization of some excision sites.

Direct repeats are used in the excision of spontaneous petite genomes

Because of the sequence conservation in the mitochondrial genomes of spontaneous petites, excision sequences can be determined by comparing the primary structure around the excision sites, or that of the entire repeat unit, of given petite genomes and of other petite genomes encompassing them.
Using this approach, a comparison of the sequences around the excision sites of a^{x}-1/7/8 and of a-3/1 indicated (fig. 2) that the excision of the repeat unit of petite a^{x}-1/7/8 took place at two direct repeats, 23 nucleotides long,located in two GC clusters (3). On the other hand, a comparison of the repeat units of petites a-1/1R/Z1 (4) and a-1/1R/14 (5) with that of a-1/1R/1 (6) showed(fig.3) that the excision sequences are two direct repeats, 13 nucleotides long (with

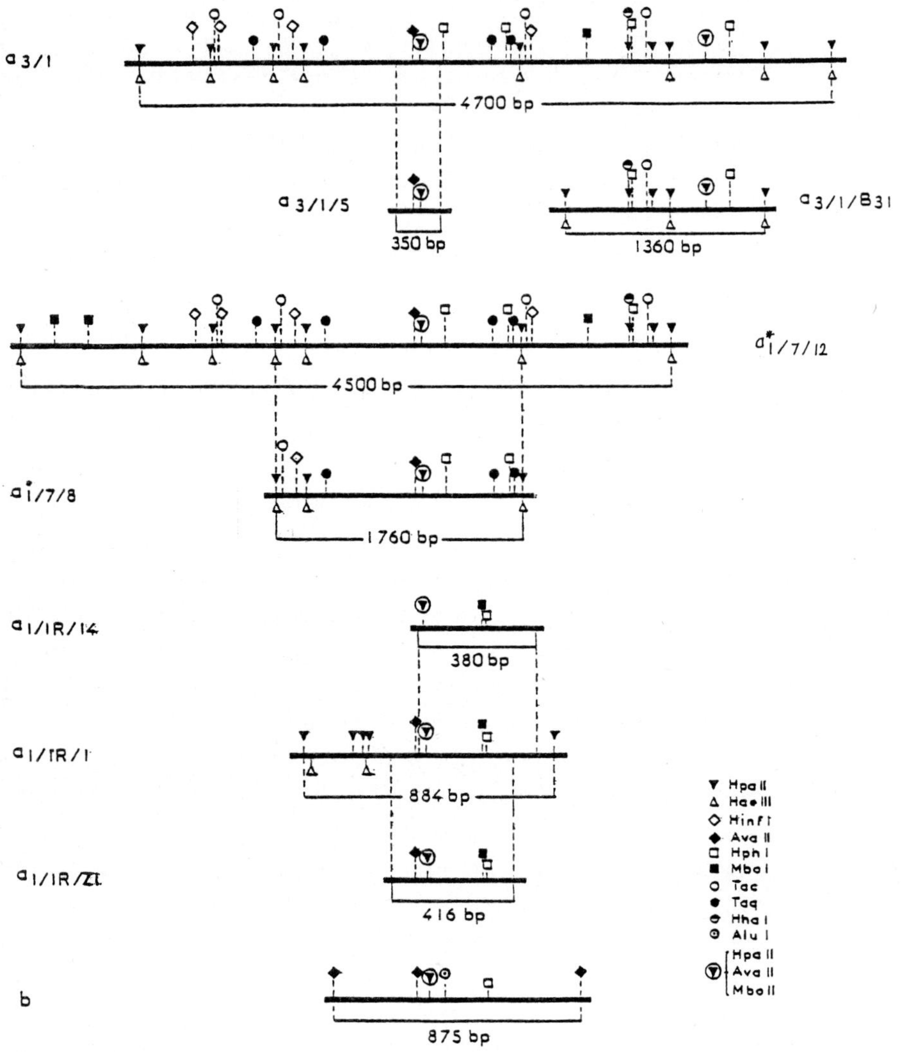

Fig. 1. Restriction enzyme maps of the repeating units of mitochondrial genomes of spontaneous petite mutants. The molecular weights, in base pairs, of the repeat units are indicated. Only some Mbo II and Ava II sites are indicated. The previous maps of a-3/1, a*-1/7/12 and a*-1/7/8 (1) were revised in the light of recent work. Sequenced excision sites are indicated by vertical broken lines. All sequences are aligned on Hpa II, Ava II, Mbo II site clusters.

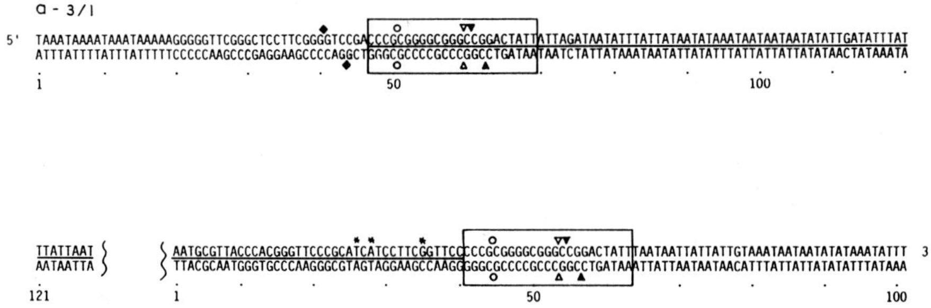

Fig. 2. Primary structure around the putative excision sequences used in the formation of the repeat unit of ă-1/7/8. The ă-1/7/8 sequence is identical with that of a-3/1 in the region indicated by the continuous line between base pairs, except for two base pair changes (A:T and G:C replace T:A and A:T) and one deletion (G:C) at the positions indicated by asterisks.

one mismatch in the case of a-1/1R/Z1), located in the AT spacers. The longer sequence of the excision repeat and its higher GC content appear to account for the higher frequency of excisions at GC clusters as opposed to AT spacers. Obviously, more excision sequences need to be studied in order to generalize this conclusion, as well as to check whether inverted repeats are also involved in the excision process.

These results fully confirm a hypothesis according to which repeated sequences in the mitochondrial genome of yeast are responsible for the excision of petite genomes by a site-specific illegitimate recombination process (7). The very high frequency of long repeats in the wild-type genome (8) accounts, again as predicted, for the very high rate of the spontaneous petite mutation.

Genomes without genes

An analysis of the sequence of a-1/1R/1 (fig. 3) fails to reveal the presence of any gene or gene segment in its repeat unit. Very recent results (to be presented at this Meeting) indicate that the genome of a-1/1R/1 is transcribed. Genomes like a-1/1R/1 are, therefore, able to replicate and to be transcribed, and yet they do not have any apparent usefulness to the cell. In this respect, and in their capacity to spread to other cells via crosses (see below), these genomes are the best examples of what has been called, rightly or wrongly, selfish DNA (9,10).

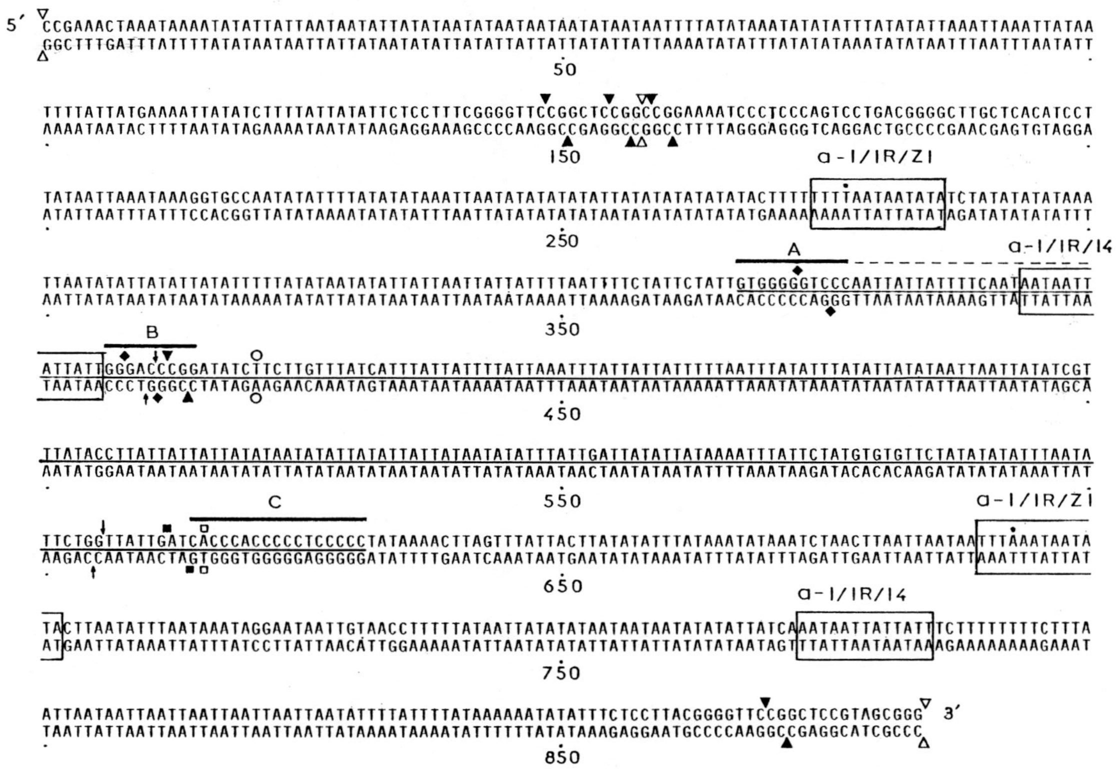

Fig. 3. Nucleotide sequence of the repeat unit of the mitochondrial genome of spontaneous petite mutant a-1/1R/1. Restriction sites : ▼ Hpa II, ▽ Hae III, ◆ Ava II, ○ Mbo II recognition sites; □ Hph I recognition site; (arrows indicate the cutting sites). Heavy lines indicate GC clusters; a broken line indicates a 23-nucleotide palindrome; boxes indicate excision sites of a-1/1R/Z1 and a-1/1R/14. The origin of replication is indicated by the continuous line between base pairs.

Origins of replication

A search for an origin of replication on the repeat unit of the genome of a-1/1R/1 (fig. 3) has focused our attention on the region comprised between positions 370 and 630. This is characterized by (4-6) two short GC clusters, A and B, flanking a 23-nucleotide AT palindrome, and one long GC cluster, C; the first two clusters are mainly formed by two symmetrical heptanucleotides, GGGTCCC and GGGACCC, and the third one by three penta-C repeats (the first one of which contains an A), each one of which is preceded by an A or a T. It should be noted that cluster A contains an Ava II site, cluster B contains a Hpa II, an Ava II and an Mbo II cutting site (the recognition site of the latter enzyme is a pentanucleotide located 8 base pairs away from the cutting site); cluster C contains an Mbo I, a Mnl I and a Hph I recognition site (the cutting site of the latter enzyme is located 7 nucleotide away from the recognition site). It is of interest to remark that the region comprised between clusters A and B can be folded in a way (fig. 4) similar to that reported for a region of the replication origin of HeLa cell mitochondrial DNA (11).

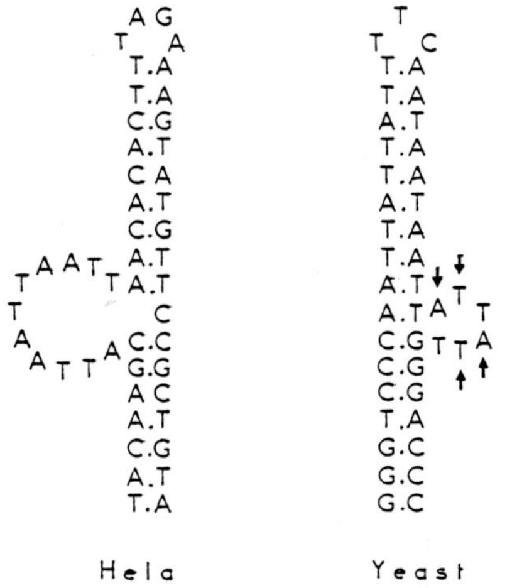

Fig. 4. Hypothetical secondary structure of the left end of the replication origins of mitochondrial DNA; the region corresponds to the sequence comprised between GC clusters A and B (see fig. 3); arrows indicate base changes found in different petite mutants. The palindromic sequence found in the replication origin of HeLa cell mitochondrial DNA (11) is shown for the sake of comparison.

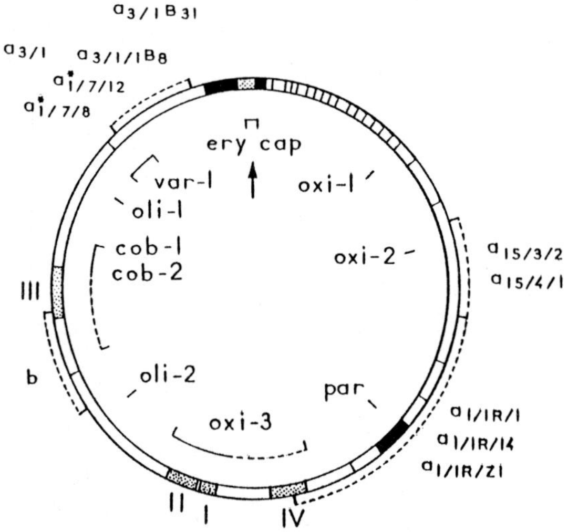

Fig. 5. Regions of excision of the spontaneous petite genomes studied here are indicated on a genome map of the mitochondrial genome of *Saccharomyces cerevisiae* (adapted from ref. 18).

Considering that the region just described might correspond to an origin of replication, we decided to look for it in several petite genomes having arisen from three different regions of the parental wild-type genomes (fig. 5) : the var-1-ery, the cob-oli 2, and the 15 S RNA regions. We first looked for the presence of the most characteristic cluster B in the repeat units of these petite genomes. Having found it, we sequenced both sides of the cluster, to discover that a stretch of about 80 nucleotides centered on the cluster was present in all cases, with only a few base pair changes (19,12 and present work; fig. 6), which were localized in the loop of fig. 4, or a short deletion, in the case of a-1/1R/14 (see below). On the left of the 80-base pair stretch, sequence homology became patchy after the Ava II site of cluster A; on the right, sequence work on the repeat unit of a-3/1 and a*-1/7/12 showed that homology extended until cluster C with some base pair changes in the long AT stretch between B and C; in the repeat unit of a-3/1/5, the whole right end of this region was absent (see also below). The whole replication origin is therefore at least 260 nucleotide long. This region has also been partially sequenced or at least mapped in the repeat units of other petite genomes. Our present knowledge is summarized in Table 1.

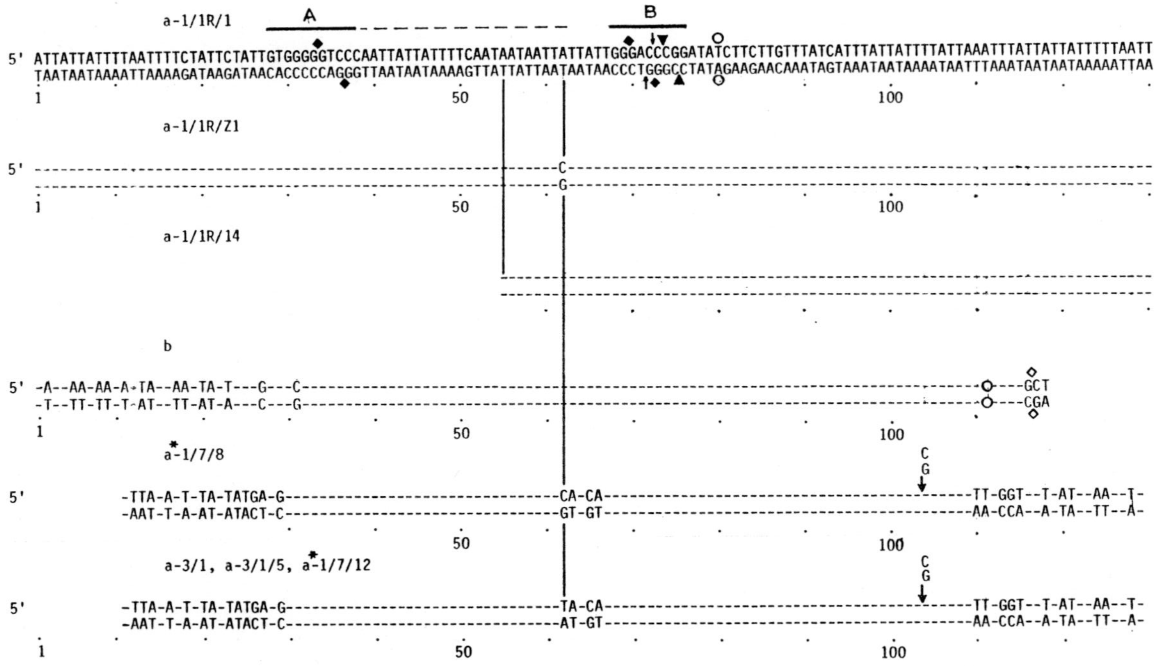

Fig. 6. Primary sequence of the left end of the replication origin of the mitochondrial genomes from several spontaneous petite mutants (see also fig.3). The sequence found in the DNA of a-1/1R/1 is taken as a reference; dashes indicate nucleotides identical to those of a-1/1R/1. Base pair changes are indicated; a double empty circle indicates a base-pair deletion, an arrow indicates an insertion. Restriction sites and other symbols as in fig.3.

Recent experiments have provided additional information on the origins of replication. The repeat units of two spontaneous petite genomes,a-15/3/2 and a-15/4/1 were shown to have arisen from the oxi-2 region and to contain at least the Hpa II, Ava II, Mbo II cluster and the Hph I site at the expected position. It is of interest that hybridization of these petite genomes took place on the same or on corresponding restriction fragments originated from the genome of four different wild-type strains; this result is reminiscent of similar ones found in the hybridization of the genome of petite b (1) and stresses the conservation of the localization of the origins of replication in wild-type genomes. A second series of results obtained with nick-translated DNA from a-1/1R/Z1 (the repeat unit of this mitochondrial genome is only slightly larger than the origin of replication) showed that hybridization of this DNA took place on the repeat unit of petite b, and on the Hae III and Hpa II fragments from the DNAs of a-3/1, a-3/1/B31, $\overset{x}{\text{a}}$-1/7/8, a-15/3/2,a-15/4/1 which contained the Ava II, Hpa II, Mbo II site cluster. A very interesting finding was that in the case of a-3/1, two Hae III fragments bound the probe; one of these two fragments is present on the repeat unit of $\overset{x}{\text{a}}$-1/7/8, the other one on that of a-3/1/B31. This indicates that a-3/1 contains in fact not one but two origins of replication, as characterized by hybridization and restriction map. The results available so far show, therefore, that at least five origins of replication are present on the mitochondrial genome of wild-type cells.

Suppressivity

Crosses of the petite mutants of fig. 1 with wild-type cells have proved extremely useful in two respects : a) in providing a functional evidence that the sequences discussed above are indeed the origins of replication of the mitochondrial genome ; b) in helping to understand the phenomenon of suppressivity.

Three years ago we observed, and reported at the Symposium on the Biochemistry and Genetics of Yeast held in Sao Paulo, Brazil, that crosses of petites a-1/1R/1 and b with wild-type cells produced diploids harboring the mitochondrial genome of the petite used in the cross. Both petites are very highly suppressive (95 % suppressivity) and have genomes characterized by repeat units corresponding to only 1 % of the wild-type genome. We proposed to call supersuppressive these petites and suggested that the reason for the petite genome being the only one found unaltered in the progeny was that these petite genomes contained multiple copies of the origin of replication and could,

TABLE 1.

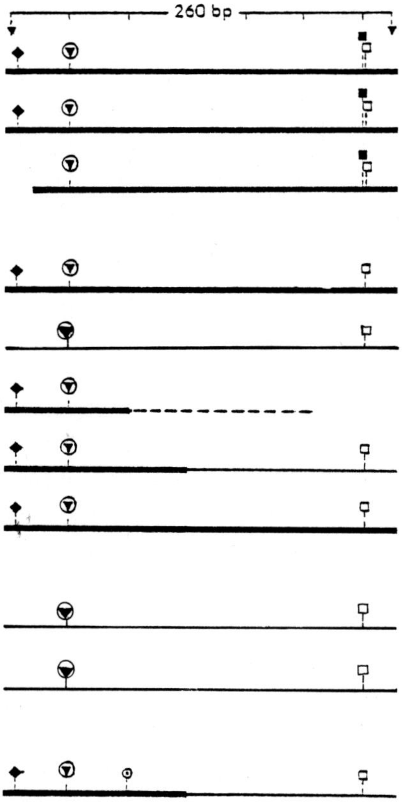

Petite Strain	Repeat Unit (bp)	Origin of replication[c]	Hybrid.[b]	Supp., %	P:M ratio[a]
a-1/1R/1	884			95	4 : 0
a-1/1R/Z1	416		+		
a-1/1R/14	380			80	
a-3/1	4700		++	60-70	28 : 4
a-3/1/B31	1360		+		
a-3/1/5	350			<5	5 : 0
a^{x}-1/7/8	1760		+	85	14 : 0
a^{x}-1/7/12	4500			60-80	36 : 0
a-15/3 /2	4300		+	50	12 : 0
a-15/4/1	4800		+	<10	
b	875		+	95	10 : 0

(a) This is the ratio of genomes of petite diploids having the repeat unit of the parental petite to those having modified repeat units; the latter may either be recombinant genomes or minority genomes (excised from the parental petite) transmitted to the progeny.

(b) Hybridization of nick-translated DNA from a-1/1R/Z1 on Hae III (or Alu I in the case of petite b) fragments.

(c) Thick lines indicate sequenced stretches. In the case of a-3/1/5 the right end of the repeat unit is located within the broken line. Restriction site symbols as in Fig. 1. The Ava II and Mbo I sites on a-3/1/B31, a-15/3/2 and a-15/4/1 have not been mapped.

therefore, compete out the genomes of wild-type cells (12), a view similar to that originally hypothesized by Rank (13).

We have now extended these investigations to other petite mutants characterized by moderate or low degrees of suppressivity, to understand which relationships, if any, exist between supersuppressive and suppressive petites on one hand, and between suppressivity and the putative origins of replication discussed above, on the other. The results are summarized in Table 1 and lead to two main conclusions. The first one is that even when petites of moderate or low suppressivity are used in crosses with wild-type cells, their mitochondrial DNA is transmitted unaltered in its restriction map to the petite diploids issued from the cross. The number of petite genomes showing evidence of recombination is very small and certainly does not account for suppressivity, ruling out explanations based on this notion (14, 15). The second conclusion is that there is a correlation between conservation or alteration of the origin of replication and suppressivity, and probably also between length of the repeat unit and suppressivity. The first point is illustrated by the case of a-3/1/5 where the loss of the whole right end of the origin of replication is associated with an almost complete loss of suppressivity; another case, though less striking, is that of a-1/1R/14 where the loss of the leftmost 25 base pairs of the origin of replication is accompanied by a decrease in suppressivity. The second point seems to be supported by the suppressivities of a*-1/7/8 and a*-1/7/12, two genomes with origins of replication probably identical; in this case, the genome with a longer repeat unit is less suppressive than that with a shorter repeat unit. The case of a-3/1 is more complex because of the presence of two close origins of replication.

Two final points should be made. The first one is that, if the sequence of the origin of replication is the most important factor in determining the degree of suppressivity as suggested above, and the length of the repeat unit is another possible factor, other features may also play a role, like the amount of mitochondrial DNA in the petite used in the cross. The second one is that petites exist, essentially among ethidium-induced ones, which do not contain an origin of replication in every repeat unit. This is the case of neutral petite RD1/A (16,17). In this case we postulate that the repeat units forming the majority of each genome carry with them an origin of replication picked up from an excision event different from that concerning the majority repeat units. Current investigations should clarify this point.

REFERENCES

1. Faugeron-Fonty G., Culard F., Baldacci G., Goursot R., Prunell A. and Bernardi G. (1979) J. Mol. Biol., 134. 493-537.
2. Lewin A., Morimoto R., Rabinowitz M., and Fukuhara H. (1978) Mol. Gen. Genet., 163. 257-275.
3. Baldacci G., de Zamaroczy M., and Bernardi G. (1980) FEBS Letters, 114. 234-236.
4. Gaillard C., and Bernardi G. (1979) Mol. Gen. Genet., 174. 335-337.
5. Bernardi G., Baldacci G., Culard F., Faugeron-Fonty G., Gaillard C., Goursot R., Strauss F., and de Zamaroczy M. (1980) in Mobilization and Reassembly of Genetic Information (Scott W.A., Werner, R., and Joseph D.R., eds.), Academic Press, New York, (in press).
6. Gaillard C., Strauss F., and Bernardi G. (1980) Nature, 283. 218-220.
7. Bernardi G. (1979) Trends Biochem. Sci., 4. 197-201.
8. Bernardi G., and Bernardi G. (1980) FEBS Letters (in press).
9. Doolittle W.F. and Sapienza C. (1980) Nature, 284. 601-603.
10. Orgel L.E., and Crick F.H.C. (1980) Nature, 284. 604-607.
11. Crews S., Ogala D., Posakony J., Nishiguchi J., and Attardi G. (1979) Nature, 277. 192-198.
12. Goursot R., de Zamaroczy M., Baldacci G., and Bernardi G. (1980) Current Genet., 1. 173-176.
13. Rank G.H. (1970) Can J. Genet. Cytol., 12. 129-136.
14. Perlman P.S., and Birky C.W., Jr. (1974) Proc. Natl. Acad. Sci. USA., 71. 4612-4616.
15. Slonimski P.P., and Lazowska J. (1977) in Mitochondria 1977 (Bandlow W. et al., eds.), pp. 39-52, de Gruyter, Berlin, 1977.
16. Van Kreijl C.F. and Bos J.L. (1977) Nucleic Acids Res., 4. 2369-2388.
17. Moustacchi E. (1972) Biochim. Biophys. Acta, 277. 59-60.
18. Borst P. and Grivell L.A. (1978) Cell, 15. 705-723.
19. de Zamaroczy M., Baldacci G., Bernardi G. (1979) FEBS Letters, 108. 429-432

The Organization and Expression of the Mitochondrial Genome
A.M. Kroon and C. Saccone, editors

YEAST MITOCHONDRIA MINILYSATES AND THEIR USE TO SCREEN A COLLECTION OF HYPERSUPPRESSIVE ρ^- MUTANTS

BERNARD DUJON and HUGUES BLANC*
The Biological Laboratories, Harvard University, Cambridge, Massachusetts and Centre de Génétique Moléculaire du C.N.R.S. Gif sur Yvette, France

INTRODUCTION

Suppressiveness is a characteristic property of ρ^- mutants such that, in crosses to a ρ^+, a given fraction of the zygotic clones are entirely composed of ρ^- cells[1,2]. In these clones therefore, although the ρ^+ mitochondrial DNA (mtDNA) was originally present in the zygote, it is not transmitted to its progeny. To study the mechanism(s) of this "elimination" we have constructed a collection of ρ^- mutants showing the highest observable degree of suppressiveness (called hypersuppressive or HS). Using one HS mutant as a model we have concluded that the hypersuppressiveness results from the preferential replication of the mtDNA molecules of the HS ρ^- parent during the transient heteroplasmic stage that follows zygote formation[3]. By heteroduplex mapping[4] of 4 different HS ρ^- mutants of this collection, it was shown that they all share a 300 bp common sequence (called *rep*) believed to represent regions that control or initiate mtDNA replication[3]. DNA sequencing has revealed that several *rep* sequences exist in the wild type mitochondrial genome, probably as the result of duplication(s) of an ancestral sequence followed by independent mutations[3]. One of them (*rep1*) has been precisely localised on the mitochondrial map *ca* 3 kb upstream of the 21 S rRNA gene.

In this paper we show, using a rapid method for mtDNA preparation that we describe first, that all HS ρ^- mutants constitute a homogeneous and discrete class characterized by common sequences and organizations of mtDNAs.

YEAST MITOCHONDRIA MINILYSATES, A RAPID SMALL SCALE PREPARATION OF mtDNA

Several methods allowing the rapid preparation of total nucleic acids from medium or small scale cultures have recently been published[5-7]. We report here a simple method (table I) in which, from a small scale culture, a crude mitochondrial pellet is prepared and used to extract DNA. The preparation is suitable for restriction enzyme analysis and hybridization to specific probes. mtDNA can also be visualized by ethidium bromide if one uses ρ^- of relatively low complexity (see fig. 1.).

---

* Present address: Stanford University Medical Center, Stanford, California

TABLE 1

YEAST MITOCHONDRIA MINILYSATE PROCEDURE

Cultures: *Grow the yeast cells until stationnary phase in 1.5 ml of a complete medium containing glucose, galactose or glycerol as appropriate. Transfer the culture into a 1.5 ml Eppendorf tube. All following steps will be performed in the same tube.*

Protoplasts formation: *Pellet the cells by centrifugation in an Eppendorf microcentrifuge (one second is sufficient) and wash once with* H_2O*. Resuspend the pellet in 1 ml of 1.2 M sorbitol, 50 mM phosphate buffer pH 7.5, 20 mM EDTA and 1% (v/v) 2-mercaptoethanol, containing 0.5 mg/ml of Zymolyase 60000 (Kirin Brewery Co., Ltd, Japan). Incubate at 37°C for 5 to 10 minutes.*

Preparation of mitochondria: *Centrifuge 2 sec., discard the supernatant (should remain clear) and add 1 ml of 0.7 M sorbitol, 50 mM Tris/Cl pH 7.8 and 10 mM EDTA. Mix well using a vortexer at maximum speed for a few sec. to break open the protoplasts. After this step a one sec. spin should not produce a sizable pellet (if it does add 0.1 ml of* H_2O *and use vortexer again). Spin for 5 minutes at 4°C and discard the supernatant. The pellet contains mitochondria and often appears brown-reddish. It can be washed in the same buffer if necessary. Then add 0.2 ml of 100 mM NaCl, 50 mM Tris/Cl pH 7.8, 10 mM EDTA and 1% Sarkosyl. Mix with vortexer until the pellet is solubilized.*

DNA preparation: *Extract twice with 0.2 ml of phenol saturated with 50 mM Tris/Cl pH 7.8 and 1 mM EDTA, each time removing carefully the phenol phase and the interphase. Traces of phenol can then be removed by two extractions with 0.5 ml diethyl ether. Then evaporate the traces of ether at 65°C for a few min. Add 0.2 ml of 5 M ammonium acetate and 1 ml of ethanol, chill at -70°C for 5 min. and centrifuge 5 min. at 4°C. Dissolve the pellet in 0.25 ml of 300 mM sodium acetate and add 0.75 ml of ethanol. Chill, centrifuge and dry the pellet under vacuum. Redissolve in 0.05 ml of 10 mM Tris/Cl pH 7.5 and 1 mM EDTA.*

CONSTRUCTION OF A REPRESENTATIVE COLLECTION OF HYPERSUPPRESSIVE ρ^- MUTANTS

To define the characteristic properties of HS ρ^- mutants we have constructed a large collection of homoplasmic ρ^- clones (≈2000) issued from a large number of independent ρ^+ to ρ^- mutational events (≈800) without *a priori* selection for any character or genetic markers in order to obtain a representative sample of all events leading to stable ρ^- clones[4]. Each clone of the collection has been tested for its degree of suppressiveness as well as for the loss or retention of several genetic loci. In total 41 clones were found HS while respectively 5, 9, 85, 24 and 34 clones have retained the loci *rib1, oxi1, par1, box4* and *oli1*. Interestingly none of the HS clone has retained any of the markers tested. This mutual exclusion suggests either that the regions responsible for hypersuppressiveness are different from those carrying the known genetic markers or that the suppressiveness of the HS ρ^- mutants is absolute (i.e. the few % of zygotic clones still containing ρ^+ cells -see table 2- may result in fact from the presence of ρ^0 cells in the HS cultures) or even both. The frequency of HS ρ^- mutants among the collection (2.3%) represents a relatively important fraction (10-15%)

of the clones which have retained mtDNA (estimated around 300-400 from the co-retention of independant genetic markers). Ten of the HS ρ^- mutants are studied in the present paper (table 2).

TABLE 2

SUPPRESSIVENESS OF THE HS ρ^- MUTANTS AND COMPLEXITY AND ARRANGEMENT OF THE REPAT UNITS OF THEIR mtDNAs.

Mutant #	% S	bp	arrangement	Mutant #	% S	bp	arrangement
HS137	96.2	640	tandem	HS1948	98.3	640	tandem
HS416	97.5	700	tandem	HS2319	98.3	760	tandem
HS634	97.5	850	tandem	HS3324	99.0	990	tandem
HS1037	98.0	750	tandem	HS3330	90.2	910	tandem
HS1826	97.9	>1150	?	HS3748	90.0	900	tandem

Suppressiveness is the % of zygotic clones entirely composed of ρ^- cells in crosses of each ρ^- mutant to the ρ^+ tester strain 777-3A (α ade_1 op_1). The sizes of the repeat units are estimated from electrophoretic mobilities of MspI, HphI and double digests as in fig. 1. In each case the sizes of the MspI and the HphI fragments were identical except for HS3748 containing two MspI sites distant by 650 bp and for HS1826 showing several fragments in addition to the one that hybridizes with HS416.

HYPERSUPPRESSIVE ρ^- MUTANTS SHARE A COMMON SEQUENCE AND ORGANIZATION OF mtDNA

The mtDNA of the ten HS ρ^- mutants have been analyzed by restriction digest with the enzymes MspI and HphI (two previoulsly known restriction sites of the *rep* sequences distant of 190 bp from each other[3]) and hybridization to HS416 used as a reference (for the complete characterization and sequence of HS416 see ref. 3). Examples of such experiments are shown in fig. 1. For each HS ρ^- mutant the MspI digestion gives rise to a single fragment (varying in size from 600 bp to 1200 bp) that cross hybridizes with the probe carrying the *rep1* sequence. Similar results are obtained if HphI is used,whereas the double digests give rise, for each mutant, to a 190 bp long fragment that hybridizes to the probe as expected if all HS ρ^- carry a *rep* sequence (data not shown). From these results we conclude that HS ρ^- mutants are tandem repeats (single and same size fragments with two enzymes cutting 190 bp from each other) of a short segment of the wild type genome (ca. 1%) containing a *rep* sequence (table 2). Implications of these findings for the mechanisms involved are discussed in ref. 3.

CONCLUDING REMARKS

Recent developments of mitochondrial genetics in yeast often involve the screening or characterization of mtDNA from a large number of clones, strains, mutants or recombinants. The minilysate method has also been found suitable for experiments in which the natural polymorphism of yeast strains (particularly the optional introns) provides the molecular markers for specific sequences in ρ^+ strains (Dujon et al. in prep.).

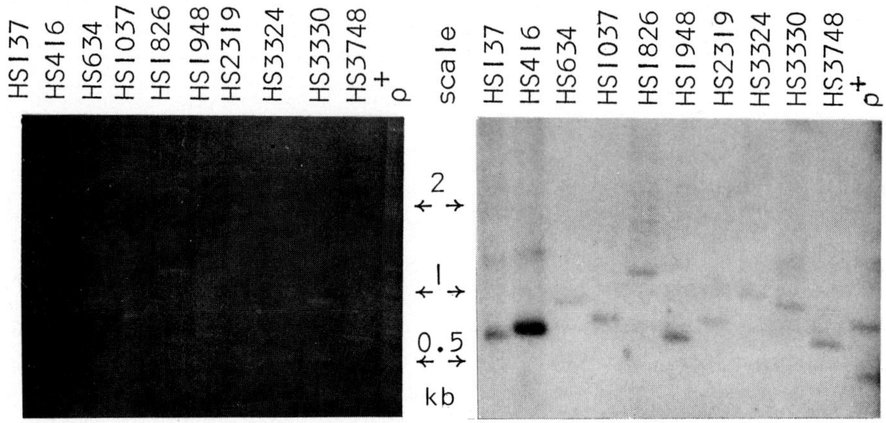

Fig. 1. Analysis of the mtDNAs of the HS ρ^- mutants and their ρ^+ parental strain (KL14-4A/121). Each minilysate was restricted with MspI, the fragments separated on a 1.5% agarose gel and transferred to nitrocellulose filters[8]. Hybridization was carried out at 55°C for 20 h in a buffer containing 1.2 M NaCl/120 mM Na citrate using nick-translated DNA of the recombinant plasmid pSCM107 (carrying one repeat unit of HS416) as a probe.

ACKNOWLEDGMENTS: We thank Professors W.Gilbert and P.Slonimski for useful discussions and support, and G.Brefort, G.Church, G.Dujardin, C.Grandchamp, R.Morimoto and D.Morisato for their discussions and informations. The excellent technical assistance of S.Robineau was highly appreciated. B.D. is the recipient of a N.S.F.-C.N.R.S. exchange fellowship, H.B. was a boursier D.G.R.S.T. This work has been supported by grants to W.Gilbert and to P.Slonimski.

REFERENCES

1. Ephrussi, B., and Grandchamp, S. (1965) Heredity, 20, 1-7

2. Ephrussi, B., Jakob, H. and Grandchamp, S. (1966) Genetics, 54, 1-29

3. Blanc, H. and Dujon, B. (1980) Proc.Nat. Acad. Sc. U.S.A. (in press)

4. Blanc, H. Thesis, Institut National Agronomique, Paris-Grignon, France (1979)

5. Hsiao, C.L. and Carbon, J. (1979) Proc. Nat. Acad. Sc. U.S.A., 76, 3829-3833

6. Nasmyth, K.A. and Reed, S.I. (1980) Proc. Nat. Acad. Sc. U.S.A., 77, 2119-2123

7. Davis, R.W., Thomas, M., Cameron, J., StJohn, T.P., Scherrer, S. and Padgett, R.A. (1980) in "Methods in Enzymology" vol. 65, pp 404-411

8. Southern, E. (1975) J.Mol.Biol., 98, 503-517.

The Organization and Expression of the Mitochondrial Genome
A.M. Kroon and C. Saccone, editors

SPLIT GENES ON YEAST MITOCHONDRIAL DNA: ORGANISATION AND EXPRESSION

L.A.GRIVELL, A.C.ARNBERG*, L.A.M.HENSGENS, E.ROOSENDAAL, G.J.B.VAN OMMEN[α] AND E.F.J.VAN BRUGGEN*

Section for Molecular Biology, Laboratory of Biochemistry, Univer University of Amsterdam, Animal Physiology Institute, Kruislaan 320, 1098 SM Amsterdam (The Netherlands) and *Biochemical Laboratory, State University Groningen, Nijenborgh 16, 9747 AG Groningen (The Netherlands)

Introduction

In the brief period that has elapsed since the discovery by Bos et al.[1] that the gene for the large ribosomal RNA (rRNA) on the mtDNA of omega$^+$ yeast strains contains an intervening sequence, evidence has accumulated that at least two other genes on this DNA are multiply split. For the gene for apocytochrome b we have a detailed picture of an intricate mosaic organisation with elaborate processing controls[2-7]. For the gene for subunit I of cytochrome c oxidase, the magnitude of its complex organisation is only now being realised. Both genes have been the object of study in our laboratory and this article briefly reviews what is known about their structure and mode of expression.

The gene for apocytochrome b

a) Structure - In the *Saccharomyces cerevisiae* strain KL14-4A the structural gene for apocytochrome b contains at least four introns and extends over approximately 7500 base pairs (bp) of the mitochondrial genome (the COB region, Fig. 1). The major transcript of the region is an 18S RNA (2200 nucleotides), which directs the synthesis *in vitro* of antigenic determinants of cytochrome b[3] and which is thus probably the functional messenger RNA (mRNA) *in vivo*. The positions of the five exons (α-ε) were established by electron microscopy of DNA-RNA hybrids formed between the 18S RNA and COB DNA fragments[3]. Both restriction mapping and electron microscopy show that two of the intervening sequences - $I_{\alpha\beta}$ and

---

[α] Present address: Pediatrics Clinic, Academic Hospital, University of Amsterdam, Grimburgwal 10, 1012 GA Amsterdam (The Netherlands).

Abbreviations: rRNA, ribosomal RNA; mRNA, messenger RNA; $I_{\alpha\beta}$, $I_{\beta\gamma}$ etc., introns between coding sequences α and β, β and γ, respectively; kb, kilo base pair(s); bp, base pair(s).

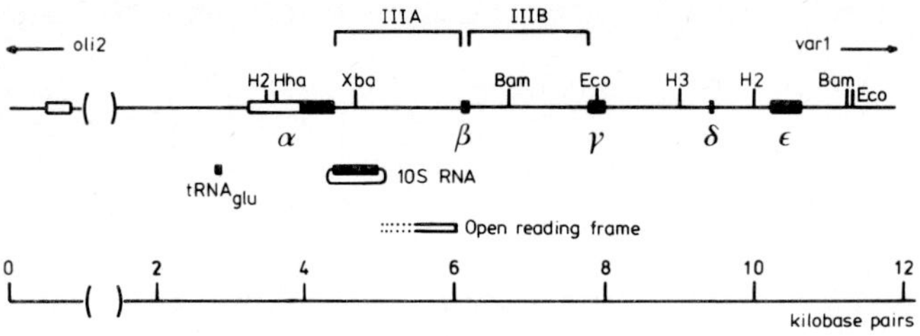

Fig. 1. The COB region in S. cerevisiae KL14-4A. The figure shows a portion of the restriction map of KL14-4A mtDNA, with the directions of the OLI2 and VAR1 genetic markers included for the purpose of orientation. Coding sequences for the 18S COB mRNA are designated α-ε and are shown as thickened bars. They were localised primarily by electron microscopy (ref. 3), with additional information from Southern blotting analysis (Crusius, J.B.A., unpublished) and DNA sequence analysis (Grivell, L.A., unpublished). The direction of transcription is from left to right. In the α exon protein-coding sequences (about 400 bp) are shown shaded; the remaining 700 bp form part of a 5'-leader. Additional sequences, coding for 250-350 bp of the leader are located leftwards of α, but these have not been mapped (see text). The open reading frame in insertion IIIA ($I_{\alpha\beta}$) is linked to the α exon by excision of 10S RNA (refs 10-13,31). The restriction sites shown are: HindII (H2); HindIII (H3); BamHI (Bam); EcoRI (Eco); HhaI (Hha); XbaI (Xba).

$I_{\beta\gamma}$ - are optional in that they are absent from the mtDNA of several yeast strains. These segments of DNA have previously been defined by Sanders et al.[8] as insertion III, one of the major differences in size and sequence between the mtDNAs of related yeast strains. In strains like S. cerevisiae D273-10B, which lacks insertion III, DNA sequence analysis[9] and shows that the gene is continuous at the positions corresponding to the insertion sites of IIIA and IIIB, so that only $I_{\gamma\delta}$ and $I_{\delta\varepsilon}$ remain. Both $I_{\alpha\beta}$ and $I_{\beta\gamma}$ are inherited together and so far no yeast strains have been found in which only part of insertion III is present.

b) Processing - Like the mRNAs of other eukaryotic split genes, the 18S RNA arises by a series of cut-and-splice events and details of this process have been studied by several groups[5,6,10]. The scheme shown in Fig. 2 is based on hybridization of intron-specific probes with splicing intermediates in both wild-type and splicing-deficient mutants[10]. It incorporates the following features:

1. The intervening sequence $I_{\alpha\beta}$ is excised in two steps. The first of these is a possibly obligate, early event in overall processing and results in the formation of a stable and abundant

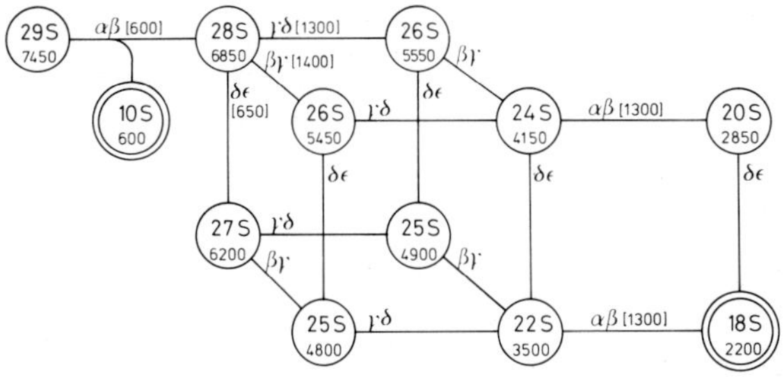

Fig. 2. Processing scheme for COB transcripts. The scheme is based on hybridization data obtained with intron-specific probes in RNA blotting analysis (ref. 10). The transcripts so identified are shown within the circles and are designated by both an approximate SE value as well as a size in nucleotides. Stable RNAs are indicated by double concentric circles; the existence of the 29S RNA is predicted on the basis of the composition of the 28S RNA and has since been demonstrated directly (ref. 14). Minor transcripts with compositions other than those shown have also been detected and imply that other pathways for splicing exist, but are infrequently used. See text for discussion. From ref. 10.

10S RNA (650 nucleotides)[11], together with a 28S RNA in which the α exon is linked to a long, open reading frame in the remainder of $I_{\alpha\beta}$. This RNA is capable of specifying a 42 kilo Dalton read-through protein[12], sharing antigenic determinants with cytochrome b. This protein has been proposed to play a role in splicing (refs 10,12,13). The 28S RNA undergoes further processing, but excision of the remainder of $I_{\alpha\beta}$ is frequently a late event and possibly even a holding point in processing. Whether the 10S RNA serves any function is as yet unknown. Curiously, this RNA - like a number of OXI3 transcripts (see below) - is circular (Arnberg, A.C., unpublished). It may be an example of an intervening sequence whose ends are ligated. So far, the only mutations known to affect 10S RNA are those which hinder its excision from 29S RNA[14] and this might argue that the bulk of its sequences are not involved in any essential function. 10S RNA must be excised by a nuclear-coded enzyme, since any petite mutant retaining this part of the COB region produces an apparently normal 10S RNA (Crusius, J.B.A., unpublished). In view of such an apparently passive role in cytochrome b gene expression, it would be interesting to verify whether yeast strains specifically deleted for this part of $I_{\alpha\beta}$ can exist.

2. Apart from excision of 10S RNA there does not appear to be an obligate order of splicing of other intermediates, although a

preferred pathway cannot be ruled out. A strictly linear progression of excision in the order $I_{\alpha\beta} \rightarrow I_{\beta\gamma} \rightarrow I_{\gamma\delta}$ inferred from studies with splicing-deficient mutants[6], is not observed but this does not necessarily contradict models which postulate the sequential production by splicing of trans-acting elements (spligase, RNA guide), which trigger subsequent splicing steps[15].

c) The mRNA for apocytochrome b contains a long 5'-leader - The 18S RNA, with a length of about 2200 nucleotides, is around twice the size minimally necessary to specify apocytochrome b and the correlation of genetic with physical mapping data shows that most of the extra sequences must be located in an unusually long 5'-leader. DNA sequence analysis (ref. 9 and Grivell, L.A., unpublished) confirms this and shows that only about 400 bp at the 3'-end of the α exon code for cytochrome b (Fig. 1). Thus, maximally 1150 bp of the combined exons are used to specify protein and the remaining 1050 nucleotides of the 18S RNA must constitute a 5'-leader. Of this, 600 nucleotides should be specified by DNA sequences located leftwards of the HindII site in α (Fig. 1), as judged from the lengths of RNA tails extending from DNA-RNA hybrids (ref. 3). So far, however, neither S_1 nuclease analysis nor Southern blot hybridization with a partially-purified 18S fraction has provided evidence for more than 250-350 bp of these sequences immediately leftwards of the HindII site (Fig. 1; Crusius, J.B.A., unpublished). The remaining 250-350 bp do not appear to be within 6000 bp of α and have so far eluded detection. At present, we have no good explanation for this discrepancy. Trivial explanations might be that the length of the 18S RNA has been seriously over--estimated in the electron microscope or that DNA-RNA hybrids, involving this part of the RNA, have an extremely low stability (the region immediately downstream of the gene for $tRNA_{glu}$ is AT-rich[9]). Alternatively, the sequences specifying the remaining 250-350 nucleotides of the leader may be located elsewhere on the genome. Thus, although a search for further 18S RNA sequences within 6000 bp of α has so far proved negative, we cannot yet exclude the possibility that the leader is spliced. If this is indeed the case, then the minimum size of the precursor RNA containing all introns will be at least 15.000 nucleotides and the leader intron will contain at least one essential gene - that for $tRNA_{glu}$ - located on the same strand as the cytochrome b gene, about 370 bp upstream of the α exon[9].

Splicing of leader sequences is a feature of gene expression in several viruses[16-18]. Adenovirus mRNA leaders can be multiply spliced[16], while in SV40 alternative splicing gives rise to a multiplicity of leader sequences which may be involved in differential translational control[17]. Resolution of the situation for the 18S mRNA leader is thus of the greatest interest.

The OXI3 region

a) Structure - The gene for subunit I of cytochrome *c* oxidase is contained within the OXI3 region, which spans - dependent on strain - minimally some 6400-11.400 bp of mtDNA[19]. It contains three sites at which major insertions or deletions can occur, without affecting either the structure of subunit I or its synthesis. More recently, one of the insertions (II; see ref. 8) has been found by electron microscopy to consist of two coordinately-inherited insertions of 1100 bp (IIA) and 1400 bp (IIB), separated by a conserved sequence of approximately 150 bp (Arnberg, A.C. and Garritsen, H.W.G.A., unpublished). Further, fine-structure restriction mapping has revealed a fifth insertion (XI; Enthoven, P.M.M., unpublished). Early indications that the gene for subunit I might be split came from restriction mapping of mit$^-$ mutations affecting the synthesis of this subunit and from the results of mtDNA-directed protein synthesis *in vitro*[19]. Definite proof that this is indeed so is only now emerging from DNA sequence analyses (Tzagoloff, A., this volume).

b) Transcripts, splicing and circular RNAs - One of the most intriguing aspects of the OXI3 region is the large variety of transcripts derived from it, suggesting that processing may play an important role in the regulation of its expression and even that it may contain other genes in addition to that for subunit I (ref. 20). A preliminary transcript map was constructed by Van Ommen *et al.*[11] and this has been refined by use of cloned fragments of mtDNA as probes in RNA blotting analysis. Our findings are outlined in Fig. 3 and can be summarised as follows:

1. In *S. cerevisiae* KL14-4A major transcripts are found migrating at 11S, 18S, 19S and 19.2S. These RNAs map at distinct sites and are probably the products of processing of longer RNAs which migrate in the range 21S-32S and which show overlapping

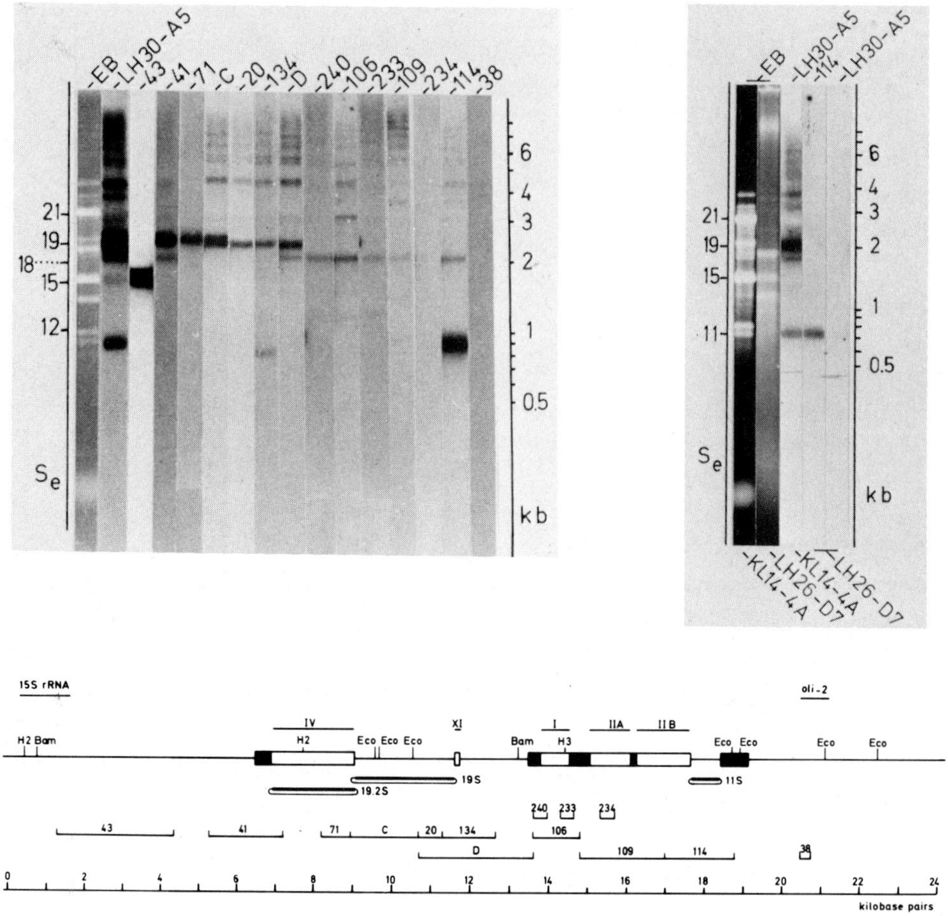

Fig. 3. Transcripts of the OXI3 region. Upper panel: Blot hybridization analysis of mtRNAs from *S. cerevisiae* KL14-4A, mit^- mutant M11-125 (refs 19,27,28) and petite mutant LH-26D7. OXI3 transcripts were visualized by hybridization with labelled probes specific for the region. For KL14-4A and M11-125, these were BclI or MboI fragments of KL14-4A mtDNA, cloned in pBR322 and labelled with ^{32}P by nick-translation. Their construction and characterization will be described elsewhere. For LH-26D7, which retains a 32.000 bp colinear segment of the wild-type genome containing all genetic markers of the OXI3 region, results with only two probes are shown. First, mtDNA from petite LH-30H5, which also retains the complete OXI3 region, was used to detect all OXI3 transcripts simultaneously. Second, clone 114 was used to specifically detect 11S RNA sequences. Results obtained with both probes were confirmed by hybridizations with other cloned fragments (data not shown). RNA electrophoresis, blotting and hybridization were carried out as described previously (ref. 11). Lower panel: The restriction map of the OXI3 region in KL14-4A with the gene for 15S rRNA and the OLI2 genetic locus included for the purpose of orientation. The approximate positions of sequences coding for the 18S OXI3 RNA are shown by thickened black bars, while the inserted sequences identified by Sanders *et al*. (ref. 8) are shown by open

hybridization. The 18S RNA shows the most complex sequence organisation, hybridizing with at least five non-contiguous DNA segments extending from the left-hand side of insertion IV, through the OXI3 region and into the OLI2 locus (*cf*. ref. 11). It is, on this basis, perhaps the most likely candidate for the mRNA of subunit I.

2. The strain-specific insertions are transcribed and their sequences are excised in discrete steps. Thus, 19.2S RNA is found only in strains containing insertion IV and is presumably a complete transcript of it. Longer RNAs still containing 19.2S sequences are present at extremely low levels, suggesting that its excision is an early event in processing. Excision of 19S, 11S RNAs and sequences of insertions I and IIA are generally speaking late events, which do not seem to occur in any specific order, although there may be a preference for early excision of IIA and 11S. A minor 13S RNA is correlated with the presence of insertion I, but so far no stable separate transcripts of either insertions IIA, IIB or of the 1500 bp segment downstream of 19S RNA sequences have been detected.

3. No 18S RNA coding sequences are detectable in the 6 kb long DNA segment, which includes insertion IV and extends just beyond the BamHI site in Fig. 3. The fact that this segment gives rise to two discrete RNAs (19.2S and 19S), each of which is removed at different stages of processing, leads to the conclusion that its excision involves at least three separate steps. The first of these is probably excision of the 19.2S RNA; the other events are not strictly ordered. The situation is reminiscent of that involving $I_{\alpha\beta}$ in the COB region[10,12,13] and it is conceivable that alternative splicing events function to link different reading frames to the first exon of 18S RNA.

(← Fig. 3 - continued)

bars and are indicated by horizontal lines above the figure. Under the main map, discrete transcripts originating from introns in 18S coding sequences are shown as black bars, together with the map positions of the cloned fragments used as labelled probes. The positions of mutations in M11-125 are indicated by Δ (refs 19,28). Restriction fragments are identified as described in ref. 8, while restriction sites are abbreviated as given in the legend to Fig. 1.

4. Of all OXI3 transcripts only the 11S RNA is present in a petite mutant (LH-26D7) spanning the complete OXI3 region (Fig. 3). Processing of other RNAs must, therefore, be dependent on mitochondrially-coded proteins and/or products specified by sequences outside the region. Similar observations, made for petite mutants retaining the cytochrome b gene, have provided support for the idea that mtDNA encodes proteins involved in splicing[13,15] and such proteins may well be required in processing of OXI3 transcripts.

The discovery that at least three major OXI3 transcripts - the 19.2S, 19S and 11S RNAs - behave as covalently-closed circular molecules (ref. 21 and Arnberg, A.C., unpublished) adds a further element of complexity to this region. Like the 10S RNA in COB, these could arise as products of splicing, but this obviously does not preclude a separate function. A clue to how the 19.2S and 19S circles may be formed has been given by electron microscopy of the 25S-28S fraction of mtRNA, which consists predominantly of long OXI3 transcripts. As Fig. 4 shows, this fraction contains tailed circular molecules (neck-ties) formed by internal base pairing[21]. Various types of these structures have been observed and these can be grouped according to the lengths of their circular portions. Their identity as OXI3 transcripts has been established by hybridization to selected restriction and cloned DNA fragments and this has also permitted correlation with transcripts mapped by other means.

In *S. cerevisiae* KL14-4A, the most abundant class of neck-tie structure is one in which the circular portion contains sequences of the 19S RNA (Fig. 5). As can be inferred from Fig. 3, sequences involved in circle formation by base pairing are thus probably located at the rightward edge of insertion IV in KL14-4A and in the immediate vicinity of insertion XI. However, some variation is seen in the position of the marker fragments in relation to the junction with insertion IV, so that it is possible that circularisation can occur at more than one site in this region. RNA blotting suggests that these neck-ties constitute a 24S RNA, which contains 19S plus 18S RNA sequences.

Two of 30 neck-ties examined possessed a circular portion with a contour length of about 4800 nucleotides, approximately equivalent to the combined lengths of 19S and 19.2S RNAs. A

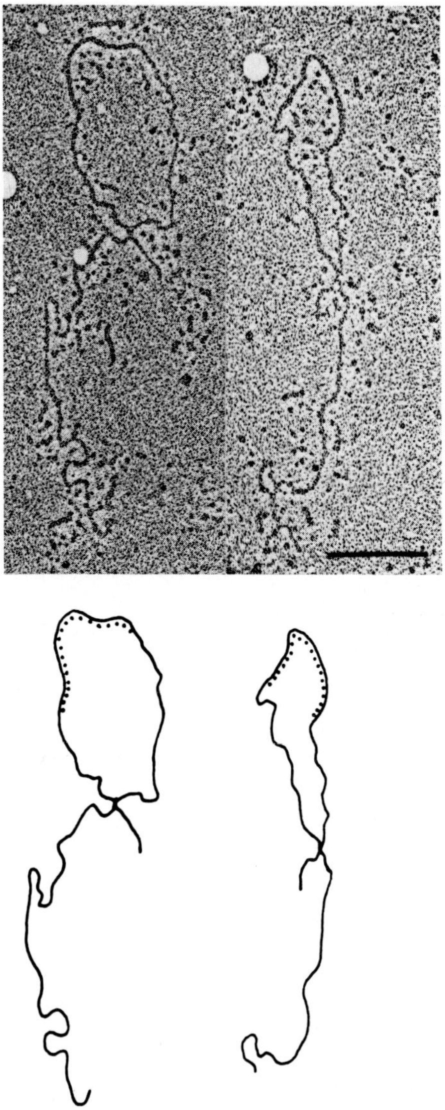

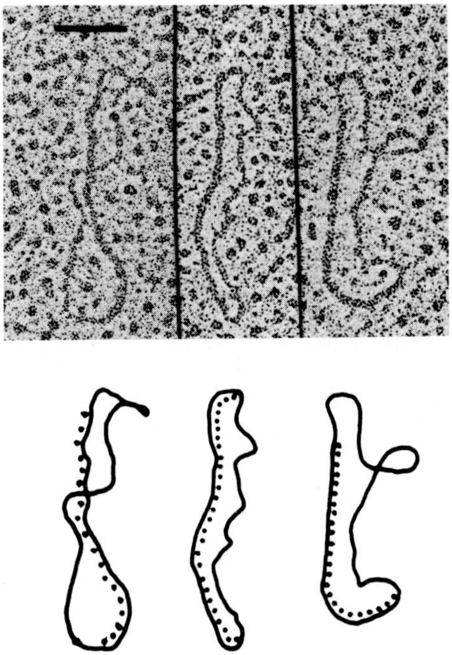

Fig. 4. Tailed circles in the 25S-28S fraction of mtRNA. The 25S-28S fraction of KL14-4A mtRNA was isolated by preparative agarose gel electrophoresis as described (ref. 11). The electron micrographs show two of the tailed circle structures which abound in this fraction. They were identified as transcripts of OXI3 by hybridization to a 950 bp EcoRI fragment of *S. carlsbergensis* mtDNA (RR8: ref. 8). Hybridization was carried out exactly as described by Thomas *et al*. (ref. 32). The bar is 0.2 µm.

Fig. 5. Identification of one class of 19S circular RNAs. The electron micrograph shows hybrids formed between a 19S fraction of KL14-4A mtRNA and EcoRI fragment RR8 from *S. carlsbergensis* mtDNA. Experimental details were as described in the legend to Fig. 4. The bar is 0.1 µm.

circularisation site must, therefore, also exist at the leftward edge of insertion IV and this could mean either that 19S and 19.2S RNAs are occasionally excised together or that an extra mechanism is involved to ensure correct excision of 19.2S RNA first.

It seems likely that GC-rich sequences are involved in the formation of neck-ties, since despite the shortness of the internally base-paired regions ($\leq$50 bp), the structures survive spreading even from 84% formamide-2.8 M urea. GC-rich clusters occur widespread in the mtDNAs of *S. carlsbergensis* and *S. cerevisiae*[22] and DNA sequence analysis shows that some of these are homologous with each other[23,24]. Intercluster base pairing within a transcript would create the tailed circles observed. What the exact role of the tailed circles may be, however, is difficult to define. In *Escherichia coli*, base pairing between short, inverted repeats at the ends of the precursor to 16S and 23S rRNAs forms a cleavage site for ribonuclease III (ref. 25), but in the case of the neck-ties it is not at all clear why such extensive base pairing should be required to achieve splicing. Certainly, neck-tie formation is not a pre-requisite for splicing in the 21S rRNA gene, since sequences around the splice point are incapable of extensive base pairing and are not GC-rich[24,26].

c) Mutations affecting processing in OXI3 - As in the case of the cytochrome *b* gene, analysis of mutants disturbed in processing should provide useful insight into the sequences involved. We have made a study of a number of such mutants (ref. 22 and Hensgens, L.A.M., unpublished) and results obtained with one of these is shown in Fig. 3. The mutant (M11-125) was isolated by Tzagoloff (see ref. 27) and has been shown to contain two widely-separated, minor changes[19,28] (Fig. 3). The first of these is the loss of a HhaI site well inside insertion IV and the second, the gain of a HpaI site within the sequences coding for 11S RNA. This mutant accumulates extremely long RNAs still containing 19.2S and 19S RNA sequences, indicating a block at an early stage in processing and in this respect it resembles 10S or $I_{\alpha\beta}$ mutants which prevent the excision of $I_{\alpha\beta}$, $I_{\beta\gamma}$ and $I_{\gamma\delta}$ in COB. Wild-type functions can only be restored to M11-125 by petite strains retaining the whole OXI3 region (*cf*. ref. 27), so that both changes seen in the DNA probably contribute to the mutant phenotype. The mutant thus defines, to within a few base pairs, sequences within two introns which are of importance for processing and the further study of these is obviously of great interest.

d) Interactions between COB and OXI3: The BOX effect - Evidence for the existence of regulatory interactions between the COB and OXI3 loci is provided by the observation that certain of the splicing-deficient mutants mapping in $I_{\alpha\beta}$ and $I_{\gamma\delta}$ lack subunit I of cytochrome *c* oxidase (the BOX effect[29]). This interaction is predominantly unidirectional since a variety of mutations in OXI3, including large deletions, have no effect on COB gene expression (refs 5,10). It is a positive regulatory effect since deletion of most of the COB region also results in an inability to synthesize subunit I (refs 5,7) and finally, it is independent of translation of COB since in exon-intron double COB mutants the effect of the intron mutation on OXI3 is epistatic regardless of its position[7]. Initially, the finding that both $I_{\alpha\beta}$ and $I_{\gamma\delta}$ mutants are disturbed in their pattern of OXI3 transcripts[5,10] provided an attractive explanation for the BOX effect and further suggested how it might be mediated: both groups of mutant accumulate long RNAs still containing $I_{\gamma\delta}$ sequences, so that the minimum hypothesis is that the excised $I_{\gamma\delta}$ intron provides an element involved in processing. This simple picture is complicated, however, by the finding that the changes in OXI3 transcript pattern cannot be directly correlated with the presence or absence of subunit I (ref. 10), so that the BOX effect seems unlikely to arise solely from changes at the level of RNA processing. How exactly it is achieved remains an intriguing problem for the future.

Prospects

It would scarcely seem necessary to enlarge on the exciting possibilities that lie ahead. The existence of introns within this apparently simple genome is in itself sufficient to fire the imagination and it is possible that a study of their origins, evolution and their relationship to the genes which they interrupt will give useful clues to the role of introns in nuclear genes[30]. However, the observation that some introns in yeast mtDNA can code for protein[24], can influence gene expression in other parts of the genome[5,7,10] and can thus perhaps control functions which are essential only in those strains possessing them is even more intriguing. The characterisation of such intron proteins and of the even more bizarre exon-intron fusion proteins, produced by translation across splice junctions, are just two of the tasks for the immediate future.

Acknowledgements

We are extremely grateful to Prof. C.P.Hollenberg (Universität Düsseldorf, Institut für Mikrobiologie, Germany) for his hospitality and help in the initial phases of recombinant clone construction; Dr R.J.Schweyen (Genetisches Institut der Universität München, Germany) and Dr A.Tzagoloff (Columbia University, Fairchild Center for Life Sciences, New York, N.Y., USA) for generous gifts of some of the yeast mutants used; Mrs G.Van der Horst and Mr M.De Haan for expert technical assistance; Mr P.H.Boer, Mr J.B.A.Crusius and Mr P.M.M.Enthoven for communication of unpublished results and Prof. P.Borst for much constructive criticism. We also thank Mr K.Gilissen for printing the electron micrographs. This work was supported in part by grants to P.Borst/LAG and EFLVB from The Netherlands Foundation for Chemical Research (SON) with financial aid (including a short-term study grant to LAMH) from The Netherlands Organization for the Advancement of Pure Research (ZWO).

References

1 Bos, J.L., Heyting, C., Borst, P., Arnberg, A.C. and Van Bruggen, E.F.J. (1978) Nature 275, 336-338.

2 Slonimski, P.P., Claisse, M.L., Foucher, M., Jacq, C., Kochko, A., Lamouroux, A., Pajot, P., Perrodin, G., Spyridakis, A. and Wambier-Kluppel, M.L. (1978) in: Biochemistry and Genetics of Yeast: Pure and Applied Aspects (Bacila, M., Horecker, B.L. and Stoppani, A.O.M., Eds), Academic Press, New York, pp. 391-401.

3 Grivell, L.A., Arnberg, A.C., Boer, P.H., Borst, P., Bos, J.L., Van Bruggen, E.F.J., Groot, G.S.P., Hecht, N.B., Hensgens, L.A.M., Van Ommen, G.J.B. and Tabak, H.F. (1979) in: Extrachromosomal DNA: ICN-UCLA Symposia on Molecular and Cellular Biology (Cummings, D.J., Borst, P., Dawid, I.B., Weissman, S.M. and Fox, C.F., Eds), Vol. 15, Academic Press, New York, pp. 305-324.

4 Haid, A., Schweyen, R.J., Bechmann, H., Kaudewitz, F., Solioz, M. and Schatz, G. (1979) Europ.J.Biochem. 94, 451-464.

5 Church, G.M., Slonimski, P.P. and Gilbert, W. (1979) Cell 18, 1209-1215.

6 Halbreich, A., Pajot, P., Foucher, M., Grandchamp, C. and Slonimski, P.P. (1980) Cell 19, 321-329.

7 Alexander, N.J., Perlman, P.S., Hansen, D.K. and Mahler, H.R. (1980) Cell 20, 199-206.

8 Sanders, J.P.M., Heyting, C., Verbeet, M.Ph., Meijlink, F.C.P.W. and Borst, P. (1977) Mol.Gen.Gen. 157, 239-261.

9 Nobrega, F.G. and Tzagoloff, A. (1980) FEBS Lett. 113, 52-54.

10 Van Ommen, G.J.B., Boer, P.H., Groot, G.S.P., De Haan, M., Roosendaal, E., Grivell, L.A., Haid, A. and Schweyen, R.J. (1980) Cell 20, 173-183.

11 Van Ommen, G.J.B., Groot, G.S.P. and Grivell, L.A. (1979) Cell 18, 511-523.

12 Van Ommen, G.J.B. (1980) Ph.D.Thesis, RNA Synthesis in Yeast Mitochondria, Krips Repro, Meppel.

13 Jacq, C., Lozowska, J. and Slonimski, P.P. (1980) C.R.Acad.Sci.Paris, Ser. D, 290, janvier 14.

14 Haid, A., Grosch, G., Schmelzer, C., Schweyen, R.J. and Kaudewitz, F. (1980) Curr.Gen. 1, 155-161.

15 Church, G.M. and Gilbert, W. (1980) in: Mobilization and Reassembly of Genetic Information (Joseph, D.R., Schultz, J., Scott, W.A. and Werner, R., Eds), Academic Press, New York, in press.

16 Berget, S.M., Moore, C. and Sharp, P.A. (1977) Proc.Natl.Acad.Sci.U.S. 74, 3171-3175.
17 Piatak, M., Ghosh, P.K., Reddy, V.B., Lebowitz, P. and Weissman, S.M. (1979) in: Extrachromosomal DNA: ICN-UCLA Symposia on Molecular and Cellular Biology (Cummings, D.J., Borst, P., Dawid, I.B., Weissman, S.M. and Fox, C.F., Eds), Vol. 15, Academic Press, New York, pp. 199-215.
18 Robertson, D.L. and Varmus, H.E. (1979) J.Virol. 30, 576-589.
19 Grivell, L.A. and Moorman, A.F.M. (1977) in: Mitochondria 1977: Genetics and Biogenesis of Mitochondria (Bandlow, W., Schweyen, R.J., Wolf, K. and Kaudewitz, F., Eds), De Gruyter, Berlin, pp. 371-384.
20 Borst, P. and Grivell, L.A. (1978) Cell 15, 705-723.
21 Arnberg, A.C., Van Ommen, G.J.B., Grivell, L.A., Van Bruggen, E.F.J. and Borst, P. (1980) Cell 19, 313-319.
22 Prunell, A. and Bernardi, G. (1977) J.Mol.Biol. 110, 53-74.
23 Cosson, J. and Tzagoloff, A. (1978) J.Biol.Chem. 254, 42-43.
24 Dujon, B. (1980) Cell 20, 185-197.
25 Bram, R.J., Young, R.A. and Steitz, J.A. (1980) Cell 19, 393-401.
26 Bos, J.L., Osinga, K.A., Van der Horst, G., Hecht, N.B., Tabak, H.F., Van Ommen, G.J.B. and Borst, P. (1980) Cell 20, 207-214.
27 Slonimski, P.P. and Tzagoloff, A. (1976) Europ.J.Biochem. 61, 27-41.
28 Morimoto, R., Lewin, A. and Rabinowitz, M. (1979) Mol.Gen.Gen. 170, 1-9.
29 Pajot, P., Wambier-Kluppel, M.L., Kotylak, Z. and Slonimski, P.P. (1976) in: Genetics and Biogenesis of Chloroplasts and Mitochondria (Bücher, Th., Neupert, W., Sebald, W. and Werner, S., Eds), North-Holland, Amsterdam, pp. 443-451.
30 Borst, P. (1980) in: Proceedings of Mosbach Colloquia (Bücher, Th., Sebald, W. and Weiss, H., Eds), Springer Verlag, Berlin, in press.
31 Kreike, J., Bechmann, H., Van Hemert, F.J., Schweyen, R.J., Boer, P.H., Kaudewitz, F. and Groot, G.S.P. (1979) Europ.J.Biochem. 101, 607-617.
32 Thomas, M., White, R.L. and Davis, R.W. (1976) Proc.Natl.Acad.Sci.U.S. 73, 2294-2298.

© 1980 Elsevier/North-Holland Biomedical Press
The Organization and Expression of the Mitochondrial Genome
A.M. Kroon and C. Saccone, editors

SEQUENCE HOMOLOGIES BETWEEN THE MITOCHONDRIAL DNAs OF YEAST AND *NEUROSPORA CRASSA*

ETIENNE AGSTERIBBE[*], JOHN SAMALLO[*], HANS DE VRIES[*], LAMBERT A.M. HENSGENS[+] AND LESLIE A. GRIVELL[+]

[*]Laboratory of Physiological Chemistry, State University, Bloemsingel 10, 9712 KZ GRONINGEN, The Netherlands.

[+]Section of Molecular Biology, Laboratory of Biochemistry, University of Amsterdam, Kruislaan 320, 1098 SM AMSTERDAM, The Netherlands.

INTRODUCTION

Recent DNA and protein sequence analysis has revealed that many mitochondrially coded proteins have been strongly conserved during evolution[1-5]. In *Neurospora crassa*, in contrast to the situation in yeast, the lack of suitable mutants has hampered attempts to identify and map genes on mtDNA. However, if the high degree of homology found for mt proteins also exists in the mtDNAs of these two organisms, it should allow the mapping of protein genes on *Neurospora* mtDNA by use of DNA fragments from defined genetic loci on yeast mtDNA as probes in hybridization. The results of such hybridization experiments enabled us to present a tentative gene map of *Neurospora* mtDNA.

MATERIALS AND METHODS

Neurospora crassa strains. Strain ANT-1[6] was used for the isolation of mtDNA. Nuclear DNA was prepared from the slime mutant resolved from the heterokaryon FGSC 327. Strain ANT-1 was kindly provided by Dr. D.L. Edwards, La Jolla, Cal.

Yeast strains and mtDNA clones. A15-1 is a ρ^- mutant derived from *S. cerevisiae* RM511-4A (a ural trpl). It covers part of the exon of the cytochrome *b* gene[8]. RP6 is a ρ^- mutant[9], that has retained in its mtDNA the gene for subunit 9 of the mt ATPase. BA2 and RP6 were generous gifts from Dr. R.J. Schweyen, Genetic Inst., Munich and Dr. P.P. Slonimski, CNRS, Gif-sur-Yvette resp. 30B3 and 30H2-1 are ρ^- mutants, derived from AMR34-17B (a trpl tsp25), that have retained the oxi-1 and oxi-2 markers and the oxi-2 marker respectively. Recombinant clones are all derived from KL14-4A. The Mbo I and Bcl I fragments were inserted into the Bam site of pBR322 and selected by hybridization to fine diges-

---

Abbreviations: mt=mitochondrial; SSC=0.15 M NaCl, 0.015 M Na_3-citrate; SDS= sodium dodecylsulphate; (k)bp=(kilo)base pairs, DCCD=dicyclohexylcarbodiimide; pvp=polyvinylpyrrolidone; BSA=bovine serum albumin.

tions of yeast mtDNA (L.A.M. Hensgens, unpublished).

Preparation of *Neurospora* and yeast DNA. *Neurospora* mt[10] and nuclear DNA[11] was prepared according to procedures described previously. Yeast mtDNA was prepared according to Moorman *et al.*[12]

Restriction enzyme digestion, electrophoresis and isolation of DNA fragments from gels. Restriction enzymes Eco RI, Hinc II and Bgl II were purchased from Boehringer Mannheim, Hap II, Hae III, Hinf I and Mbo I from New England Biolabs. Digestion conditions were as prescribed by the manufacturers. Agarose gel electrophoresis was in 40 mM Tris-HCl pH 7.8, 20 mM sodium acetate and 2 mM EDTA[10]. DNA fragments were eluted from gel slices by buffer treatment and centrifugation.

Nick translation of DNA and hybridization conditions. Cloned DNA fragments and ρ^- mtDNAs were uniformly labeled with ^{32}P by nick translation[13]. Hybridization to *Neurospora* DNA fragments fixed to nitrocellulose strips[14] was performed in 3xSSC, 0.1% SDS, 0.2% BSA, 0.2% pvp and 0.2% Ficoll for 40 hours at 55° unless otherwise stated.

RESULTS AND DISCUSSION

Only the genes for the rRNAs and the tRNAs have so far been identified on *Neurospora* mtDNA[15-17]. A restriction map indicating the position of these genes is shown in fig. 1. In an attempt to locate sequences specifying other known genes, we have hybridized *Neurospora* mt- and nuclear-DNA with defined segments of yeast mtDNA, containing genes for subunits 1, 2 and 3 of cytochrome aa_3, for cytochrome *b* and for subunit 9 of the mt ATPase (see figs. 2 and 5). Positive hybridization was obtained with the probes for oxidase and ATPase genes, no hybridization was found with the cytochrome *b* gene probes, presumably as a result of low homology. These sequences were not further investigated.

The genes for subunits 2 and 3 of cytochrome aa_3 (OXI1 and OXI2 region resp.) In order to locate these genes we used the labeled mtDNAs from the ρ^- 30B3 and 30H2-1 as probes. MtDNA from 30B3 contains the OXI1 and OXI2 regions as well as a number of tRNA genes and probably part of the gene of the 15S rRNA (fig.2). MtDNA from 30H2-1 has retained the oxi-2 marker only (fig. 2). In principle it should be possible to discriminate between the regions on the *Neurospora* mtDNA, which correspond to the OXI1 and 2 regions in yeast by comparing the hybridization patterns of labeled 30B3 and 30H2-1 mtDNA with *Neurospora* mtDNA. The results of these experiments are shown in fig. 3. MtDNA from 30H2-1 hybridizes to *Neurospora* Eco RI fragment 1 and to Hind III fragments 1 and 7. Since the concomitant hybridization to Hinc II fragment 2 limits the region of homology to the DNA segment between the rRNA genes, it can be inferred that the hybridization

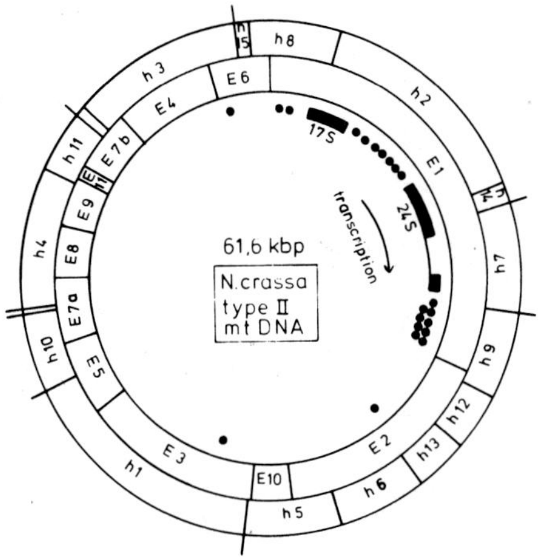

Fig. 1.

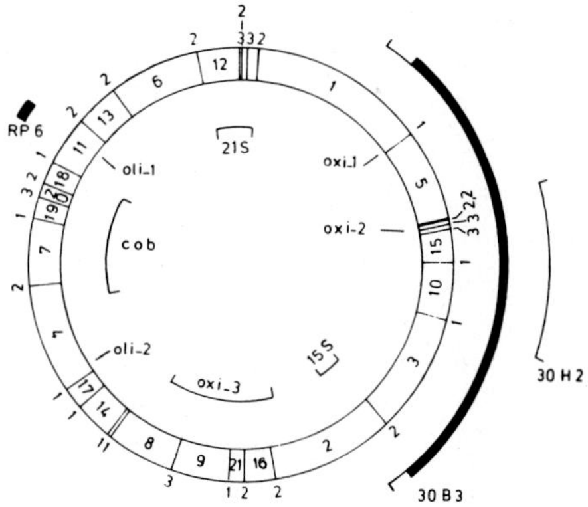

Fig. 2.

Fig.1. Physical and genetic map of mtDNA from *N. crassa*[15-17]. E and h are Eco RI and Hinc II restriction fragments. The inner ring shows the positions of the 24S and 17S rRNA genes, tRNAs genes are represented by black dots.

Fig.2. Physical and genetic map of mtDNA from *S. cerevisiae* KL14-4A[21]. 1,2 and 3 represent Eco RI, Hind II and Hind III restriction sites. On the inner ring the loci for mt genes are indicated; oxi-1,2,3 for the subunits of cyt. aa_3, cob for cytochrome *b*, 15S and 21S for rRNA, oli-1,2 for resistance markers for oligomycin The two outer rings show the locations of the mtDNAs from the petites RP6, 30B3 and 30H2-1 with respect to the wild type genome.

with the Hind III digest is restricted to fragment 7A and part of fragment 1. See fig. 4 for the Hinc II and Hind III restriction sites in Eco RI fragment 1. No hybridization could be detected with Hind III fragment 12, situated between fragment 1 and 7A. It cannot be excluded that the hybridization with fragment 1 is caused by cross-homology between rRNA genes on *Neurospora* mtDNA and yeast nuclear DNA, present as a contaminant in the petite mtDNA preparation, although this would imply that the small rRNA gene is stronger conserved than the large rRNA gene. So it seems probable that the sequence homologous to the yeast OXI2 region is on Hind III fragment 7A. MtDNA from petite 30B3 hybridizes to Eco RI fragments 1 and 4 and to the overlapping Hinc II fragments 2 and 3 (see fig.3). If we assume that the hybridization with Eco RI fragment 1 is due to complementarity with the yeast OXI2 region, and perhaps also with the gene of the small rRNA, it seems likely then that Eco RI fragment 4 contains a sequence homologous to the yeast OXI1 region.

The petite mtDNAs also hybridize to a DNA fragment of approximately 15,000 bp, an example is indicated with an asterisk in fig. 3. A possible explanation will be given in the following section.

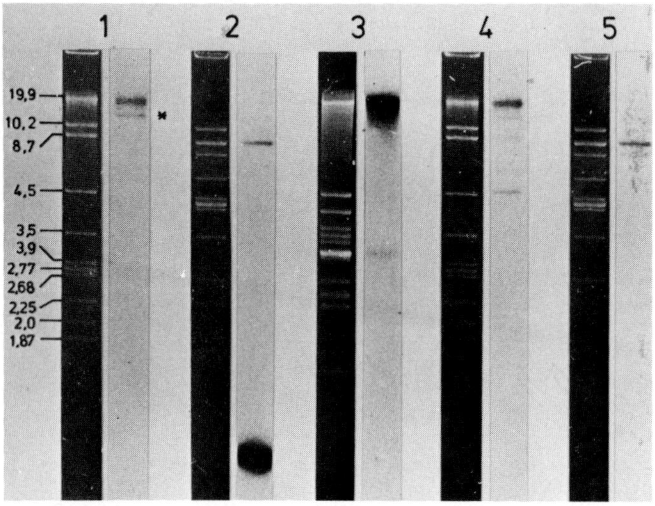

Fig.3. Hybridization of mtDNAs from the petites 30H2-1 (lanes 1,2 and 3) and 30B3 (lanes 4 and 5) to Eco RI (lanes 1 and 4), Hinc II (lanes 2 and 5) and Hind III (lane 3) digests of *Neurospora mtDNA*. The length of the Eco RI fragments is given in kbp.

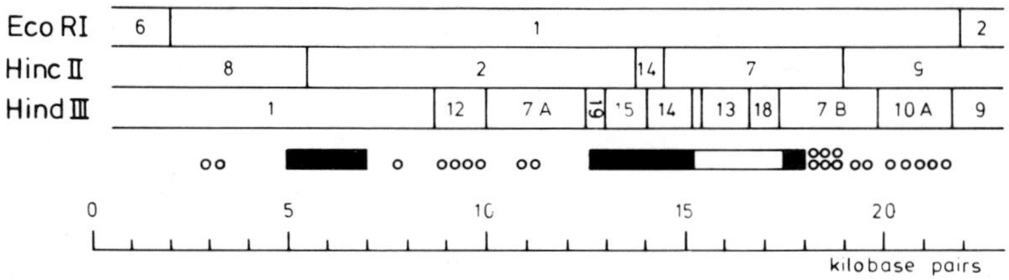

Fig.4. Physical and genetic map of Eco RI fragment 1 from *Neurospora* mtDNA[18,19]. The closed bars represent the rRNA genes, the open bar the intervening sequence in the gene for 24S rRNA. The tRNA genes are indicated by open circles.

The gene for subunit 1 of cytochrome aa_3 (OXI3 region). The OXI3 region in yeast mtDNA is large, spanning approx. 15% of the total mt genome, and also incompletely characterized. The coding sequences for subunit 1 of cytochrome aa_3 have not been localized accurately[18]. Several major insertions and deletions can occur within the region[19,20] and it is further not excluded that the OXI3 region contains other as yet not identified genes[21]. Hence we have screened a series of Mbo I and Bcl I fragments from the yeast OXI3 region for homology. The arrangement of these fragments on yeast mtDNA is shown in fig. 5. Hybridization was positive for the fragments D, A, 111 and 108, and not detectable for the fragments C, 20 and 134. Complementarity of these four fragments was restricted to Eco RI fragment 3, and to Hinc II fragment 1 which largely overlaps Eco RI fragment 1 (see fig. 6). Sometimes a very weak hybridization was observed with Eco RI dragment 7A and Hinc II fragment 3. The temperature chosen for most hybridization was 55°. In the case of the OXI3 fragments the optimum was at 58°. No hybridization was observed at 65°, the temperature used for yeast DNA-DNA hybridizations. The cloned fragments, the ρ^- mtDNAs and even pBR322 also hybridize to a minor DNA fragment of 15,000 bp (see figs. 3 and 6) which is not cut by Eco RI and Hinc II. This hybridization is stronger at 65° than at 55°. The reason for this hybridization as well as the origin of this band are unknown.

We have tried to construct a more detailed map of the sequences within Eco RI fragment 3, homologous to the OXI3 region in yeast by hybridization of the yeast fragments D, A, 111 and 108 to Bgl II and Hap II digests of *Neurospora* mtDNA. Eco RI fragment 3 comprises the Bgl II fragments 4,5,10,12,13 and probably 15 and the Hap II fragments 1,6,16 and a not exactly known number of smaller fragments. Bgl II fragments 4 and 13 and Hap II fragments 1 and 6 are situated at terminal positions in the Eco RI fragment 3, since they are cut by Eco RI. The

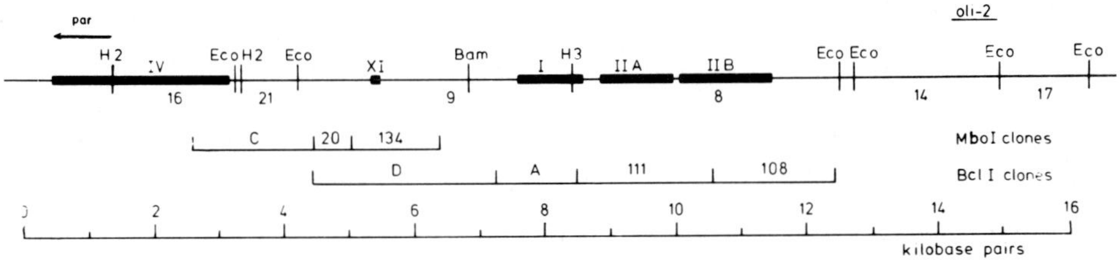

Fig.5. Physical map of the OXI3 region on mtDNA from yeast strain KL14-4A. Sites for Bam HI (Bam), Eco RI (Eco), Hind II (H2) and Hind III (H3) and the positions of the cloned fragments are shown. The closed bars represent insertions not present in *S. carlsbergensis*.

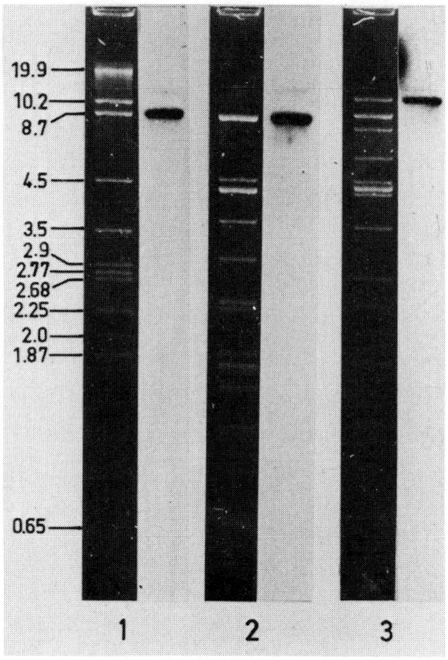

Fig.6. Hybridization of cloned Bcl I fragment 108 (see fig.5) with Eco RI (1), Eco RI + Hinc II (2) and Hind II (3) digests of *Neurospora* mtDNA. The length of the Eco RI fragments is given in kbp.

exact order of the remaining fragments has not yet been established. From the hybridization results shown in fig. 7, it can be concluded that discrete locations with each of the yeast fragments are found on *Neurospora* mtDNA. A tentative map is shown in fig. 8. We conclude that fragment A hybridizes to two non-contiguous fragments, implying either sequence duplication, or possibly the presence of inserts in *Neurospora* mtDNA in a region which is continuous in yeast. The segment of *Neurospora* mtDNA hybridizing to the yeast fragments contains 2500 -3000 bp. If we assume that the inserts I, IIA and IIB are largely devoid of homologous sequences, then the size of the yeast OXI3 segment homologous to *Neurospora* mtDNA is also approx. 3500 bp. It is important to notice that the segment of the yeast mtDNA encompassed by fragments D,A,111 and 108, is complementary to the smallest linear transcript of the OXI3 region, namely an 18S RNA. This RNA may be the mRNA for subunit 1 of cytochrome aa_3[20], so that we may have localized the coding sequences for this protein in *Neurospora*.

The gene for subunit 9 of the mitochondrial ATPase (OLI1). The gene for subunit 9, the DCCD-binding protein, of the mt ATPase is located in the OLI1 region

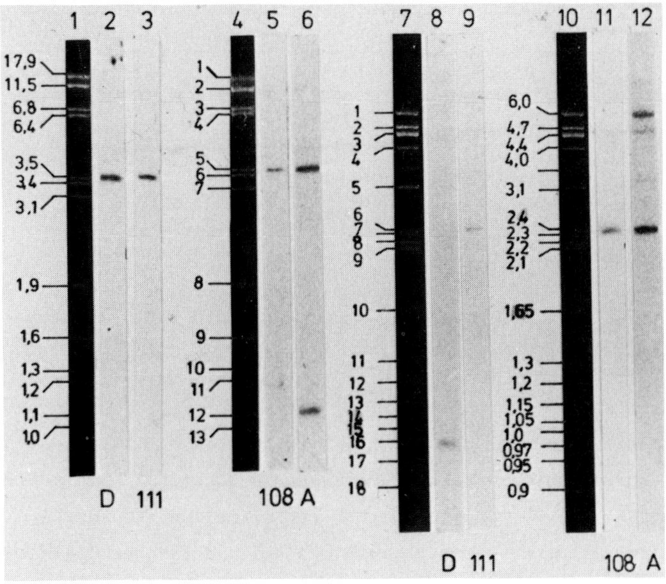

Fig.7. Hybridization of cloned Bcl I fragments D,111,108 and A of the yeast OXI3 region (see fig.5) to Bgl II (1-6) and Hap II (7-12) digests of *Neurospora* mtDNA. The length of the Bgl II and Hap II fragments is given in kbp next to lanes 1 and 10 resp., the fragment numbers next to lane 4 (Bgl II) and 7 (Hap II).

of the yeast mtDNA. Conclusive evidence has been presented that in *Neurospora* this particular gene is located in the nuclear genome[22]. Because of the high homology of these proteins an attempt was made to identify the *Neurospora* nuclear gene by using nick-translated mtDNA from petite RP6 as a probe. The mtDNA of this strain consists of a 1025 bp repeating unit, containing the structural gene for subunit 9[9]. It was of interest to look also at *Neurospora* mtDNA for two reasons, first the possibility exists that this gene was transferred to the nucleus in the course of evolution and this event may have been accompanied by gene duplication, so that a silent copy could still be present in the mtDNA. Second there is the unexplained observation of Machleidt *et al.*[23] that in the nuclear cni-1 mutant of *Neurospora* a polypeptide, strongly resembling the DCCD binding protein, appears to be a mitochondrial product. Labeled mtDNA from petite RP6 hybridized weakly to 2 or 3 bands in Eco RI digest of *Neurospora* nuclear DNA. However, these hybridizations disappeared after competition with total RNA from ρ^{o} petite mutants, indicating that the hybridizations are the result of homology between rRNA genes. (Results not shown). With *Neurospora* mtDNA hybridization was

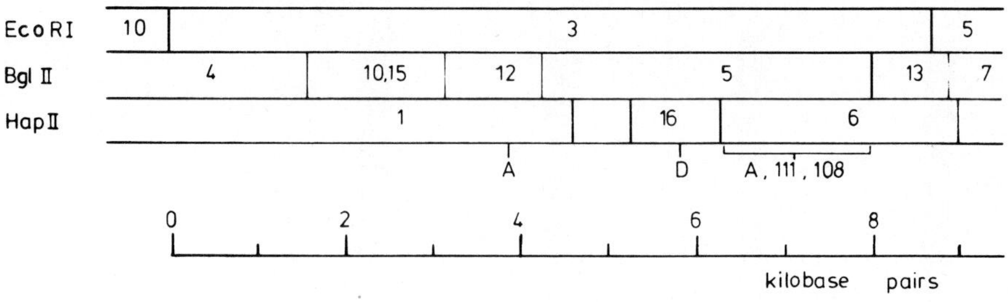

Fig.8. Tentative map of the base sequences in *Neurospora* mtDNA with homology to the Bcl I fragments D,111,108 and A from the OXI3 region in yeast (see fig.5).

found at discrete locations, namely to Eco RI fragment 4 (see fig.9) and the overlapping Bam HI fragment 2 (result not shown). With digests of *Neurospora* nuclear DNA sometimes hybridization was observed with a band at the same position as was found in mtDNA, provably the result of contamination with mtDNA. We have isolated Eco RI fragment 4 from a preparative gel and further digested with Mbo I, Hinf I and Hap II. From the hybridization results shown in fig. 9 it can be concluded that the smallest fragment with sequence homology to mtDNA from RP6 is an Hinf I fragment of approx. 1700 bp. The implication for this finding is that possibly a silent gene for the DCCD-binding protein or part of this gene is present on *Neurospora* mtDNA.

CONCLUSIONS

Our experiments show that parts of the *Neurospora* mt genome are homologous with defined genetic loci on yeast mtDNA. Their locations on the restriction map are shown in fig. 10. Assuming that the homologous regions indeed represent protein-coding sequences, then gene order has obviously not been conserved between the two organisms. It thus seems likely that gene position is not an important element in gene expression, or its regulation. The finding of sequence homology in *Neurospora* mtDNA to the yeast subunit 9 gene is the most intriguing of the results shown. Sequence analysis of this gene and the flanking regions is of greatest importance, since it may give an answer to the question of how and when a mt gene is expressed.

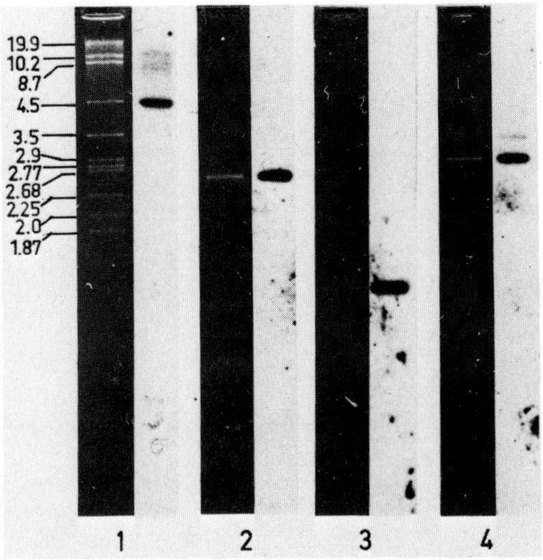

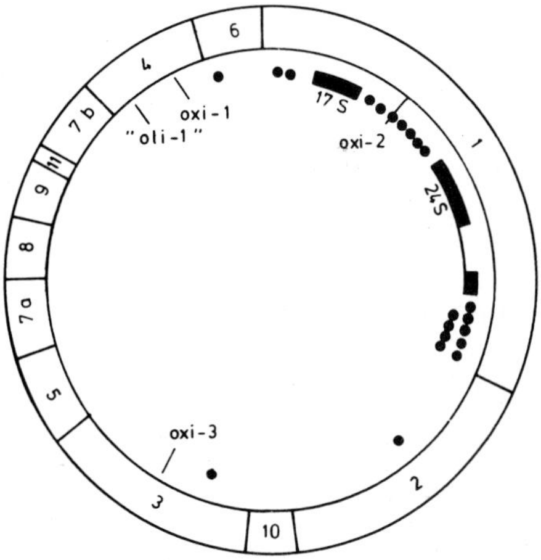

Fig. 9.

Fig. 10.

Fig.9. Hybridization of mtDNA from the petite RP6 to an Eco RI digest of total *Neurospora* mtDNA (1) and to Mbo I (2), Hinf I (3) and Hap II (4) digests of isolated Eco RI fragment 4. The length of the Eco RI fragments is given in kbp.

Fig.10. Gene map of *Neurospora* mtDNA. The tentative loci for oxi 1, 2 and 3 and oli-1 are indicated in the Eco RI restriction map of *Neurospora* mtDNA.

ACKNOWLEDGEMENTS

We thank Dr. A.M. Kroon and Dr. P. Borst for advice and criticism and Dr. C.P. Hollenberg for hospitality to L.A.M.H. in the initial phases of recombinant clone construction. Thanks are also due to Mrs. G. v.d. Horst and M. de Haan for expert technical assistance, to H. Schokkenbroek for drawing the figures, to B. Tebbes for photography and to Mrs. R. Kuperus for preparing the manuscript. This work was supported in part by grants to A.M. Kroon, P. Borst and L.A.G. from the Netherlands Foundation for Chemical Research (SON), with financial aid from the Netherlands Organization for the Advancement of Pure Research (ZWO).

REFERENCES

1. Steffens, G.J. and Buse, G. (1979) Hoppe-Seyler's Z. Physiol. Chemie, 360, 613-619.
2. Coruzzi, G. and Tzagoloff, A. (1979) J. Biol. Chem., 254, 9324-9330.
3. Fox, T.D. (1979) Proc. Natl. Acad. Sci. USA, 76, 6534-6538.
4. Machleidt, W. and Werner, S. (1979) FEBS Letters, 107, 327-330.
5. Barrell, G.B., Bankier, A.T. and Drouin, J. (1979) Nature, 282, 189-194.
6. Edwards, D.L., Chalmers, J.H., Guzik, H.J. and Warden, J.T. (1976) in: Genetics and Biogenesis of Chloroplasts and Mitochondria (Bücher, T. *et al.* Eds.) Elsevier/North-Holland, Amsterdam, pp. 865-872.
7. Van Ommen, G.-J.B., Boer, P.H., Groot, G.S.P., De Haan, M., Grivell, L.A., Haid, A. and Schweyen, R.J. (1980) Cell, 20, 173-183.
8. De Haan, M. and Grivell, L.A., unpublished observations
9. Hensgens, L.A.M., Grivell, L.A., Borst, P. and Bos, J.C. (1979) Proc. Natl. Acad. Sci. USA, 76, 1663-1667.
10. Terpstra, P., Holtrop, M. and Kroon, A.M. (1977) Biochim. Biophys. Acta, 475, 571-588.
11. Van 't Sant, P., Mak, J.F.C. and Kroon, A.M., these proceedings.
12. Moorman, A.F.M., Grivell, L.A., Lamie, F. and Smith, H.L. (1978) Biochim. Biophys. Acta, 518, 351-365.
13. Jeffreys, A.J. and Flavell, A.R. (1977) Cell, 12, 1097-1108.
14. Southern, E.M. (1975) J. Mol. Biol., 98, 503-517.
15. De Vries, H., De Jonge, J.C., Bakker, H., Meurs, H. and Kroon, A.M. (1979) Nucl. Acids Res., 61, 1791-1803.
16. Hahn, U., Lazarus, C.M., Lünsdorf, H. and Küntzel, H. (1979) Cell, 17 191-200.
17. Heckman, J.E. and RajBhandary, U.L. (1979) Cell, 17, 583-595.
18. Van Ommen, G.-J.B., Boer, P.H., Groot, G.S.P., De Haan, M., Roosendaal, E., Grivell, L.A., Haid, A. and Schweyen, R.J. (1980) Cell, 20, 173-183.
19. Sanders, J.P.M., Heyting, C., Verbeek, M.Ph., Meijlink, F.C.P.W. and Borst, P. (1977) Mol. Gen. Genet., 157, 239-261.
20. Grivell, L.A., Arnberg, A.C., Hensgens, L.A.M., Roosendaal, E., Van Ommen, G.-J.B. and Van Bruggen, E., these proceedings.
21. Borst, P. and Grivell, L.A. (1978) Cell, 15, 705-723.
22. Sebald, W., Sebald-Althaus, M. and Wachter, E. (1977) in: Mitochondria 77 (Bandlow, W. *et al.*, Eds.) De Gruyter, Berlin and New York, pp. 433-440.
23. Machleidt, W., Michel, R., Neupert, W. and Wachter, E. See ref. 8, pp. 195-198.

GENETIC ORGANISATION OF MITOCHONDRIAL DNA OF *KLUYVEROMYCES LACTIS*

GERT S.P. GROOT and NEL VAN HARTEN-LOOSBROEK
Biochemical Laboratory, Vrije Universiteit, de Boelelaan 1083,
1081 HV Amsterdam (The Netherlands)

INTRODUCTION

The mitochondrial DNA of the petite-negative yeast *Kluyveromyces lactis* consists of circular molecules with a contour length of 11.4 μm and a kinetic complexity of 20×10^6, which corresponds well with the size[1]. This DNA is therefore about half the size of the mtDNA of *Saccharomyces*[2]. Although the exact size of the mtDNA of different *Saccharomyces* species varies, the gene order is essentially constant[3]. Moreover the localisation of the rRNA genes on *Saccharomyces* mtDNA is rather unique in nature. The two genes are diametrically opposed on the mtDNA molecule and separated by approximately 30 000 bp[4]. In order to compare the genetic organisation of the mtDNA's of the two yeast species we have constructed a restriction fragment map of the mtDNA of *K. lactis* using the enzymes Hind II and Hind III[5]. Gene localisation was performed by hybridisation of separated restriction fragments to homologous rRNA or heterologous *Saccharomyces* DNA fragments of defined genetic content.

RESULTS AND DISCUSSION

Digestion of *K. lactis* mtDNA with the restriction enzymes Hind II, Hind III and Hind II+III leads to the production of 8, 3 and 10 fragments respectively (Table I). The sum of the lengths of the fragments 36 000 bp is in good agreement with the size and the previously estimated kinetic complexity. Since the number of fragments obtained in the double digest does not equal the sum of the numbers obtained in the single digests, one double digest fragment must be so small that it escaped detection. Using partial digests and redigestion of isolated fragments we have constructed the restriction fragment map of *K. lactis* mtDNA (Fig. 1). The genes for the large and small RNA were localized by hybridising [^{32}P]-labelled rRNA to separated restriction fragments after transfer to nitrocellulose filters[5]. The large rRNA hybridizes mainly to D_2 and for about 15% to D_3. Assuming a length of 3200 nucleotides (as for *Saccharomyces* 21S rRNA) this RNA can be placed on the map stretching from map position 0.82 to 0.91, provided that no intervening sequence is present in this gene[6]. The small rRNA hybridizes exclusively to D_2 or T_2 and T_7. Again

TABLE I

FRAGMENTS OF *K. LACTIS* mtDNA OBTAINED BY CLEAVAGE WITH HIND II AND HIND III
The experiment was carried out as described before[5].

Hind II		Hind III		Hind II+III	
Fragment	Length(bp)	Fragment	Length(bp)	Fragment	Length(bp)
T_1	11 600	D_1	18 100	TD_1	6 500
T_2	6 500	D_2	14 400	DT_2	6 200
T_3	5 100	D_3	3 800	TD_3	5 400
T_4	4 900			TT_4	5 100
T_5	3 800			TT_5	4 900
T_6	2 800			TD_6	3 500
T_7	1 650			TT_7	2 800
T_8	180			TT_8	1 650
				DT_9	295
				TT_{10}	180
	Σ 36 350		Σ 36 300		Σ 36 525

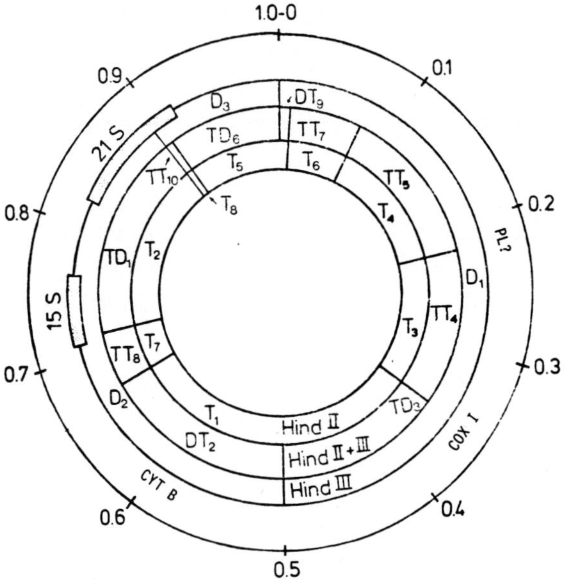

Fig. 1. The physical map of *K. lactis* mtDNA. The nomenclature used is that of Sanders *et al.*[3]. 21S and 15S: genes for the large and small rRNA; CYT B, COX I and PL are sequences homologous to those of the genes coding for cytochrome b, subunit I of cytochrome c oxidase and subunit 9 of the ATPase complex.

assuming the same length as *Saccharomyces* 15S rRNA and the absence of an intervening sequence this RNA can be placed on the map from position 0.71 to 0.76.

In order to localize other genes we have used heterologous *Saccharomyces* probes. We have shown before that *Saccharomyces* mtDNA has a small, but significant homology with *K. lactis* mtDNA of about 20%[7]. We have used mtDNA from the rho^- strains RP6 and A15-1 as probes for the genes for ATPase subunit 9 (PL) and cytochrome *b* (CYT B)[8,9]. As a probe for the gene for subunit I of cytochrome *c* oxidase (COX I) we have used a cloned Mbo fragment of the OXI 3 region. Hybridisation of these DNA's to *K. lactis* restriction fragments was very specific (Fig. 2). RP6 mtDNA hybridizes to TT_4 and TT_5, A15-1 mtDNA to DT_2

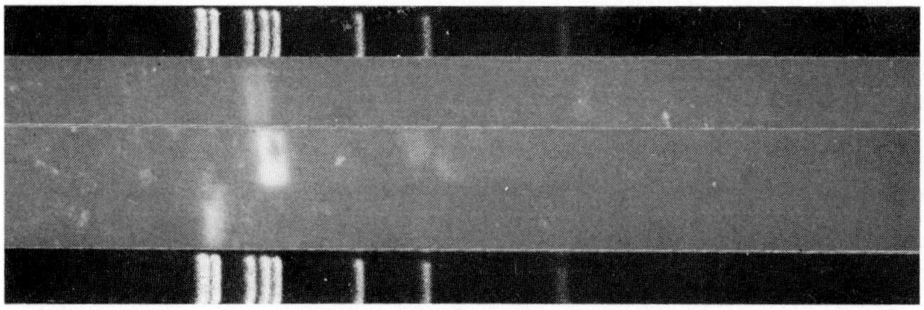

Fig. 2. Stripfilter hybridization of ^{32}P-labelled probes for sequences homologous to the COB, OXI 3 and OLI 1 region of *Saccharomyces*. Hind II+III: gels stained with ethidium bromide; COB, OXI 3 and OLI 1: autoradiograms (see text).

and the cloned Mbo-fragment to TD_3 and TT_4 localizing tentatively the respective genes as indicated in Fig. 1.

With respect to the genetic organisation on mtDNA of *K. lactis* there are several interesting observations. In the first place the rRNA genes map quite closely together. With the assumptions made, the maximal distance between the two genes is 2250 bp. The organisation of these two genes therefore is much more like that of *N. crassa*[10] and *A. nidulans*[11]. Since the latter do have intervening sequences in the gene for the large rRNA it would be interesting to see whether this is also the case for *K. lactis*. Secondly one might wonder whether the observed hybridisation with the *Saccharomyces* protein probes indeed reflects similar genes on *K. lactis* mtDNA. In view of the sequence homology and hybridisation to specific fragments,we feel that our approach is as a first approximation justified. A special case is the tentative localisation of the gene for subunit 9 of the ATPase complex. In *N. crassa* the gene for this proteolipid is located on the nuclear DNA. However sequences homologous to RP6 mtDNA are still present in the mtDNA of this organism . Unfortunately, the mitochondrial translation products of *K. lactis* have not yet been described. Since however extrachromosomally inherited oligomycin resistant mutants are known in *K. lactis* (A. Bronner-L., personal communication) it is most likely that the gene for this subunit is localised on mtDNA. Identification of the mt translation products of *K. lactis* mtDNA however is necessary before a definite conclusion can be drawn.

Finally the order of the genes on mtDNA of *K. lactis* is different from that on *Saccharomyces* mtDNA. It would therefore be interesting to investigate the nature of the differences, especially in respect with the elaborate gene organisation in *Saccharomyces* mtDNA.

ACKNOWLEDGEMENTS

We like to thank Drs L.A. Grivell and L.A.M. Hensgens for their cooperation and Dr H. van Heerikhuizen and Mrs. P.G. Brink for their help in the preparation of this manuscript.

REFERENCES

1. Sanders, J.P.M., Weyers, P.J., Groot, G.S.P. and Borst, P. (1974) Biochim. Biophys. Acta, 374, 136-144.
2. Hollenberg, C.P., Borst, P. and Van Bruggen, E.F.J. (1970) Biochim. Biophys. Acta, 209, 1-15.
3. Sanders, J.P.M., Heyting, C., Verbeet, M.Ph., Meylink, F.C.P.W. and Borst, P. (1977) Molec. gen. Genet., 157, 239-261.
4. Sanders, J.P.M., Heyting, C. and Borst, P. (1975) Biochem. Biophys. Res. Commun., 65, 699-707.
5. Groot, G.S.P. and Van Harten-Loosbroek, N. (1980) Curr. Genetics, 1, 133-135.
6. Bos, J.L., Heyting, C., Borst, P., Arnberg, A.C. and Van Bruggen, E.F.J. (1978) Nature, 275, 336-338.
7. Groot, G.S.P., Flavell, R.A. and Sanders, J.P.M. (1975) Biochim. Biophys. Acta, 378, 186-194.
8. Sanders, J.P.M., Heyting, C., DiFranco, A., Borst, P. and Slonimski, P.P., (1976) The Genetic Function of Mitochondrial DNA, Elsevier/North-Holland, Amsterdam, pp. 259-272.
9. Van Ommen, G.J.B., Boer, P.H., Groot, G.S.P., de Haan, H., Haid, A., Roozendaal, E., Schweyen, R.J. and Grivell, L.A. (1980) Cell, 20, 173-183.
10. Hahn, U., Lazarus, C.M., Lünsdorf, H. and Künzel, H. (1979) Cell, 17, 191-200.
11. Lazarus, C.M., Lünsdorf, H., Hahn, U., Stepien, P.P. and Künzel, H. (1980) Molec. gen. Genet., 177, 389-397.

© 1980 Elsevier/North-Holland Biomedical Press
The Organization and Expression of the Mitochondrial Genome
A.M. Kroon and C. Saccone, editors

MITOCHONDRIAL MIT$^-$ MUTATIONS AND THEIR INFLUENCE ON SPORE FORMATION IN SACCHAROMYCES CEREVISIAE

ELKE PRATJE, SUSANNE SCHNIERER, AND GEORG MICHAELIS
Universität Bielefeld, Fakultät für Biologie, Postfach 8640,
D-4800 Bielefeld 1, Federal Republic of Germany

INTRODUCTION

Wild type yeast cells under anaerobic conditions and cytoplasmic petite mutants are unable to sporulate, suggesting a role of the mitochondrial system during meiosis and sporulation[1]. To characterize this role we have studied the sporulation ability of 216 mitochondrial point mutants (mit$^-$) defective in electron flow or ATP synthesis. 215 of these mit$^-$ mutations were isolated in the op1 strain 777-3A[2] and assigned to the four mitochondrial genes cob, oxi1, oxi2, and oxi3 coding for cytochrome b or the three largest subunits of cytochrome c oxidase[3]. One mitochondrial ATPase mutant (pho1) was included in this study. This mutant (M28-81/2) lacking mitochondrial rutamycin sensitive ATPase activity was isolated in strain D273-10B/A by Foury and Tzagoloff[4].

RESULTS

All 216 haploid mit$^-$ mutants were crossed with a ρ^0 tester. After incubation of the diploids for seven days in sporulation medium the occurrence and frequency of tetrads were determined with a light microscope. 56 of the respiratory deficient mutants are blocked in sporulation whereas 160 mutants produced asci (Table 1). In order to distinguish whether the asci observed are formed by mit$^+$ revertants or by mit$^-$ cells about ten tetrads were isolated with a micromanipulator from each of the sporulating cultures. Colonies derived from the undissected asci were assayed for the presence of respiratory competent revertants by replica plating to glycerol medium. In some mutants all asci isolated have retained the mit$^-$ phenotype. These mutants exhibit a new phenotype and, therefore, are of special interest[3,5]. The sporulation efficiency of non-revertible mit$^-$ mutants ranges from 0.1% to 20% compared to 30-40% of the wild type. Sporulation-positive

TABLE 1

THE ABILITY OF DIPLOID MIT$^-$ MUTANTS TO SPORULATE AND THE PHENOTYPE OF THE RESULTING TETRADS

Mitochondrial genotype of 216 diploid mit$^-$ mutants	Sporulation	Phenotype of undissected tetrads		
		mit$^-$ tetrads	mit$^-$ and mit$^+$ tetrads	mit$^+$ tetrads
17 oxi1	12 positive	3	3	6
	5 negative			
3 oxi2	3 positive		3	
	0 negative			
155 oxi3	111 positive	43	42	26
	44 negative			
40 cob	33 positive	15	8	10
	7 negative			
1 pho1	1 positive	1		
	0 negative			

About ten tetrads were isolated by micromanipulation from each sporulating mutant culture. The undissected tetrads grew up on glucose medium to heterogeneous colonies which were replica plated on glycerol medium. Mit$^-$ phenotype: no growth on glycerol; mit$^+$ phenotype: growth on glycerol of at least one sector of the colony.

and sporulation-negative mit^- mutations map in the same genes of the mitochondrial genome[3].

The mitochondrial translation products of some sporulating mit^- mutants were analyzed. The absence of one major polypeptide does not prevent sporulation as illustrated in Fig.1. The cob mutant GM50/374 is especially interesting (Fig.1 lane 6). This mutant does not revert, shows no trace of residual growth on non-fermentable carbon sources after three weeks of incubation and, nevertheless, produces about 14% tetrads. Four viable mit^- spores were found in each of the five dissected asci. The nuclear markers segregated 2:2. This mutant exhibits a pleiotropic phenotype. Cytochrome b and subunit I of cytochrome c oxidase are missing and three new polypeptides are observed.

The one pho1 mutant which has lost mitochondrial rutamycin sensitive ATPase activity produces as many tetrads as the wild type. No pho mutation has been found so far in an op1 strain which is defective in the adenine nucleotide translocase[2,3]. Are pho mutations lethal in op1 strains? To answer this question we have crossed the pho1 mutant with an op1 strain. Several diploid pho colonies were isolated, purified and sporulated. Two spores of each mit^- tetrad were of the op1 pho1 genotype as shown by test crosses with appropriate ρ^o and op1 ρ^+ testers. The op1 pho1 spores grew well on glucose medium. This result demonstrates that the nuclear op1 mutation can coexist with mitochondrial pho1 mutations in one haploid yeast cell.

CONCLUSIONS

1. Respiratory deficient mit^- mutants are either defective or capable of sporulation. The two classes of mutations have been assigned to the same genes (oxi1, oxi2, oxi3, and cob) of the mitochondrial genome.
2. Sporulation can proceed at an extremely low level of respiration. Mutants in which one major mitochondrial translation product is not detectable are still able to sporulate. Among the non-revertible mit^- mutants one cob mutant produces nearly

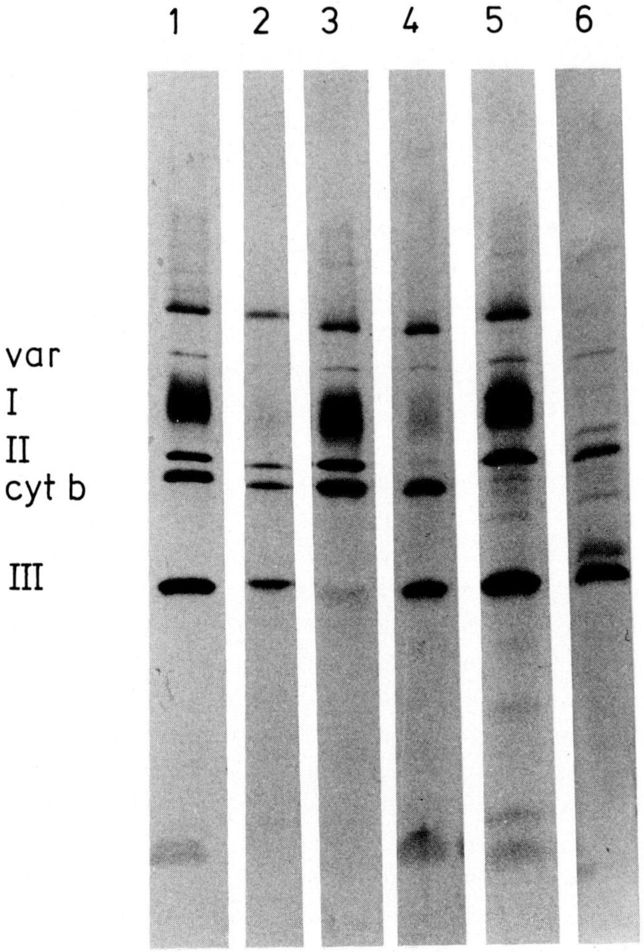

Fig. 1. Mitochondrial translation products of some sporulating mit$^-$ mutants. Mitochondrially translated polypeptides of wild type GM50 and the mutants were labeled in vivo with $^{35}SO_4^{--}$ in the presence of cycloheximide[6], separated electrophoretically on exponential lithium dodecylsulfat polyacrylamide gels and visualized by autoradiography.
Strains, mutations, and frequency of tetrads:
1. GM50, wild type, 40% asci
2. GM50/332, oxi3 locus, 7% asci
3. GM50/385, oxi2 locus, 4% asci
4. GM50/116, oxi1 locus, 5% asci
5. GM50/144, cob locus, 21% asci
6. GM50/374, cob locus, 14% asci

I,II,III: subunit I to III of cytochrome c oxidase; cyt b: cytochrome b; var: var polypeptide.

as many tetrads as the wild type. Cytochrome b and subunit I of cytochrome c oxidase are missing and three new mitochondrial translation products are present. This pleiotropic cob mutant probably belongs to the class of intron mutants.

3. Mit$^-$ asci contain four viable mit$^-$ spores. Meiotic segregation of nuclear markers is undisturbed.

4. The one ATPase mutant (pho1) studied produces as many asci as the wild type. We conclude from the occurrence of op1 pho1 spores that mitochondrial ATPase mutations are viable in op1 strains with a defective adenine nucleotide translocase.

ACKNOWLEDGEMENTS

This work was supported by the Deutsche Forschungsgemeinschaft.

REFERENCES

1. Fowell, R.R. (1969) In: The Yeasts, Rose, A.H., Harrison, J.S. (eds.), Vol.I, Academic Press, London and New York, pp. 303-383.

2. Kotylak, Z. and Slonimski, P.P. (1976) In: The Genetic Function of Mitochondrial DNA, Saccone, C., Kroon, A.M. (eds.), Elsevier/North-Holland, Amsterdam, pp. 143-154.

3. Pratje, E., Schulz, R., Schnierer, S., Michaelis, G. (1979) Molec. gen. Genet. 176, 411-415.

4. Foury, F. and Tzagoloff, A. (1976) Eur. J. Biochem. 68, 113-119.

5. Hartig, A. and Breitenbach, M. (1980) Current Genetics 1, 97-102.

6. Douglas, M.G. and Butow, R.A. (1976) Proc. Natl. Acad. Sci. 73, 1083-1086.

© 1980 Elsevier/North-Holland Biomedical Press
The Organization and Expression of the Mitochondrial Genome
A.M. Kroon and C. Saccone, editors

SELECTION OF A NEW CLASS OF CYTOPLASMIC DIURON-RESISTANT MUTATIONS IN *Saccharomyces cerevisiae* : TENTATIVE EXPLANATION FOR UNEXPECTED GENETIC AND PHENOTYPIC PROPERTIES OF THE MITOCHONDRIAL CYTOCHROME b SPLIT GENE IN THESE MUTANTS

ANNE-MARIE COLSON* and LILIANE WOUTERS
Laboratoire d'Enzymologie, University of Louvain, 1, Place Croix du Sud,
1348 Louvain-la-Neuve (Belgium)

INTRODUCTION

Selection for diuron-resistance using glycerol as carbon source yielded diuron-resistant mutations belonging to two loci of the mitochondrial cytochrome b split gene, diu1 and diu2[1]. The diu2 mutations were located in the vicinity of the cytochrome b deficient box4 locus and the diu1 mutations were found to be closely linked to the box1 locus[2].

We report, here, the isolation and characterization of a new class of cytoplasmic diuron-resistant mutants, Diu3, which were obtained after selection for resistance using DL-lactate instead of glycerol as carbon source. The diu3 mutations represent a third diuron-resistant locus and show some unexpected genetic and phenotypic properties in relation with the cytochrome b split gene. We make the hypothesis that the diu3 locus is carried by a mitochondrial episome and that its physiological role could be correlated with the splicing specificity of the cytochrome b premessenger. The present report summarizes and discusses briefly our results. They will be published in a full paper (Colson and Wouters in preparation).

RESULTS

The Diu3 mutants which have been selected using DL-lactate instead of glycerol do not show marked differences about their spontaneous frequencies of occurence (about 1 x 10^{-8}) from that observed with the Diu mutants which have been selected on glycerol plus diuron. The majority of the mutants exhibits a weak resistance to diuron on DL-lactate (six to seven days of incubation) and is sensitive on glycerol. The majority of the mutations selected on DL-lactate plus diuron (19 out of 23) belongs to a third diuron-resistant locus, diu3, which is clearly unlinked to diu1 and diu2 mutations. The diu3 locus seems to be carried by a cytoplasmic DNA since ethidium bromide can induce the loss of its genetic transmission. The diuron-resistance segregates out in Diu3 mutants, independantly of the rho^+ mit DNA factor indicating that the diu3 mutation is carried by a cytoplasmic DNA distinct from the mit DNA. The Diu3 resistance exhibits high transmission frequency in crosses with Diu^S testers (± $95D^R/5D^S$) suggesting that the diu3 locus is amplified. The Diu3 resistance presents uncoordinated transmission with two unlinked markers carried by the mit DNA (oli1 and cap1). This result shows that the diu3 locus should be specifically amplified without concurrent amplification of mit DNA markers. Surprisingly, the diu3 mutations exhibit low recombinants frequencies with box3 alleles (an average of 0.5% out of 12 crosses) whereas high recombinants frequencies were observed between diu3 mutations and box mutations belonging to other loci of the split gene (average of 7% out of 12 crosses). These results suggest that diu3 and box3 loci are closely linked. Strong resistance, (expression in three days) is selected at high frequency (1×10^{-3} to 1×10^{-2}) on glycerol plus diuron from the weak resistant Diu3 mutants. Thus a "second step mutation" occurs at high frequency conferring an increased resistance to diuron. In crosses with Diu^S testers, a double mutant, named Diu3/3' exhibits the segregation of two phenotypes : a) strong resistance which suggests the presence in the segregant of the second mutation diu3'; b) weak resistance indicating the loss of the second mutation diu3' and the retention of the first mutation diu3. The diu3' mutation shows mitochondrial segregation ($43D^R/56D^S$) whereas the diu3 mutation still presents high transmission frequency in the double mutant. In crosses using discriminating Rho^- mutants, we found that the second mutation diu3'is included in the cytochrome b split gene probably between box1 and box4 loci.

DISCUSSION

An hypothesis for the origin of the diu3 and diu3/3' mutations : " a diu3 episome".

We postulate that the diu3 locus is carried by an episome resulting from the excision and amplification of a mit DNA segment probably located in the vicinity of the box3 locus. The mechanism of excision and amplification could be similar to that of rho^- formation. In the Diu3/3' mutant, we postulate that the "second step mutation" is the reinsertion of a part of the amplified diu3 episome back in the cytochrome b split gene.

The "diu3 episome" postulate is based on the following observations : 1° the diu3 locus seems to be carried by a cytoplasmic DNA (sensitivity to ethidium bromide); 2° the diu3 locus seems to be independant from the mit DNA (independant segregation from Rho^+, not included in the cytochrome b split gene, high frequency of transmission, uncoordinated transmission with mit DNA markers); 3° in contradiction with the two first sets of data, the diu3 locus seems to be closely linked to the box3 locus which is an intron of the cytochrome b gene[3 4]. This apparent close linkage could, however, be interpreted if one assumes that one end of the diu3 episome excision site is linked to the box3 locus. In crosses between $box3^{+}diu3$ (deletion in cyt b gene + diu3 episome) and $box3^{-}diu^{S}$ (intact cyt b gene) the absence of homology between the cyt b genes of the parents in the box3 region would reduce the recombinants frequencies between $box3^{+}$ and diu^{S}. As a consequence, the diu3 episome might have more homology with the intact cyt b gene of the tester strains than with the deleted cyt b gene.

The "diu3' reinsertion" hypothesis is based on the following results :

1° The diu3' mutation is genetically distinct from the diu3 mutation since it arises at a high frequency compared to the diu3 mutation; also it exhibits, in a double mutant Diu3/3', mitochondrial segregation whereas the diu3 mutation appears to be still highly amplified.

2° The diu3' mutation is included in the cytochrome b gene most likely between the box1 and box4 loci. The reinsertion of a diu3' segment in the cytochrome b gene should reconstitute a region of homology in the diu3/3' cyt b gene with the undeleted diu^{S} cyt b DNA of the tester strains.

An hypothesis for the role of the diu3 episome : a diu3 RNA splicer.

DL-lactate is known to give its reducing equivalents to the respiratory chain through at least two paths[5]. In the first path, the electron flow bypasses the cytochrome b while in the second path, it includes the cytochrome b. In the presence of diuron, it is reasonable to assume that the second path only is blocked at the level of the cytochrome b. Since diuron inhibits growth on DL-lactate, the first path does not seem to support growth by itself. Therefore, we postulate that a leak of electron flow through cytochrome b is necessary for growth on DL-lactate using mainly the first path.

In the Diu3 mutants, a weak resistance to diuron is expressed on DL-lactate and sensitivity to the drug is observed on glycerol. This weak resistance could be interpreted by a leak of electron flow through cytochrome b allowing slow growth on DL-lactate plus diuron and insufficient to permit growth on glycerol plus diuron.

Since diu3 is genetically related to box3, an intron locus controlling a splicing step of the cytochrome b premessenger[3 4], it is tempting to postulate that the diu3 mutation is expressed by interfering in someway with the normal splicing. The hypothesis that small RNA's could act in the processing of pre-messenger RNA lead to the notion of "splicer RNA"[6]. The diu3 episome could produce a new splicer RNA competing with the normal splicer RNA. A poor efficiency of the new splicer to interact with the splicing complex would result in the synthesis of few diuron-resistant cytochrome b molecules (new splicer) and a majority of diuron-sensitive cytochrome b molecules (preexisting splicer).

Such unequal proportion of the two types of cytochrome b could account for the weak resistance of the Diu3 mutants on DL-lactate plus diuron.

In the Diu3/3' double mutant, the reinserted diu3' DNA would still code for the new splicer RNA. Increased interaction efficiency of the new splicer RNA with the splicing complex and possible inactivation of the preexisting splicer would lead to the synthesis of a majority or a totality of diuron-resistant cytochrome b molecules. This can explain the strong diuron-resistance of the double mutant Diu3/3' on DL-lactate and on glycerol.

In our model, the splicer RNA interacts with the splicing complex in order to determine the specificity of the clivage(s) to be performed by nucleases, possibly "maturases"[7] or "spligases"[8] enzymes.

Tests for the validity of the model.

1° Mitochondria isolated from Diu3 and Diu3/3' mutants should contain a new DNA beside the mit DNA.

2° The postulated deletion in mutant Diu3 could be detected and sized by a restriction map analysis of its mit DNA. A similar analysis with the Diu3/3' mit DNA should reveal the reinsertion of diu3 episome DNA.

REFERENCES

1. Colson, A.M., Luu The Van, Convent, B., Briquet, M. and Goffeau, A. (1977) Eur. J. Biochem. 74, 521-526.
2. Colson, A.M. and Slonimski, P.P. (1979) Molec. gen. Genet. 167, 287-298.
3. Church, G.M., Slonimski, P.P. and Gilbert, W. (1979) Cell 18, 1209-1215.
4. Halbreich, A., Pajot, P., Foucher, M., Grandchamp, C. and Slonimski, P.P. (1980) Cell 19, 321-329.
5. Pajot, P. and Claisse, M.L. (1974) Eur. J. Biochem. 49, 275-285.
6. Murray, V. and Holliday, R. (1979) FEBS Letters, 106, 5-7
7. Jacq, C., Lazowska, J. and Slonimski, P.P. (1980) Co. Re. Acad. Sc. Paris, 291 D p 89-92.
8. Church, G.M. and Gilbert, W. in : Mobilization and reassembly of genetic information eds. Joseph, D.R., Schultz, J., Scott, W.A., Werner, R. Acad. Press, New York (in press).

* Chercheur qualifié du F.N.R.S.

The authors wish to express their gratitude to Drs C. Colson and A. Goffeau for stimulating discussions.

© 1980 Elsevier/North-Holland Biomedical Press
The Organization and Expression of the Mitochondrial Genome
A.M. Kroon and C. Saccone, editors

PHENOTYPIC AND GENETIC CHANGES IN YEAST CELLS TRANSFORMED WITH MITOCHONDRIAL DNA SEGMENTS JOINED TO 2 MICRON PLASMID DNA

PHILLIP NAGLEY, BENTLEY A. ATCHISON, RODNEY J. DEVENISH, PAUL R. VAUGHAN AND ANTHONY W. LINNANE
Department of Biochemistry, Monash University, Clayton, Victoria, 3168, Australia

INTRODUCTION

Transformation of *Saccharomyces cerevisiae* with yeast nuclear genes cloned in a bacterial plasmid vector was first described by Hinnen et al[1]. The frequency of transformation of yeast cells has been greatly improved by incorporating segments of the 2 μm yeast plasmid DNA in these yeast-*E.coli* vectors[2-4]. We report here that mtDNA sequences joined to 2 μm yeast plasmid DNA can be used to achieve the transformation of yeast. Cells transformed by mtDNA can be identified using two different approaches: (i) direct selection of transformants in which phenotypic changes in the recipient cells themselves are observed, such as conversion of an antibiotic sensitive *rho*$^+$ strain to antibiotic resistance, and (ii) indirect selection of transformants, in which a *rho*° recipient strain is used; in this case the transformants into which the mtDNA segment has been introduced are identified by their ability to rescue particular *mit*$^-$ mutations.

SUITABILITY OF YEAST STRAINS FOR TRANSFORMATION WITH mtDNA

Transformation of yeast involves conversion of cells to spheroplasts, which are incubated with DNA in the presence of Ca^{2+} and polyethylene glycol[1]. The spheroplasts are regenerated to yield vegetative cells[5]. The use of mtDNA in transformation experiments imposes extra conditions on potential recipient strains which are not of immediate relevance to experiments involving nuclear genes. For example, the recipient antibiotic sensitive *rho*$^+$ cells to be used in a direct selection approach would need to regenerate efficiently into *rho*$^+$ colonies. Accordingly a number of laboratory strains were tested for this property by regenerating them in agar media containing glucose or ethanol as carbon source (some examples are listed in Table 1). Surprisingly, some strains failed to regenerate at all (e.g. STX-1C-1A) whereas others (e.g. J69-1B) regenerated on glucose but not on ethanol. The failure to regenerate using ethanol as carbon source was found to be due to extensive petite induction (Table 1),

TABLE 1

REGENERATION OF SOME YEAST STRAINS

Strain	Percent regeneration[a]		Petite frequency (%)[b]	
	Glucose	Ethanol	Initial	Final
STX-1C-1A	0	0	4.6	-
J69-1B	6.0	0	12.4	99
SC3	18.5	12.5	1.3	20
X4005-11A	58.4	57.8	8.3	10.4

[a] Spheroplasts of each strain (made using zymolyase) were regenerated in 3% agar/1M sorbitol/yeast extract/peptone medium[6] containing 2% glucose or 1% ethanol.

[b] Petite frequencies were determined on cells just before conversion to spheroplasts (initial), or on cells recovered from regeneration agars containing glucose (final).

possibly occurring when spheroplasts were mixed with molten agar at 45°. Of the many strains tested in this way, strain X4005-11A was found to regenerate efficiently and not be converted to the petite state during regeneration (Table 1). This strain was selected for further study as it was also sensitive to several anti-mitochondrial drugs.

PREPARATION OF TRANSFORMING mtDNA LINKED TO 2 μm DNA

The mtDNA we have used was obtained from a petite strain 70M-J which retains an 8.6 kb segment of mtDNA carrying the *oli2* gene coding for the 20,000 dalton subunit 6 protein of the mitochondrial ATPase[7] together with an oligomycin resistance mutation located within this gene. The 70M-J mtDNA was first cleaved at its unique *Pst*I site (which is outside the ATPase subunit gene), then ligated *in vitro*[6] to 2 μm plasmid DNA (6.4 kb) previously linearized with *Pst*I.

DIRECT SELECTION OF OLIGOMYCIN RESISTANT TRANSFORMANTS

Spheroplasts of X4005-11A were treated with 70M-J mtDNA linked to 2 μm DNA and were regenerated under non-selective conditions. Regenerated cells were recovered from the agar *en masse* and were spread onto plates containing oligomycin. It was found in three such experiments[6] that oligomycin resistant transformants could be detected at frequencies ranging from 25 to 60 per 10^6 cells recovered from regeneration plates, whereas few if any oligomycin resistant cells were obtained from spheroplasts of X4005-11A treated with 2 μm DNA or 70M-J mtDNA alone under the same conditions (<2 per 10^6 cells).

TABLE 2

INDIRECT SELECTION OF TRANSFORMANTS OF THE *RHO°* STRAIN EX-O

Treatment of EX-O cells[a]	*mit*$^-$ tester	*rho*$^+$ cells obtained (per 10^6 diploids)
With DNA	M44	28
No DNA	M44	<3
With DNA	M18-5	17
No DNA	M18-5	<3
With DNA	37-16-6	<5
No DNA	37-16-6	<5

[a] See text for details.

A priori, the segment of 70M-J mtDNA linked to 2 μm DNA can lead to the expression of oligomycin resistance in the transformed cells in two ways. Firstly the mtDNA could physically recombine with the host mtDNA in a manner analogous to the events in a regular *rho*$^-$ x *rho*$^+$ cross. Alternatively, the mtDNA segment of 70M-J may be replicating in the cytosol or nucleus as part of a 2 μm plasmid replicon. In this case the ATPase 20,000 dalton subunit gene in 70M-J mtDNA may be transcribed and translated outside the mitochondrion. This ATPase subunit could then be incorporated into the mitochondrial membrane to become assembled into the ATPase complex and thus produce an oligomycin resistant cell. It was, therefore, of interest to analyze the oligomycin resistant transformants for their genetic properties, in terms of possibly locating the oligomycin resistance determinant within the transformants.

It was found that there was more than one class of transformants. One class is stable for the oligomycin resistant phenotype and does not segregate oligomycin sensitive cells. Moreover, the genetic inheritance of the oligomycin resistance is consistent with some of the 70M-J sequences having been integrated into the mitochondrial genome of the recipient cells. A second class of cells is moderately unstable, and continues to segregate oligomycin sensitive cells at low frequency. However, the genetic properties of this class of transformants are not consistent with the oligomycin resistance behaving as a genetic determinant integrated into the mitochondrial genome; specifically, extensive segregation of oligomycin resistance takes place in diploids formed on crossing this class of transformants with a *rho°* haploid strain. It is conceivable that competition between different 2 μm DNA based replicons is being observed in the diploid cells formed in this type of cross.

INDIRECT SELECTION OF TRANSFORMANTS OF *RHO*° CELLS

Strain X4005-11A was converted to the *rho*° state by treatment with ethidium bromide. The *rho*° strain (denoted EX-0) was converted to spheroplasts and was incubated with 70M-J ligated *in vitro* to 2 μm DNA at the respective *Pst*I sites. The treated spheroplasts were regenerated, and mated to a series of *mit*$^-$ tester strains. Control spheroplasts not treated with DNA were also regenerated. It is evident (Table 2) that the EX-0 population that had been treated with 70M-J/2 μm DNA contained transformed cells able to rescue the *mit*$^-$ mutations at the *oli2* locus (M44 and M18-5) which affect the 20,000 dalton ATPase subunit, but did not rescue a *mit*$^-$ mutation (37-16-6) in the cytochrome *b* gene which is not included in the 70M-J segment of mtDNA.

In general, it has now become possible to convert a *rho*° petite into a *rho*$^-$-like strain replicating a particular segment of mtDNA. The genetic content of mtDNA segment can be readily tested by marker rescue tests in crosses with many different *mit*$^-$ and temperature sensitive tester strains. Shorter restriction fragments of a mtDNA segment could be introduced into recipient *rho*° cells using 2 μm DNA itself, or using other autonomously replicating yeast vectors[2-4] carrying genes for which a direct selection can be made. It should thus be possible to map genetic loci with respect to restriction sites in mtDNA in a manner that overcomes the random deletions now available in petite mutants, and to study in a novel way the functional properties of individual mtDNA segments.

REFERENCES

1. Hinnen, A., Hicks, J.B. and Fink, G.R. (1978) Proc. Natl. Acad. Sci. USA, 75, 1929-1933.
2. Beggs, J.D. (1978) Nature, 275, 104-109.
3. Struhl, K., Stinchcomb, D.T., Scherer, S. and Davis, R.W. (1979) Proc. Natl. Acad. Sci. USA, 76, 1035-1039.
4. Gerbaud, C., Fournier, P., Blanc, H., Aigle, M., Heslot, H. and Guerineau, M. (1979) Gene, 5, 233-253.
5. Van Solingen, P. and Van Der Plaat, J.B. (1977) J. Bacteriol. 130, 946-947.
6. Atchison, B.A., Devenish, R.J., Linnane, A.W. and Nagley, P. (1980) Biochem. Biophys. Res. Commun. (in press).
7. Roberts, H., Choo, W.M., Murphy, M., Marzuki, S., Lukins, H.B. and Linnane, A.W. (1980) FEBS Lett. 108, 501-504.

The Organization and Expression of the Mitochondrial Genome
A.M. Kroon and C. Saccone, editors

THE MITOCHONDRIAL GENOME OF ASPERGILLUS NIDULANS

H. Küntzel, N. Basak, G. Imam, H. Köchel, C.M. Lazarus, H. Lünsdorf
Abteilung Chemie, Max-Planck-Institut für experimentelle Medizin,
Hermann-Rein-Str. 3, D-3400 Göttingen, FRG,

E. Bartnik, A. Bidermann and P.P. Stępień
Department of Genetics, Warsaw University, Warszawa, Poland

Recent mapping studies have shown that the mitochondrial genomes of Neurospora crassa (about 60 kb) and Aspergillus nidulans (35 kb) are similar in their gene organization: the split gene for L-rRNA* is separated from the contiguous S-rRNA gene by relatively long spacer[1-4], and most tRNA genes are clustered in two regions flanking the L-rRNA gene[2,5]. There is also significant sequence homology between other DNA regions at corresponding map positions (unpublished data).

We have cloned restriction fragments covering 90 % of A. nidulans mtDNA into E. coli using the vector pBR 322 and sequenced large parts of tRNA gene clusters. This report presents a map of genes coding for rRNAs, tRNAs and mRNAs, and of sequences inserted in a related strain (A. echinulatus). We further present the partial amino acid sequence of a major mitochondrially synthesized proteolipid of unknown function.

MOLECULAR CLONING

Figure 1 shows a circular map of fragments (boxed) which were ligated with pBR 322 in the respective restriction sites and cloned into E. coli. The cloned fragments were identified by restriction fine mapping and by molecular hybridization with nick-translated mtDNA. For unknown reasons the Hind III fragment H4 could not be recovered as stable recombinant DNA.

TRANSCRIPTION MAP

The rRNA genes have been mapped by gel transfer hybridization and electron micoroscopy[4]. Figs. 1 and 2 show that the S-rRNA gene (1.4 kb) is separated

---

*Abbreviation: L-rRNA, large ribosomal subunit RNA.
S-rRNA, small ribosomal subunit RNA.

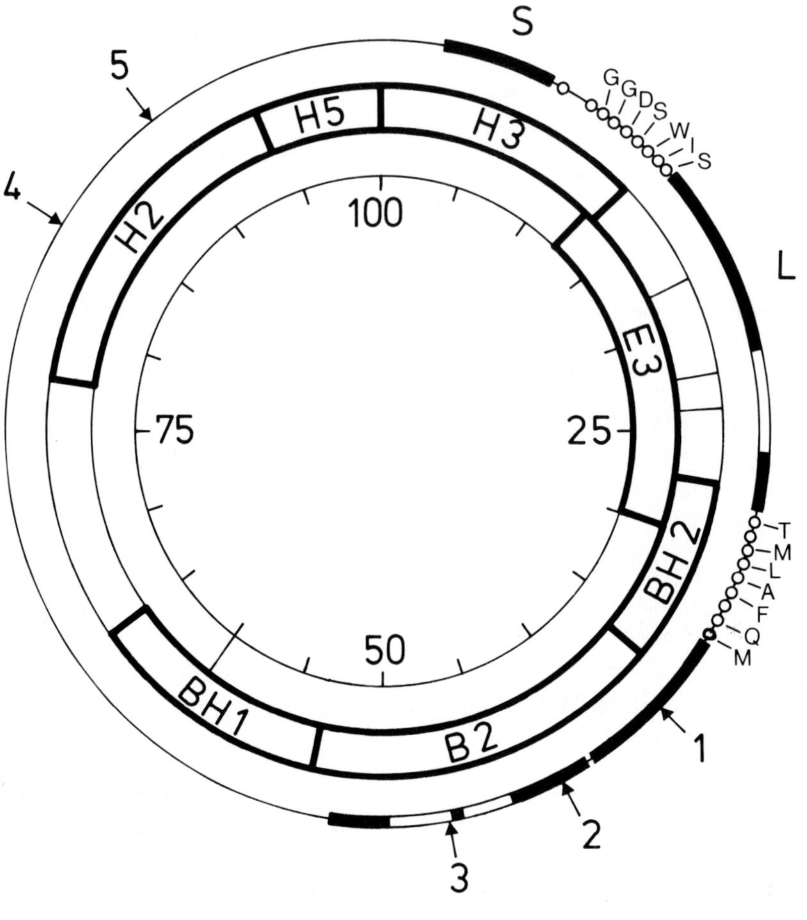

Fig. 1. Circular map of the 35 kb mtDNA of Aspergillus nidulans. Inner ring: Eco RI fragments. Middle ring: Hind III fragments (the largest fragment H1 is presented as the three Bam HI subfragments BH1, B2 and BH2). The boxed fragments were cloned into E.coli. Outer ring: transcription map. The positions of rRNA and mRNA genes (black boxes: exons, open boxes: introns) were determined by electron microscopy of DNA/RNA hybrids, the positions of tRNA genes (circles) by gel transfer hybridization and by sequencing. The amino acid specificities as deduced from DNA sequencing are given in single letter code. The arrows 1-5 denote the positions of inserted sequences in the related mtDNA of A. echinulatus, as determined by electron microscopy of DNA/RNA heteroduplexes.

from the split L-rRNA gene (exon I = 2.4 kb, intron = 1.8 kb, exon II = 0.5 kb) by a 2.8 kb spacer sequence. Gel transfer hybridization with bulk mt tRNA labelled at the -CCA end by nucleotidyl transferase showed that about 19 out of 25 tRNA genes are equally divided between two clusters mapping in the ribosomal spacer and within a 2 kb region following the 3' terminus of the L-rRNA gene. This rRNA-tRNA gene region (total length about 10 kb) appears to be a condensed copy of the corresponding N. crassa region (total length about 17 kb)[2], which contains about the same number of tRNA genes but longer intron, spacer and rRNA sequences.

Another large transcribed region has been detected on Bam H1 fragment B2. This fragment was isolated from the cloned plasmid pan B2 by digestion with Hha I which does not cleave B2[4], hybridized with high salt insoluble mitochondrial RNA under DNA/RNA duplex conditions and analyzed by electron microscopy[6].

About 50 % of all measured hybrids (total number 80) exhibited two intervening loops whereas the other molecules appeared to be fully transcribed. This suggests that the two introns are transcribed and more slowly removed than the intron of the L-rRNA gene.

Most of the spliced hybrid molecules had a total length of 6.5 kb, which exceeds that of B2 by 1 kb. We therefore interpret the single-stranded end of the long hybrid arm as RNA. The same single stranded end appears as a fork if hybridization is performed with Pst I-linearized plasmid pan B2 DNA. These hybrids were also used to determine the polarity of the spliced transcript (data not shown). Some of the spliced hybrid molecules had the length of B2 and contained a much shorter hybrid region at the longer arm.

The data are interpreted as follows: Fragment B2 binds two transcripts. The contiguous transcript (mRNA 1, Fig. 1) extends into the neighbouring fragment BH2 by about 1 kb and covers about 1.7 kb of the left end of B2. This transcript is separated by a small spacer region (not visible by EM) from the spliced transcript (mRNA 2) containing three exons (1 kb, 0.15 kb and 0.4 kb) and two introns (1 kb each). The presence of a small spacer is also indicated by sequencing: a fragment mapping in that region contains stop codons in all reading frames. Sequencing also indicates that the BH2-B2 region is transcribed in the same direction as the rRNA and tRNA genes.

INSERTS IN A. ECHINULATUS mtDNA

Fig. 2 shows the linearized restriction maps of mtDNA from A. nidulans and the related species A. echinulatus. The latter DNA contains five inserts of around 1.3 kb lengths which were mapped by electron microscopy of heteroduplexes between the corresponding Bam HI (B2) and Hind III (H2) fragments of the two genomes. Otherwise the two mtDNA molecules show strong sequence homology as studied by interspecific DNA/RNA hybridization and restriction mapping. A few restriction sites outside the inserted sequences are found only in one of the two DNAs and were used, together with the inserts, as physical markers for the analysis of recombinant mtDNA (Earl, Turner, Lünsdorf, Lazarus and Küntzel, manuscript in preparation). It is interesting to note that inserts 1 and 2 map within the hybrid region of the B2 fragment (Fig. 1): possibly the B2 transcripts of A. echinulatus are interrupted by additional introns.

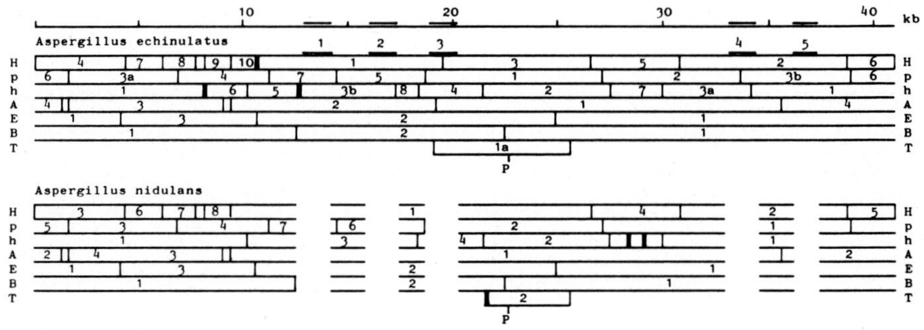

Fig. 2 Linearized restriction maps of mtDNA from A. echinulatus and A. nidulans. H = Hind III, p = Pvu II, h = Hind II, A = Hha I, E = Eco RI, B = Bam HI, T = Taq I, P = Pst I. The thick bars represent restriction sites unique to one genome (excluding sites within inserts).

NUCLEOTIDE SEQUENCES OF tRNA GENE CLUSTERS

We have determined the nucleotide sequences of large parts of the two tRNA gene clusters (Fig. 1) by using the Maxam-Gilbert technique[7].

All 14 tRNA genes so far sequenced are on the same strand as the L-rRNA gene and are transcribed clockwise, as drawn in Fig. 1.

The first cluster in the ribosomal spacer contains two genes for tRNA Gly (reading the codons GGR and GGU), two genes for tRNA Ser (AGY and UGR), and genes for tRNA Asp (GAY), Ile (AUY) and Trp (UGR). The second cluster contains two genes for tRNA Met (AUG), and genes for tRNA Thr (ACR), Leu (YUR), Ala (GCR), Phe (UUY) and Gln (CAR). A few exceptional features are notable: In tRNA Ser (AGY) the base at position 9 (generally a purine[8]) is deleted. tRNA Thr (ACR) contains A and G in positions 18 and 55, respectively, as already found for mitochondrial initiator tRNA Met from Neurospora crassa[9], and in the two tRNAs Met and Gly (GGU) the G·C pair at positions 53·61 is replaced by an A·T pair.

The tRNA genes are separated from each other by AT-rich spacers of lengths between a single base and 150 basepairs.

The 5'- and 3'-termini of the L-rRNA gene have tentatively been located by gel transfer hybridization and by comparing DNA sequences with that of the *E. coli* 23S rRNA gene[10]. Fig. 3 shows that the mtDNA region at the junction

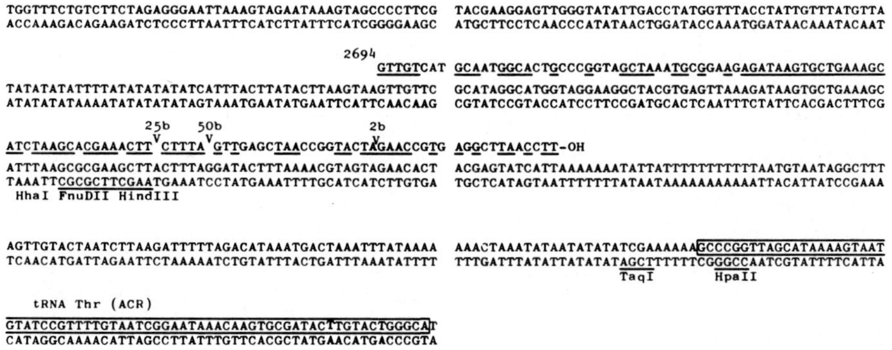

Fig. 3 Nucleotide sequence of a section of mtDNA around the junction H3/BH2 (see Fig. 1). The superimposed sequence is that of the 3'-end of the *E. coli* 23S rRNA gene (bases 2694-2904, deletions indicated). Sequence homologies are underlined.

H8/BH2 (see Fig. 1) contains a 74 nucleotide sequence highly homologous to bases 2694-2768 of the E. coli sequence. The homologous region can be extended to the 3'-end of the E. coli sequence by deleting three regions of 25, 50 and 2 bases lengths. The putative 3' end of the mt L-rRNA gene is followed by an AT-rich spacer sequence of 130 bp, and then by a threonyl-tRNA gene. There is no evidence for the presence in this region of a gene coding for 5S rRNA or any small RNA functionally equivalent to 5S rRNA.

A MAJOR MITOCHONDRIALLY SYNTHESIZED PROTEOLIPID OF UNKNOWN FUNCTION

Two proteins are removed from an aqueous suspension of A. nidulans mitochondria by extraction with 20 volumes of chloroform: methanol (2:1, v/v) and precipitation with diethyl ether: a cytoplasmically synthesized proteolipid identified as the DCCD-binding ATPase-subunit[11] and a mitochondrially synthesized fMet-starting proteolipid which is partially associated with the functional oligomycin-sensitive ATPase complex[12] but absent from the immunoprecipitated ATPase[11]. The two proteins are separated from each other by their differential solubility in organic solvents (Table 1). Similar fMet-starting proteolipids

Table 1. Properties of A. nidulans proteolipids extracted from aqueous mitochondria by chlororform: methanol (2:1, v/v).

	DCCD-proteolipid	fMet-proteolipid
Yield per mg mitochondrial protein	~10 μg	~20 μg
solubility in		
a) dry C:M	soluble	insoluble
b) DMF	soluble	insoluble
c) 5 % Formic acid	soluble	insoluble
d) 90 % ethanol	insoluble	soluble
residues	84	67
polarity	30 %	38 %
N-terminus	Tyr	fMet
C-terminus	Ala	Val

have been found in Neurospora crassa and human mitochondria, whereas only the fMet-starting DCCD-binding ATPase subunit is extracted from yeast mitochondria under identical conditions.

fMet-Gln-Leu-Val-Leu-Ala-Ala-Lys-Tyr-Ile-Gly-Ala-Gly-Ile-Ser-
fMet-

Thr-Ile-Gly-Leu-Leu-Gly-Ala-Gly-Ile-Gly-Ile-Ala-Ile-Val-Phe-
Ala-Met-Gly-Phe-Leu-Gly-Val-His-Tyr-Asp-Ile-Gln-Arg-Asp-Glu-

Ala-Ala-Leu-Ile- (yeast ATPase proteolipid)
-Ala-Leu-Ser- (A. nidulans fMet-proteolipid)

Fig. 4 N-terminal amino acid sequences of the A. nidulans fMet-proteolipid and the yeast ATPase proteolipid[13].

The sequence of the first 19 N-terminal amino acids (Fig. 4), as determined by manual Edman degradation, shows a certain degree of homology to a near N-terminal region of the yeast ATPase proteolipid[13] and it appears to be possible that both A. nidulans proteolipids had a common evolutionary precursor related to the yeast proteolipid.

Acknowledgement

We thank Dr. Hans Kössel for teaching us DNA sequencing methods, and Ina Ahlborn and Marion Dornwell for technical assistance. This study was partially supported by the Volkswagenstiftung.

REFERENCES

1. Hahn, U., Lazarus, C.M., Lünsdorf, H. and Küntzel, H. (1979) Cell, 17, 191-200.
2. Heckman, J.E. and RajBhandary, U.L. (1979) Cell, 17, 583-595.
3. Manella, C.A., Collins, R.A., Green, M.R. and Lambowitz, A.M. (1979) Proc. Natl. Acad. Sci. USA, 76, 2635-2639.
4. Lazarus, C.M., Lünsdorf, H., Hahn, U., Stepien, P.P. and Küntzel, H. (1980) Molec. Gen. Genet., 177, 389-397.
5. de Vries, H., de Jonge, J., Bakker, H., Meurs, H. and Kroon, A. (1979) Nucleic Acids Res., 6, 1791-1803.
6. Wellauer, P.K. and Dawid, I.B. (1977) Cell, 10, 193-212.
7. Maxam, A.M. and Gilbert, W. (1977) Proc. Natl. Acad. Sci. USA 74, 560-564.
8. Gauss, D.H., Gruter, F. and Sprinzl, M. (1979) Nucleic Acids Res., 6, r1-r19.

9. Heckman, J.E., Hecker, L.I., Schwartzbach, S.D., Barnett, W.E., Baumstark, B. and RajBandhari, U.L. (1978) Cell, 13, 83-95.

10. Brosius, J., Dull, T.J. and Noller, H.F. (1980) Proc. Natl. Acad. Sci. USA 77, 201-204.

11. Turner, G., Imam, G. and Küntzel, H. (1979) Eur. J. Biochem. 97, 565-571.

12. Marahiel, M.A., Imam, G., Nelson, P., Pieniazek, N.J., Stepien, P.P. and Küntzel, H. (1977) Eur. J. Biochem., 76, 345-354.

13. Sebald, W., Wachter, E. and Tzagoloff, A. (1979) Eur. J. Biochem., 100, 599-607.

© 1980 Elsevier/North-Holland Biomedical Press
The Organization and Expression of the Mitochondrial Genome
A.M. Kroon and C. Saccone, editors

AMPLIFICATION OF A COMMON MITOCHONDRIAL DNA SEQUENCE IN THREE NEW RAGGED MUTANTS OF *ASPERGILLUS AMSTELODAMI*

Colin M. Lazarus and Hans Küntzel
Max-Planck-Institut für experimentelle Medizin,
Abt. Chemie, Hermann-Rein-Str. 3, D-3400 Göttingen, FRG

INTRODUCTION

"Ragged" is a cytoplasmically-inherited, suppressive, loss of viability condition affecting mitochondrial function in *Aspergillus amstelodami*[1,2]. Ragged mutants (Rgd) are easily recognized by their uneven radial growth on solid medium (Handley and Caten, manuscript in preparation). We recently reported that a small sequence of the mtDNA of ragged mutant Rgd1 is amplified in the form of multiple head-to-tail repeats (rgd1DNA)[3]. The extent of amplification is extremely high and variable but non-random, since, on agarose gel electrophoresis, discrete bands of very high molecular weight are observed in addition to the intact mitochondrial genome in untreated mtDNA, or all the wild-type restriction fragments in restricted mtDNA. *Eco*RI and *Hpa*II each reduce the multimers to the monomeric, approx. 870 bp, repeat unit. Observations on a single mutant can only *imply* a causal relationship between DNA amplification and ragged growth, and raises the question of whether all ragged isolates have amplified mtDNA sequences, and, if so, whether the same or different sequences are involved. Investigations of three further ragged mutants are described in this paper.

MATERIALS AND METHODS

All techniques used have been previously described or cited[3].

Strains. Rgd3 and Rgd4 were obtained from Dr. C.E. Caten, University of Birmingham. Rgd3 arose spontaneously among the progeny of an *A. amstelodami* inter-isolate cross. Rgd4 arose spontaneously from a derivative of a wild isolate of *A. heterocaryoticus*, which has been described as a natural variant of *A. amstelodami*, differing (as far as has been determined) only in spore-colour[4]. Rgd5 was isolated as a spontaneous ragged-growth mutant of wild-type strain 37 (Rgd^+), which was also the progenitor of Rgd1.

RESULTS

"Additional" bands of very high molecular weight were observed in undigested and *Hha*I-digested mtDNA of all ragged strains. *Hha*I restriction fragments of

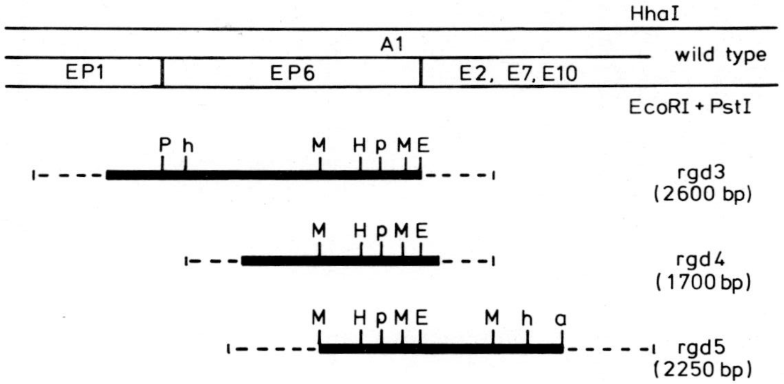

Figure 1. Restriction maps of amplified mtDNAs of Rgd3, Rgd4 and Rgd5, aligned at the common *Eco*RI site. Three additional *Mbo*I sites within rgd3DNA have not been mapped. The total length of each rgdDNA is the solid line plus one dashed line, and the points where each amplification begins and ends with respect to the wild-type genome lie within the dashed lines. Restriction endonuclease single-letter code: A, *Hha*I; a, *Hae*III; E, *Eco*RI; H, *Hin*dIII; h, *Hin*dII; M, *Mbo*I; P, *Pst*I; p, *Pvu*II.

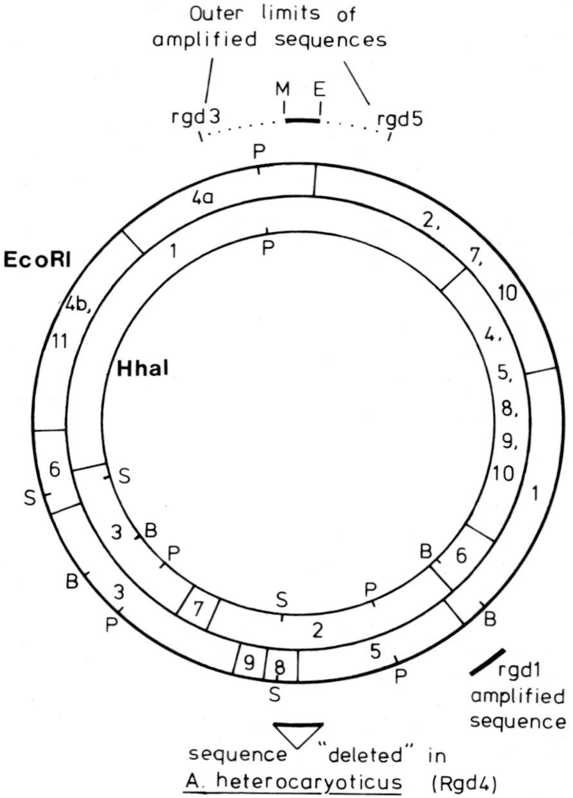

Figure 2. Restriction map of the *A. amstelodami* mitochondrial genome (*Eco*RI outer circle, *Hha*I inner circle; B, *Bam*HI; P, *Pst*I; S, *Sal*I). Located on the outside are the regions amplified in ragged mutants and the deleted region in *A. heterocaryoticus*. M-E denotes the 680 bp sequence bordered by one *Mbo*I site and the single *Eco*RI site shared in common by rgd3DNA, rgd4DNA and rgd5DNA (Figure 1).

Rgd3 and Rgd5 were identical with those of Rgd1 and Rgd^{+}[3], but fragment A2 of Rgd4 was about 1000 bp shorter than the corresponding fragment of the other strains. The slight shortening of *Eco*RI fragment E5 and absence of E8 roughly locate this "deletion" on the *A. amstelodami* mtDNA map (shown in Figure 2). This deletion indicates a second point of difference between wild isolates of the *A. amstelodami* and *A. heterocaryoticus* types (see Strains, above).

Screening with a range of restriction endonucleases identified enzymes having sites within the amplified sequences. Monomeric-repeat-unit sizes were thus established, and in a series of double-digestion experiments the repeat units were partially mapped (Figure 1). Five restriction sites covering approx. 680 bp have identical locations in rgd3-, rgd4- and rgd5DNA (= "common amplified sequence", indicated by ME in Figure 2). Gel transfer hybridizations of nick-translated rgd3- and rgd5DNA to *Hha*I-digested Rgd^{+} mtDNA resulted in the labelling of fragment A1 in both cases (data not shown) in contrast to the labelling of A2 by rgd1DNA[3]. Figure 2 shows the locations of the amplified sequences and the *A. heterocaryoticus* deletion on the *A. amstelodami* wild-type mitochondrial genome.

DISCUSSION

Spontaneous mutation to ragged growth is correlated in all cases tested with amplification of mtDNA sequences, and it is proposed that the latter causes the former. The involvement of (at least) two distinct segments of the mitochondrial genome implies that amplification *per se*, rather than reiteration of a particular nucleotide sequence, is the causative factor. This does not assume that all mtDNA sequences are equally likely to be amplified; indeed the presence of an homologous sequence in three out of four amplified segments suggests that this sequence is a "hot spot" of amplification. How rgdDNAs arise is unknown, but the mechanism could be similar to the process of excision and amplification proposed for the generation of spontaneous petite genomes in yeast[5]. The presence of a common internal sequence and differing termini in rgd3-, rgd4- and rgd5DNAs, however, makes it unlikely that nucleotide sequences at or around the termini are responsible for triggering the process[6]. The behaviour of the rgdDNAs as independent replicons suggests that they all contain an origin of replication, which for rgd3-, rgd4- and rgd5DNAs may be included in the common amplified sequence, and which may also be used by wild-type mtDNA.

Amplification of mtDNA sequences features among a number of recent reports, indicating that the phenomenon is by no means rare. Electron microscopy of mtDNA from senescent mycelium of *Podospora anserina*[7] and from Rgd1 (G. Turner, unpublished results) has in both cases revealed multimeric series of circular DNA molecules. Major differences are, however, that the amplified DNA of ragged mycelium exists mainly in very high molecular weight form[3], and that the generation of small circles in *Podospora* may leave no copy of the amplified sequence in the mitochondrial genome[7,8]. In a comparable situation low molecular weight amplified sequences present in mtDNA from certain male-sterile strains of *Zea mays* were found to be present in the high molecular weight mtDNA of fertile strains, but not in that of male-sterile strains. They are thought to have arisen by excision from the mitochondrial genome. In *Neurospora crassa* tandem reiteration of mtDNA sequences occurs both within the intact mitochondrial genome (thus increasing its size) or with excision[10]. In the latter case the intact mitochondrial genome was also present, albeit in low concentration, (compare ragged mtDNA), but no mutant traits have as yet been correlated with mtDNA amplification in this organism[10,11].

REFERENCES

1. Caten, C.E. (1972) *J. Gen. Microbiol. 72*, 221-229.
2. Handley, L. (1975) Ph.D. Thesis, University of Birmingham.
3. Lazarus, C.M., Earl, A.J., Turner, G. and Küntzel, H. (1980) *Eur. J. Biochem. 106*, 633-641.
4. Caten, C.E. (1979) *Trans. Brit. Mycol. Soc. 73*, 65-74.
5. Gaillard, C., Strauss, F. and Bernardi, G. (1980) *Nature 283*, 218-220.
6. Faugeron-Fonty, G., Culard, F., Baldacci, G., Goursot, R., Prunell, A. and Bernardi, G. (1979) *J. Mol. Biol. 134*, 493-537.
7. Cummings, D.J., Belcour, L. and Grandchamp, C. (1979) *Mol. Gen. Genet. 171*, 239-250.
8. Esser, K., Tudzynski, P., Stahl, U. and Kück, U. (1980) *Mol. Gen. Genet. 178*, 213-216.
9. Thompson, R.D., Kemble, R.J. and Flavell, R.B. (1980) *N. A. Res. 8*, 1999-2008
10. Mannella, C.A., Goewert, R.R. and Lambowitz, A.M. (1979) *Cell 18*, 1197-1207.
11. Mannella, C.A. and Lambowitz, A.M. (1980) *Genetics 93*, 645-654.

© 1980 Elsevier/North-Holland Biomedical Press
The Organization and Expression of the Mitochondrial Genome
A.M. Kroon and C. Saccone, editors

SENESCENCE SPECIFIC DNA OF *PODOSPORA ANSERINA* : ITS VARIABILITY AND ITS RELATION WITH MITOCHONDRIAL DNA.

CORINNE VIERNY, ODILE BEGEL, ANNE-MARIE KELLER, ALAIN RAYNAL AND LEON BELCOUR.
Centre de Génétique Moléculaire, C.N.R.S., 91190 Gif-sur-Yvette (France)

In the ascomycete *Podospora anserina*, all cultures maintained in continuous growth eventually reach a senescent state characterized by progressive reduction and finally complete arrest of growth [1, 2]. Each strain displays a genetically defined life span. Both the senescent state [3] and the life span [4] are maternally inherited. The senescent state is moreover infectious through anastomoses between mycelia [2].

A mitochondrial basis of senescence was first suggested by the observation that mitochondrial mutations [5, 6] and inhibitors of mitochondrial functions[7] modified the life span. Strong support of this hypothesis was provided by the identification in senescent cultures (1) of modifications in the restriction pattern of mitochondrial DNA (2) of a senescence specific DNA (SEN-DNA) constituted of circular molecules arranged in a multimeric set of contour lengths [8].

Experiments to establish the origin of this SEN-DNA and to determine whether this SEN-DNA represents the cytoplasmic factor responsible for the establishement of the senescent state have been undertaken. Most of the results presented here are detailed in two papers [9, 10].

RESULTS

Variability of the SEN-DNA'S

In any young culture , senescence is triggered by a random primary event leading to the appearance and spreading of the cytoplasmic factors which eventually will provoque the arrest of growth [2]. Independent senescent cultures were obtained either from different subclones of the same strain or from different strains and their total DNA was extracted. Depending on the culture, the SEN-DNA was recovered, after DAPI-ClCs centrifugation, either in the mitochondrial DNA band (density = 1,694 g/cm^3) or in an additionnal band of higher density (1,699 g/cm^3). In all cases, the SEN-DNA consisted in circular DNA molecules distributed in multimeric sets of contour lenghts [10] (Fig. 1).

The restriction enzyme patterns (Fig. 2) and electron microscopic analysis of different SEN-DNA's have shown that: (1) these DNA's consist of tandemly repeated sequences; (2) the size of the repeat unit varies from 0,89 µm to

2,05 µm; (3) there is a good correlation between the size of the repeat unit estimated by restriction analysis and the length of the monomer circles measured on electron micrographs; (4) the products of Hae III digestion of SEN-DNA's which have the mitochondrial DNA density all include a 2.3 KB fragment. The results obtained by the Southern technique [11] (Fig. 2) show that SEN-DNA's of identical density cross-hybridize whereas no hybridization is observed between SEN-DNA's of different densities. Therefore the SEN-DNA's so far obtained belong two classes which have different sequence origins [19].

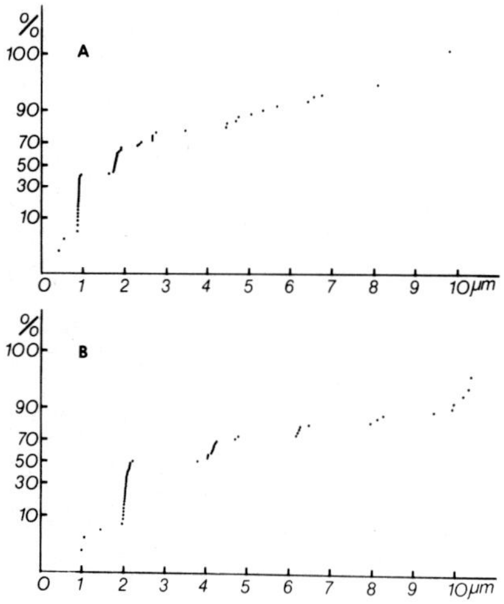

Fig. 1. Cumulative distribution analysis of the lengths of circular DNA molecules from two independent senescent cultures of strain s mt⁻.
A : SEN-DNA of density 1,699 g/cm^3.
B : SEN-DNA of mitochondrial density 1,694 g/cm^3.
Abcissae : circles lengths.
Ordinates : cumulative %, gaussian scale.
DNA recovered after DAPI-ClCs centrifugation was prepared for electron microscopy as described by Cummings, Belcour and Grandchamp [8].

Origin of the SEN-DNA's.

A mitochondrial origin for the SEN-DNA's was demonstrated by Southern hybridization experiments using ^{32}P cRNA probes obtained from one SEN-DNA of each density class. These experiments [9, 10] led to the following conclusions: (1) neither of the density class of SEN-DNA hybridize with nuclear DNA; (2) SEN-DNA's of both density classes hybridize with mitochondrial DNA in a fashion which indicates that a) their monomeric sequence is an integral part of the mitochondrial chromosomes and b) the monomeric sequence of the SEN-DNA of mitochondrial density (1,694 g/cm^3) is present in young mycelia in its integrated state only, while SEN-DNA of density 1,699 g/cm^3 seems to exist not only in integrated form but also as a free repeated sequence.

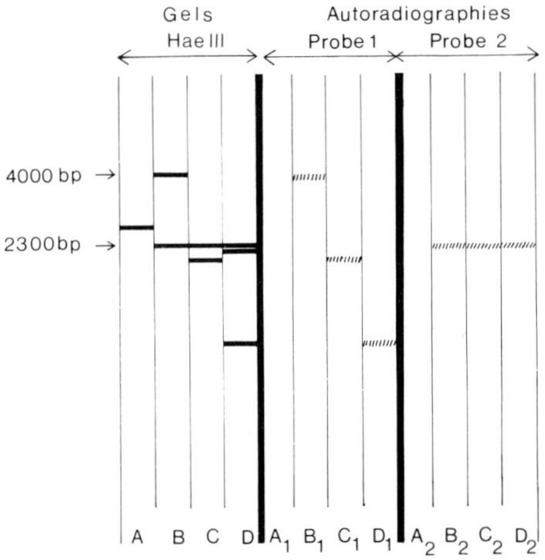

Fig. 2. Schematic representation of Southern hybridization experiments between different SEN-DNA's.
Each of the two fragments resulting from cleavage of SEN-DNA B by Hae III was transcribed into ^{32}P cRNA. They were used to hybridize the Hae III restriction fragments of four different SEN-DNA's. A : SEN-DNA from strain s mt^- Cap^s (d = 1,699 g/cm^3). B : SEN-DNA from strain s mt^- Cap^s (d = 1,694 g/cm^3). C : SEN-DNA from strain s mt^- Cap^r (d = 1,694 g/cm^3). D : SEN-DNA from strain s mt^+ Cap^s (d = 1,694 g/cm^3). Probe 1 = ^{32}P cRNA obtained from the 4000 bp fragment of SEN-DNA B. Probe 2 = ^{32}P CRNA obtained from the 2300 bp fragment of SEN-DNA B.

Relationship between SEN-DNA and senescence.

To investigate the possible cause-effect relationship between SEN-DNA and senescence, contamination experiments were carried out as described in fig. 3. It was observed that the SEN-DNA extracted from the contaminated culture strongly resembled that of the contaminating culture. Although repetitions of similar experiments are needed, this observation supports the idea that the SEN-DNA is the infectious cytoplasmic factor causing senescence.

DISCUSSION.

On the basis of previous genetical and physiological studies of senescence [5, 6, 7], it was proposed that senescence results from a mechanism similar to that involved in the appearance and suppressivness of some "petite" mutants of yeast. This hypothesis is supported by the existence and properties of the senescence specific DNA's, *i.e:* (1) mitochondrial origin [9, 10]; (2) repeated arrangement [8, 9, 10]; (3) variability in the sequence depending on the senescent subclone studied [10]. However no definitive evidence has yet been provided that the SEN-DNA is the cause of senescence.

Important modifications of the mitochondrial DNA organization have only been observed in yeast because of the capability of this organism to dispense with its mitochondrially encoded functions. In *Podospora* senescent cultures, we were able to detect a senescence specific modified mitochondrial DNA, resulting, like the DNA of "petite" mutants, from the amplification of a sequence of the mitochondrial chromosome [12, 13]. This is not so surprising since modifications such as those found in "petite" mutants are expected to be either conter-selected or to cause death in strictly aerobic organisms. If definitive evidence that SEN-DNA has a suppressive effect and is responsible for senescence could be obtained in *Podospora*, it would then be of particular interest to examine the possible mitochondrial origin of senescence in other aerobic organisms.

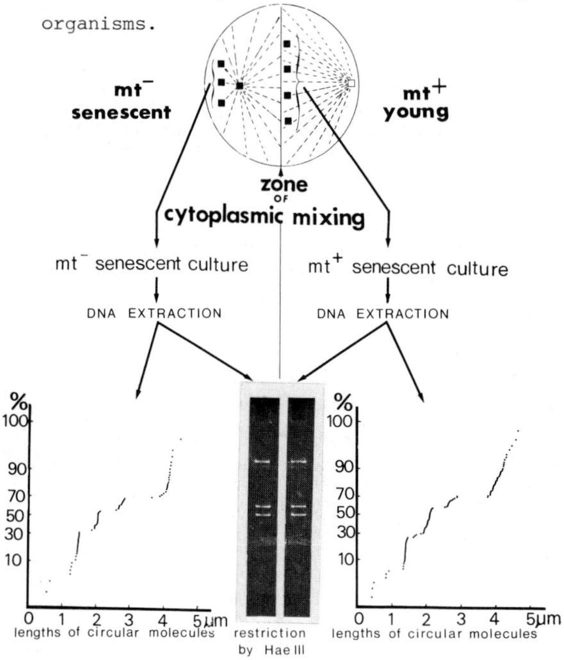

Fig. 3. Contamination of a young culture by a senescent one.
A senescent culture of strain s mt$^-$ Cap^r was used to contaminate, through anatomoses, a young s mt$^+$ Cap^S culture. 6 days after establishement of cytoplasmic contact between the two mycelia, implants of the contaminated mt$^+$ Cap^S strain were transferred onto fresh medium dishes and were shown (1) to display immediate senescence (2) to carry only the mt$^+$ nuclear genome and (3) to contain Cap^r mitochondrial determinants. Total DNA was extracted from both the contaminating and the contamined mycelia and purified through DAPI-ClCs gradient. The DNA's present in bands of mitochondrial DNA density were analyzed a) by electrophoresis after Hae III digestion b) by electron microscopy: the diagrams presented correspond to the cummulative distribution analysis of the lengths of circular molecules found in SEN-DNA's.

REFERENCES

1. Rizet, G. (1953a). C.R. Acad. Sci. (Paris) 237, 838-840.

2. Marcou, D. (1961). Ann. Sci. Natur. Botan. 11, 653-764.

3. Rizet, G. (1957). C.R. Acad. Sci. (Paris) 244, 663-665.
4. Smith, J.R. and Rubenstein, I.J. (1973). Gen. Microbiol. 76, 297-304.
5. Belcour, L. and Begel, O. (1977). Gen. Genet. 153, 11-21.
6. Belcour, L. and Begel, (1978). Mol. Genet. 163, 113-123.
7. Belcour, L. and Begel, O. (in press) O.J. of Gen. Microbiol.
8. Cummings, D.J., Belcour, L., and Grandchamp, C., (1979b). Molec. Gen. Genet. 171, 239-250.
9. Jamet-Vierny, C., Begel, C. and Belcour, L. (in press) Cell.
10. Belcour, L., Begel, O. and Jamet-Vierny, C. (in preparation).
11. Southen, E.M. (1975). J. Mol. Biol. (1975). 98, 503-517.
12. Locker, J., Rabinowitz, M., Getz, G.S., (1974). J. Mol. Biol. 88, 489-507.
13. Lazowska, J, and Slonimski, P.P., (1976). Molec. Gen. Genet. 146, 61-78.

© 1980 Elsevier/North-Holland Biomedical Press
The Organization and Expression of the Mitochondrial Genome
A.M. Kroon and C. Saccone, editors

CLONING OF SENESCENT MITOCHONDRIAL DNA FROM PODOSPORA ANSERINA: A BEGINNING

DONALD J. CUMMINGS, JANE L. LAPING and PETER E. NOLAN
University of Colorado Health Sciences Center, Department of Microbiology and Immunology, Denver, Colorado 80262, U.S.A.

INTRODUCTION

All races of Podospora anserina exhibit vegetative death or senescence[1]. Senescence has been described in terms of life span, incubation distance, transformation rate from wild-type to the senescent state, median length of growth, etc. and each race has unique values for these parameters[1]. Several studies have indicated that senescence is maternally inherited[1,2] and a transmissible cytoplasmic factor has been implicated[3]. An obvious candidate for this cytoplasmic factor is the mitochondrion and this has been supported by the effects of mitochondrial inhibitors on life span[4], cytochrome spectra of wild-type and senescent mycelia[5,6] as well as by studies with mitochondrial mutants[5]. Belcour and Begel (1978) indicated that in many respects, senescence in Podospora anserina resembles the mitochondrial rho$^-$ suppressive mutation in the yeast, S. cerevisiae.

Clearly, studies on mitochondrial (mt) DNA were essential for the understanding of maternal inheritance of senescence in Podospora anserina. In 1979, we[7] reported that wild-type mt DNA consisted of closed circular molecules, 31 μm in length but that senescent cultures yielded mt DNA of quite a different character. A multimeric set of closed circular DNA molecules were observed ranging in size from about 0.9 μm to about 15 μm[6,8]. Moreover, in some cases, two densities of mt DNA were observed[6], 1.694 g/cm^3, the same as wild-type and 1.699 g/cm^3. Separation of these density populations indicated that the "heavy" DNA contained the greatest proportion of the multimeric set of closed circular molecules. Restriction enzyme analysis of the so-called heavy mt DNA revealed that this DNA had no EcoRI sites. Digestion with HaeIII restriction endonuclease yielded one predominant fragment of 2600 bp[6], quite close to that expected for a 0.9 μm length molecule. We concluded that senescent mycelia from Podospora anserina contained mt DNA composed of tandemly repeated subunits. Such molecules were observed earlier[9,10,11] in rho$^-$ petite mutants, thus extending to the molecular level, the similarities between these seemingly dissimilar phenomena.

Three major questions arose as a result of these studies[7]. First, would other senescent events give rise to the same or a different set of tandemly repeated closed circular molecules. In yeast, independent rho$^-$ mutants yielded different sets of repeats[9,10,11]. Second, what was the origin of this DNA? This question could only be answered by hybridization studies with wild-type DNA. And third, what was the causal relationship of this multimeric set of DNA molecules with senescence? Were defective mitochondria the transmissible cytoplasmic factors involved with senescence? Transformation studies with either senescent mt or DNA could be most revealing. To answer all these questions, we have begun to clone mt DNA obtained from several independent senescent events.

MATERIALS AND METHODS

Wild-type and senescent cultures of Podospora anserina race s$^-$ and s$^+$ were grown as described previously[5,6,7]. Mitochondria were isolated by either grinding with glass beads[6,7] or by homogenization of mycelia in a Tekmar Tissumizer. Mt DNA was separated from nuclear DNA by density centrifugation in DAPI-CsCl gradients[6,7]. Only occasionally was a "heavy" fraction observed in senescent cultures and never in wild-type cultures.

Restriction enzymes were obtained from Bio Labs, or Boehringer-Mannheim and reaction conditions were as specified by the supplier. DNA fragments were isolated from Sea-Plaque agarose gels as previously described[12].

Isolated HaeIII fragments of senescent mt DNA were cloned into the EcoRI site of plasmid pBR 325[13] using EcoRI molecular recombination linkers[14] (Obtained from Collaborative Research, Waltham, Mass.) on blunt-ended HaeIII DNA fragments. Transformants were selected for and amplified with spectinomycin as described[13]. Cloned senescent mt DNA fragments were nick translated with the Nick Translation System supplied by New England Nuclear[15].

Gel electrophoresis and Southern blotting hybridizations were done under standard conditions[12,16].

RESULTS

Mitochondrial DNA from five independent senescent events was isolated. In each case, as before[6] about 10 percent of the molecules consisted of a set of closed circles. Detailed analysis of these circles has not yet been completed. Purified DNA was digested by HaeIII and examined on Sea Plaque agarose for the isolation of specific fragments (Figure 1). Lanes 3,4,7 and 8 represent separate isolates of the same senescent event. For lane 3, this DNA was the only mt DNA which had the so-called heavy density, and the one predominant HaeIII fragment had the same molecular weight, about 2560 bp as before[6]. Other senescent events yielded DNA having the same density as wild-type (young) mt DNA. In some thirty experiments, it should be pointed out that wild-type mt DNA never

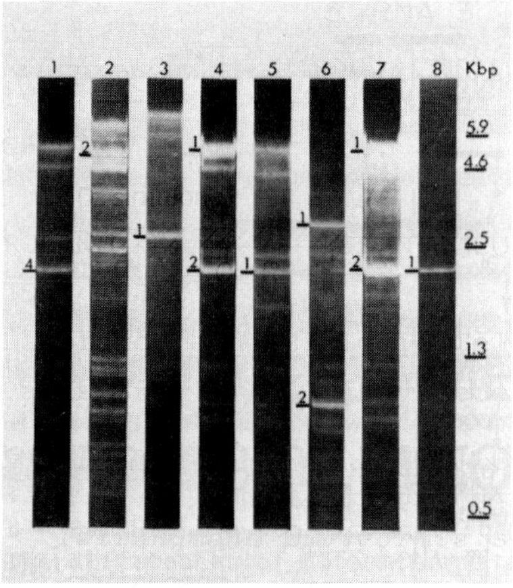

Figure 1. HaeIII digest of mt DNA obtained from several senescent events. 1) s^- SD2; 2) s^+ SD4; 3) s^- SD1; 4) s^- SD1; 5) s^- SD3; 6) s^+ SD5; 7) s^- SD1; 8) s^- SD1. s^- or s^+ refer to race s of + or - mating type. SD designates Senescent event Denver.

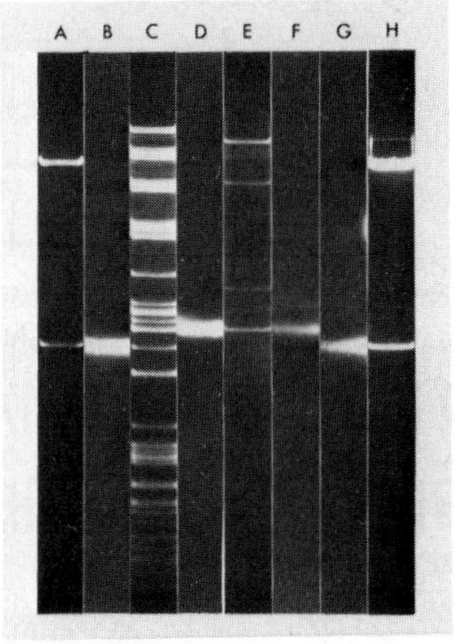

Figure 2. Hybridization analysis with ^{32}P 1_4 probe. A) EcoRI digest of 1_4 clone; B) Radioautograph of A; C) HaeIII digest of wt mt DNA; D) Radioautograph of C; E) HaeIII digest of 1; F) Radioautograph of E; G) Radioautograph of H; H) EcoRI digest of 8_1 clone.

contained a set of small circular molecules. The DNA indicated by numbers in each lane was isolated and subjected to cloning techniques. Thus far, we have not yet cloned the three interesting fragments $3_1, 5_1, 6_1$, or 6_2 but have cloned successfully the DNA labeled $1_4, 2_2, 4_1, 4_2, 7_1, 8_1$.

In Figure 2 is shown the data for the 1_4 clone. This insert hybridized to wt mt DNA HaeIII fragment 13 which has a molecular weight of 2000 bp, slightly higher than the insert of 1860 bp but the same as that fragment isolated (see Figure 1). In addition, the 1_4 insert is homologous with the insert in the 8_1 clone (lanes G and H). These DNA fragments originated from two independent senescent events. Figure 3 contains information on the 4_2 clone. Here, the insert had the same molecular weight 2000 bp, as that isolated and also hybridized to wt HaeIII 13. This clone was not homologous with the 1_4 insert however (not shown) indicating, as is obvious from lane E, that wt mt DNA HaeIII 13 contained at least two co-migrating species of DNA. This 4_2 clone is homologous to the 5_1 fragment, another independent senescent event. Figure 4 is particularly interesting in that it shows that the high molecular weight HaeIII fragment isolated (lanes G and H, about 5500 bp) contained EcoRI sites (lanes I and

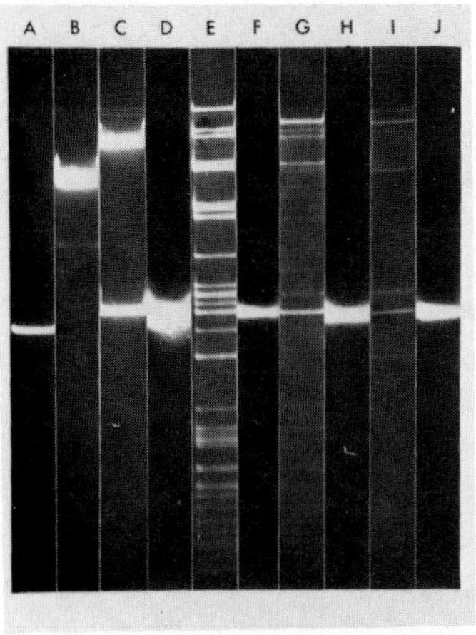

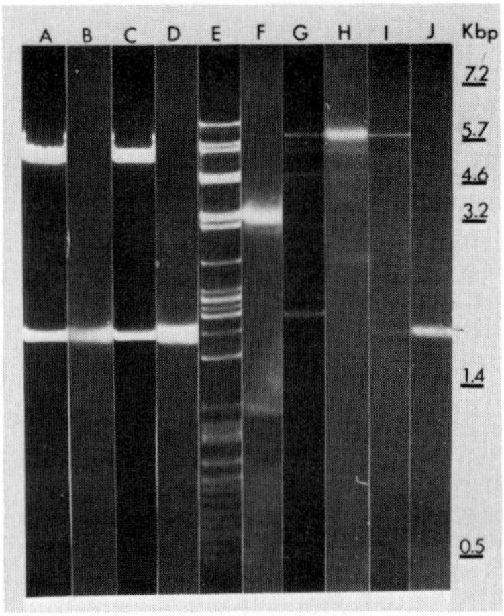

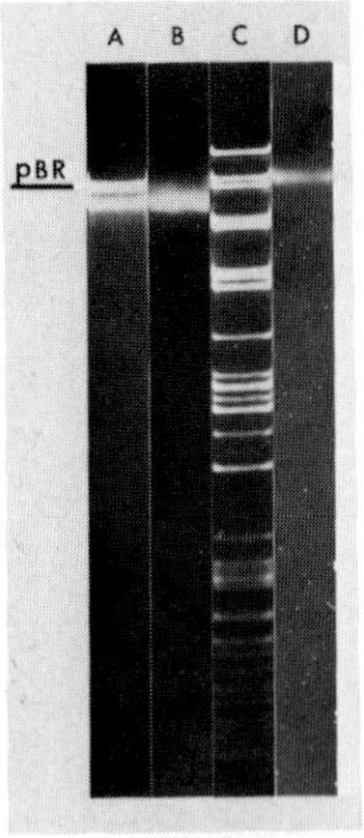

Figure 3. Hybridization Analysis with $^{32}P4_2$. A) Insert I_4; B) EcoRI digest of 4_{2a} "clone". Note the decreased molecular weight; C) EcoRI digest of 4_2; D) Radioautograph of C; E) HaeIII digest of wt mt DNA; F) Radioautograph of E; G) HaeIII digest of 4; H) Radioautograph of G; I) HaeIII digest of 5; J) Radioautograph of I.

Figure 4. Hybridization Analysis with $^{32}P4_{1a}$. A) EcoRI digest of 7_1 clone; B) Radioautograph of A; C) EcoRI digest of 4_{1a} clone; D) Radioautograph of C; E) HaeIII digest of wt mt DNA; F) Radioautograph of E; G) HaeIII digest of 4; H) Radioautograph of G; I) EcoRI digest of 4; J) Radioautograph of I.

Figure 5. Hybridization Analysis with $^{32}P4_{1f}$: A) EcoRI digest of 4_{1f}; B) Radioautograph of A); C) HaeIII digest of wt mt DNA; D) Radioautograph of C.

J). Our cloning procedure was such that bluntended HaeIII fragments were linked with EcoRI recognition sequences and then treated with EcoRI to generate sticky ends. We had assumed that, based on our earlier results[6], senescent DNA would not have EcoRI sites. Examination of several senescent events has indicated that some mt DNA had EcoRI sites (2,4 and 8) and some did not (1,3). Statistically, this is not unexpected. Figure 4 also shows that this clone 4_{1a} is homologous with the 7_1 clone, derived from the same senescent event. The homologous wt HaeIII fragment 7 has a molecular weight of 3250 bp, much greater than the fragment cloned. Finally, another fragment (4_{1f}) from senescent event 4 was cloned (Figure 5) and this was cloned from the same fragment isolated for the 4_{1a} clone. This insert of about 5100 bp hybridized with wt HaeIII fragment 2, of about 5500 bp. No homologies of this 4_{1f} insert were observed with the clones or senescent events discussed here but recently another race s^+ senescent event yielded two high molecular weight mt DNA fragments homologous with 4_{1f}.

DISCUSSION AND SUMMARY

Two of the major questions raised concerning the DNA isolated from senescent Podospora anserina have been answered by the results presented here. DNA isolated from several independent senescent events can give rise to sometimes similar but often times different DNA. This is in accord with rho^- petites[9,10,11]. Most important, we have shown that this senescent DNA is homologous with specific fragments of wt mt DNA. No homologies were observed with either wt or senescent nuclear DNA (unpublished observations). Similar data have also been obtained by L. Belcour and his colleagues (personal communication). Thus we conclude that the multimeric set of tandemly repeated closed circular molecules isolated from senescent mycelia is of mitochondrial origin and not of either nuclear or plasmid[8] sequences. One could suggest that perhaps we have inadvertently isolated residual wt mt DNA. The data (see Summary Table) argue

SUMMARY TABLE OF CLONED SENESCENT mt DNA

Senescent Event	Fragment Isolated (bp)	Fragment Cloned (bp)	wt Fragment Hybridized	Homologous DNA
1 (4)	2000 (HaeIII)	1860	2000 (HaeIII 13)	8_1 (not to 4_2)
2 (2)	5500 (HaeIII)	5500	5500 (HaeIII 3)	None
4 (1)	5750 (HaeIII)	5100	5600 (HaeIII 2)	None
(1)	1860 (EcoRI)	1860	3250 (HaeIII 7)	7_1
(2)	2000 (HaeIII)	2000	2000 (HaeIII 13)	5_1 (not to 1_4)
8 (1)	2000 (HaeIII) (1860 EcoRI)	1860	2000 (HaeIII 13)	1_4 (not to 4_2)

against this since some fragments isolated from senescent cultures hybridized to a higher molecular weight fragment from wt mt DNA. This is as expected if a particular sequence within a larger sequence is selected for amplification. Why some isolated fragments yield a lower molecular weight upon cloning (1_4, 4_1) is not clear. It is possible that a deletion mechanism is in operation. Certain senescent fragments (4_2) for example, led to the complete deletion of the chloramphenicol gene in the pBR 325 plasmid (unpublished data). In terms of interrelationships, there appears to be at least five senescent classes discussed here: 4_{1a} and 7_1; 1_{4b} and 8_1; 4_2 and 5_1; 2_2; and finally 4_{1f}. Clearly other senescent events are needed and fine structural analysis of the DNA is required before we can establish a further correlation with senescence or with petite suppressiveness.

No answer is yet known regarding the causal relationship of senescence and these tandemly repeated DNA molecules. The availability of these DNA clones will be of great value in transformation studies.

ACKNOWLEDGMENT

The work reported here was supported by grants from the National Institutes of Health (AG01367) and the National Science Foundation (PCM-79 01805).

REFERENCES

1. Marcou, D. (1961). Ann. Sci. Natur., 11, 653-764.
2. Smith, J.R. and Rubenstein, I. (1973). J. Gen. Microbiol. 76, 297-304.
3. Marcou, D. and Schecroun, J. (1959). Comptes Rendus Acad. Sci. 248, 280-283.
4. Tudzynski, P. and Esser, K. (1977). Molec. gen. Genet. 153, 111-113.
5. Belcour, L. and Begel, O. (1978). Molec. gen. Genet. 163, 113-123.
6. Cummings, D.J., Belcour, L. and Grandchamp, C. (1979). Molec. gen. Genet. 171, 239-250.
7. Cummings, D.J., Belcour, L. and Grandchamp, C. (1979). Molec. gen. Genet. 171, 229-238.
8. Stahl, U., Lemke, A., Tudzynski, P., Kück, U. and Esser, K. (1978). Molec. gen. Genet. 162, 341-343.
9. Locker, J., Rabinowitz, M. and Getz, G.S. (1974). J.Mol.Biol. 88, 489-507.
10. Faye, G., Fukuhara, H., Grandchamp. C., Lazowska, J., Michel, F., Casey, J., Getz, G.S., Locker, J., Rabinowitz, M., Bolotin-Fukuhara, M., Coen, D., Deutsch, J., Dujon, B., Netter, P., and Slonimski, P.P. (1973). Biochimie 55, 779-792.
11. Lazowska, J. and Slonimski, P.P. (1976). Molec. gen. Genet. 146, 61-78.
12. Cummings, D.J., Maki, R.A., Conlon, P.J. and Laping, J. (1980). Molec. gen. Genet., in press.
13. Bolivar, F. (1978). Gene 4, 121-136.
14. Maniatis, T., Hardison, R.C., Lacy, E., Lauer, J., O'Connell, C. Quon, D., Sim, G.K. and Efstratiadis, A. (1978). Cell 15, 687-701.
15. Rigby, P.W.J., Dieckmann, M., Rhodes, C. and Berg, P. (1977). J. Mol. Biol. 113, 237-251.
16. Southern, E.M. (1975). J. Mol. Biol. 98, 503-517.

The Organization and Expression of the Mitochondrial Genome
A.M. Kroon and C. Saccone, editors

THE REMARKABLE FEATURES OF GENE ORGANIZATION AND EXPRESSION OF HUMAN MITOCHONDRIAL DNA

GIUSEPPE ATTARDI, PALMIRO CANTATORE, EDWIN CHING, STEPHEN CREWS, ROBERT GELFAND, CHRISTIAN MERKEL, JULIO MONTOYA and DEANNA OJALA
Division of Biology, California Institute of Technology, Pasadena, California 91125, U.S.A.

Introduction

About ten years ago, one of us gave a lecture at the California Institute of Technology entitled "One thousand circles in search of a function". This quasi-Pirandellian title illustrated the state of confusion and ignorance that at that time existed concerning the informational role of mitochondrial DNA (mtDNA) in animal cells. During the past ten years, the function of this DNA has become increasingly apparent, conforming more and more to the picture which was unfolding in the more easily analyzable yeast system. At the same time, the distinctive features of the mitochondrial genetic system of animal cells have gradually emerged. Thus, the unusual characteristics of the processes of replication and transcription of animal cell mtDNA were recognized quite early. It is, however, only in the last two or three years that a detailed structural analysis of this system, made possible by the recent technological developments in the area of DNA and RNA sequencing and gene mapping, has revealed the astounding features of the mitochondrial genetic code and of the gene organization and expression in animal cell mtDNA, that have been or will be discussed in this Symposium.

Our main approach in the last few years has been to identify the transcription products of HeLa cell mtDNA and to characterize them in detail in their structural, metabolic and mapping properties. Correlation of this information with the DNA sequencing data obtained in F. Sanger's laboratory and, in small part, in our laboratory, and with protein sequence data has allowed us to develop a coherent picture of gene organization in human mtDNA and a plausible model of expression of this genome.

Isolation and fractionation of mitochondrial RNA

An important development from the point of view of the analysis of the structural and metabolic properties of mitochondrial RNA has been the recent introduction in our laboratory of a procedure for the isolation in pure form of mitochondrial RNA species. This procedure involves treatment of the mitochondrial fraction with high concentrations of micrococcal nuclease to destroy extramitochondrial nucleic acids.[1] Figure 1b shows the autoradiogram, after high resolution CH_3HgOH-agarose slab gel electrophoresis, of the oligo(dT)-bound RNA fraction from micrococcal nuclease treated mitochondria of HeLa cells labeled with [^{32}P]orthophosphate for 2.5 hr in the absence of inhibitors of nuclear RNA synthesis. The pattern is identical in every detail to that obtained with mitochondrial oligo(dT)-bound RNA from the crude mitochondrial fraction (untreated) of cells labeled for 3 hr with [^{32}P]orthophosphate in the presence of 20 µg/ml camptothecin (to block all high mol. wt. nuclear RNA synthesis) (Fig. 1a). One recognizes the 18 discrete components, with molecular weights covering the range from 9.0×10^4 to 3.4×10^6 daltons, which were previously described.[2] The sequence complementarity to the heavy (H) and light (L) mtDNA strands of the individual components, determined in earlier[2] and more recent experiments, is indicated in Figure 1b. With the exception of the three largest RNA species (#1, 2 and 3) and of the smallest one (#18), all these components are coded for by the H-strand.

The possibility of obtaining mitochondrial RNA species not detectably contaminated by other components has opened the way to the sequencing analysis of these RNA species. Figure 1c shows the electrophoretic pattern, after ethidium bromide staining, of a large scale preparation of mitochondrial oligo(dT)-bound RNA from micrococcal nuclease treated organelles (from 20 gm of cells). One clearly recognizes the characteristic set of components shown in Figure 1a and 1b, with the exception of RNA species 1, which is extremely short lived and present in mitochondria in very small amount.[3,4] There is a previously undetected component, of unknown origin, migrating between species 16 and 17.

As to the nature of the above described poly(A)-containing RNA species, components 5, 7, 9, 11 to 16 and 17, because of their relative abundance, their enrichment in partially purified polysomal structures, and their relatively long half-life,[3,4] are probably specific mRNAs. For several of these RNA species such functional characterization has been corroborated, as will be discussed below, by the correlation of RNA sequences with protein sequence data and with known yeast gene sequences. RNA species 1 to 4 and 6, because of their presence in only marginal amounts in partially purified polysomal structures and their relatively short half-life, are presumably either precursors or intermediates in the pathway of maturation of the functional species. The mapping data to be discussed below support this conclusion for RNAs 4 and 6. The nature of the L strand transcripts (RNAs 1, 2, 3 and 18) will be discussed below.

General organization of mtDNA transcripts

A preliminary mapping of the poly(A)-containing RNAs was done using the Southern[5] and the Alwine et al.[6] procedures. This analysis allowed a gross localization in the mtDNA physical map of the sequences complementary to these RNA species. A more precise mapping was done using the S1 protection technique.[7,8] The results of this study, as well as the previous results on the fine mapping of the mitochondrial rRNA genes,[9] are presented in a diagrammatic form in Figure 2. In this diagram, the positions and identities of the tRNA genes were derived from the DNA sequence (F. Sanger and B. Barrell, personal communication, and ref. 1). The identification of the putative mRNAs, and thereby of the corresponding genes, for cytochrome c oxidase subunit I (COI) and cytochrome b was made by correlating, as will be described below, RNA sequences with protein sequence data for human COI (Chomyn, Hunkapiller, and Attardi, in preparation) or with the yeast cytochrome b sequence (A. Tzagoloff, personal communication); the identification of RNA 16 as presumptive mRNA for cytochrome c oxidase subunit II (COII) was based on the prior recognition of the complementary DNA sequence as the COII gene.[10]

The results illustrated in Figure 2 allow several general conclusions to be drawn. First, with the exception of the D-loop region and of a stretch of about 170 nucleotides between the 5'-end of the H-strand coded $tRNA^{Phe}$ and the 5'-end of the L-strand coded RNA 18, the HeLa cell mtDNA sequences are completely saturated by the discrete poly(A)-containing RNAs, rRNAs and tRNAs coded for by the two strands. No evidence for intervening sequences has been found in the mitochondrial genes coding for the various poly(A)-containing RNAs. This observation extends the previous similar results concerning the human mitochondrial rRNA cistrons[9] and is in agreement with the evidence presented for the *Xenopus* mitochondrial genome.[11] A second conclusion of the present studies is that, at the level of resolution allowed by this approach, and with the exception of RNA 4 and RNA 6, there is no apparent overlapping in the H-strand of the sequences coding for the various poly(A)-containing RNA, rRNA and tRNA species. There are good reasons to believe that the two RNAs representing an exception to the "non-overlapping" rule may be precursors of functional, mature species. In particular, RNA 4 is a possible precursor of the two rRNA species and RNA 6 a possible precursor of RNA 9. Both the correspondence in mapping position with that of the mature species, and the relatively short half-life of these presumptive precursors[3,4] support the above conclusion.

As concerns the relative mapping position of the H-strand coding sequences

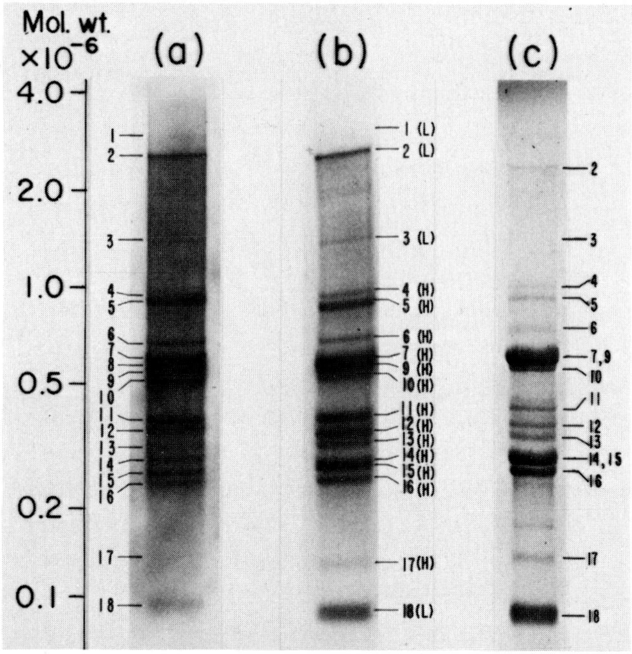

Fig. 1. Autoradiograms (a and b) and ethidium bromide stained pattern (c) of oligo(dT)-bound mitochondrial RNA from HeLa cells, after electrophoresis through agarose-CH_3HgOH slab gels. See text for details.

and the L-strand sequences which specify what are presumably mature RNA products (i.e., RNA 18 and tRNAs), some of the latter (RNA 18, $tRNA^{Gln}$, $tRNA^{Pro}$) appear to fall within noncoding regions of the H-strand, others correspond to terminal regions of H-strand coded poly(A)-containing RNAs. In particular, the $tRNA^{Ala}$, $tRNA^{Asn}$, $tRNA^{Cys}$ and $tRNA^{Tyr}$ are complementary to the 5'-terminal segment of RNA 6 preceding the initiator codon of RNA 9 (COI mRNA); $tRNA^{Ser(L)}$ and $tRNA^{Glu}$ would appear from the mapping data to be complementary to the 3'-end proximal segments of RNA 9 (and of its putative precursor RNA 6) and RNA 5, respectively; both of these segments lie outside of the polypeptide coding stretches of these RNAs (Barrell and Sanger, personal communication). The functional significance of the L-strand coded large transcripts 1, 2 and 3, which overlap many poly(A)-containing RNA coding sequences on the H-strand, is not known. Their very short half-life suggests that they are precursors or intermediates in the processing of L-strand coded tRNAs; that they may contain polypeptide coding stretches cannot, however, be excluded. Therefore, while it

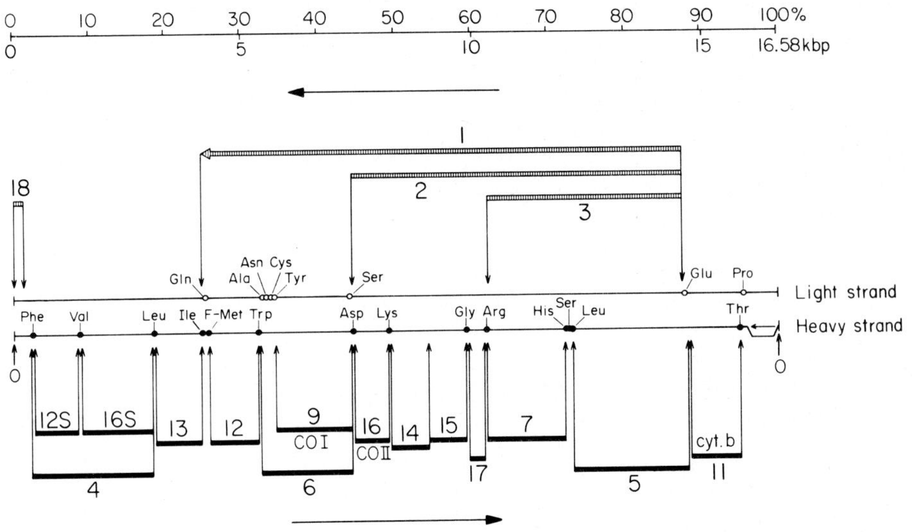

Fig. 2. Transcription map of HeLa cell mtDNA. The positions and identities of the tRNA genes on the H-strand (●) and on the L-strand (O) were derived from the DNA sequence (F. Sanger and B. Barrell, personal communication, and ref. 1). The upper and lower arrows indicate the direction of L- and H-strand transcription, respectively. See text for details.

appears from the available information that there is no significant overlapping of coding sequences within the H-strand domain, nothing conclusive can be said as yet concerning the possible overlapping of information between the H-strand and L-strand domains.

The most important conclusion of the results summarized in Figure 2 is that the sequences coding for the poly(A)-containing RNA species on the H- and L-strand are immediately contiguous on one side, and most frequently on both sides, to tRNA coding sequences. These observations extend the results of the earlier E.M. mapping experiments which had demonstrated a similar relationship with 4S RNA genes of the two rRNA coding sequences.[12,13] There are a few exceptions to the rule that each end of a coding sequence for an rRNA or poly(A)-containing RNA species is immediately flanked by a tRNA gene. Thus, the RNA 9 coding sequence does not have a tRNA gene on its 5'-side; this RNA, however, is probably not a primary transcription product, but derives, as mentioned above, from RNA 6 by removal of a 5'-terminal stretch containing sequences complementary to four L-strand coded tRNA species. The RNAs 14 and 15 coding sequences are apparently not separated by a tRNA gene; the mapping data indicate that their coding sequences are immediately contiguous on the H-strand, suggesting the possibility

```
5' CCCCCCAAAGCTGGTTTCAAGCCAACCCCATGGCCTCCATGACTTTTTCAAAAAGGTATT
3' GGGGGGTTTCGACCAAAGTTCGGTTGGGGTACCGGAGGTACTGAAAAAGTTTTTCCATAA
  -120                                                      -61

                        5'—— tRNA^Asp ——>3'
   AGAAAAACCATTTCATAACTTTGTCAAAGTTAAATTATAGGCTAAATCCTATATATCTTA
   TCTTTTTGGTAAAGTATTGAAACAGTTTCAATTTAATATCCGATTTAGGATATATAGAAT
  -60                                                          -1

   5'          RNA #16
   AYGGYAYAYGYAGYGYAAGYAGGYYYAYA ——————>
   ATGGCACATGCAGCGCAAGTAGGTCTACAAGACGCTACTTCCCCTATCATAGAAGAGCTT 3' L strand
   TACCGTGTACGTCGCGTTCATCCAGATGTTCTGCGATGAAGGGGATAGTATCTTCTCGAA 5' H strand
   1                                                          60

   ———————————————— COII ————————————————>
```

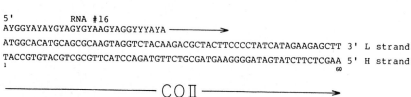

Fig. 3. Alignment of the nucleotide sequence of the 5'-end proximal region of poly(A)-containing RNA 16 with the DNA sequence of the COII gene.[10] Y: pyrimidine.

of a common processing intermediate giving rise to the two mature species by a precise endonucleolytic cleavage. A similar situation seems to occur for RNAs 5 and 11. However, no processing intermediate corresponding to joined RNA 14 and RNA 15 or joined RNA 11 and RNA 5 has been identified as yet.

Sequence analysis of mitochondrial poly(A)-containing RNAs

It has been recently reported by Barrell et al.[10] that the $tRNA^{Asp}$ gene in human mtDNA is juxtaposed to the 5'-end of the COII coding sequence. Since all eukaryotic mRNAs so far analyzed have been shown to have on the 5'-side of the coding sequence a non-coding stretch, which is presumably used for ribosome attachment,[14,15] it was reasonable to ask whether, in the case of the human COII mRNA, the above function is performed by the $tRNA^{Asp}$ sequence or a portion of it, or whether on the contrary this mRNA completely lacks a 5' non-coding stretch. The identification, in the above described mapping analysis, of the presumptive COII mRNA (RNA 16) has made it possible to address the above question. In particular, the 5'-end proximal region of RNA 16 was sequenced by partial enzymatic digestions of the 5'-end $[^{32}P]$labeled RNA followed by electrophoretic fractionation of the products on polyacrylamide/urea gels; the sequence thus determined was aligned with the COII coding sequence. As shown in Figure 3, the entire 29 nucleotide stretch of RNA 16 which has been sequenced appears to be colinear with the DNA sequence.[10] The striking result is that the 5'-end of the RNA corresponds precisely to the first nucleotide of the COII coding sequence.

The above described finding raised the question of whether the lack of a 5' non-coding stretch is a frequent or even general feature of mitochondrial mRNAs in HeLa cells. In order to answer this question, a 5'-end proximal segment of the majority of the polysome-associated poly(A)-containing RNAs has been sequenced and the sequences thus obtained have been compared with the DNA sequence (ref. 16, and Barrell and Sanger, personal communication). Figure 4 summarizes in a schematic form these results, as well as the sequencing data on the COII mRNA and on the 12S and 16S rRNAs.[1] It is clear that almost all the putative mRNAs analyzed so far either start directly with an AUG or AUA triplet (which is a methionine codon in human mitochondria[10]), or have a few nucleotides (1 to 8) preceding the AUG or AUA. Protein sequencing data (Chomyn, Hunkapiller and Attardi, in preparation) and comparison with the yeast cytochrome b gene sequence have indicated that the AUG of RNA 9 (COI mRNA) and of RNA 11 (cytochrome b mRNA), besides that of COII mRNA, are initiator codons for the corresponding polypeptides. Thus, it seems reasonable to extrapolate from these results and interpret the AUGs of the other mRNAs likewise as initiator codons. There is the possibility that AUA too may function as initiator codon in human mitochondria (Sanger and Barrell, personal communication). A single mRNA (#12), among those analyzed, has a long stretch (108 nucleotides) preceding the first AUG.

The observation that most mitochondrial mRNAs either start directly at the initiator codon or have only a few nucleotides (1 to 8) preceding this codon raises interesting questions concerning the mechanism whereby the mitochondrial ribosomes attach to these messengers. Both in eukaryotic and prokaryotic mRNAs, there is a stretch of variable length preceding the initiator codon, and at least in E. coli mRNAs there is evidence that this stretch contains a ribosome-binding site.[17,18] A single exception to this rule has been described, namely that of the lambda repressor mRNA involved in repressor maintenance.[19] In this case, it was argued that the lack of a strong ribosome binding site could account for the low efficiency of translation of this mRNA. In the case of mitochondria from human cells (and probably from other animal cells), it is reasonable to think that the special features of their ribosomes make them suitable for binding directly to the initiator codon.

A significant observation illustrated in Figure 4 is that the coding sequence for each mRNA is almost always immediately adjacent to the 3'-end of a tRNA gene. This situation thus reproduces that previously described for the relationship of the 12S rRNA gene and the $tRNA^{Phe}$ gene.[1] The two exceptions

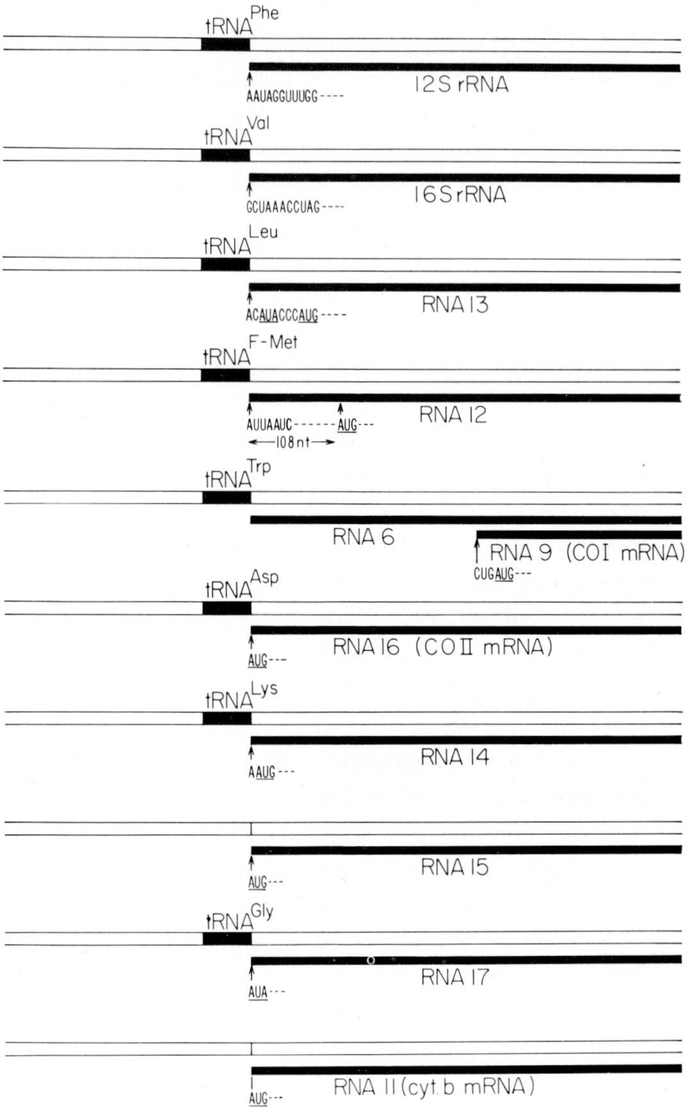

Fig. 4. Diagram illustrating the juxtaposition of the rRNA or poly(A)-containing RNA coding sequences and the flanking tRNA genes on their 5'-side. The sequence of the 5'-end proximal region of each RNA was aligned with the DNA sequence (ref. 1, 10 and 16, and F. Sanger and B. Barrell, personal communication). The identification of the pyrimidines as U or C was made on the basis of the DNA sequence. The arrows indicate the first nucleotide of the RNAs sequenced; this nucleotide corresponds in the DNA to the residue immediately adjacent, on the 3'-side, to the tRNA gene flanking the rRNA or poly(A)-containing RNA coding sequence. See text for details.

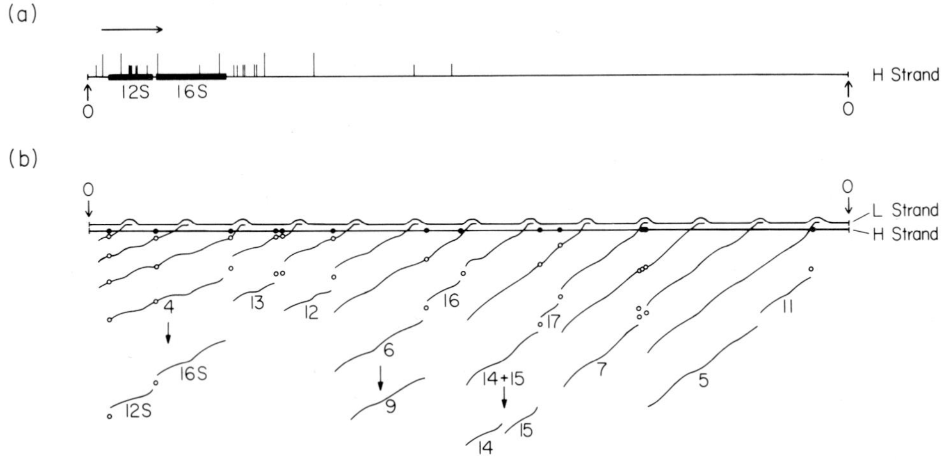

Fig. 5. Proposed model for mtDNA H-strand transcription. (a) Distribution in the mtDNA H-strand of the hybrids formed with nascent RNA chains isolated from transcription complexes (modified from ref. 20). Each tick marks the position in the HpaII map (not shown) of one end of each hybrid, while the other end is arbitrarily localized at the closest, origin-proximal HpaII site. Long and short ticks refer to abundant and rare hybrids, respectively. (b) Diagrammatic representation of the processing of nascent mitochondrial RNA chains in the transcription complexes ●, tRNA gene; O, mature tRNA.

to the rule of a tRNA gene immediately preceding, on the 5'-side, the mRNA coding sequence, are those previously mentioned concerning the RNA 9 (COI mRNA) and RNA 15. We do not have yet any information concerning the precise relationship between the 3'-ends of the sequences coding for the mitochondrial RNAs and the adjacent tRNA genes. Sequencing experiments are in progress to elucidate this relationship.

A model for mtDNA H-strand transcription

The above described sequencing and mapping results strongly suggest that the H-strand sequences coding for the rRNA species, the poly(A)-containing RNAs and the tRNAs are immediately contiguous to each other, extending continuously from coordinate 2/100 to coordinate 95/100 (relative to the origin taken as 0/100). This arrangement is consistent with a model of transcription of the H-strand in the form of a single molecule which is processed by precise endonucleolytic cleavages into functional, mature RNA species (Fig. 5). In the processing of the primary transcripts, the tRNA sequences may play an important role as recognition signals, providing the punctuation in the reading of mtDNA information. It is conceivable that the processing enzyme(s) recognizes the cloverleaf structure of the tRNA sequence or some portion of it. A similar structure (although not a bona fide tRNA sequence) may exist at the site where RNA 9 is cleaved out of RNA 6, and at the border between RNA 14 and RNA 15 and between RNA 11 and RNA 5. Analysis of the sequences near the 5'- and 3'-termini of the various poly(A)-containing RNAs should reveal any common sequence or secondary structure which may represent the recognition site for the putative processing enzyme(s).

Previous work, in which nascent RNA molecules isolated from mtDNA transcription complexes were mapped by the S1 protection technique on the HeLa cell mitochondrial genome, had indicated a concentration of H-strand nascent transcripts in the quadrant of the mtDNA HpaII map which is adjacent to the origin of replication in the direction of H-strand transcription (Fig. 5a).[20] These results strongly suggested that the region of mtDNA around the origin of replication contains an initiation site for H-strand transcription. Other types of evidence had indicated that the H-strand nascent transcripts are on the average considerably shorter than the L-strand nascent transcripts.[2,20,21] This observation, combined with the failure to detect any giant-size H-strand transcripts, in contrast to the occurrence of discrete giant-size L-strand transcripts, pointed to the possibility that the processing of the H-strand transcripts proceeds while they still reside on the mtDNA transcription complexes.[20] The previous and the present results can be brought together to support a model in which transcription of the H-strand starts near the origin of replication and proceeds uninterruptedly at least up to the distal end of the D-loop, while the growing chains are processed to yield the mature rRNAs, poly(A)-containing RNAs and tRNAs (Fig. 5b). For simplicity's sake, the diagram in Figure 5 presents a sequential processing scheme; in reality, the release from the primary transcripts of the individual mature mRNAs and tRNAs may deviate appreciably from a regular temporal pattern reflecting their mapping positions. Given the extremely high degree of packing of genetic information in human mtDNA revealed by the results reported here and by the DNA sequencing analysis carried out in F. Sanger's laboratory, a mechanism of transcription of the H-strand from one single promoter, as proposed here, with the generation of the individual mature RNA species by precise endonucleolytic cleavages involving no or only minimal wastage, has indeed attractive features of simplicity, efficiency and economy.

If a single transcript gives rise to all mitochondrial mature RNA species coded for by the H-strand, one has to account for the quite different amounts in which these species are present in mitochondria. Estimates made in our laboratory (R. Gelfand and J. Montoya, personal communication) indicate that the 12S rRNA is present in HeLa cell mitochondria in a molar amount which is at least 100 times higher than the molar amounts of most of polysome-associated, presumably mature poly(A)-containing RNAs; fairly wide variations in amount also occur among the latter, as revealed by the electrophoretic patterns of long-term [^{32}P]labeled mitochondrial oligo(dT)-bound RNA after autoradiographic exposure, or of unlabeled RNA after ethidium bromide staining (Fig. 1).

Two factors can be invoked to account for the wide variations in amount of the individual RNA species. First, premature termination of the transcripts beyond the rRNA cistrons may give a larger molar yield of the rRNA species relative to that of the mRNAs or the majority of the tRNAs. Indeed, the concentration of nascent transcripts in the quadrant of the map adjacent to the origin of replication in the direction of H-strand transcription (Fig. 5a) is consistent with a much higher rate of transcription of this portion of the genome relative to the other portions. A premature termination of transcription, which produces a 4 to 6-fold greater molar amount of RNA synthesized from the first 2000 nucleotides of the transcriptional unit than from any other region late in adenovirus-2 infection, has been well documented.[22,23] It is attractive to think that the location of the mitochondrial rRNA cistrons close to the initiation point of H-strand transcription in animal cell mtDNA, as has emerged in the course of evolution, may have proven to be advantageous to the cell as one which gave the rRNA cistrons a transcription advantage. The second factor which probably plays a role in determining the differences in steady state amount of the various RNA species coded by the H-strand are differences in their metabolic stability. Recent measurements of the half-life of HeLa cell mitochondrial poly(A)-containing RNAs and rRNAs by a variety of approaches have indeed shown a considerable variation in metabolic stability both within the set of poly(A)-containing RNAs and between these and the rRNAs.[3,4]

The large L-strand coded transcripts

Although the picture of L-strand transcription is still very preliminary, the available evidence suggests that it may follow the same pattern as the H-strand transcription. In particular, the tRNA sequences may also play a role in the processing of L-strand transcription products. The three large discrete L-strand transcripts (RNAs 1, 2 and 3) have an apparently common 5'-end in correspondence of an L-strand coded tRNA gene ($tRNA^{Glu}$). The above mentioned mapping study on the nascent mitochondrial RNA molecules had pointed to the existence of an initiation site for L-strand transcription near the origin

```
                                                                 (A)
                                                               5' AAGAY
5'  GAGGGGAAAATAATGTGTTAGTTGGGGGGTGACTGTTAAAAGTGCATACCGCCAAAAGAT
3'  CTCCCCTTTTATTACACAATCAACCCCCCACTGACAATTTTCACGTATGGCGGTTTTCTA
    360       350       340       330       320       310      301

            7S RNA (#18)
    AAAAYYYGAAAYYYGGYYAGGYYGGYG ------------------------------->
    AAAATTTGAAATCTGGTTAGGCTGGTGTTAGGGTTCTTTGTTTTTGGGGTTTGGCAGAGA
    TTTTAAACTTTAGACCAATCCGACCACAATCCCAAGAAACAAAAACCCCAAACCGTCTCT
    300       290       280       270       260       250      241

    ------------------------------------------------------------>
    TGTGTTTAAGTGCTGTGGCCAGAAGCGGGGGGAGGGGGGGGGTTTGGTGGAAATTTTTTG
    ACACAAATTCACGACACCGGTCTTCGCCCCCCTCCCCCCCCCAAACCACCTTTAAAAAAC
    240       230       220       210       200       190      181

    ------------------------------------------------------------>
    TTATGATGTCTGTGTGGAAAGCGGCTGTGCAGACATTCAATTGTTATTATTATGTCCTAC
    AATACTACAGACACACCTTTCGCCGACACGTCTGTAAGTTAACAATAATAATACAGGATG
    180       170       160       150       140       130      121
                                     O
                                     ↓        7S DNA
    ------------------- - - - - - -> 5'TGTTCGCCTGTAATATTGAACGT-->
    AAGCATTAATTAATTAACACACTTTAGTAAGTATGTTCGCCTGTAATATTGAACGTAGGT
    TTCGTAATTAATTAATTGTGTGAAATCATTCATACAAGCGGACATTATAACTTGCATCCA
    120       110       100       90        80        70        61

    GCGATAAATAATAGGATGAGGCAGGAATCAAAGACAGATACTGCGACATAGGGTGCTC   3' H strand
    CGCTATTTATTATCCTACTCCGTCCTTAGTTTCTGTCTATGACGCTGTATCCCACGAGGC 5' L strand
    60        50        40        30        20        10        1
```

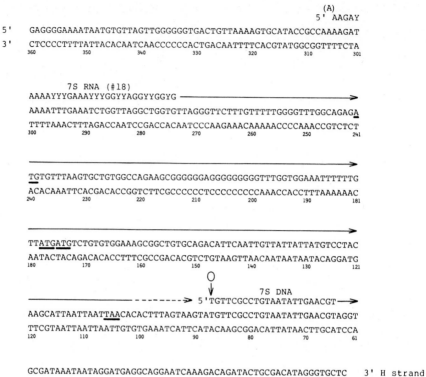

Fig. 6. Alignment of the nucleotide sequence of the 5'-end region of poly(A)-containing RNA 18 (7S RNA) with the DNA sequence of HpaII fragment 8.[1] The sequence of the 5'-terminal segment of 7S DNA and its alignment with the HpaII-8 sequence[26] is also shown.

of replication; therefore, it is possible that the common 5'-end of RNAs 1, 2 and 3 represents a processing point of the primary transcripts. It has been possible to map precisely the 3'-end in two of these three large RNAs: in one (RNA 2), the 3'-end appears to correspond to an L-strand tRNA gene, in the other (RNA 3), to an H-strand tRNA gene. The three transcripts may result from termination of transcription at alternative fixed points or may represent successive steps in a processing scheme. The physiological significance of these large RNAs is not clear. They may be precursors of L-strand coded tRNAs. On the other hand, the possibility that these transcripts, in particular RNA species 2 and 3 which are the more abundant and the relatively more stable among the three, may themselves function as mRNAs or be precursors of more stable mRNAs as yet unidentified cannot be excluded.

A small L-strand coded RNA (7S RNA) mapping near the origin of replication

Among the L-strand coded RNAs, there is a small polyadenylated species (RNA 18 according to the classification of Amalric et al.[2]), whose existence had been recognized early in this work and which had been called 7S RNA on account of its sedimentation constant in the native state.[24] Mapping experiments have localized the sequences complementary to this RNA in the region immediately preceding the origin of replication in the direction of L-strand transcription (Fig. 2). Because of the intriguing possibilities suggested by this mapping position, it was desirable to obtain a more precise localization of the coding sequences for this RNA. For this purpose, the 5'-end proximal region of 7S RNA was sequenced and aligned with the previously determined sequence of mtDNA HpaII fragment 8,[1] which contains the origin of replication.[25,26] As shown in Figure 6, the 5'-end of 7S RNA corresponds to a residue in the L-strand at 219 nucleotides from the origin of replication. The size of the RNA (after subtraction of the poly(A) contribution) has been estimated to be ∿215 nucleotides on the basis of the mapping experiments by the S1 protection technique, and ∿230 nucleotides on the basis of its electrophoretic mobility in denaturing gels.[24] Therefore, according to these size estimates, the 3'-end of 7S RNA would map at or very near to the origin of replication. Sequencing experiments are in progress to localize more precisely the 3'-end of this RNA.

The significance of the 7S RNA is unknown. There is a reading frame starting at 65 nucleotides and ending at 199 nucleotides from the 5'-end of the 7S RNA, which would code for a polypeptide 45 amino acids long. However, this frame contains two arginine codons (AGA and AGG) which have been suggested to represent termination codons in human mitochondria.[27] If this codon assignment is correct, then the only significant reading frame remaining in 7S RNA would be that starting at nucleotide 128 or 131 and ending at nucleotide 199 from the 5'-end, which would code for a polypeptide 24 or 23 amino acids long. respectively. Both the large and the small hypothetical polypeptides would have a considerable proportion of hydrophobic amino acids. It should be emphasized that, although small polypeptides are synthesized in HeLa cell mitochondria (see below), there is no evidence that the 7S RNA functions as mRNA. The immediate proximity of the sequence coding for this RNA to that coding for the H-strand initiation sequence, the 7S DNA, suggests the possibility that the 7S RNA or some derivative of it may function as a primer for mtDNA H-strand synthesis.

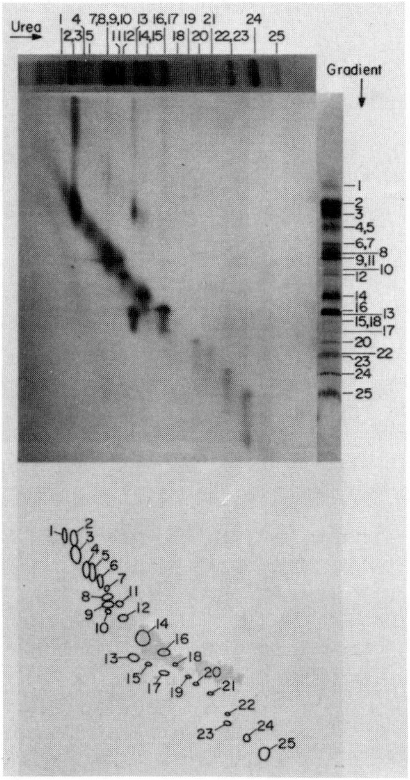

Fig. 7. Fractionation of the products of HeLa cell mitochondrial protein synthesis, labeled for 2 hr with [^{35}S]-methionine in the presence of emetine, by two-dimensional 12.5% polyacrylamide-SDS/8M urea and 15 to 25% polyacrylamide-SDS slab gel electrophoresis.[31,32] The products were first run on an SDS-urea polyacrylamide slab gel, then a track was cut out and overlaid on an SDS-polyacrylamide gradient slab gel and the products run into the gel. An SDS-urea gel pattern and an SDS-gradient gel pattern are shown alongside to aid in recognition.

Evidence suggesting that RNA priming is utilized at the origin for H-strand synthesis in mouse cells has been reported.[28]

Human mitochondrial genes and gene products

On the basis of the reasonable assumption of a functional analogy with the mitochondrial genetic systems from lower eukaryotes, human mtDNA would be expected to code for three subunits of the cytochrome c oxidase, three to four subunits of the oligomycin sensitive ATPase, one subunit of the cytochrome bc_1 complex and one protein associated with mitochondrial ribosomes.[29] Unfortunately, direct information on the nature of the mitochondrial translation products in human, and in general in animal cells, has been very sluggish in coming forth. Among the above mentioned polypeptides, only the three subunits of cytochrome c oxidase have been identified in human mitochondria.[30] In spite of the lack of genetic analysis, it has been possible to start mapping structural genes for specific proteins in the human mitochondrial genome by correlating DNA and RNA sequences with protein sequence data or with known yeast gene sequences. The observation, made in our laboratory, that almost all human mitochondrial mRNAs have their presumptive initiator codon at or very close to the 5'-end suggested that the sequence of the 5'-end proximal region of an mRNA should with high probability yield the expected NH_2-terminal sequence of the polypeptide coded for by that mRNA. Thus, by comparing the 5'-end proximal sequences of the mitochondrial mRNAs with the NH_2-terminal amino acid sequence of human COI recently determined here (Chomyn, Hunkapiller and Attardi, in preparation) and with the NH_2-terminal sequence of yeast cytochrome b (Tzagoloff, personal communication) we have been able to identify the mRNAs for these polypeptides and to map the corresponding genes on the genome. A similar gene assignment, utilizing for comparison the human mtDNA sequences instead of the mRNA sequences, has been made in F. Sanger's laboratory (F. Sanger and B. Barrell, personal communication). Previously, the COII gene had been similarly mapped.[10]

The statement has often been made that the mtDNA gene products so far detected in yeast and the corresponding genetic loci completely saturate the yeast mitochondrial genetic map. By extrapolation, the same statement should apply to the human mitochondrial genome. However, the situation may be more complicated. We have previously reported the resolution among the mitochondrial translation products from HeLa cells of 18 or 19 discrete components, in the molecular weight range between 3500 and 51,000 daltons, after electrophoresis through an SDS-polyacrylamide/8M urea slab gel, or an SDS-polyacrylamide gradient slab gel, respectively.[3,31] An even higher resolution has been recently obtained by a bidimensional combination of the urea and the gradient gel systems.[32] Figure 7 shows such bidimensional fractionation of $[^{35}S]$methionine labeled mitochondrial protein products, so defined on the basis of the emetine resistance and chloramphenicol sensitivity of their labeling: 25 discrete polypeptide components can be identified in the autoradiogram. Control experiments tend to argue against the possibility of degradation artifacts as being responsible for the large number of components detected here. Furthermore, the similarity in the patterns observed after an $[^{35}S]$methionine pulse and a pulse-chase indicates that the great majority of these components are not related by obvious precursor to product relationship. One certainly cannot exclude alternative pathways of processing or secondary modifications, nor the possibility that some polypeptides may overlap in their coding sequences either in the same or in different reading frames. It is interesting, however, to mention that the sum of the molecular weights of the mitochondrial products detected here, together with the two rRNA species and 22 tRNA species, just saturates the single-stranded non-overlapping coding capacity of the HeLa cell mitochondrial genome.[32] The analysis of the nature of these polypeptides and of their relationship represents a formidable challenge for the future, and may reveal unexpected facets of the expression and regulation of this remarkable genome.

Acknowledgements

This work was supported by a research grant from the U.S.P.H.S. (GM-11726). We are very grateful to Drs. F. Sanger and B. Barrell for communicating to us human mtDNA sequence data and for sending us two manuscripts prior to publication.

References

1. Crews, S. and Attardi, G. (1980) Cell, 19, 775-784.
2. Amalric, F., Merkel, C., Gelfand, R. and Attardi, G. (1978) J. Mol. Biol. 118, 1-25.
3. Attardi, G., Cantatore, P., Ching, E., Crews, S., Gelfand, R., Merkel, C. and Ojala, D. (1979) In Extrachromosomal DNA, ICN-UCLA Symposia on Molecular and Cellular Biology, 15 (D. Cummings et al., eds.), Academic Press, New York, pp. 443-469.
4. Gelfand, R. (1980) Ph.D. Thesis, California Institute of Technology, Pasadena, California.
5. Southern, E. M. (1975) J. Mol. Biol. 98, 503-517.
6. Alwine, J. C., Kemp, D. J. and Stark, G. R. (1977) Proc. Nat. Acad. Sci. USA 74, 5350-5354.
7. Berk, A. J. and Sharp, P. A. (1977) Cell 12, 721-732.
8. Berk, A. J. and Sharp, P. A. (1978) Proc. Nat. Acad. Sci. USA 75, 1274-1278.
9. Ojala, D. and Attardi, G. (1980) J. Mol. Biol. 138, 411-420.
10. Barrell, B. G., Bankier, A. T. and Drouin, J. (1979) Nature 282, 189-194.
11. Rastl, E. and Dawid, I. B. (1979) Cell 18, 501-510.
12. Wu, M., Davidson, N., Attardi, G. and Aloni, Y. (1972) J. Mol. Biol. 71, 81-93.
13. Angerer, L., Davidson, N., Murphy, W., Lynch, D. and Attardi, G. (1976) Cell 9, 81-90.
14. Hagenbüchle, O., Santer, M., Steitz, J. A. and Mans, R. J. (1978) Cell 13, 551-563.
15. Kozak, M. (1978) Cell 15, 1109-1123.
16. Sanger, F., Coulson, A. R., Barrell, B. G., Smith, A.J.H. and Roe, B. A. (1980) J. Mol. Biol., in press.

17. Shine, J. and Dalgarno, L. (1974) Biochem. J. 141, 609-615.

18. Steitz, J. A. and Jakes, K. (1975) Proc. Nat. Acad. Sci. USA 72, 4734-4738.

19. Ptashne, M., Backman, K., Humayun, M. Z., Jeffrey, A., Maurer, R., Meyer, B. and Sauer, R. T. (1976) Science 194, 156-161.

20. Cantatore, P. and Attardi, G. (1980) Nucleic Acids Research, in press.

21. Aloni, Y. and Attardi, G. (1971) Proc. Nat. Acad. Sci. USA 68, 1957-1961.

22. Evans, R., Weber, J., Ziff, E. and Darnell, J. E. Jr. (1979) Nature 278, 367-370.

23. Fraser, N. W., Sehgal, P. B. and Darnell, J. E. Jr. (1979) Proc. Nat. Acad. Sci. USA 76, 2571-2575.

24. Ojala, D. and Attardi, G. (1974) J. Mol. Biol. 88, 205-219.

25. Ojala, D. and Attardi, G. (1978) J. Mol. Biol. 122, 301-319.

26. Crews, S., Ojala, D., Posakony, J., Nishiguchi, J. and Attardi, G. (1979) Nature 277, 192-198.

27. Barrell, B. G., Anderson, S., Bankier, A. T., de Bruijn, M.H.L., Chen, E., Coulson, A. R., Drouin, J., Eperon, I. C., Nierlich, D. P., Roe, B. A., Sanger, F., Schreier, P. H., Smith, A.J.H., Staden, R. and Young, I. G. (1980) Proc. Nat. Acad. Sci. USA, in press.

28. Bogenhagen, D., Gillum, A. M., Martens, P. A. and Clayton, D. A. (1978) Cold Spring Harbor Symp. Quant. Biol. 43, 253-262.

29. Borst, P. and Grivell, L. A. (1978) Cell 15, 705-723.

30. Hare, J. F., Ching, E. and Attardi, G. (1980) Biochemistry 19, 2023-2030.

31. Attardi, G. and Ching, E. (1979) In Methods in Enzymology, Vol. 56 (Biomembranes), Academic Press, New York, pp. 66-79.

32. Ching, E. (1979) Ph.D. Thesis, California Institute of Technology, Pasadena, California.

© 1980 Elsevier/North-Holland Biomedical Press
The Organization and Expression of the Mitochondrial Genome
A.M. Kroon and C. Saccone, editors

TWO STUDIES ON MAMMALIAN mtDNA: EVOLUTIONARY ASPECTS; ENZYMOLOGY OF REPLICATION

FRANK J. CASTORA, GREGORY G. BROWN, AND MELVIN V. SIMPSON
Dept. of Biochemistry, State University of New York, Stony Brook, N.Y. (U.S.A.)

Intraspecific mtDNA polymorphism within mammalian populations, including the laboratory rat, is now a well established phenomenon, and restriction enzyme analysis of such DNA's is beginning to prove useful in evolutionary studies on wild populations[1-3] and on the mtDNA molecule itself[2-4]. We report such results here along with cloning and sequencing studies[2-4] which show, with De Vos et al.[5], that variant mtDNA's differ by base substitution rather than methylation.

In a project to develop an enzyme system for mtDNA replication, we have isolated a mitochondrial type II topoisomerase which we plan to use in conjunction with mtDNA polymerase[6] and mitochondrial nicking-closing enzyme[7].

RESULTS AND DISCUSSION

Evolutionary Aspects

mtDNA types and subtypes. We have surveyed wild *R. norvegicus* and *R. rattus* populations from various parts of the world[2,3] (see legend, Fig. 2). Using 6 restriction enzymes whose cleavage sites were mapped, plus Hinf I which makes many cuts and is thus useful in detecting minor variants, we detected several mtDNA types and subtypes (variants whose sequences diverge less than 1% from the type). Four types were revealed in *R. norvegicus* with type NC, the common U.S. type, comprising 5 subtypes. One of these, NC_1, is notable because it exists only in a small discrete area in New York City, suggesting that some rat populations are localized in individual pockets despite the absence of physical or distance barriers. Of special interest are subtypes NC_2, NC_3, and NC_5, which differ from one another only in 1 small Hinf I fragment. The gel pattern (Fig. 1) shows an alteration in only the 140 bp band which differs by about 4 bp in each of the 3 subtypes, suggesting the occurrence of occasional deletions/insertions. In *R. rattus*, variation among the morphs (i.e., restriction patterns) exceeds that in *norvegicus* but confirmation of this seemingly greater mtDNA diversity awaits results on more widely distributed *norvegicus* samples. Five types of rattus mtDNA were revealed, RD being the most common in the U.S. While one of its three subtypes, RD_1, is concentrated in Alameda County, Cal., a sample of RD_1 was also found 3000 miles away in Florida. The Sri Lanka rats (RA) yielded a unique morph for each enzyme suggesting less sequence conservation relative to the other rattus mtDNA's.

Fig. 1. Hinf 1 restriction morphs of mtDNA subtypes. PAGE was from left to right.

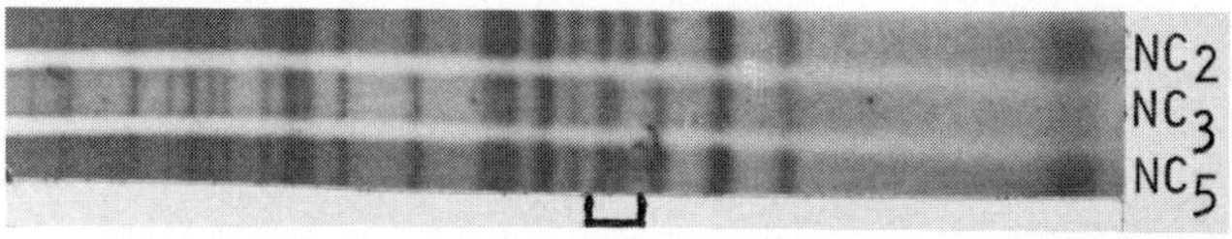

Restriction Maps; Evolutionary Features. We find no evidence for variant site clustering in the R. rattus or the R. norvegicus maps (Fig. 2). However, the rRNA region at 82.5-100 map units (u) is almost free of such sites. Ten sites are conserved in all the DNA's and these cluster in the rRNA and 64-75 u regions. While the H strand replication origin is not exceptionally conserved, the light strand origin at position 67[8] contains a Hind II site which is conserved in all these DNA's and in mouse DNA as well, and this is also true of a site at 73 u and one in the large rRNA gene (83 u). Interestingly, some sites vary intraspecifically but are conserved interspecifically among certain DNA's.

Cloning and sequencing. An assumption in the above studies is that site variations result primarily from base substitution rather than methylation, rearrangements, or deletions/insertions. This view was tested on types NA and NB mtDNA's[9,10] by cloning the two Bam HI fragments in E. coli (K803 (RecA$^-$), Eco RI-methylase free; pBR 322) and sequencing corresponding Hind III fragments containing a variant Eco RI site[2,4]. The cloned and native mtDNA's yielded identical analyses with Eco RI, Hha I, Hinf I and Hae III. Three base substitutions were found in the Hind III fragments, one at the Eco RI site (Fig. 3). Thus, both types of results support the point mutation assumption.

The preferred reading frame (starting at position 169, upper strand) was inferred from the known coding function of the heavy (upper) strand, the presence of multiple stop codons in all lower strand frames and the upper frame starting at 168, and the uncharacteristically high G content in the third positions of upper frame 167. The three mutations are in third positions in the preferred frame and would introduce no changes in amino acids encoded. Further sequencing will show the extent to which such intraspecific changes are silent.

Sequence Divergence. The calculated[11] sequence divergences of the DNA's are shown in Fig. 4. The wide spectrum of intraspecific divergences in R. rattus, 0.4% to 10.5%, is notable, particularly when compared to the norvegicus range of 0.4-1.8%. All interspecific divergences were higher than any intraspecific difference, ranging from 13.7-18.4%. With respect to the Hinf I subtypes (data not shown), the most closely related DNA's from both species diverge only 0.1%-0.2%.

Relatedness of the DNA's. One type of computer-generated tree, based on the application of the unweighted pair group method to the data in Fig. 4, shows the present day relationship among these DNA's (discussed elsewhere[2,3]) (Fig. 5).

Evolutionary pathways can be inferred only if constancy of evolutionary rate and absence of convergence are assumed, both assumptions premature for mtDNA.

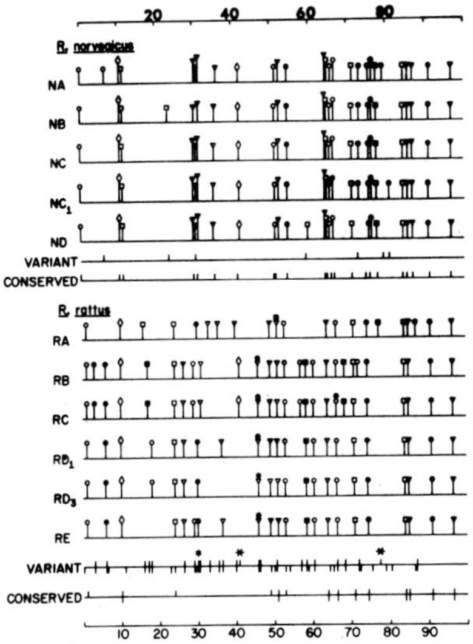

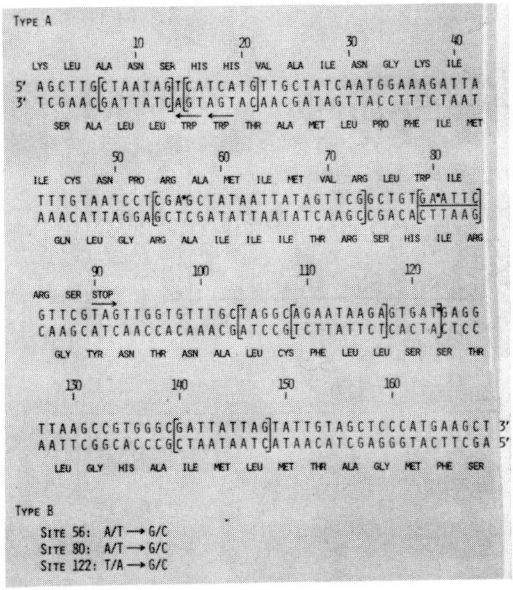

Fig. 2. Restriction maps of rat mtDNA's. Scale begins at origin of H strand replication. Indicators extending downward from R. rattus axes represent composites of variant and conserved sites in both species. *, sites in all R. norvegicus but only in some R. rattus mtDNA's; , sites in certain mtDNA's of both species. ●, Eco R1; ▼, Sst II; ■, Sst I; o, Hind II; ∇, Hind III; □, Hha I; ◊, Bam HI. NA and NB, laboratory types; NC, U.S. type found in 14 widely distributed cities; NC_1, small area in N.Y.C.; ND, Japan; RA, Sri Lanka; RB, Hong Kong; RC, Japan; RD, common U.S. type; RE, Fort Lauderdale, Florida.

Fig. 3. Sequences of the 169 bp Hind II fragments from type NA and NB DNA's. *, sites of nucleotide replacement; GAATTC, Eco RI recognition site.

Enzymology of mtDNA replication in rat liver

Isolation of topoisomerase. Type II topoisomerases can reversibly break the double helix permitting the double strand of the same or of another molecule to pass through[12]. Thus, reversible reactions such as knotting, catenation, and supercoiling are catalyzed. Inhibitor studies on intact mitochondria indicated the presence of a gyrase[13] but its direct detection was unsuccessful. Recently, using a novel assay[12], with pBR322 as substrate, we have detected ATP-stimulated type II activity in mitochondrial extracts. The assay gel shows the formation of catenates of varying complexity, the simplest (Fig. 6, D) running slower than monomeric forms (I, II, III) while the most complex is immobile (E). These forms (Fig. 7) resemble the complex DNA's we have found in intact mitochondria[14]. The insensitivity of the partially purified enzyme to novobiocin or Berenil[7] does not suggest a relationship to gyrase or mt-nicking-closing enzyme. (We thank Drs. Leroy Liu and Rolf Sternglanz for helpful discussions.)

	R.rattus						R.norvegicus				
	RA	RB	RC	RD_1	RD_3	RE	NA	NB	NC	NC_1	ND
RA	-										
RB	10.5	-									
RC	10.5	0.7	-								
RD_1	7.3	4.5	4.4	-							
RD_3	6.9	4.2	4.0	0.4	-						
RE	7.3	3.4	3.4	1.0	1.4	-					
NA	14.7	18.4	18.0	16.4	16.1	16.4	-				
NB	15.4	17.8	17.3	17.1	16.8	17.1	1.6				
NC	13.7	17.5	17.1	15.4	15.1	15.4	0.7	0.9	-		
NC_1	14.5	18.2	17.8	16.1	15.9	16.1	1.2	1.4	0.4		
ND	14.1	17.8	16.7	15.7	15.2	15.7	1.6	1.8	0.9	1.3	-

Fig. 4. Sequence divergence matrix of mtDNA's. See legend to Fig. 2 for nomenclature.

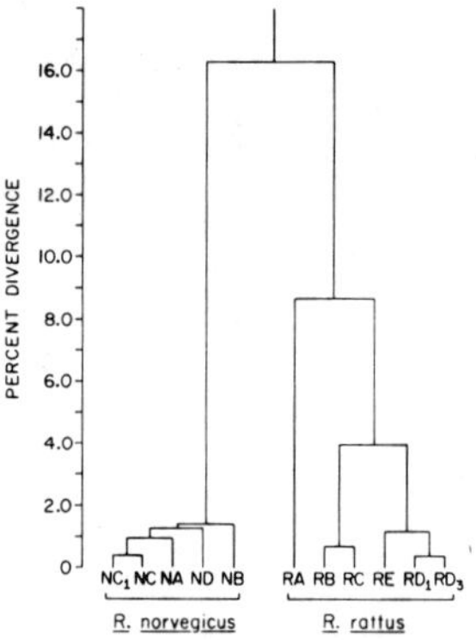

Fig. 5. Phenogram of mtDNA's.

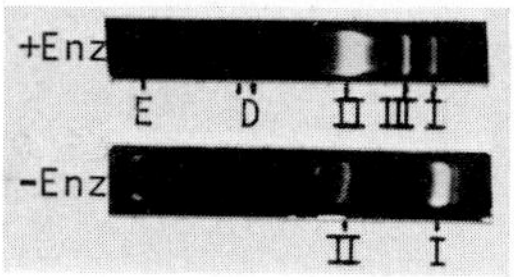

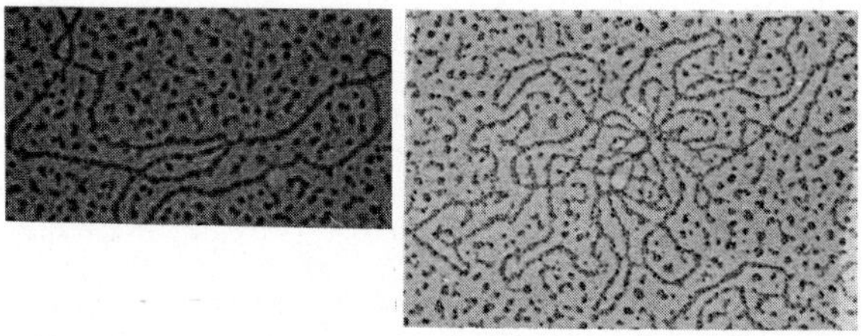

Fig. 6. Assay gel   Fig. 7. Topoisomerase products

REFERENCES

1. Hayashi, J., Yonekawa, H., Gotoh, O., Tagashira, Y., Moriwaki, K., and Yosida, T.H. (1979) Biochim. Biophys. Acta 564, 202-211.

2. Brown, G.G., Castora, F.J., Frantz, S.C., and Simpson, M.V. (1980) Ann. N.Y. Acad. Sci. In press.

3. Brown, G.G., Frantz, S.C., and Simpson, M.V. In press.

4. Castora, F.J., Arnheim, N., and Simpson, M.V. (1980). In press.

5. De Vos, W.M., Bakker, H., Saccone, C., and Kroon, A.M. (1980) Biochim. Biophys. Acta 607, 1-9.

6. Meyer, R.R. and Simpson, M.V. (1970) J. Biol. Chem. 245, 3426-3435.

7. Fairfield, F.R., Bauer, W.R., and Simpson, M.V. (1979) J. Biol. Chem. 254, 9352-9354.

8. Martens, P.A. and Clayton, D.A. (1979) J. Mol. Biol. 135, 327-351.

9. Francisco, J.F. and Simpson, M.V. (1977) FEBS Lett. 79, 291-294.

10. Francisco, J.F., Brown, G.G., and Simpson, M.V. (1979) Plasmid 2, 426-436.

11. Upholt, W.B. (1977) Nucl. Acids Res. 4, 1291-1299.

12. Liu, L.F., Liu, C-C., and Alberts, B.M. (1980) Cell 19, 697-707.

13. Castora, F.J. and Simpson, M.V. (1979) J. Biol. Chem. 254, 11193-11195.

14. Francisco, J.F., Vissering, F.F., and Simpson, M.V. (1977). In "Mitochondria 1977" (W. Bandlow et al., eds.) de Gruyter, Berlin. pp. 25-37.

VARIATION IN BOVINE MITOCHONDRIAL DNAs BETWEEN MATERNALLY RELATED ANIMALS

PHILIP J. LAIPIS AND WILLIAM W. HAUSWIRTH
Departments of Biochemistry and Molecular Biology and Immunology and Medical Microbiology, J. Hillis Miller Health Center, University of Florida, Gainesville, Florida 32610.

INTRODUCTION

A number of laboratories have reported that mitochondrial DNAs from related species and from individuals within a species show sequence variability by restriction enzyme analysis (1-12). The conclusion has been drawn that mitochondrial DNA may be evolving more rapidly than nuclear DNA sequences (10, 12). We wish to report here that examination of mitochondrial DNAs from maternally related Holstein cows reveals an extremely rapid and specific variation in a single Hae III restriction site, mapping at 37.1% on the bovine mitochondrial map (13).

MATERIALS AND METHODS

Mitochondrial isolation, DNA preparation, restriction enzyme digestion, gel electrophoresis, 5' end labeling, and Southern blotting were as previously described (13,14). Liver biopsies of 20-60g were done in collaboration with Dr. K. Braun, Department of Preventive Medicine, University of Florida, College of Veterinary Medicine.

RESULTS

The maternal descendents of Holstein Cow Remcrest Ovation Lucky (Holstein-Fresian Association of American registration number 3797669, barn number H15) are shown in Figure 1. When mitochondrial DNAs from the livers of seven representative animals of this maternal lineage were compared by Hae III restriction enzyme analysis, a variation in the size of a single restriction fragment was observed. As can be seen in Figure 2 fragment L (525 base pairs) is present in four animals whereas it is missing and a larger fragment (L', 563 base pairs) is present in three animals. Because all other Hae III fragments visible on the gel are identical for all animals, it is most probable that fragments L and L' arise from the same genome location.

We have previously shown that Hae III fragment L maps between 37.1 and 40.6 map units (Figure 3) because it contains the Eco RI site at 39.3 and the Bgl II

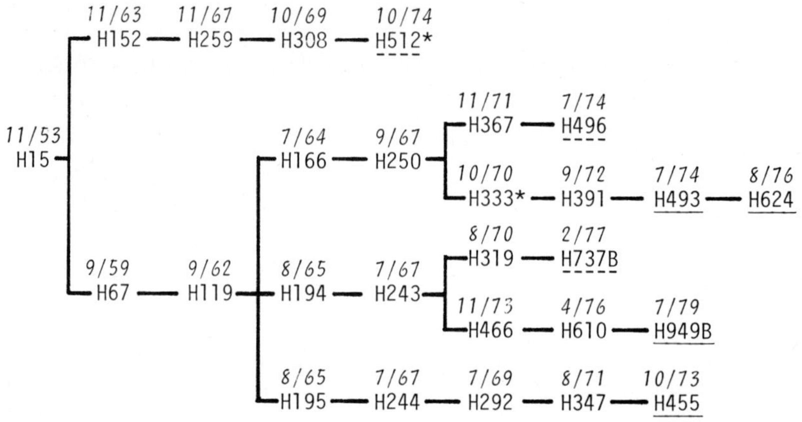

Figure 1. Maternal descendents of Holstein Cow Remcrest Ovation Lucky (H15). Animals are denoted by barn number (e.g. H15 etc.) month and year of birth (superscript). Solid underlines indicate mitochondrial DNA samples having the Hae III-L fragment; dashed underlines indicate those having the Hae III L' fragment; * denotes animal is still alive. For clarity, only the relevant relationships are shown; nineteen living descendents which have not yet been analyzed are still alive in the herd.

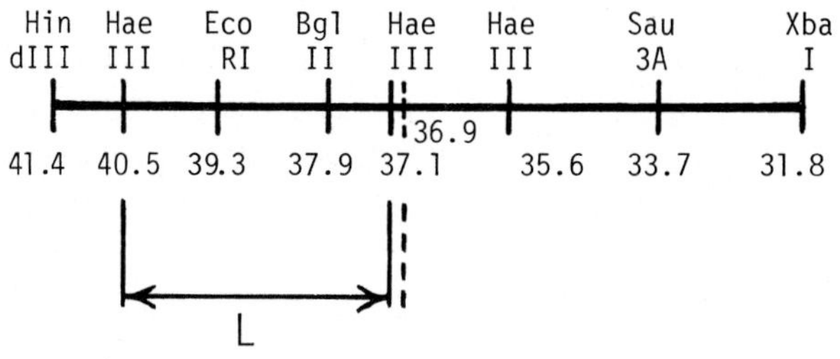

Figure 3. Physical map of restriction sites between 31.8 and 41.4 map units for Holstein H493, H624, H455, and H949B. The dashed line indicates the position of Hae III-L' found in H512, H496, and H737B.

site at 37.9 (Cole, V. C., Crawford, P. C., Laipis, P. J., and Hauswirth, W. W., manuscript submitted). To confirm the relationship between L and L', H493 and H512 mitochondrial DNA (data not shown) or cloned DNA from these two animals (Figure 4) was digested with Hae III, and then either with Bgl II or Eco RI. All resulting fragments longer than 200 base pairs were identical in size, except that the Hae III + Eco RI digest of H493 DNA contained a fragment of 368 base pairs whereas the same digest of H512 DNA was missing this fragment and contained a new fragment of 406 base pairs, the same length difference as between L and L'. Hence L and L' map to the same genome location, and the additional 38 base pairs of L' must map to the right of 37.9 map units (Fig. 3).

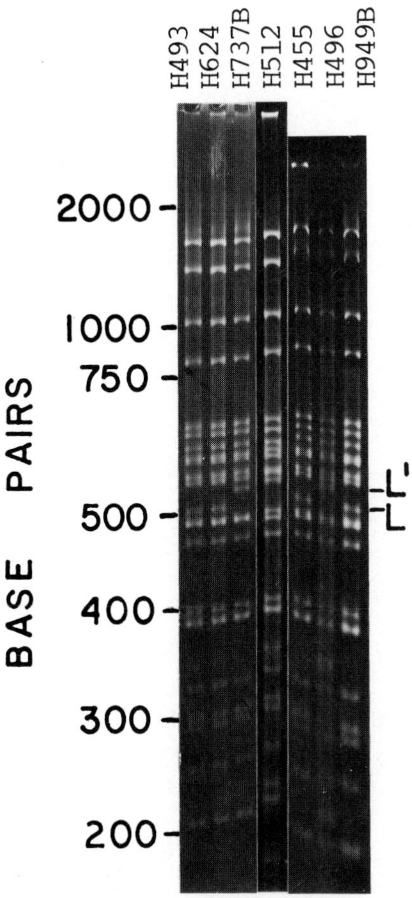

Figure 2. Hae III digestions of mitochondrial DNA from maternal descendents of Holstein H15 compared by electrophoresis on a 6% polyacrylamide gel. Size calibration is based on Hae III digestion products of SV40 DNA run on the same gels.

Two general possibilities remain for the structure of L' relative to L; either the extra segment in L' has been inserted between 37.1 and 37.9 or the extra length of L' resulted from loss of the Hae III site at 37.1 and subsequent joining of a small, unmapped Hae III fragment to L. If an insertion of 38 base pairs occurred between 37.1 and 37.9, this size difference would be apparent in other restriction fragments mapping to this region. Sau 3A

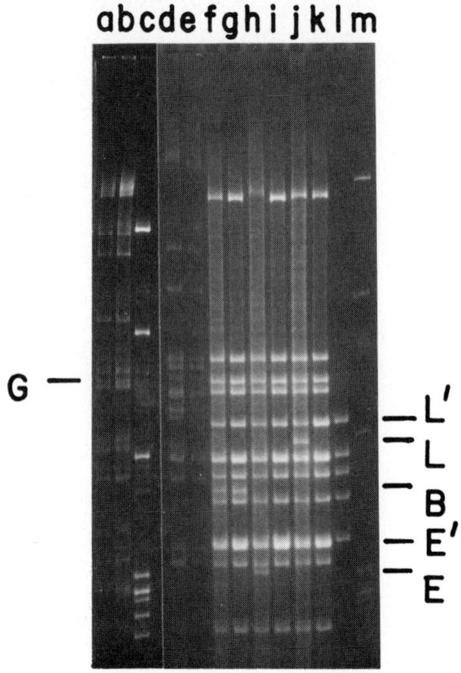

Figure 4. Polyacrylamide gels of restriction enzyme digests of H493 and H512 mitochondrial DNA and Pst I (19.2/62.2) cloned fragments from those DNAs. Lanes a and b: Sau 3A digests of H493 and H512 DNA, respectively. "G" indicates Sau 3A (33.7/37.9). See Figure 3. Lane c: Hae III digest of SV40 DNA. Lanes d and e: Hae III digests of H493 and H512 DNA, respecitvely. "L" and "L'" indicate the positions of Hae III L (37.1/40.5) and Hae III L' (36.9/40.5). Lanes f and g: Bgl II digests of clones of the Pst I (19.2/62.2) fragments of H493 (lane f) and H512 (lane g) in pBR 322. "B" denotes the Hae III/Bgl II (37.9/40.5) fragment. Lanes h and i: Eco RI digests of the same clones in lanes f and g. "E" and "E'" indicate the positions of Hae III/Eco RI (37.1/39.3) and (36.9/39.3), respectively. The Eco RI/Hae III (39.3/40.5) fragment was the same in both digests (data not shown). Lanes j and k: Hae III digests of the same clones in lanes f and g. "L" and "L'" denote the positions of Hae III L (37.1/40.5) and Hae III L' (36.9/40.5). Lanes l and m: Hae III digests of PBR 322 and SV40 DNA, respectively.

digestion produces a fragment (Sau 3A G) of 685 base pairs mapping between 33.7 and 37.9 (J. Alger and P. J. Laipis, unpublished results). A comparison of Sau 3A digests of H493 and H512 mitochondrial DNA shows no differences in Sau 3A G or any other fragment (Figure 4). Additionally a Hinf I digest comparison between H455, H496 and H949B shows no differences in the 46 largest fragments comprising >95% of the genome (data not shown). These results are inconsistent with the insertion model and suggests that L' has lost the Hae III restriction site at 37.1 and is 38 base pairs longer than L for that reason.

Loss of the Hae III site at 37.1 could occur either by a limited alteration in base sequence or perhaps by base modification of an unaltered sequence. To test whether H493 and H512 mitochondrial DNA differ in their level of modification, the Pst I (19.2/62.2) fragment of H493 and H512 were each cloned via the plasmid pBR322 into *E. coli*. Hae III digests of the cloned DNA from these two animals are compared in Figure 4. Because the L, L' difference is retained when both DNAs are grown in the same host (and in presumably the identical modification background), it appears that the difference is due to a limited base sequence difference affecting the Hae III restriction site rather than a difference in base modification.

DISCUSSION

We have observed a rapid change in the restriction patterns of mitochondrial DNA isolated from direct maternal descendants. This variation appears to involve changes in one or a few bases affecting a Hae III restriction site which maps at 37.1 map units (18.0% clockwise from the 3' end of the D-loop). This change appears to be extremely rapid, since differences are present in animals separated by a very few generations (Figure 1). We have compared 95 restriction sites between H493 and H496 and only this one Hae III site is altered. Assuming that the alteration results from a point mutation, the calculated mutation rate is 1.4×10^{-4} changes/base pair/year using either the method of Upholt (15) or Brown et al. (10). This rate is 10,000 times more rapid than that calculated from evolutionary divergence between primate species (10). Even if this Hae III site difference is the only change present between the mitochondrial DNA of these two animals, the calculated mutation rate is still about 200 times more rapid than evolutionarily observed rates. In addition, inspection of the lineage (Figure 1) shows that this event occurred independently at least three times, at the same apparent site, to give the observed distribution of descendants. Hence, in addition to its rapid rate, this change appears to be sequence specific. The fact that it does not appear to be a large insertion, or loss of a restriction site through base modification eliminates the two most obvious mechanisms. We cannot rule out amplification, in some animals, of a mitochondrial DNA species present at low relative levels. We can rule out any obvious paternal influence from inspection of breeding records.

Because of the rapid rate and type of variation we have observed in this lineage, this mutational process probably does not make a major contribution to evolutionary divergence. However, its mechanism should be of importance in understanding rapid genetic variation in eukaryotes.

ACKNOWLEDGEMENTS

We appreciate the constant assistance of Dr. C. Wilcox and Mr. A. Green, Dairy Research Unit, University of Florida. Supported by grants AG 01636 from the National Institutes of Health and PCM 78-23221 from the National Science Foundation.

REFERENCES

1. Potter, S. S., Newbold, J. E., Hutchison, C. A., III and Edgell, M. H. (1975) Proc. Natl. Acad. Sci. U.S., 72, 4496-4500.
2. Jakovcic, S., Casey, J. and Rabinowitz, M. (1975) Biochemistry, 14, 2043-2050.
3. Upholt, W. B. and Dawid, I. B. (1977) Cell, 11, 571-583.
4. Francisco, J. F. and Simpson, M. V. (1977) FEBS Letters, 79, 291-294.
5. Buzzo, K., Fouts, D. L. and Wolstenholme, D. R. (1978) Proc. Natl. Acad. Sci. U.S., 75, 909-913.
6. Ramirez, J. L. and Dawid, I. B. (1978) J. Mol. Biol., 119, 133-146.
7. Hayashi, J.-I., Yonekawa, H., Gotoh, O., Motohashi, J. and Tagashira, Y. (1978) Biochem. Biophys. Res. Commun., 81, 871-877.
8. Hayashi, J.-I., Yonekawa, H., Gotoh, O., Watanabe, J. and Tagashira, Y. (1978) Biochem. Biophys. Res. Commun., 83, 1032-1038.
9. Kroon, A. M., DeVos, W. M. and Bakker, H. (1978) Biochim. Biophys. Acta, 519, 269-273.
10. Brown, W. M., George, M., Jr. and Wilson, A. C. (1979) Proc. Natl. Acad. Sci. U.S., 76, 1967-1971.
11. Shah, D. M. and Langley, C. H. (1979) Nature, 281, 696-699.
12. Brown, W. M. (1980) Proc. Natl. Acad. Sci. U.S., in press.
13. Laipis, P. J., Hauswirth, W. W., O'Brien, T. W. and Michaels, G. S. (1979) Biochim. Biophys. Acta, 565, 22-32.
14. Hauswirth, W. W., Laipis, P. J., Gilman, M. E., O'Brien, T. W., Michaels, G. S. and Rayfield, M. A. (1979) Gene, 8, 193-209.
15. Upholt, W. B. (1977) Nucl. Acids Res., 4, 1257-1265.

The Organization and Expression of the Mitochondrial Genome
A.M. Kroon and C. Saccone, editors

AVIAN mtDNA: STRUCTURE, ORGANIZATION AND EVOLUTION

K. R. GLAUS, H. P. ZASSENHAUS, N. S. FECHHEIMER AND P. S. PERLMAN
Department of Genetics, Ohio State University, Columbus, Ohio 43210, U.S.A.

INTRODUCTION

Recent studies have indicated that the physical organization of mtDNA from different animals is evolutionally conserved[1,2,3,4], despite considerable divergence of primary sequence. We have begun a physical analysis of avian mtDNA with two goals in mind: 1) to compare the organization of avian mtDNA to that of other higher animals, and 2) to study the structural and sequence evolution of mtDNA from a number of closely related species in hopes of gaining some insight into the nature and mechanism(s) of mtDNA evolution in animals. We have chosen for our studies mtDNA isolated from the livers of chicken and four other closely related birds in the avian order Galliforme.

RESULTS

Physical Map of Chicken mtDNA. Figure 1 shows the relative positions of the replicative origin and rRNAs, as well as the map positions of several additional mtDNA transcripts, aligned with the physical map of chicken mtDNA. The circular restriction map, deduced from partial, double and triple digests of mtDNA, has been linearized at one of two Xho I sites.

The replicative origin (D-loop) has been localized on the restriction map by two criteria. First, the D-loop DNA was labelled with ^{32}P-dNTPs in a chain extension reaction of closed circular mtDNA with DNA Polymerase I[5]. The DNA was subsequently restricted and labelled restriction fragments identified by autoradiography. The results, shown in Figure 2a, indicate that the D-loop is contained entirely within a 2,000 base pair (bp) fragment generated by a Bam-Xba double digest of mtDNA. Secondly, the D-loop, 800 bp in average size, was mapped at a position 1,000 bp from the unique Xba I site by electron microscopy of restricted DNA. We have not as yet identified the direction of replication of the D-loop vis-a-vis the physical map but additional electron microscopic studies should clarify this point.

We are presently mapping the transcripts of chicken mtDNA by Northern hybridization of isolated, nick-translated restriction fragments to DBM-immobilized mtDNA. Initial hybridization studies identify 6 to 10 different mtDNA transcripts. Two of these transcripts are present in vast molar excess over

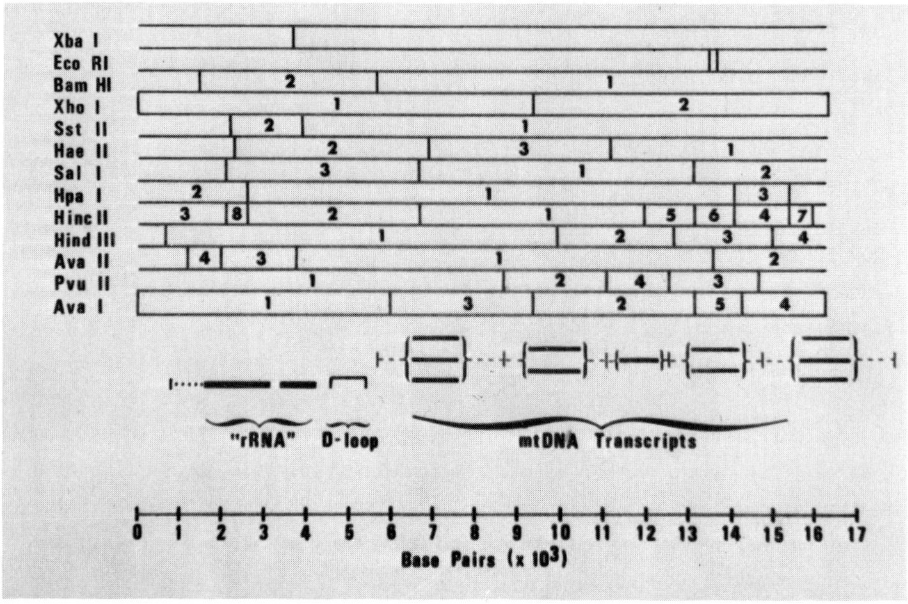

Fig. 1. The physical map of chicken mtDNA, linearized at one of two Xho I sites and aligned with the D-loop and rRNA genes. The numbers and approximate positions of additional less abundant mtDNA transcripts have been positioned relative to the Bam-Pvu II map (dark lines within parentheses).

any other mtDNA transcript and presumably represent the large and small rRNAs of chicken mitochondria. These two RNAs, with apparent sizes of 1,750 and 1,100 bp (as determined from a comparison of the electrophoretic mobility of glyoxylated RNA and single stranded restriction fragments of λ DNA[6]) map within the same 4,100 bp Bam fragment as the D-loop. In order to determine the relative orientation of these two RNAs within the Bam fragment we have used as probes three restriction fragments generated by a Hpa I - Xba I double digest of the Bam fragment. Figure 2B shows the results of this experiment. It can be seen that the larger RNA maps distal to the D-loop whereas the smaller RNA spans the Xba I site, and is proximal to the D-loop. Figure 1 shows the relative position of these two RNAs in the mtDNA restriction map. The other less abundant mtDNA transcripts map throughout the mitochondrial genome, and we are presently constructing a detailed transctiption map for these RNAs.

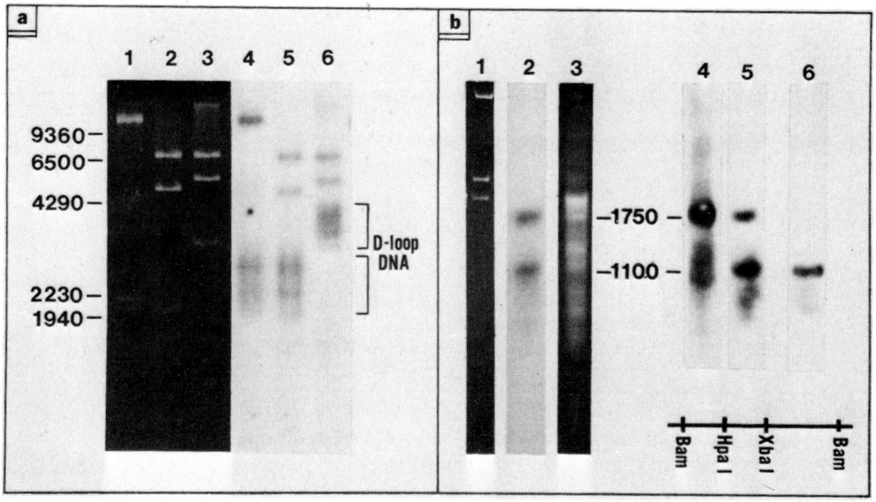

Fig. 2. a) Mapping of D-loop by chain extension with Pol I. Lanes 4-6 are autoradiographs of the digests shown in lanes 1-3. b) Orientation of the rRNAs in Bam-2. Lane 2 is autoradiograph of mtRNA using Bam-2 as probe. Lanes 4-6 are autoradiographs of mtRNA using as probes those fragments shown in the map below each lane. Lane 1 is glyoxylated λ-Hind III digest and lane 3 is photograph of the EB-stained agarose gel of glyoxylated RNA.

Sequence Heterogeneity of Avian mtDNA. We have investigated the sequence evolution of mtDNA from Galliforme birds by comparing detailed restriction maps of mtDNA from chicken, guinea fowl, Ring-neck pheasant, turkey and Japanese quail. In order to determine the fraction of restriction sites common to two mtDNAs we have aligned the different restriction maps relative to one another by two criteria: 1) restriction sites common to all or most birds (e.g. the Sst II fragment pattern was identical in all birds) were placed in register, and 2) the alignment of maps was confirmed by Southern hybridization of nick-translated restriction fragments from chicken mtDNA to appropriate digests of other mtDNAs. From the comparison of the fraction of common restriction sites between any two DNAs we calculated the amount of sequence divergence between any two DNAs using a mathematical model recently developed by Nei and Li[7] (Table 1). In Table 1, numbers below the diagonal indicate the fraction of total restriction sites common to any two mtDNAs whereas the

TABLE 1

THE SEQUENCE DIVERGENCE OF AVIAN mtDNA

	C	G	P	T	Q
C	X	0.097	0.091	0.133	0.10
G	0.56	X	0.153	0.115	0.118
P	0.58	0.40	X	0.126	0.175
T	0.45	0.50	0.47	X	0.157
Q	0.55	0.49	0.35	0.39	X

Abbreviations: C = Chicken, G = Guinea Fowl, P = Ring-neck Pheasant, T = Turkey, Q = Japanese Quail.

numbers above the diagonal indicate the calculated estimates of the amount of sequence divergence between any two mtDNAs.

The estimates of sequence divergence, presented in terms of average nucleotide substitutions per base pair, suggest that the mtDNA of these species are indeed evolutionally closely related. While it is possible to construct, qualitatively, a phylogenetic history of galliforme mtDNA using these estimates we feel that the present mathematical models for estimating sequence divergence from a comparison of common restriction sites are not adequate for a rigorous quantitative treatment of these data. Furthermore, these estimates of sequence divergence provide no information on the randomness of sequence evolution around the mitochondrial genome. We have noted, qualitatively, that a large portion of those sites which are common to any two mtDNAs map within the region of the genome that we have tentatively identified as encoding the rRNAs in chicken. Such nonrandom conservation of sequence has also been seen in closely related species of Xenopus[8].

We should note in passing that, in contrast to the sometimes extensive intraspecific polymorphism of restriction sites seen in most other animal mtDNAs studied to date, chicken mtDNA shows little intraspecific variation in restriction sites. We have examined a number of chickens from seven diverse racial and geographical backgrounds and have noted only two chickens which have apparently lost a single Eco RI site. A plausible explanation for the low level of intraspecific polymorphism is that chicken mtDNA has undergone a relatively recent evolutionary bottleneck since the time of its domestication.

In support of this hypothesis, we have found that the mtDNA from jungle fowl, the presumed modern ancestor of the domestic chicken, is identical to the more common form of chicken mtDNA. We have noted intraspecific polymorphism in restriction sites in some pheasants and turkeys, suggesting that low-level polymorphism is not a general characteristic of avian mtDNA.

DISCUSSION

On the basis of the relative positions of the D-loop and the putative rRNA genes, the organization of chicken mtDNA appears similar to that of all other animals studied to date. Furthermore, among birds from the Galliforme order, we have been unable to find evidence of structural rearrangements of mtDNA (insertions, deletions, etc.) by restriction analysis, suggesting most differences among the primary sequences of these mtDNAs are due to nucleotide base substitutions. We are now in a position to extend our studies on the evolution and organization of avian mtDNAs to a detailed analysis of gene structure and arrangement of avian mtDNA.

REFERENCES

1. Ramirez, J.L. and Dawid, I.B. (1978) J. Mol. Biol., 119, 133-146.
2. Ojala, D. and Attardi, G. (1977) Plasmid, 1, 78-105.
3. Kroon, A.M., Pepe, G., Bakker, H., Holtrop, M., Bollen, J.E., Van Bruggen, E.F.J., Cantatore, P., Terpstra, P., and Saccone, C. (1977) Biochim. Biophys. Acta, 478, 128-145.
4. Hanswirth, W.W., Laipis, P.J., Gilman, M.E., O'Brien, T.W., Michaels, G.S. and Rayfield, M.H. (1980) Gene, 8, 193-209.
5. Ter Schegget, J., Flavell, R.A. and Borst, P. (1971) Biochim. Biophys. Acta, 254, 1-14.
6. McMaster, G.K. and Carmichael, G.G. (1977) Proc. Natl. Acad. Sci. U.S.A., 74, 4835-4838.
7. Nei, M. and Li, W. (1979) Proc. Natl. Acad. Sci. U.S.A., 76, 5269-5273.
8. Dawid, I.B. (1972) J. Devel. Biology, 29, 139-151.

MITOCHONDRIAL GENE CHARACTERIZATION

© 1980 Elsevier/North-Holland Biomedical Press
The Organization and Expression of the Mitochondrial Genome
A.M. Kroon and C. Saccone, editors

CYTOCHROME b MESSENGER RNA MATURASE ENCODED IN AN INTRON REGULATES THE EXPRESSION OF THE SPLIT GENE : I. PHYSICAL LOCATION AND BASE SEQUENCE OF INTRON MUTATIONS

JACQ, C., LAZOWSKA, J., and SLONIMSKI, P.P.
Centre de Génétique Moléculaire du C.N.R.S., F-91190 Gif-sur-Yvette
Université P. et M. Curie, Paris.

Two recent advances have allowed a rapid increase in our understanding of the organisation and expression of mitochondrial genes. The first one is general and results from the development of modern techniques in molecular biology like cloning, DNA sequencing and selective probing of polynucleotide segments by molecular hybridizations. The second one is specific to yeast mitochondria. It is essentially genetic and takes advantage of the possibility of generating thousands of point and deletion mutants, investigating them by genetic recombination, phenotypic complementation and analyses of mutated gene products. The $\underline{rho}^-$ petites constitute in this respect a rather unique tool, methodologically equivalent to the bacterial transformation with purified DNA fragments.

During the last Bari meeting[1], it became apparent that a number of obvious questions dealing with the identification, localization and function of the most conspicuous mitochondrial genes was going to be solved rapidly. At the same time a number of non-obvious observations was made and it was less apparent whether they constituted interesting problems or obscure puzzles.

One of these non-obvious observations dealt with the genetic control of cytochrome b and cytochrome-oxidase synthesis and was called "the BOX paradox"[2]. We have found that four unlinked genetic loci with pleiotropic effects were involved in the control of cytochrome b synthesis[2]. As pointed out in the summary[3] : "the major question remains, therefore : what should all these genes specify ? Specification for regulatory functions is one possibility. The BOX loci discussed by Kotylak and Slonimski may represent examples of this type".

The aim of our presentation to day is to bring a solution to the "BOX paradox". Although the solution is not yet a complete one, it permits already to develop new concepts of gene expression in eukaryotes and to uncover a novel class of proteins endowed with regulatory functions.

It was demonstrated[4,5,6], confirmed and extended[7,8,9], that the mitochondrial gene (referred to as the <u>cob</u>[10]-<u>box</u>[2] region), which specifies the amino-acid sequence of cytochrome b and controls the formation of cytochrome oxidase subunit I, displays a mosaic organisation composed of exons and introns. An essential piece of evidence came from the physical location of different clusters of mutational sites (the box genetic loci) on the restriction map of the wild type mtDNA. Although this genetic and physical map[4] has been widely used in the subsequent studies (see[11] for literature quotations), its full description has not been published until now. In this paper we shall first present a detailed restriction map and a refined physical location of the box genetic loci, which serves as a basis for any further molecular analysis of the gene structure and function. In the second part, we shall summarize briefly our more recent results on the base sequence analysis of the N-terminal part of the gene in two different wild type strains (a "long" strain which has four introns and a "short" strain which has only the two last introns) and in a few different mutants of the first intron. We shall then discuss some concepts which emerge from these studies. A full account of this work will appear elsewhere[12].

I. FINE RESTRICTION MAP AND PHYSICAL LOCATION OF GENETIC LOCI

1. Strategy employed, general comments and precautions.

In order to establish a fine restriction map of the cob-box region, we took advantage of the large number of rho^- clones genetically characterized[4,13]. It is now well documented that rho^- clones are deletions mutants which have retained and amplified a part of rho^+ mtDNA sequence. The rho^- clones whose mtDNA was physically analyzed were chosen according to the following criteria : i) They must be genetically stable. ii) The rho^- clone considered must restore a contiguous set of mit^- mutations like in Fig. 2. iii) The frequency of appearance of various rho^- clones should follow a simple frequency of separation rule[4,14]. This permits to separate rho^- clones which have retained a unique wild type segment from those which are due to multiple deletions and/or rearrangements. iv) The frequency of wild type recombinants in rho^+ mit^- by rho^- mit^+ crosses should be similar for mit^- mutations located inside the corresponding sequence of mtDNA rho^- repeat unit ; this frequency may decrease near the boundaries of this sequence[4].

Having fulfilled these genetic requirements the rho^- clone must contain a mtDNA which presents the following physical properties :

i) The restriction patterns must fall in one of the two classes.

- all fragments are in stoechiometric amounts and only one of these fragments is different in size from those of the wild type mtDNA restriction pattern ; the rho$^-$ mtDNA has a straight head-to-tail arrangement ;

- two fragments are different in size from those of the wild type mtDNA restriction pattern and they are in sub-stoechiometric amounts ; the rho$^-$ mtDNA is composed of inverted repeats whose boundaries are in the middle (or close to it) of the junction fragments.

ii) The non-junction restriction fragments must be present in the wild type genome.

Out of the 35 clones derived from this region, 4, which fulfilled the genetic requirements, did not fulfill the physical ones : they were discarded. It might be worth noticing that the obtention of stable and non rearranged rho$^-$ clones seems to be dependent on the region of the genome. Among rho$^-$ clones retaining short DNA repeating units in the C terminal part of the gene (box6) none was rearranged whereas several of those containing only the penultimate exon (box2) appeared to be rearranged. Similarly, the occurence of inverted repeats among petites does not seem to be random. We obtained the majority of palindrome arranged clones from the N terminal region (box4/5) whereas none was obtained from the C terminal region (box6). These particularities may be related to the existence, in the mitochondrial genome, of certain sequences specifically involved in recombination mechanisms[15,16].

2. General restriction map and gross localization of genetic loci.

Having selected a genetically characterized set of rho$^-$ clones (Fig. 2) from different parts of cytochrome b gene and showing unambigous restriction patterns with different restriction enzymes, we established their restriction map. We used classical methods such a double or triple digestion, elution of different fragments to redigest them with a second enzyme and precise evaluation of the size of restriction fragments. The restriction maps of the rho$^-$ clones could then be compared and the different repeat units were localized relatively to each other using, in particular, the HaeIII sites (A) which appeared to be scattered in this part of the genome. This allowed us to reconstitute a restriction map of the wild type genome in this region (Fig. 1). The restriction maps derived from the strains KL14-4A and 777-3A were indistinguishable in the region and found to be in a very good agreement with the general map already established directly from the wild type strain KL14-4A[17]. The relative position (Fig. 1) of the different rho$^-$ mtDNA repeat units can be used in conjunction with the data of Fig. 2, to localize the gentic loci. For instance the wild type alleles corresponding to the intervals 6-8 are kept in mtDNA of the rho$^-$ clone PK1

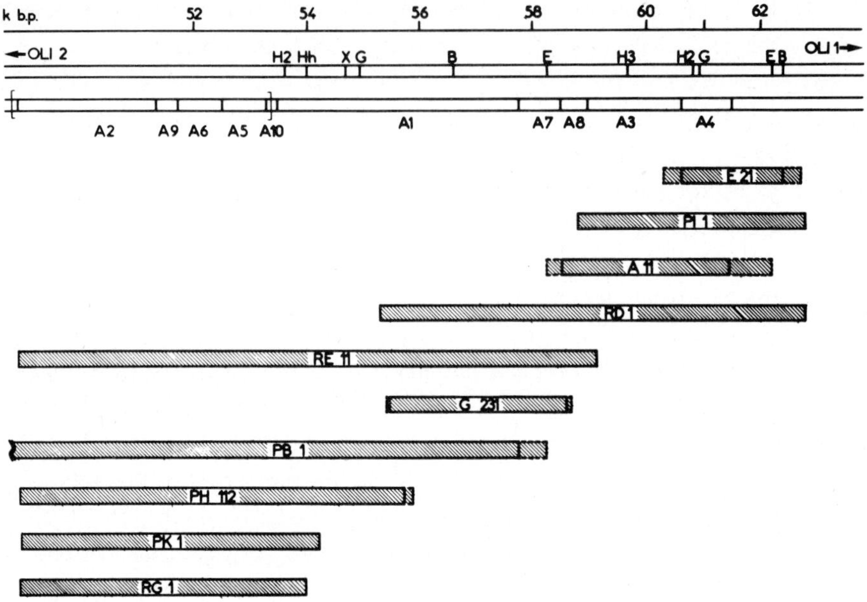

Fig.1. Gross restriction map of the cob-box region. The upper lines represent the rho^+ mtDNA sequence as deduced from the structure of different rho^- mtDNAs shown in the lower lines. The symbols used for restriction enzymes are : A = HaeIII, H2 = HindII, H3 = HindIII, Hh = HhaI, X = Xba, G = BglII, B = BamHI, E = EcoRI, M = MboI, MII = MboII, U = Alu, F = HinfI, V = AvaII. The map position on mitochondrial genome is indicated on the upperline assuming that the size of the genome is 75 kbp, with the conventional origin placed at the SalI site. When the boundaries of the junction fragments in repeat units arranged head-to-tail could not be precisely localized, the uncertainties are represented by dotted areas. Bracketed HaeIII restriction fragments are not ordered.

while they are deleted in RG1. Physically, the mitochondrial genomes of these two clones are very similar (Fig. 1) : they are composed of palindrome arranged repeats units, one junction (on the Oli2 side) is identical while there is a slight difference in the size of the other junction fragment. This additional sequence, present in PKI genome, can be located between the restriction sites HhaI and Xba, respectively present and absent in PK1 and RG1. It contains the wild type alleles of the intervals 6-8. Such an observation has important implications if one wishes to establish the base sequence of a mutation. It eliminates spurious changes which could result from a genetic

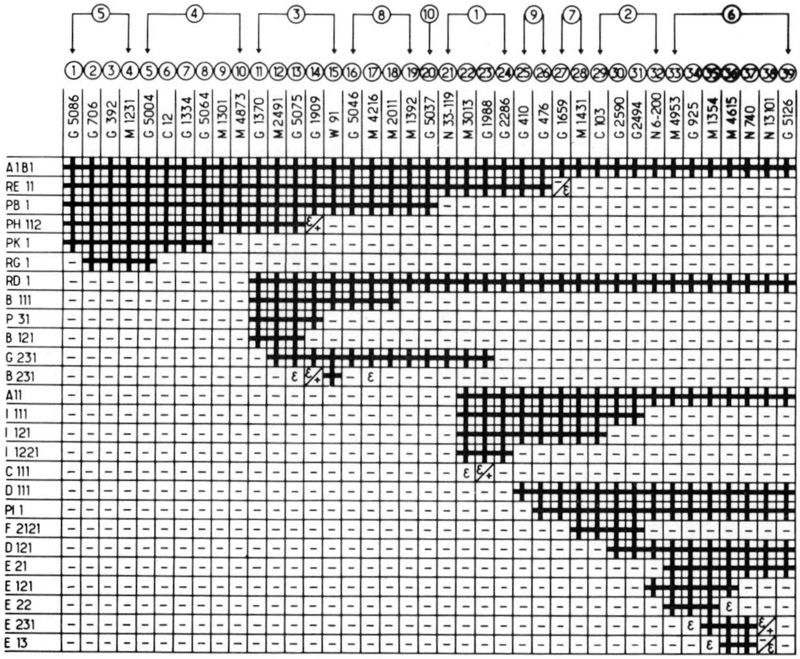

Fig. 2. Genetic map of the cob-box region deduced from the restoration tests in rho+ box- x rho- box+ crosses[13]. The map is divided into intervals (second line, 1 to 39) based on the fact that, at least, one rho- clone differentiates between mutants allocated to different intervals. One typical box mutation is represented per each interval (third line). Successive intervals are grouped into box loci (1 to 10, top line) on the basis of phenotype and complementation analyses. The rho- clones (first column) obtained by ethidium bromide treatment of the strain KL14-4A, rho+ box+, have been crossed with different rho+ box- mutants. The ability of a rho- clone to restore the respiration of a box mutant is indicated by a (+) while (ε) denotes very rare wild type recombinants. A full description of 142 mutants mapped in this region will appear elsewhere[13].

drift, polymorphic variations, cloning, or human errors, because only those modifications in the base sequence which correlate with the physical location of the wild type allele are genetically meaningful. Thus, the finer is the physical location of a genetic marker, the more certain is the establishment of its base sequence.

3. Refined physical location of mutational clusters in the middle part of the gene.

The gross localization of box1 on the restriction map (Fig. 3) was obtained from the comparison of rho⁻ clones RE11, G231 and I1221 which have retained the corresponding wild type sequence (Fig. 2) whereas PI1 and D111 are deleted for it. A more precise localization of intervals 22 and 23 could be directly deduced from the position of the rho⁻ clone C111. This remarkable clone has an extremely short repeat unit, 55 bp long only. It is the smallest genome ever found. Nevertheless, this short sequence replicates true to type, contains the wild type alleles of seven box1 mutations and is able to recombine faithfully with the rho⁺ genome. The wild type alleles of box9, which are neither in mtDNA sequence of I1221 nor in P11 (Fig. 3), should be near the edge of the repeat unit of D111. box2 mutations are neither restored by rho⁻ clone I1221 nor by E121 but are partially restored by I111 (Fig. 3). The locus could be mapped next to the HindIII site. The localization of box7 is more tentative and relies upon the boundary effect[4] exhibited by the rho⁻ clone RE11.

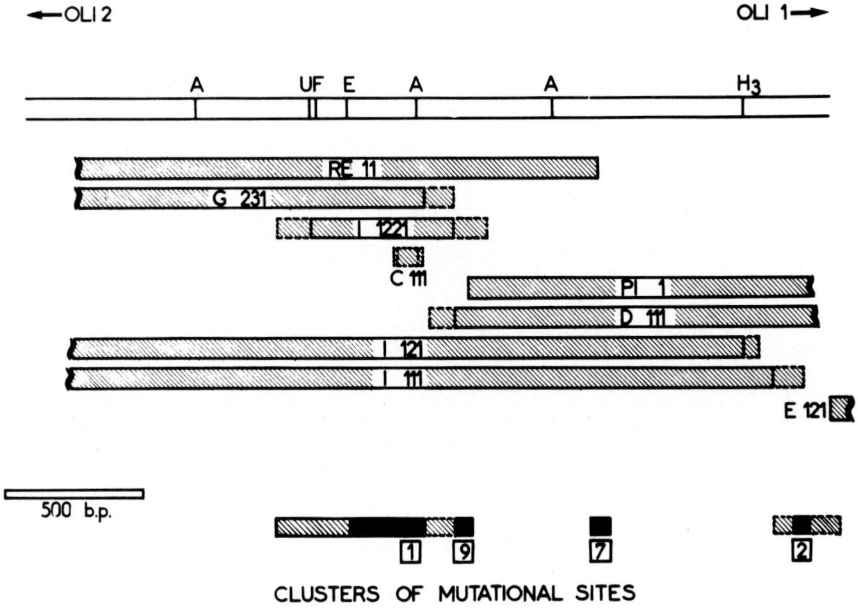

Fig. 3. Refined physical location of mutational clusters in the middle part of the gene (Symbols like in Fig. 1).

4. Refined physical location of mutational clusters in the C terminal part of the gene.

The set of rho⁻ clones analyzed in this region allowed us to get a very precise localization of different intervals of the box6 locus. For instance, the wild type sequence of interval 33, present in the repeat unit of rho⁻ clone E22, is absent in E231 and should be localized in a small fragment next to the HindII site. On the other hand, intervals 37 and 38 could be localized on the other edge of E231. The physical location of different genetic intervals on this 350 to 400 bp long fragment is in a good agreement with the size of this terminal exon as determined by electron-microscopy studies of heteroduplexes[9].

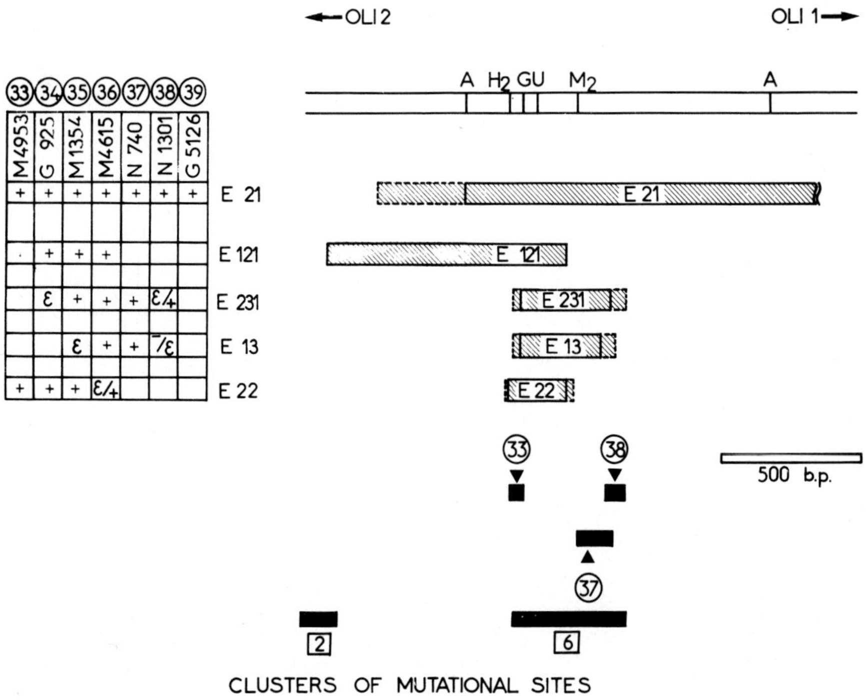

Fig.4. Refined physical location of mutational clusters in the C terminal part of the gene (symbols like in Fig. 1 and 2).

5. General conclusions concerning the physical location of genetic loci.

Physical mapping of mutations reported here confirms our previous results[4] and goes further into detail. Moreover, it permits to allocate the box8 cluster of mutational sites[13] to the second exon found by RNA-DNA hybridization[9]. The mutations have allowed to study the functions of various parts of the gene. Fig. 5 summarizes the data shown in Fig. 1 to 4 and integrates the salient properties of different segments which are described in detail elsewhere[5,7,8,13,18,19,29,30]. Five exons code for the aminoacid sequence of a unique cytochrome b polypeptide. Three different complementation groups distinct between themselves and in respect to the exon sequences are located in introns. Mutations in these segments do block cytochrome b mRNA processing and arrest cytochrome oxidase synthesis. Some 30 mutations are located in a segment (box3) of the first intron. Only one and 9 mutations are actually known in the next two introns (box10 and 7). A few mutations which, by complementation, behave like the exon ones but display a complex pattern of numerous polypeptides synthesized are also found. These mutations may be located at or affect the exon-intron junctions.

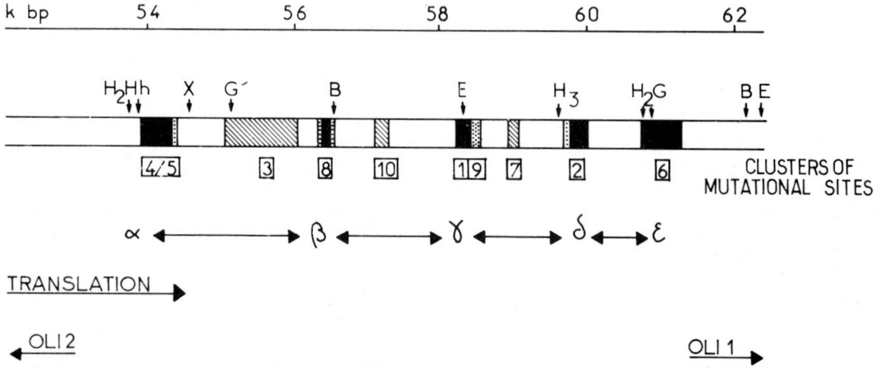

Fig. 5. Physical map, genetic loci and functionally different parts of cytochrome b gene in strains 777-3A and KL14-4A. Five exons (black areas) are separated by introns. Three main classes of mutations have been found : one located in exons and two different ones (hatched or dotted areas) in introns and/or in intron/exon junctions (see text). Clusters of mutational sites correspond to different box loci numbered in historical and not logical order. Density of mutations is quite different in various introns : high and moderately high in the first and the third intron ; very low and nil in the second and the fourth one, respectively.

II. SEQUENCE ANALYSIS OF THE INITIAL PART OF THE GENE.

The first steps of splicing of cytochrome b pre-mRNA occur in the N terminal part of the gene[18,19], where a large number of intron mutations has been mapped. This region is therefore of special interest.

1. Location of two exons flanking the first intron.

Analyses of rho⁻ clones have permitted us to locate the genetic loci box4/5, box3 and box8 in the N terminal part of the gene (Fig. 5). The physical location of the loci box4/5 and box8 should correspond to the location of the first two exons. In order to check it and to get a better insight in the exact positions of these exons, we have sequenced a part of a rho$^-$ clone derived from a "short" strain D273-10B in which the first three exons of a "long" strain are supposed to be fused[9,20]. We compared the sequences of the two polymorphic strains D273-10B and 777-3A in the N terminal part of the gene. As expected, the two sequences are homologous in the first exon and they diverge about 50 nucleotides downstream the Alu site (Fig. 6). Another homologous region, 70 to 80 nucleotides long was found in the vicinity of the BamH1 site (10 to 15 nucleotides upstream this site) ; this second region, corresponding to our localization of box8 represents certainly the second exon.

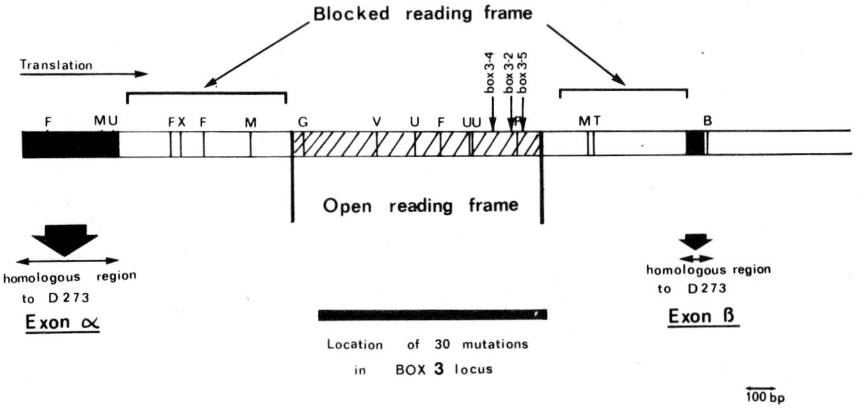

Fig.6. Structure of the N terminal part of cytochrome b gene determined by the nucleotide sequence of two wild types and three intron mutants. Sequence determinations[21] have been carried out on rho$^-$ clones or on DNA fragments cloned in pBR322. Two regions in the "long", 777-3A strain which contain the first intron have been found homologous to the "short" D273-10B strain which lacks it. A long open reading frame in the middle of the intron was discovered and the sequence of three mutants placed in it. Other symbols like in Fig.1.

2. Significance of a remarkably long open reading frame found in the first intron. Sequence of three mutations located in it. An intron encoded protein.

Base sequence studies[22] disclosed an interesting organization of the first intron in the wild type strain 777-3A (Fig. 6). The intron sequence is not of an almost pure AT type : its overall GC content (18%) is practically identical to that of the average yeast mtDNA and contrary to what has been observed in other yeast mitochondrial protein genes, which are embedded in AT rich segments, the exon sequences, flanking the intron are not conspicuous at the first sight. The most remarkable finding concerning this intron is the presence of a very long open reading frame. This uninterrupted sequence, assuming that the UGA codons are not stop codons in the mitochondrial genetic code, cf[23] for literature quotations, is located in the middle of the intron. It is able to code for a polypeptide 322 aminoacids long (or 318 aminoacids long if the translation would begin with the first AUG codon present in the open reading frame) and is terminated with two adjacent UAA codons. This long open reading frame is unique in the whole intron. In the five remaining frames numerous (17 to 45 UAA or UAG stop codons are present on the coding or on the opposite strand), in agreement with a theoretical calculation, which predicts that a UAA or UAG stop codon should occur every 30 to 50 nucleotides. The probability for such a long open reading frame to occur at random is extremely low (less than 10^{-8}). Its existence cannot be therefore fortuitous. Furthermore, the base sequences of the intron, surrounding the central open reading frame, are full of UAA and UGA stop codons in all the different frames (Fig. 6). In conclusion, a protein can be translated from the intron. Base sequence determinations of several box3 mutants demonstrates that it is actually translated.

Some 30 mutations belonging to the box3 locus have been mapped in the intervals 11 to 15 of the genetic map (Fig. 2). Using rho^- clones the wild type alleles of these mutations have been accurately localized on the restriction map of the gene (Fig. 5). We have sequenced the mutated alleles present in three box mutants : 3-2, 3-4 and 3-5 and compared their base sequence to the isomitochondrial wild type strain 777-3A from which they derive. All sequence changes occur within the open reading frame : two of them, box3-4 and box3-5 generate UAA codons at positions -215 and - 62 respectively (0 being the first base of the UAAUAA termination signal of the open reading frame). The third mutant, box3-2, carries simultaneously two non adjacent base substitutions at positions - 180 and - 92 corresponding respectively to the VAL → PHE and to the ILE → ASN substitutions. Sequence studies of mutants are in agreement with the previous genetic and physical mapping : box3-5 (- 62) was allocated to the last interval of the box3 locus and box3-2 (- 180) and 3-4 (- 215) were allocated to the preceding interval (compare Fig. 2 and 6). Furthermore, these mutations are very closely linked by genetic recombination[24] and our finding that the box3-2 mutant actually carries two mutations explains the fact that it does not revert[25].

Analyses of mitochondrially synthesized polypeptides prove that the open reading frame is translated. It has been found previously[5] that numerous box3 mutants synthesize novel polypeptide chains, not observed in the wild type and significantly longer than the wild type cytochrome b. In particular, the box3-2 mutant synthesizes a 41.7 Kd, box3-5 a 38.5 Kd and box3-4 a 34 Kd polypeptide[5]. Given the fact that the box3-2 mutant carries only the missense mutations while the two other mutants carry nonsense mutations, one can calculate from the translated base sequence of the open reading frame the expected length of polypeptides : one finds 39.3 Kd for box3-5 and 33.3 Kd for box3-4. Expectation is in an excellent agreement with observation.

We would like to make a few remarks concerning the open reading frame i) it corresponds to the general sense of transcription and translation of all the mitochondrial genes examined so far ; ii) The protein translated from it should be highly basic, rich in lysine and poor in arginine, which may be related to its function dealing with RNA processing (vide infra) ; iii) The density of mit$^-$ mutations in the box3 locus (30 found already[13]) is at least as high as that observed in a typical cytochrome b exon (eg the last exon contains 16 mutations[13]). As the two open reading frames, box3 and box6, are of comparable lengths (Fig. 5 and 6) it follows logically that both should be translated into proteins the function of which is essential for mitochondrial biogenesis. Quite a different situation is found for the second and fourth introns of the "long" split gene (cf. Fig. 5) ; iv) The codon usage exhibits a strong bias in favor of A or U in the third position as it has been already observed for other mitochondrial genes[26,27,28]. Some codons which have not been found as yet in other yeast mitochondrial genes coding for proteins are found in the open reading frame. Interestingly enough, most of these unusual codons are located in the 5' part of the sequence, between the BglII and AvaII restriction sites (Fig. 6). The significance of this finding will be discussed below.

3. The intron encoded protein is translated after an initial splice which ligates the first cytochrome b exon with the open reading frame of the first intron.

Several lines of evidence argue in favor of the idea that only the C terminal part of the 41.7 Kd protein is encoded by the intron open reading frame while its N terminal part is encoded totally or partially by the first cytochrome b exon : i) The open reading frame is too short to code for a 41.7 Kd polypeptide, especially if one considers that numerous unusual codons (eg CGU) are present in the first part of it ; ii) Nonsense but not missense mutants in the box4/5 exon fail to complement, in trans, box3 mutants[4,30] ; iii) Chain termination mutants in the box4/5 exon are epistatic on the formation of novel polypeptide chains characteristic of box3 mutants[31] ; iv) Anti-cytochrome b serum cross reacts with novel polypeptide chains of box3 mutants[32].

Our sequence data (Fig. 6) rule out the hypothesis[33] that a protein could be translated in an uninterrupted reading frame from box4/5 to box3. Too many stop codons separate these two segments. We favor therefore our original hypothesis[22] (see also[19]), which requires an initial splicing step in order to bridge, in phase, the first exon and the box3 coding sequence. This initial splice, which would generate the small circular RNA[18,19] appears to be catalysed by cell sap made enzymes[18].

4. Physiological function : a self regulated messenger RNA maturase involved in splicing of the intron which encodes it.

What is the function of the exon-intron encoded protein ? It must be an essential element of the excision of the first intron and ligation of its flanking cytochrome b exons, because box3 mutants do not make cytochrome b mRNA, being unable to splice the first intron[18,19,34]. In other terms the protein must be actively involved in the maturation of messenger RNA : it is a mRNA maturase[22]. The mRNA maturase would be endowed with a remarkable "matricidal" property : it would destroy the RNA sequence (present in the cytochrome b pre-mRNA) which codes for its own synthesis. It follows that the amount of active maturase should be subject to a negative feed-back regulation :[22,33] an increase in activity would lead to a decrease in its biosynthesis, while a decrease in activity would automatically increase its biosynthesis. In the wild type, minute amounts would be present, while an inactive maturase should accumulate in abundance in box3 mutants, what is actually observed[5]. O. Groudinsky (in preparation) has proved that the active maturase must be present in very low amounts in cob-box$^+$ strains : a mutant carrying a large deletion of the oxi3 gene synthesizes cytochrome b, but makes neither COXI nor any detectable protein in the 35 to 45 Kd range ; the same mutant recombined in cis with the box3-5 mutant makes large amounts of the 38.5 Kd polypeptide.

5. Generality of the messenger RNA maturase model.

Last figure summarizes the essential features of the model[22]. We are inclined to think that the box3 maturase (symbolized MM) is related to the specificity of splicing of this particular intron, rather than responsible for a general splicing machinery which would be supplied by the cell sap (symbolized CM) made enzymes. Its activity would trigger however, all the successive steps. It is probable that other elements like the secondary intramolecular structure of pre-mRNA, RNA-RNA inter-molecular interactions are also involved in splicing. What we would like to stress is the idea that mRNA maturases, translated as hybrid proteins with at least two structural domains, may act

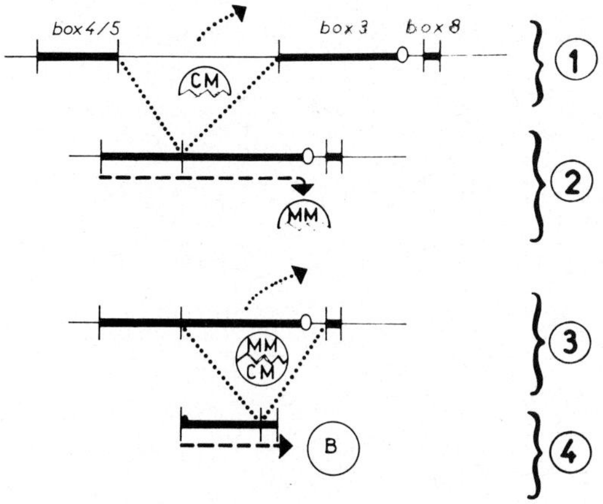

as selective signals for (or molecular "niches" where) the actual RNA processing takes place. They could act along a single split gene or between different split genes like between B and OXidase genes. Other implications are discussed in[12,30,35,36].

ACKNOWLEDGEMENTS

We thank B. Poirier for her generous help in the completion of the manuscript. This work was supported by DGRST, INSERM, Ligue Nationale contre le Cancer and Fondation pour la Recherche Médicale Française Grants.

REFERENCES

1. "The Genetic Function of Mitochondrial DNA" (1976) Eds. Saccone,C. and Kroon, A.M., North-Holland, Amsterdam, pp. 1-354.

2. Kotylak, Z. and Slonimski, P.P. (1976) in 1, pp. 143-153.

3. Kroon, A.M. and Saccone, C. (1976) in 1, pp. 343-347.

4. Slonimski, P.P., Pajot, P., Jacq, C., Foucher,M., Perrodin,G., Kochko,A. and Lamouroux,A. (1978) "Biochemistry and Genetics of Yeast". Bacila, M. et al., Eds. Academic Press, pp. 339-368.

5. Claisse, M.L., Spyridakis, A., Wambier-Kluppel,M.L., Pajot,P. and Slonimski,P.P. (1978), in 4, pp. 369-390.

6. Slonimski, P.P., Claisse, M.L., Foucher,M., Jacq,C., Kochko,A., Lamouroux,A.,Pajot,P., Perrodin,G., Spyridakis,A., and Wambier-Kluppel,M.L. (1978),in 4,pp. 391-401.

7. Hanson, D.K., Miller, H.R., Mahler, H.R., Alexander, N.J. and Perlman,P.A. (1979) J. Biol. Chem. 254, 2480-2490.

8. Haid, A., Schweyen,R.J., Kaudewitz, F., Solioz,M., Schatz, G. (1979) Eur. J. Biochem. 94, 451-465.

9. Grivell,L.A., Arnberg,A.C., de Boer,P.H., Borst,P., Bos, J.L., Groot, G.S.P., Hecht,N.B., Hengsens, L.A., Van Ommen, G., and Tabak, H.F. (1979) in "Extrachromosomal DNA". D/ Cummings et al. Eds. Academic Press, New York, pp. 305-324.

10. Slonimski, P.P. and Tzagoloff, A. (1976) Eur. J. Biochem., 61, 27-41.

11. Dujon, B. (1979) Nature, 282, 777-778.

12. Lazowska,J., Jacq, C. and Slonimski, P.P. in preparation.

13. Slonimski, P.P., Claisse, M.L., Kotylak, Z., Perlman, P.A., Schweyen,R.J., Solioz, M., and Spyridakis, A. (1980) in preparation.

14. Schweyen, R.J., Steyrer, V., Kaudewitz,F., Dujon, B. and Slonimski,P.P. (1976) Mol. Gen. Genet. 146, 117-132.

15. Lazowska, J. and Slonimski, P.P. (1977) Mol. Gen. Genet. 156, 163-175.

16. Michel, F., Grandchamp, C., Dujon, B. (1979) Biochim., 61, 985-1010.

17. Sanders,J.P.M., Heyting, C., Verbeet, M.P., Meylink, F.C., Borst, P. (1977) Mol. Gen. Genet. 157, 239-261.

18. Halbreich, A., Pajot, P., Foucher, M., Grandchamp, C., Slonimski, P.P. (1980) Cell, 19, 321-329.

19. Van Ommen, G.J.B., Boer, P.M., Groot, G.S.P., De Haan, M., Roosendal,E., Grivell, L., Haid,A., Schweyen, R.J. (1980) Cell, 20, 173-183.

20. Morimoto, R., Rabinowitz, M. (1979) Mol. Gen. Genet. 170, 25-48.

21. Maxam, A. and Gilbert, W. (1980 "Methods in Enzymology", 65, Moldave K. and Grossman, L. Eds Academic Press, New York, pp. 499-560.

22. Jacq, C., Lazowska, J., Slonimski, P.P. (1980) Compt. Rend. Acad. Sci. Paris, 290, Série D, 89-92.

23. Hall, B.D. (1979) Nature, 282, 129.

24. Kotylak, Z. and Slonimski, P.P. (1977) in "Mitochondria 1977" Genetics and Biogenesis of Mitochondria.Eds Bandlow et al., W. de Gruyter,Berlin, pp. 161-172.

25. Pajot, P., Wambier-Kluppel,M.L., Slonimski,P.P. (1977) in 24, pp. 173-183.

26. Hensgens, L.A., Grivell, L.A., Borst,P., Bos, J.L. (1979) Proc. Natl Acad. Sci. USA, 76, 1663-1667.

27. Fox, T.D. (1979) Proc. Natl Acad. Sci USA, 76, 6534-6537.

28. Macino, G. and Tzagoloff, A. (1979) J. Biol. Chem. 254, 4617-4623.

29. Claisse, M.L., Slonimski, P.P., Johnson, J., Mahler, H. (1980) Mol. Gen. Genet. 177, 375-387.

30. Lamouroux, A., Pajot, P., Kochko, A., Halbreich, A. Slonimski, P.P., this Symposium.

31. Alexander, N.J., Perlman, P.S., Hanson, D.K., Mahler, H.R. (1980) Cell, 20, 199-206.

32. Solioz, M. and Schatz, G. (1979) J. Biol. Chem. 254, 9331-9334.

33. Church, G. and Gilbert, W. (1980) in Mobilization and Reassembly of Genetic Information, Joseph, D.R. et al. Eds, in press.

34. Church, G., Slonimski, P.P. and Gilbert, W. (1979) Cell 18, 1209-1215.

35. Slonimski, P.P. (1980) Compt. Rend. Acad. Sci. Paris, 290, Série D, pp. 331-334.

36. Dujardin, G., Groudinsky, O., Kruszewska, A., Pajot, P., Slonimski, P.P. This Symposium.

The Organization and Expression of the Mitochondrial Genome
A.M. Kroon and C. Saccone, editors

CYTOCHROME b MESSENGER RNA MATURASE ENCODED IN AN INTRON REGULATES THE EXPRESSION OF THE SPLIT GENE: II. TRANS- AND CIS-ACTING MECHANISMS OF mRNA SPLICING.

LAMOUROUX, A., PAJOT, P., KOCHKO, A., HALBREICH, A., SLONIMSKI, P.P.
Centre de Génétique Moléculaire du C.N.R.S., F-91190 Gif-sur-Yvette,and
Université P. et M. Curie, Paris.

Novel functions of introns were discovered by the study of complementation between mutants in the cob-box gene. First, we have shown that mutations in the locus box3 constitute a distinct group of complementation from those allocated to the loci box1,2,4 and 6 [1]. As the box3 mutations were shown to be located in the first intron[1] while the others were located in exons[2], it became clear that the rôle of intron sequences in the expression of the gene was different from that of exon sequences. Two other complementation groups were discovered later and shown to correspond to the next two introns, box10 and box7 respectively[3,4]. Thus, all exons constitute a single complementation group, which is understandable because they specify the unique polypeptide chain of cytochrome b, whereas each intron corresponds to a different complementation group.

A few examples illustrate this conclusion (Table 1). The respiration of heteroplasmic zygotes, in μmole O_2/h/10^8 cells, was measured in a number of different combinations of exon and intron mutations, before the genetic recombination has taken place : four complementation groups were found.

TABLE 1

THE cob-box GENE CONTAINS FOUR COMPLEMENTATION GROUPS.

	1st Exon box5-4	3rd Intron box7-1	5th Exon box6-2
1st Exon	0.3	7.1	0.4
3rd Intron		0.1	6.4
5th Exon			0.5

	3rd Exon box1-7	2nd Intron box10-1	5th Exon box6-2
3rd Exon	0.4	10.2	0.6
2nd Intron		0.8	6.1
5th Exon			0.5

	3rd Exon box1-2	1st Intron box3-5	4th Exon box2-2
3rd Exon	0.5	6.6	0
1st Intron		0.3	6.3
4th Exon			0.3

	1st Intron box3-5	2nd Intron box10-1	3rd Intron box7-1
1st Intron	0.3	8.9	6.8
2nd Intron		0.8	10.3
3rd Intron			0.1

A further step towards the elucidation of the mechanism was made by the use of double mutants in which an exon was labelled with a diuron resistance marker [5,6]. The D^R marker was associated in a cis configuration either to a deficient exon (box1,2 or 6) mutation or to a deficient intron (box3 or 7) mutation. As illustrated in Fig.1 the expressed message is the one contained in the exons of the intron mutated parent[6]. The expression of the gene blocked by the intron mutation, is made possible by a trans acting product which is supplied by the information contained in the wild type intron.

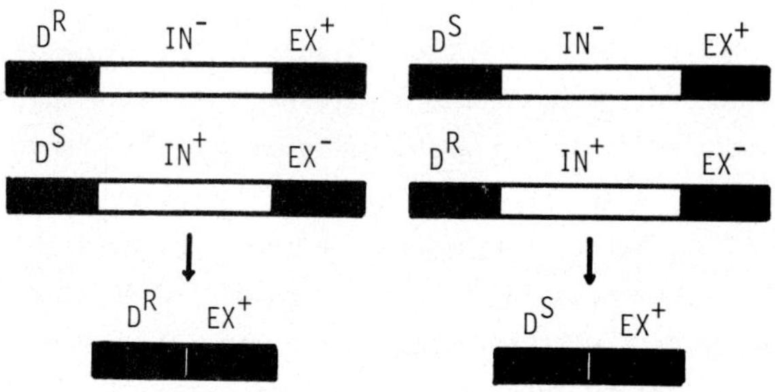

Fig. 1 : Complementation expresses a message in cis, through a trans-acting mechanism. The use of double mutants with a diuron resistance marked exon has shown that cytochrome b was translated from the exons (black bars) of the intron mutated parent.

It is now established[7-10], that intron mutations block the formation of cytochrome b mRNA. Mutations in different introns arrest splicing at different steps. A sequential pathway of the excision of introns was proposed[8]. Although the late steps of this pathway are still unclear[8,10], the progression of early steps is well documented : i) the first step consists in an excision, from the first intron, of a small single-stranded circular RNA which accumulates[8,10]. This step does not require mitochondrial protein synthesis[8]; ii) the subsequent steps may not occur in rho^- petites[11]; iii) cis-polar effects (i.e. upstream vs downstream) are observed in splicing : a box3 mutation in the first intron not only prevents splicing of the intron which carries it but also that of the downstream introns, while a box7 mutation in the third intron does not prevent splicing of the first upstream intron[7,8]. Thus, introns have not only functions which they can supply in trans, as revealed by complementation, but also functions which are normally propagated in cis along the primary transcript, 7 to 8 kb long, in the same direction as the translation of cytochrome b exons. Furthermore, some mutations in the first (box4/5) exon greatly diminish the formation of cytochrome b mRNA and lead to the accumulation of large size precursors[7].

Polar effects are also found in complementation. In the first experiments

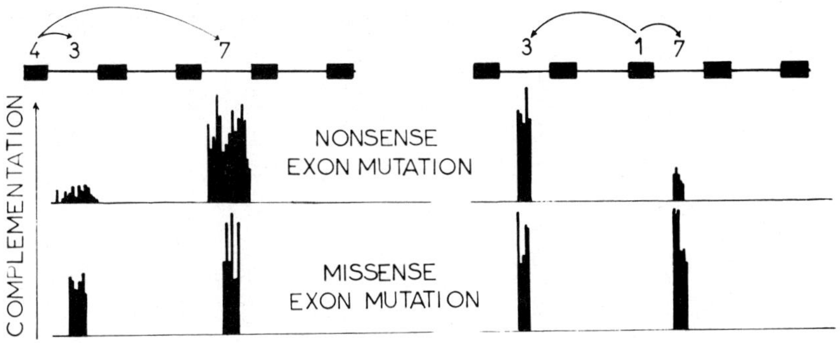

Fig. 2 : Polar effects in complementation. Each bar represents a different combination of exon by intron mutations.

performed, we have found[1] that some box4 mutations, located in the first exon complement rather poorly box3 mutations occuring in the next intron. By performing a large matrix of crosses[3], it was possible to establish a general rule, which is illustrated by a few examples in Fig.2. Two mutations, located either in the first, box3-5, or in the third, box7-1, intron were complemented one by one, by a collection of different mutants located either in the first, box4, or in the third, box1, exon. A polar effect is quite apparent. Numerous box4 mutants complement poorly the immediate downstream intron mutant whereas the complementation is good with the third intron mutant. Numerous box1 mutants complement poorly the immediate downstream intron mutant whereas the complementation is good with the upstream intron mutant. Analysis of polypeptides made by various box4 and box1 mutants[2] permits to establish a correlation : the polar effect is observed with non-sense rather than with missense mutants. Among the box4/5 mutants, a third class of mutants is found. It is unable to complement neither box3-5 nor box7-1 mutants.

The ensemble of results can be integrated in a coherent explanation in the light of the demonstration that a long open reading frame is present in the first intron and that the box3 mutants are non-sense or missense mutations of the protein encoded by that intron[12]. The active mRNA maturase produced by the sufficient intron would complement in *trans* and splice the exons in *cis*. This explains the results summarized in Fig.1. The activity of the box3^{+} maturase would be necessary also to permit the expression of the third, box7^{+}, intron. This explains the polar effect of box3 on splicing. The maturase would result from a fusion of the first exon and the translatable part of the intron. The fusion is explained by the first splice which leads to the formation of a small circular RNA[8,10] and the polar effects of nonsense mutations in the box4/5 exon on complementation (Fig.2) and splicing are easily accounted for. The third intron, box7^{+}, may also code for a maturase, but this maturase must be a different one from that encoded by the box3$^+$ sequence because they complement each other (Table 1). The two intron coded maturases may be, however, synthesised according to the same general principle of exon-intron fusions. Polar effects of box1 mutations on complementation (Fig.2) are the only evidence available at the present time.

ACKNOWLEDGEMENTS

This work was supported by DGRST, INSERM , Ligue Nationale contre le Cancer and Fondation pour la Recherche Médicale Française Grants.

REFERENCES.

1. Slonimski, P.P., Pajot, P., Jacq, C., Foucher, M., Perrodin, G., Kochko,A., and Lamouroux, A., (1978) in Biochemistry and Genetics of Yeast, M. Bacila, B.L. Horecker, and A.O.M. Stoppani, eds. (New York : Academic Press) pp. 339-368.

2. Claisse, M., Spyridakis, A., Wambier Kluppel, M.L., Pajot, P. and Slonimski P.P. (1978) in Biochemistry and Genetics of Yeast, M. Bacila, B.L. Horecker and A.O.M. Stoppani eds. (New York : Academic Press) pp. 369-390.

3. Lamouroux, A. (1979) Thèse de Doctorat de 3ème cycle, Université Paris XI.

4. Lamouroux, A., Kochko, A., Pajot, P., Colson, A.M. and Slonimski, P.P.(1979) Abstract of a paper presented at the meeting on the Molecular Biology of Yeast, Cold Spring Harbor Laboratory Aug. 1979, p. 67.

5. De Kochko, A., Colson, A.M. and Slonimski, P.P. (1979) Arch. Int. Physiol. Biochim., 87, pp. 619-620.

6. De Kochko, A. (1979) Thèse de Doctorat de 3ème cycle, Université Paris VI.

7. Church, G.M., Slonimski, P.P. and Gilbert, W. (1979) Cell 18 pp.1209-1215.

8. Halbreich, A., Pajot, P., Foucher, M., Grandchamp, C. and Slonimski, P.P. (1980) Cell, 19 pp. 321-329.

9. Haid, A., Grosch, G., Schmelzer, C., Schweyen, R.J. and Kaudewitz, F.(1980) Current Genetics, 1 pp. 155-161.

10. Van Omnen, G.J.B., Boer, P.H., Groot, G.S.P., Haan, M.D., Roosendaal, E. Grivell, L., Haid, A. and Schweyen, R.J. (1980) Cell, 20 pp. 173-183.

11. Church, G.M. and Gilbert, W. (1980) in Mobilization and Reassembly of Genetic Information, D.R. Joseph, J. Schultz, W.A. Scott and R. Werners, eds, in press.

12. Jacq, C., Lazowska, J. and Slonimski, P.P. (1980) Compt. Rend. Acad. Sci. Paris, 290, série D, pp. 89-92.

The Organization and Expression of the Mitochondrial Genome
A.M. Kroon and C. Saccone, editors

CYTOCHROME b MESSENGER RNA MATURASE ENCODED IN AN INTRON REGULATES THE EXPRESSION OF THE SPLIT GENE. III. GENETIC AND PHENOTYPIC SUPPRESSION OF INTRON MUTATIONS.

DUJARDIN, G., GROUDINSKY, O., KRUSZEWSKA, A., PAJOT, P., SLONIMSKI, P.P.
Centre de Génétique Moléculaire du C.N.R.S., F-91190 Gif-sur-Yvette

In order to uncover the functional circuitry both within the mitochondrial genome and between the mitochondrial and the nuclear genome, we have undertaken a systematic search of suppressors of mitochondrial mit^- mutations and devised a general methodology that permits to establish the genetic nature of suppressors : nuclear suppressors were called NAM (for nuclear accomodation of mitochondria) and the mitochondrial ones MIM (for mitochondrial-mitochondrial interactions). More than 500 such suppressors were isolated and genetically characterized. In a number of particularly interesting cases, the suppressor was precisely localized either in one of the nuclear chromosomes or in a specific segment of the mitochondrial genome[1].

As the first step in the investigation of the mechanism of action of these suppressors, we have tested the action-spectrum of each suppressor upon some 250 mit^- mutations located in different mitochondrial genes. Several distinct classes of suppressors implying different mechanisms in their mode of action were found. In addition to the expected allele-specific, gene non-specific suppressors of the informational type, we found novel classes of extragenic suppressors located either in the nuclear or in the mitochondrial DNA, and specific of intron mutations in the cob-box gene. As intron mutations prevent the correct splicing to occur[2,3], these intron specific suppressors must alleviate splicing and mRNA maturation defects. They permit to single out genes or gene segments involved selectively in RNA splicing. Fig. 1 summarizes different action spectra.

The nuclear dominant suppressor NAM2 and the mitochondrial one MIM2, located in the oxi3 gene, are intron-specific and allele non-specific. They act upon all mit^- mutations located in the third intron of the cob-box gene (box7) in mitochondrial variants which have four introns and suppress no other mutation in other genes. Interestingly enough, they also suppress a mutation in the first intron of mitochondrial variants which have only 2 introns, demonstrating the functional equivalence of the third and the first intron in a "long" vs a "short" gene, respectively. Furthermore, the fact that (i) mutations in these introns have pleiotropic effects preventing the expression of the oxi3 gene specifying the COXI subunit and that (ii) they can be cured by a suppressor located

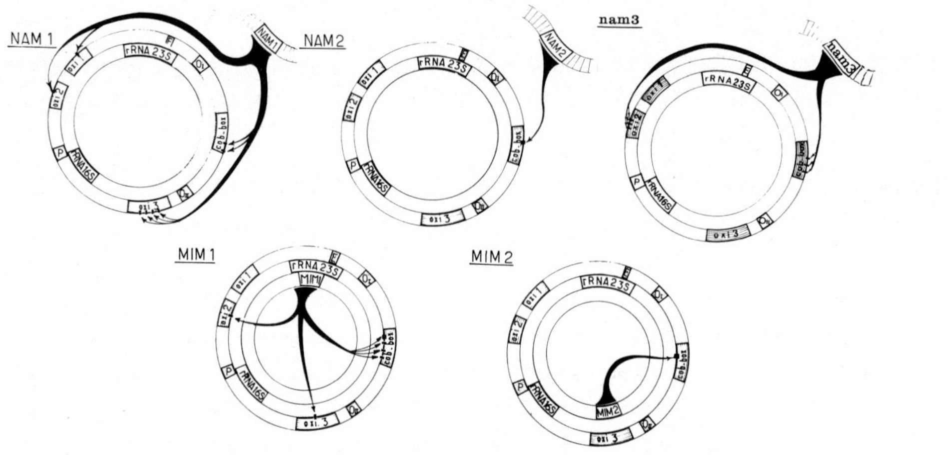

Fig. 1. Action spectra of 3 nuclear and 2 mitochondrial suppressors.

in the oxi3 region itself, strengenths the notion of regulatory interactions between these two genes. We interpret these interactions at the pre-mRNA maturation level, the product of the $box7^+$ intron being normally required for splicing of the $oxi3^+$ primary transcript, while in MIM2 mutant the product of mutated oxi3 intron being able to compensate the deficiency in splicing of the mutated box7 intron. Possibly, introns in both genes (MIM2 in oxi3 and $box7^+$ in cob-box) could be translated into similar proteins involved in splicing.

NAM1 and nam3 are dominant and recessif nuclear suppressors. MIM1 is a mitochondrial suppressor which we have located in the gene coding for the large rRNA. These 3 suppressors present a large action-spectrum, display allele specificity but no gene specificity, strongly suggesting an informational mechanism of suppression[4]. They act probably by modifying the structure of mitochondrial ribosomes either at the large rRNA level (MIM1) or at the ribosomal protein level (NAM1 and nam3). Interestingly, they suppress some mutations in the first intron (box3) of the "long" cob-box gene (see Table 1). The fact that informational suppressors alleviate box3 intron mutations is in excellent agreement with an intron coded mRNA maturation protein responsible for splicing[5].

We have asked whether mitochondrial mutations can be cured by an aminoglycoside antibiotic, paromomycin. It is known that paromomycin promotes extensive misreading of the RNA code words, in vitro, and suppresses phenotypically, in vivo, some nonsense mutations in prokaryotes and eukaryotes[6,7,8].

TABLE 1

MAJOR CLASSES OF SUPPRESSIBLE mit$^-$ MUTATIONS WITHIN THE FIRST, box3, INTRON

No of mutations tested	Phenotypic suppression by paromomycin	Genetic suppression by nam3	NAM1	MIM1	NAM2 MIM2	Examples of mutations suppressed
7	+	+	+	-	-	box3-6
3	+	+	-	-	-	box3-9
1	-	-	-	+	-	box3-11
19	-	-	-	-	-	box3-2
No of mutations suppressed	10	10	7	1	0	

We have found that paromomycin does suppress specific mitochondrial mutations. In Fig. 2, growth surrounding the disk indicates the suppression. We have verified that the cure is purely phenotypic : no permanent revertants are induced by paromomycin. Out of 142 cob-box mutations tested, 11 were clearly suppressed by paromomycin : 10 located within the first, box3, intron, one in the third box7, intron and none in the exons. This result is quite remarkable and surprising at first sight ; it becomes comprehensible in view of following considerations.

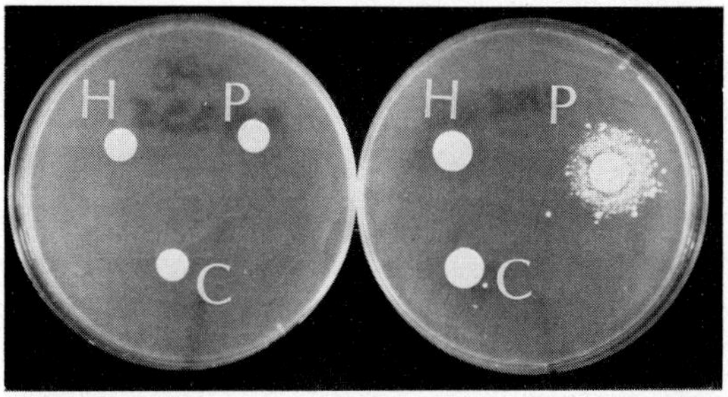

Fig. 2. Phenotypic suppression by paromomycin of mitochondrial mutations. $5\ 10^7$ mit$^-$ cells were spread on nonfermentable medium. Sterile paper disk containing 1 mg of paromomycin (P) or 100 µg of hygromycin B (H), or the control (C) buffer were placed as shown. The right disk represents an intron mutation box3-9, the left one an exon mutation box1-1.

Mutations suppressed by paromomycin are precisely those suppressed by informational suppressors. The majority of them are chain termination mutations[5,9]. Thus, mitoribosomal misreading, whether genetic or phenotypic, of some nonsense codons cures RNA splicing defects caused by intron mutations. This proves that the splicing defect must be due to an inactive protein because the defect is alleviated by a translational relief. How the preferential suppression by paromomycin of box3 intron mutation can be explained ? According to the model[5] a messenger RNA maturation protein is translated from the box3 intron. This protein, when active, is required in very small amounts in order to process pre-mRNA into mRNA. It is known, that suppression is only partial[6,7,8]. Thus, very small amounts (let us say 0.1 to 1 %, in comparison to the wild type) of active protein would be formed. If the nonsense codon is located in a cytochrome b exon, mitochondria would synthesize only 0.1 to 1% of active cytochrome b. This amount may be much to small to ensure respiration and growth. Now, if the nonsense codon is located in the cytochrome b intron, 0.1 to 1% of the active mRNA maturase would be synthesized. However a single molecule of this protein would ensure the splicing of numerous pre-mRNA molecules permitting therefore the synthesis of numerous mRNA molecules. Because in an intron mutant, all the exons are intact a relatively more abundant synthesis of active cytochrome b would occur, ensuring respiration and growth.

ACKNOWLEDGEMENTS

This work has been supported by DGRST, INSERM and Ligue Nationale Française contre le Cancer grants ; A.K. acknowledges the receipt of an EMBO fellowship.

REFERENCES

1. Dujardin, G., Pajot, P., Groudinsky, O, Slonimski, P.P. (1980) Mol. Gen. Genetics (in press).
2. Church, G.M., Slonimski, P.P., Gilbert, W. (1979) Cell, 18, 1209-1215.
3. Halbreich, A., Pajot, P., Foucher, M., Grandchamp, C., Slonimski, P.P. (1980) Cell, 19, 321-329.
4. Hawthorne, D.C., Leupold, U. (1974) Current topics in Microbiology and Immunology, 64, 1-47.
5. Jacq, C., Lazowska, J., Slonimski, P.P. (1980) Compt. Rend. Acad. Sci.,Paris, 290, Série D, 89-92.
6. Gorini, L. (1974) Ribosomes. Eds Nomura et al. Cold Spring Harbor Lab., 791-803
7. Singh, A., Ursic, D., Davies, J. (1979) Nature, 277, 146-148.
8. Palmer, E., Wilhelm, J.M., Sherman, P. (1979) Nature, 277, 148-150.
9. Claisse, M.L., Spyridakis, A., Wambier-Kluppel, M.L., Pajot,P.,Slonimski,P.P. (1978) Biochemistry and Genetics of Yeast. Eds Bacila M. et al., Academic Press, New York, 369-390 and M. Claisse, personnal communication.

© 1980 Elsevier/North-Holland Biomedical Press
The Organization and Expression of the Mitochondrial Genome
A.M. Kroon and C. Saccone, editors

ALTERNATE FORMS OF THE COB/BOX GENE : SOME NEW OBSERVATIONS

P.S. PERLMAN[a,b], H.R. MAHLER[c], S. DHAWALE[b], D. HANSON[c] and N.J. ALEXANDER[b,d]
[a]Université de Paris-Sud, 91405 Orsay, France ; [b]Genetics Department, Ohio State University, Columbus, Ohio 43210 ; [c]Chemistry Department, Indiana University, Bloomington, Indiana 47405 ; [d]Northern Regional Research Center, Peoria, Illinois 61604

INTRODUCTION

There are two different forms of the cob/box gene under investigation in several laboratories. The "long form" present in mtDNA from strains such as ID41-6/161, 777-3A and KL14-4A, has been shown to have a mosaic structure containing at least five exons and four introns spanning at least 7000 base pairs (bp) (reviewed in 1, 2). Mutants in that form of the gene have been well characterized biochemically and genetically and each has been assigned to one of ten "box" loci that loosely correspond to eight of the nine portions of that gene (Fig. 1). The "short form" present in mtDNA from strain D273-10B (D273) has not been well-defined genetically but appears to be identical to the long form except that the first two introns (box 3 and 10) including roughly 3300 bp are absent and the first three exons (box 5/4, 8 and 1) form one continuous exon[3,4].

An unexpected property of long form intron mutants is that they are simultaneously deficient in cytochrome b and cytochrome c oxidase (the BOX phenotype[5-10]). In particular the product of the cob/box gene, apocytochrome b (p30) and the product of the oxi3 gene, subunit I of cytochrome oxidase (cox I), are absent while the products of the other mitochondrial genes specifying subunits of cytochrome oxidase are present. Intron mutants (except conditional ones) fail to express oxi3 under all conditions of growth tested and so exhibit a stringent BOX phenotype. Some mutants in box 4 and 5 of strains having the long form exhibit a conditional BOX phenotype where coxI is present when the cells are cultured on media containing galactose or low concentrations of glucose but is absent when cells are grown on high concentrations of glucose (e. g., ref. 8, 10). Until now neither form of the BOX phenotype was observed when mutants of the short form (D273) were examined; the few cases where a BOX phenotype was obtained were shown to be due to the presence of two mutations, one in cob/box and another in an oxidase gene[11].

In the last year we have concentrated on trying to resolve this apparent

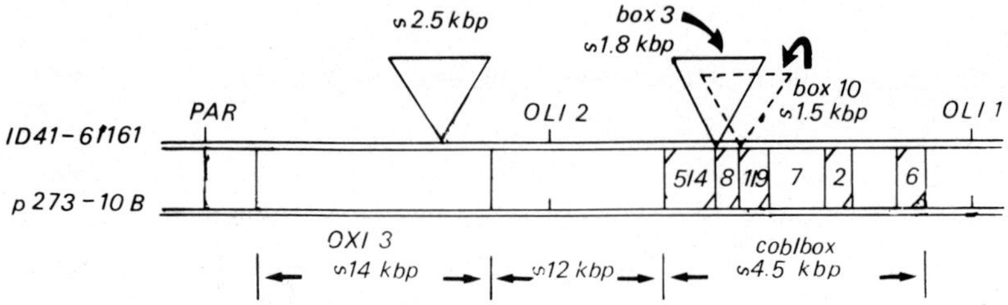

Fig. 1. Schematic diagram of cob/box and oxi3 genes of strains 161 and D273.

discrepancy between the two forms of the cob/box gene. In this paper we summarize the approaches used and the results obtained and then describe several new observations that resulted from our analysis of the short cob/box gene.

ANALYSIS OF THE BOX DISCREPANCY

Analysis of hybrid strains

Rationale. Since it was possible that the short and long forms of cob/box are functionally different with respect to oxi3 regulation we sought explanations for such a difference that could be tested. A variable content of introns is not limited to the cob/box gene of yeast MtDNA; in some strains an intron is present in the gene for the 21S rRNA species while it is absent from others (reviewed in 1, 2). Also, while the structure of the oxi3 gene has not yet been defined in great detail, it, too, is probably mosaic and is longer in strains such as 161 than it is in others such as D273 (c. f., Fig. 1 and refs. 13, 14). If a product of box3 and/or 10 were to regulate oxi3 by permitting the maturation of transcripts of the extra sequences of the long form of oxi3 then it would follow that the short form of oxi3 would be independent of mutations in those introns. Even for the case of box7, which is apparently present in both forms, it is possible that mutants there would be defective only in cytochrome b for that reason. These hypotheses are readily tested by obtaining box7 mutants of D273 or, lacking that, by constructing new strains having the short form of oxi3 but also well characterized BOX mutations that originated in a long strain and then by examining their phenotypes.

Construction of hybrid strains. Since we had no BOX mutants of D273 we undertook the second approach initially. The strategy used for those constructions is shown in Figure 2 for the case of hybrid strain D273/A103 and the same approach was used to obtain the other strains described below. We have exclu-

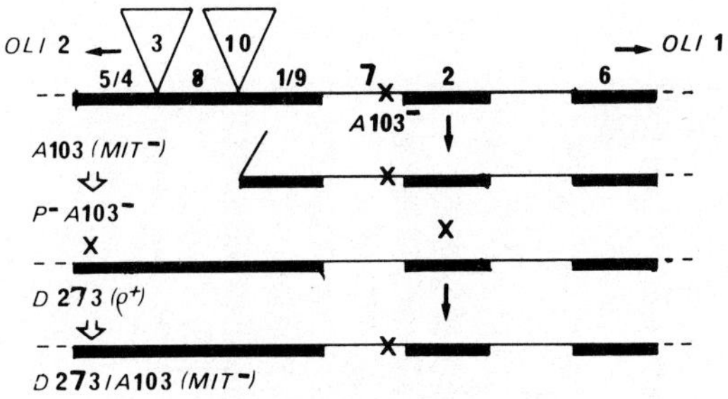

Fig. 2. Diagram of steps used to construct hybrid strain D273/A103. Bold lines indicate exons and narrow lines indicate introns. The petite used retains var1, oli1 and box6 through box1 but is deleted for the remainder of the genome. The petites used in the other constructions retain the entire cob/box region.

ded the participation of mutations in oxidase genes in the resulting phenotypes by using only petite strains that lack the oxidase regions.

The strains obtained and analyzed are the following: i) A103 ($box7^-$), long cob, long oxi3; ii) D273/A103 ($box7^-$), short cob, short oxi3; iii) PZ1 ($box3^-$), long cob, long oxi3; iv) D273/PZ1 ($box3^-$), long cob, short oxi3; v) EM25 ($box4^-$, conditional BOX), long cob, long oxi3; vi) D273/EM25 ($box4^-$), long cob, short oxi3; and vii) tr1 (rho^+), short cob, long oxi3.

For each new strain the form of the cob/box and oxi3 genes was assessed by restriction enzyme analysis of its MtDNA. This analysis for hybrid strains D273/A103 and D273/EM25 is shown in Figure 3. HincII fragment 5 (H2-5) of strain 161 contains most of the long form of the cob/box gene[1, 7, 8] while H2-9 of D273 contains most of the short form[3]. They differ in size by roughly 3300 bp and this reflects the presence or absence of box3 and 10. From Fig. 3 it is clear that D273/A103 has the short form of cob/box while D273/EM25 has the long form. The different forms of cob/box can also be evaluated in EcoR1 (R1) digests where R1-3 fragments differ in size by roughly 3300 bp. The mobility difference between R1-4 species reflects the two forms of oxi3[1, 3], that of 161 being roughly 2500 bp longer than that of D273 and the other strains shown. The size of the cob/box and oxi3 genes present in MtDNA of the other strains listed above was also established in this fashion.

In Figure 4 are shown the phenotypes of these strains. It is evident that the patterns generated are identical regardless of whether a mutation in an intron is in its original (161) or hybrid (D273) context. The stringent BOX phenotype and the nature of the intron products are independent of the form of oxi3 present; thus, both forms of oxi3 are regulated by cob/box introns. Simi-

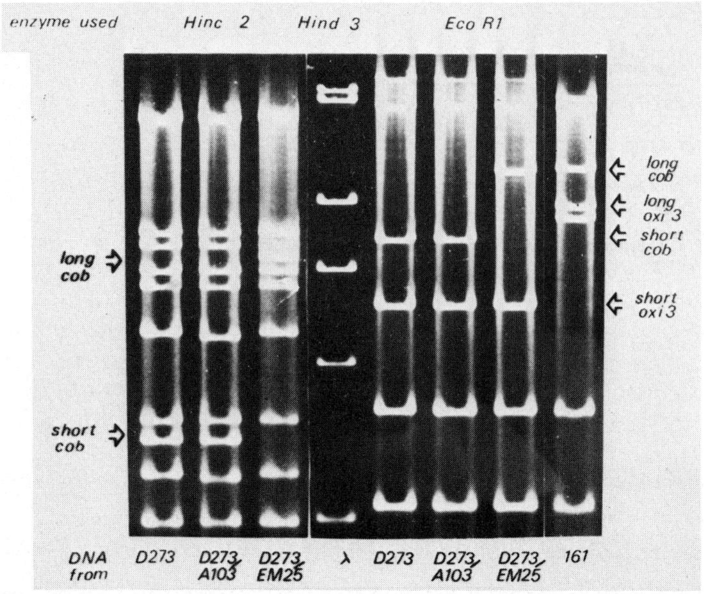

Fig. 3. Restriction enzyme analysis of MtDNA from hybrid strains D273/A103 and D273/EM25.

larly, the conditional BOX phenotype of EM25 is expressed in both contexts. Since none of the box5/4 mutants of D273 (see below) have a conditional BOX phenotype it appears that box3 and/or 10 interacts with mutations in the first exon to yield that phenotype. Strain tr1 yields a typical wild-type pattern (not shown).

Box7 directly controls oxi3 expression. Since A103 has the same phenotype in both contexts (short cob-short oxi3 or long cob-long oxi3) we conclude that box7 exerts a direct effect on oxi3 expression in a manner that is not dependent on any interaction with box3 and/or 10. These results also show that the novel proteins typical of box7 mutants (e. g., p56, p35 and p26 noted in Fig. 4) do not require the presence of box3 or 10 for their specification. Since rho^+ strain tr1 lacks box3 and 10 but has the long form of oxi3 we also conclude that box3 and 10 have no direct effect on oxi3 expression. Several groups have already shown that box3 mutants fail to process box7 as well as most other steps in the processing of the cob/box transcript[14-17] and so their failure to express oxi3 is probably a consequence of their being "phenocopies" of box7. Apparently box7 must be processed in order to permit the expression of oxi3.

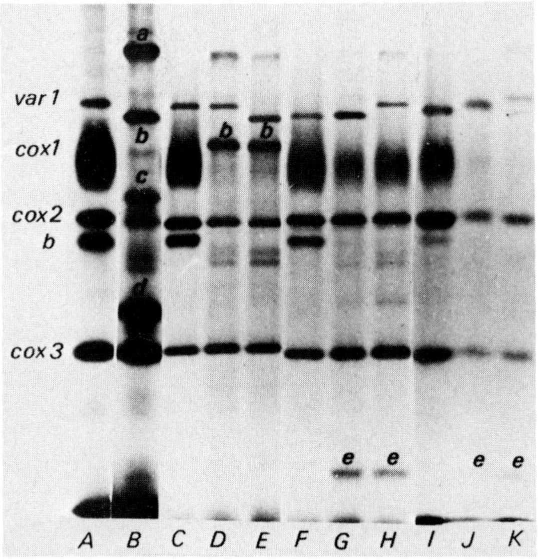

Fig. 4. Phenotypes of hybrid strains. Mitochondrial polypeptides were labelled and analyzed as described in ref. 8. A and C, D273; B, D273/A103; D, D273/PZ1; E, PZ1; F and I, 161; G and J, EM25; H and K, D273/EM25. Samples A-H were labelled in media containing 1% galactose and samples I-K, in media containing 0.3% glucose. Novel proteins are a) p56; b) p42.5; c) p35; d) p26; and e) p15.

It has been suggested that box3 codes for a protein (maturase) that participates in the splicing of part of box3[18, 19]. Our results show that such a maturase cannot play any role in the expression of oxi3.

A box7 mutant of D273

Genetic map of the short form of cob/box. Although it was already clear from restriction[3] and transcript mapping[4] studies that the common portions of the two forms of the cob/box gene are virtually identical we wanted to establish whether their genetic structures are similar. We isolated a number of new cob/box mutants from D273 and prepared a genetic map of them (Fig. 5) using the Gif collection of petite strains that had been used previously to define the ten box loci of the long form[2]. As shown, our mutants were readily assigned to box loci 5, 4, 8, 1, 9 and 6; although none of them maps in box7 we have found that one mutant from Tzagoloff's collection, M8-219, does. From phenotypic studies of long form mutants a good correlation was found between phenotype and map position[1, 2, 7, 9]. As shown in Fig. 5, a similar situation was obtained for these new mutants; some synthesize p30 and are probably missense mutants while others accumulate one novel protein smaller than p30 (arrows) that is probably a chain termination fragment of p30. While the molecular basis of the phenotypes of box9 mutants is not currently understood, the two new box9 mutants (S3 and S118) have phenotypes typical of long form mutants mapping in that region (see ref. 20).

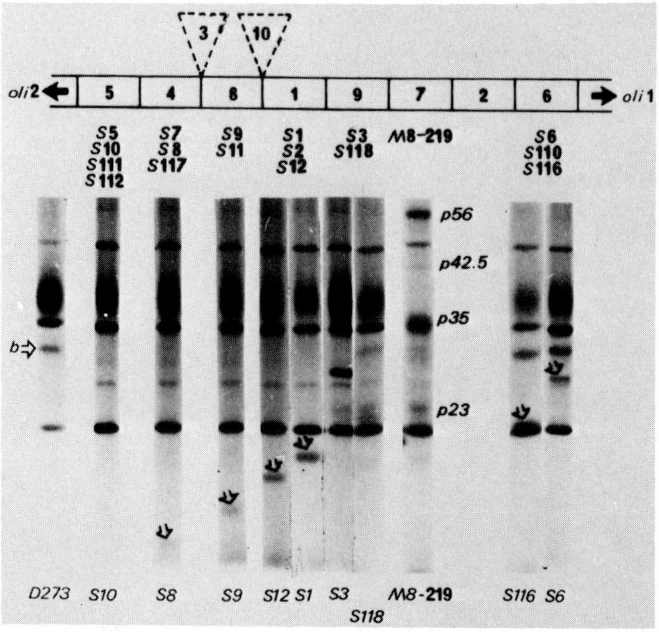

Fig. 5. Map positions and phenotypes of short form cob/box mutants. Phenotypes of 10 of the 18 strains are shown below the map. S5 and S112 resemble S10; S2 resembles S1; and S111, S7, S117, S11 and S110 have p30 and no extra bands. As was the case with some long form chain termination mutants, S116 and S6 have p30 as well as a fragment of p30. Chain termination fragments are indicated with arrows.

Mutant M8-219. Mutant M8-219 has a typical box7 phenotype including the absence of coxl and the accumulation of several novel protein species larger (p56, p42.5 and p35) and smaller (p23) than the missing p30 (Fig. 5, 6). Most long form box7 mutants accumulate a p42.5 species[20, 21] that comigrates with the p42.5 species found in many box3 mutants (such as PZ1, Figs. 4 and 6) and which are very similar upon fingerprinting (c. f., Fig. 9). While we had anticipated that those p42.5 species are, in fact, the same protein, that of box3, our results with M8-219 clearly rule out that interpretation; M8-219 has a p42.5 species even though box3 and 10 are absent from its cob/box gene. In Figure 7 we illustrate with S. aureus V8 protease digests the similarities between the protein species p56, p42.5, p35 and p23 of M8-219 with those of long form box7 mutants G1659 and A103. As noted previously, p35 and p23 species resemble p30 while p56 resembles coxl; the p26 species of A103 resembles neither p30 nor coxl[20].

A new gene near box5. Because the beginning of the cob/box gene has not

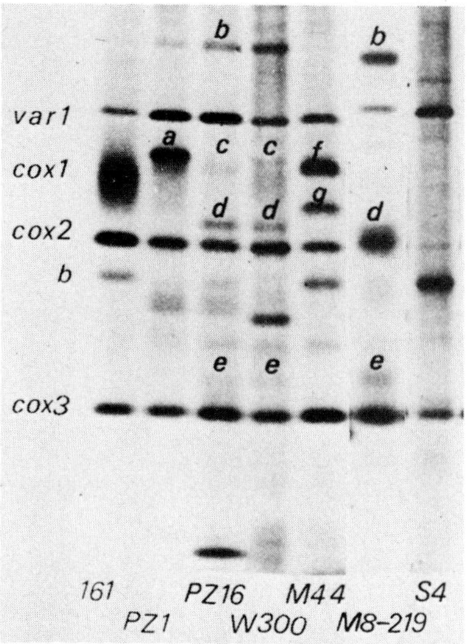

Fig. 6. Phenotypes of other mutant strains. As described in the text phenotypes are shown of PZ1 (box3), PZ16 and M8-219 (box7), W300 (box2), M44 (oxi3) and S4 (glu-tRNA(?)). G1659 (box7) closely resembles PZ16. Mutant specific proteins are: a), c) and f) p42.5; b) p56; d) p35; e) p23 and g) p40.

been localized with great precision we were especially interested in mutant S4 of D273 which maps between box5 and oli2. As shown in Fig. 6 it lacks cox1 and 2 but synthesizes var1, p30 and cox3. We have found that S4 complements cob/box exons and introns (including box3) consistent with its being in an intron of cob/box or in another gene. Since it fails to modify the phenotype of the downstream cob/box mutant PZ1 (box3) in double mutants we conclude that it lies in a separate gene. By deletion mapping we have found that the S4 mutation is located within the HhaI fragment adjacent to box5 (c. f., ref. 8). Although S4 is neither conditional for growth on glycerol nor syn$^-$ it is probably a mutant in the glu-tRNA gene that is located in that HhaI fragment (G. Macino, personal communication). While we do not know why this non-conditional mutant is mit$^-$ rather than syn$^-$ we note that we have characterized several other mutants (2-34s and PZ204) which also have that phenotype; those mutants, however, lack only cox1 (see ref. 22).

NEW FORMS OF THE COB/BOX GENE

Considering our interest in identifying the functions of cob/box introns we attempted to separate box3 from box10 by taking advantage of our collections of mutants in both forms of the gene. As diagrammed in Figure 8 we analyzed the progeny of crosses between short and long form mutants that would require recombination within box8 to yield rho$^+$ progeny. Although considerable inter-

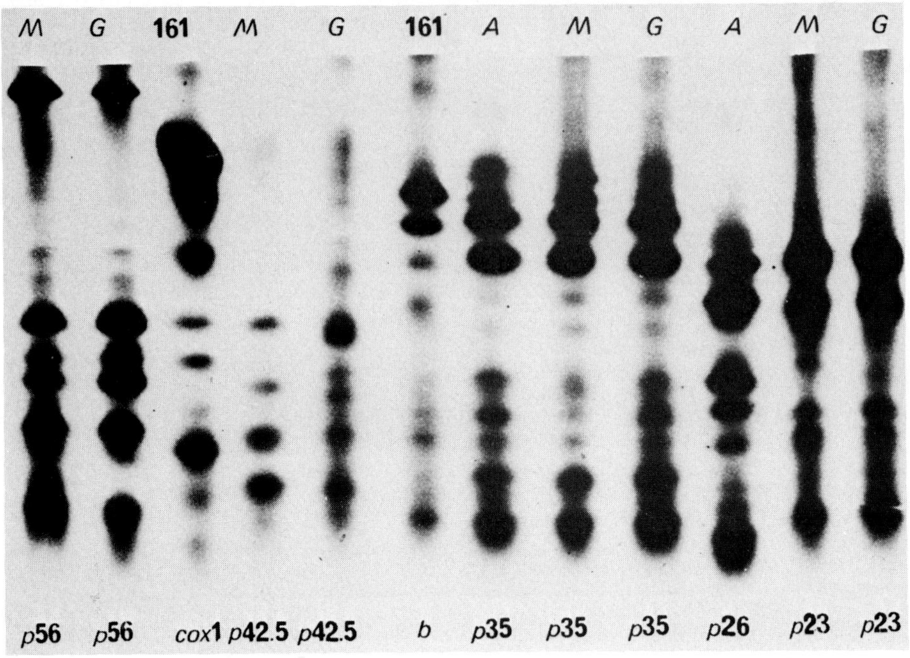

Fig. 7. S. aureus V8 protease fingerprints of novel proteins accumulated by M8-219 (M) and representative long form box7 mutants (C, G1659; A, A103).

ference with recombination was encountered due to the long regions of non-homology on both sides of box8 some rho^{+} progeny were obtained. Based on restriction analysis of MtDNA from several recombinants from each of the crosses shown, some of which is summarized in the legend to Fig. 8, we conclude that we have constructed the two new forms of the cob/box gene that are shown in the diagram. Since these strains have a rho^{+} phenotype, we conclude that the processing of box10 is independent of processing events within, or products of, box3 and vice-versa.

A FAMILY OF PROTEINS TRANSLATED FROM EXON AND INTRON SEQUENCES

Based on immunological[21, 23], fingerprinting[20] and double-mutant[10] studies we and others have concluded that species p42.5 of box3 and box7 mutants and p35 and p23 of box7 mutants are translated from adjacent exon and intron sequences. The finding that intron mutants are defective in the maturation of cob/box transcripts[14-17] is consistent with that conclusion which also implies that portions of box3 and box7 are translatable. The recent demonstration by

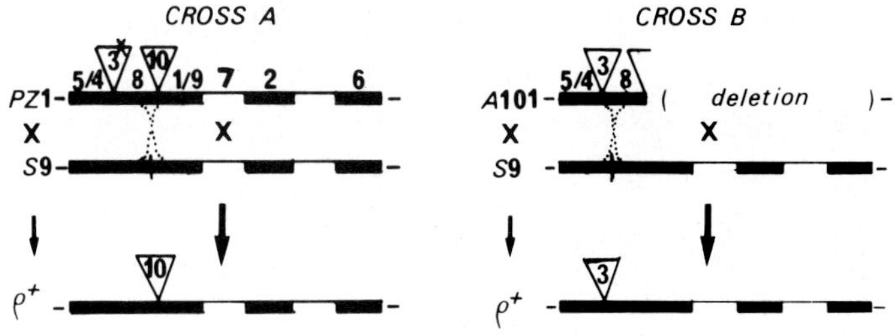

Fig. 8. Schematic diagram of crosses used to construct new forms of the cob/box gene. Restriction analysis of MtDNA from progeny of these crosses shows them to lack both parental Hinc2 and EcoR1 fragments that contain most of the cob/box gene (see Fig. 3). In Hinc2 digests of MtDNA from progeny of cross A there is an extra, non-parental, fragment comigrating with H2-8 of 161 which contains a BamH1 site; in digests of MtDNA from progeny of cross B there is an extra fragment slightly larger than H2-8 of 161 which has no BamH1 site.

Jacq et al.[19] that box3 contains a long open reading frame and that several mutations of box3 affect that sequence in a fashion consistent with the phenotype of each mutant strongly supports this conclusion.

Some years ago we reported an oxi3 mutant, M44, which, unlike most others in that gene fails to synthesize cox1 but at the same time accumulates two new proteins, p42.5 and p40[24] (Fig. 6). By fingerprinting we have found that those two species are nearly identical to each other and that they bear no resemblance to cox1 or p30 (c. f., Fig. 9). That mutant synthesizes functional cytochrome b and so it is unlikely that its p42.5 species is a product of the cob/box gene; that conclusion is supported by our analysis of double mutants containing M44 and a BOX mutation in which the large proteins typical of both mutants are synthesized.

We have completed a detailed comparison of these p42.5 species and in this section we report our finding that while these species are distinguishable in primary digests with S. aureus V8 protease they appear to be homologous over roughly half of their length. This result suggests that there are base sequences in the introns box3 and box7 of cob/box and a portion (presumably an intron) of oxi3 that are homologous. Since several groups have proposed that the p42.5 species of many box3 mutants is a defective form of a maturase needed for the splicing of at least that intron we recognize the possibility that introns of two different mitochondrial genes may each specify one of a family of closely-related, perhaps intron-specific, maturases. In the absence of evidence that each of these proteins is present in wild-type cells and has maturase activity,

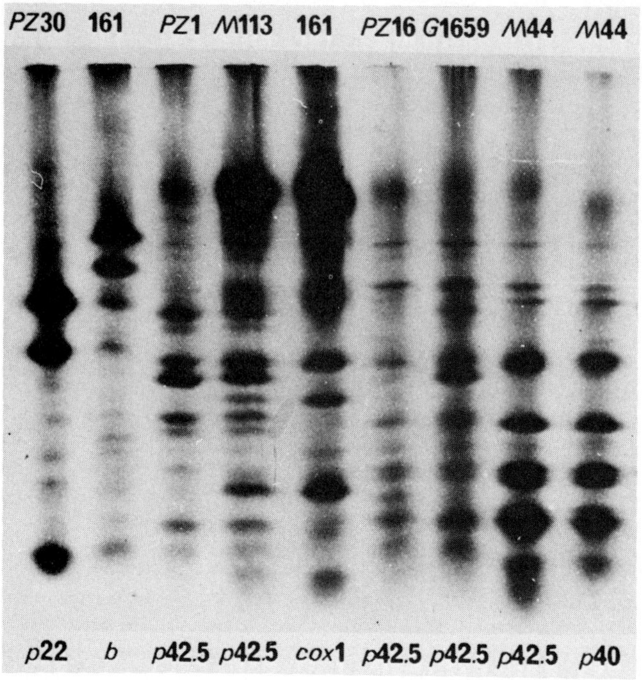

Fig. 9. S. aureus V8 protease fingerprints of p42.5 species accumulated by box3, box7 and oxi3 mutants. For comparison, fingerprints of the p22 species of box6 mutant PZ30 and wild-type b and cox1 are shown. Box3 mutant M113 accumulates p42.5 and p26 species and is probably leaky since it has some cox1; it is clear from the pattern that the p42.5 species of M113 is contaminated with some cox1.

we should also consider the possibility that one or more of them is defective, even in wild-type cells. If the oxi3 p42.5 species is defective then the BOX phenomenon could be a consequence of the dependence of oxi3 on the related processing factor specified by cob/box. In such a case it is clear that we should focus attention on the box7 protein because we have already shown in this paper that the box3 protein plays no role in the BOX phenomenon. The guide RNA model[25] can also be used to explain the BOX phenomenon and such a hypothesis would also benefit from the existence of a family of related sequences within introns.

Our results will be documented in detail elsewhere[26] but we present here a portion of the evidence supporting our conclusion. Figure 9. contains S. aureus V8 protease digests of the p42.5 species of two different box3 mutants (PZ1 and M113), two box7 mutants (PZ16 and G1659) and the oxi3 mutant (M44).

While some differences are evident in these primary digests, there are also numerous striking similarities. We have probed the similarities between these species further by excising comigrating peptides and analyzing them by isoelectric focusing or by redigestion with a second protease[26]. In each case they proved to be indistinguishable in the second dimension. Since the largest fragment analyzed in this way has an M_r of roughly 20,000 daltons we conclude that all of these species are highly homologous over roughly half of their length.

ACKNOWLEDGMENTS

These studies were supported by Grants GM12228 and GM19607 from NIH-NIGMS. HRM is a recipient of a Research Career Award K06 05060 from the same institute. We thank Kathryn Hyams for assistance in some experiments and in the preparation of the manuscript, Jean-Claude Callan, for his photographic assistance and Jean-Claude Mounolou and Mario Luzzati for stimulating discussion and encouragement.

REFERENCES

1. Perlman, P. S., Alexander, N. J., Hanson, D. K. and Mahler, H. R. (1980), in, Gene Structure and Function (Dean, D. H., Johnson, L. F., Kimball, P. C. and Perlman, P. S., eds.) Ohio State University Press, Columbus, Ohio, in press.

2. Slonimski, P. P., Claisse, M. L., Kotylak, Z., Perlman, P. S., Schweyen, R. J., Solioz, M. and Spyridakis, A. (1980), in preparation.

3. Morimoto, R. and Rabinowitz, M. (1979), Molec. gen. Genet., 170, 25-48.

4. Grivell, L. A., Arnberg, A. C., de Boer, P. H., Borst, P., Bos, J. L., Groot, G. S. P., Hecht, N. B., Hensgens, L. A. M., van Ommen, G. J. B. and Tabak, H. F. (1979), in, Extrachromosomal DNA, ICN-UCLA Symposium on Molecular and Cellular Biology, Vol. XV (Cummings, D., Borst, P., Dawid, I., Weissman, S. and Fox, C. F., eds.) Academic Press, New York, pp. 305-324.

5. Pajot, P., Wambier-Kluppel, H. L., Kotylak, Z. and Slonimski, P. P. (1976), in, Genetics and Biogenesis of Chloroplasts and Mitochondria (Bucher, Th., Neupert, W., Sebald, W. and Werner, S., eds.) Elsevier/North Holland, Amsterdam, pp. 443-451.

6. Cobon, G. S., Groot-Obbink, D. J., Hall, R. M., Maxwell, R., Murphy, M., Rytka, J. and Linnane, A. W. (1976), in, ref. 5, pp. 453-460.

7. Slonimski, P. P., Pajot, P., Jacq, C., Foucher, M., Perrodin, G., Kochko, A. and Lamoroux, A. (1978), in, Biochemistry and Genetics of Yeast (Bacilla, M., Horecker, B. L. and Stoppani, A. O. M., eds.) Academic Press, New York, pp. 339-368.

8. Alexander, N. J., Vincent, R. D., Perlman, P. S., Miller, D. H., Hanson, D.

K. and Mahler, H. R. (1979), J. Biol. Chem., 254, 2471-2479.

9. Haid, A., Schweyen, R. J., Bechmann, H., Kaudewitz, F., Solioz, M. and Schatz, G. (1979), Europ. J. Biochem., 94, 451-464.

10. Alexander, N. J., Perlman, P. S., Hanson, D. K. and Mahler, H. R. (1980), Cell, 20, 199-206.

11. Foury, A, and Tzagoloff, A, (1976), Molec. gen. Genet., 149, 43-50.

12. Dujon, B. (1979), Nature, 282, 777-778.

13. Sanders, J. P. M., Heyting, C., Verbeet, M. Ph., Meijlink, F. C. P. W. and Borst, P. (1977), Molec. gen. Genet., 157, 239-261.

14. Van Ommen, G. J. B., Boer, P. H., Groot, G. S. P., de Haan, M., Haid, A., Roosendaal, E., Schweyen, R. J. and Grivell, L. A. (1980), Cell, in press.

15. Church, G. M., Slonimski, P. P. and Gilbert, W. (1979), Cell, 18, 1209-1215.

16. Halbreich, A., Pajot, P., Foucher, M., Grandchamp, C. and Slonimski, P. P. (1980), Cell, 19, 321-329.

17. Haid, A., Grosch, G., Schmelzer, C., Schweyen, R. J. and Kaudewitz, F. (1980), Curr. Genet., in press.

18. Church, G. and Gilbert, W. (1980), in, Mobilization and Reassembly of Genetic Information (Joseph, D. R., Schultz, J., Scott, W. A. and Werners, R., eds.), in press.

19. Jacq, C., Lazowska, J. and Slonimski, P. P. (1980) Comptes rendus Acad. Sci. Paris, 290, 89-92.

20. Claisse, M., Slonimski, P. P., Johnson, J. J. and Mahler, H. R. (1980), Molec. gen. Genet., 177, 375-387.

21. Kreike, J., Bechmann, H., van Hemert, F. J., Schweyen, R. J., Boer, P. H., Kaudewitz, F. and Groot, G. S. P. (1979), Europ. J. Biochem., 101, 607-617.

22. Hanson, D. K., Miller, D. H., Mahler, H. R., Alexander, N. J. and Perlman, P. S. (1979), J. Biol. Chem., 254, 2480-2490.

23. Solioz, M. and Schatz, G. (1979), J. Biol. Chem., 254, 9331-9334.

24. Mahler, H. R., Bilinski, T., Miller, D., Hanson, D., Perlman, P. S. and Demko, C. A. (1976), in, ref. 5, pp. 857-863.

25. Kochko, A. (1979), These de 3eme Cycle, Université Pierre et Marie Curie, Paris VI, Paris, France.

26. Hanson, D. K., Mahler, H. R. and Perlman, P. S. (1980), in preparation.

© 1980 Elsevier/North-Holland Biomedical Press
The Organization and Expression of the Mitochondrial Genome
A.M. Kroon and C. Saccone, editors

PROCESSING OF THE mRNA FOR APOCYTOCHROME b IN YEAST DEPENDS ON A PRODUCT ENCODED BY AN INTERVENING SEQUENCE

H. BECHMANN, A. HAID, C. SCHMELZER, R. J. SCHWEYEN and F. KAUDEWITZ

Genetisches Institut der Universität München, Maria-Ward-Str. 1a
D-8000 München 19, W-Germany

It is well established that in yeast mtDNA apocytochrome b is encoded by a split gene, called COB or BOX[1-3]. Its physical map is shown in Fig. 1. The formation of the messenger RNA for apocytochrome b is accomplished by transcription of a large RNA, containing coding and intervening sequences of COB, followed by a series of splicing events[4-7]. The processing of the COB transcript has been shown to be arrested by mutations in intervening sequences (IVS). They either prevent the excision of the IVS mutated only or block several excision events simultaneously[5, 6]. The latter effect is exerted by many mutations in IVS α/β. We no longer regard the excision of IVS α/β as a prerequisite of further splicing. Instead we assume that the pleiotropy of α/β mutants arises from their effect on a processing function common to several IVS.

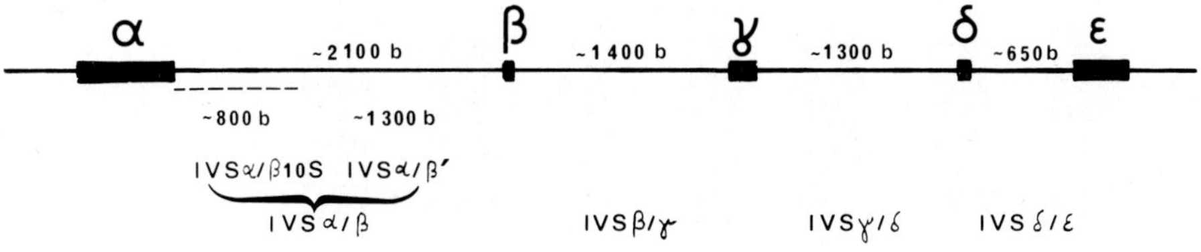

Fig. 1. Physical map of the split gene COB
The position of sequences α to ε coding for apocytochrome b are indicated. IVS = intervening sequence.

Arrest of splicing by IVS mutations

The patterns of COB transcripts presented in Fig. 2 were obtained by electrophoretic separation of mitochondrial RNA followed by transfer hybridization with DNA of the COB region. A comparison of wild type and mutant patterns reveals that the 18S RNA species, the messenger for apocytochrome b[7], is lacking in all intervening sequence mutants. Instead, larger transcripts accumulate some of which also are detected in the wild type. These RNA species have been shown to be intermediates in the processing of a transcript covering all COB sequences. A 10S species, detected in wild type and in most of the mutants, is known to result from excision of part of IVS α/β[5, 7]. It is clear from patterns shown in Fig. 2 that mutations in IVS β/γ and γ/δ allow several processing steps to proceed; hybridization of IVS specific DNA probes to these RNA species (not shown) has revealed that the smallest intermediate accumulated (24S and 23S respectivly) retains only the IVS mutated besides the coding sequences. A similar result is obtained with

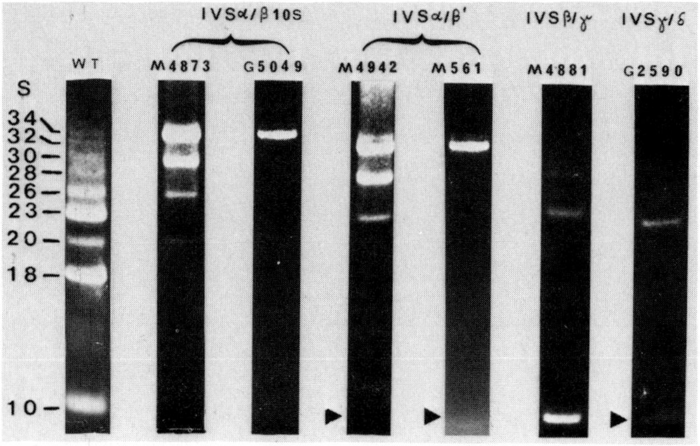

Fig. 2. COB transcripts in wild type (777-3A) and in different IVS mutants
The autoradiographs show mtRNA separated by electrophoresis on agarose/methyl mercury hydroxide gels, transfered to DBM paper and hybridized with ^{32}P labelled probes containing COB coding and intervening sequences.

some mutants of IVS α/β (M4873 and M4942 in Fig. 2): the only intervening sequences retained in the smallest intermediates observed (20S and 23S) are $\alpha/\beta\cdot 10S$ and α/β', respectively. According to these data - and consistent with studies on processing in wild type - we conclude that the various excision events <u>per se</u> are widely independent of each other. Most mutations in α/β sequences block the excision of all sequences (G5049 in $\alpha/\beta\cdot 10S$, Fig. 2), or impair the excision of IVS α/β, β/γ, and γ/δ (M561 in α/β', Fig. 2), leading to the accumulation of transcripts of 34S and 32S, respectively. This contrasts with the finding that none of the excision events necessarily has to occur prior to others.

<u>Evidence for translation of part of IVS α/β</u>

It has been concluded previously that at least part of IVS α/β can be translated and that the product may be involved in processing of the COB transcript[3, 8]. In fact, an open reading frame has been detected in a long sequence in the α distal part of IVS α/β[9]. Mutations in IVS α/β may affect this translation product and thus interfere with the excision of other intervening sequences.

The study of mitochondrial translation products in a large series of mutants provides data which allows us to identify the sequence translated (Fig. 3 and Bechmann et al., in prep.): In mutants of IVS α/β' a series of polypeptides in the range from 17000d to 42000d are observed. The increase in size of this translation product follows the order of mutational sites on the map, indicating 'colinearity'. We conclude that the series of mutations, which spread over most of IVS α/β', cause premature chain termination in a polypeptide of 42000d.

Polypeptides smaller than 17000d are observed with many mutants in α. This and genetic data[8] indicate that translation of the 42000d polypeptide starts with sequence α, which also is the first of the sequences coding for apocytochrome <u>b</u>. Consistent with this assumption is the finding that both polypeptides share antigenic determinants (see Fig. 3). Sequence α, coding for a 15000d N-terminal fragment and sequence α/β', coding for a C-terminal part from 17000 to 42000d are separated by sequence $\alpha/\beta\cdot 10S$ which is about 800bp long. It appeares to be an intervening sequence in the translation unit $\alpha + \alpha/\beta'$. Only after its excision from the COB

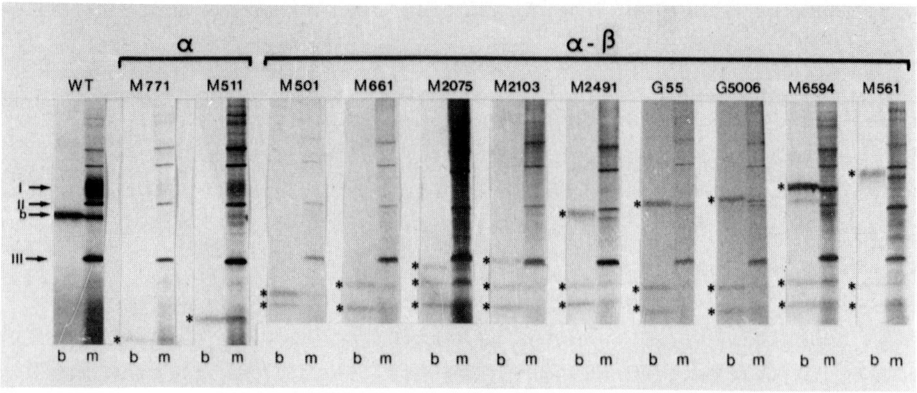

Fig. 3. Mitochondrial translation products were labelled with $^{35}SO_4^{2-}$ in the presence of cycloheximide, separated by polyacrylamid SDS-gelelectrophoresis and autoradiographed[4]. The arrows indicate subunit I, II, III of cytochrome c oxidase and cytochrome b. Polypeptides not found in wild type patterns are marked with an asterisk.
Lane m: total mitochondrial protein; lane b: immunoprecipitate obtained with anti-apocytochrome b serum prior to electrophoresis.

transcript is the 42000d polypeptide synthesized.

A COB translation product affects RNA processing

It has been hypothesized recently[9] that a protein encoded by sequences α and IVS α/β' is involved in processing of the COB transcript. If its function is lacking several excision events may be blocked. The effects which mutations described here have on transcript processing are consistent with this model:

- Mutations in IVS α/β·10S block excision of this sequence and, thereby, prevent the formation of an RNA in which sequences α + α/β' can be translated. These mutants are defective in all other processing events. In leaky mutants excision of α/β·10S may be impaired only; this would allow the production of small amounts of the gene product and a low rate of processing as observed with M4873 (Fig. 2).
- Mutations in sequence α/β' render the translation product defective; the excision of several sequences is blocked. In leaky mutants processing will proceed at a reduced rate, following the

preferred order of events as in wild type[7]. IVS α/β' then will be the last to be excised. This applies to mutant M4942 (Fig. 2).

The model predicts that mutations in sequence α will have effects on processing similar to those in α/β'. In fact, many mutations in α exert these effects; others are not or only slightly impaired in processing but lack functional apocytochrome b (not shown). Both classes of mutations map throughout sequence α. We may assume that one affects the function of both products specified by α, the other apocytochrome b only.

It is open to question whether the 42000d polypeptide or another, so far undetected, product participates in processing. However, the basic idea that there exists a product specified by sequences α and α/β' which has a vital function in the excision of COB intervening sequences, is strongly supported by the data now available.

Acknowledgements

This work was supported by the Deutsche Forschungsgemeinschaft.

REFERENCES

1. Slonimski, P.P., et al. (1978) Biochemistry and Genetics of Yeast, Academic press, New York, pp. 391-401
2. Haid, A., et al. (1979) Europ. J. Biochem., 94, 451-464
3. Kreike, J., et al. (1979) Europ. J. Biochem., 101, 607-617
4. Church, G.M., et al. (1979) Cell, 18, 1209-1215
5. Haid, A., et al. (1980) Current Genetics, 1, 155-161
6. Halbreich, A., et al. (1980) Cell, 19, 321-329
7. Van Ommwn, G.J.B., et al. (1980) Cell, 20, 173-183
8. Alexander, N.J., et al. (1980) Cell, 20, 199-206
9. Jacq, C., et al. (1980) C. R. Acad. Sci. Paris, 290

© 1980 Elsevier/North-Holland Biomedical Press
The Organization and Expression of the Mitochondrial Genome
A.M. Kroon and C. Saccone, editors

PREDICTED SECONDARY STRUCTURES OF THE HYPOTHETICAL BOX 3 RNA MATURASE

R.A. REID and L. SKIERA
University of York, Heslington, York, England

Jacq, Lazowska and Slonimski[1] have recently proposed that the open reading sequence in the Box 3 intron of the Cob-box gene of S. cerevisiae codes for an RNA maturase that is involved in splicing the transcript for the apoprotein of cytochrome b. Although there is genetic evidence for the hypothesis the protein has not been purified, there is no definitive biochemical evidence for a splicing role and it may be some time before critical tests are possible. It was therefore interesting to consider whether the amino acid sequence[1] coded in the Box 3 intron would fall into random coil configurations (which would make the hypothesis improbable) or show a degree of organised secondary structure basic to the notion of a catalytic or structural role.

The results of applying the Chou and Fasman[2] rules for predicting helical and b-sheet regions in proteins and the occurence of b-turns are shown in Fig. 1 and indicate that the secondary structure is typical of biologically active globular proteins. The predictive model, which is capable of delineating the regions with 80% accuracy, proposes at least three a-helices and seven b-sheet regions including some antiparallel hydrogen bonding sections. For comparison, ribonuclease has three helices and six b-sheets and lysozyme has six helices and four b-sheets. The incidence of random coil is relatively low and less than in a-chymotrypsin and carboxypeptidase A. Eleven b-turns are predicted, which compares favourably with carboxypeptidase (7), thermolysin (7) subtilisin (5) and an average of fourteen turns over a range of globular proteins[2]. Little can be said about tertiary structure other than that modelling has shown that the first two cys residues are close enough for disulphide bridging. The general features of the predicted secondary structure suggest that the polypeptide would form a compact globular protein. The Box 3.2 mutant shows two substitutions, the first of which (Val---Phe) would not affect secondary structure but might influence tertiary structure. The second substitution (Ile---Asn) would weaken the b-sheet concerned.

1. Jacq,C. et al. (1980) C.R.Acad.Sc.Paris t290, serie D - 1
2. Chou, P.Y. and Fasman, G.D. (1974) Biochem. 13, 222-248.

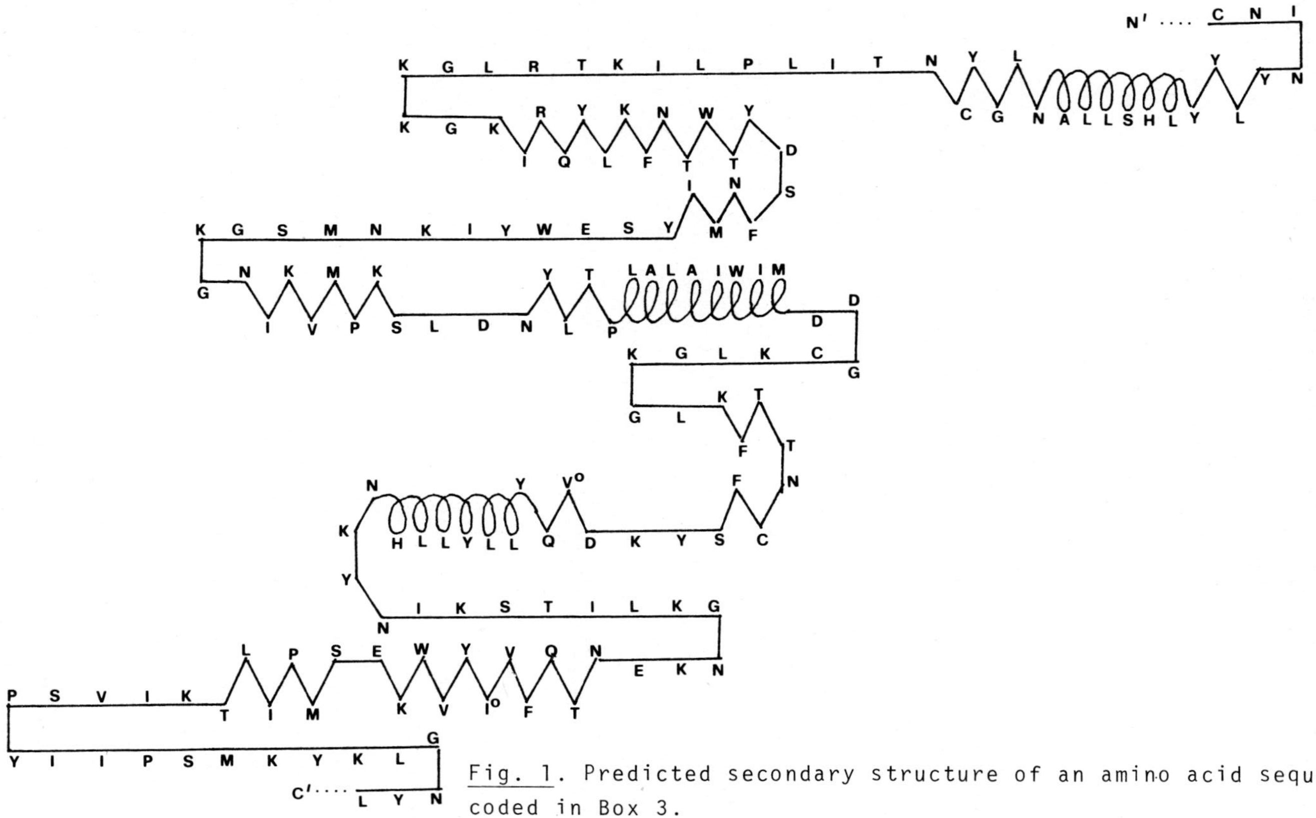

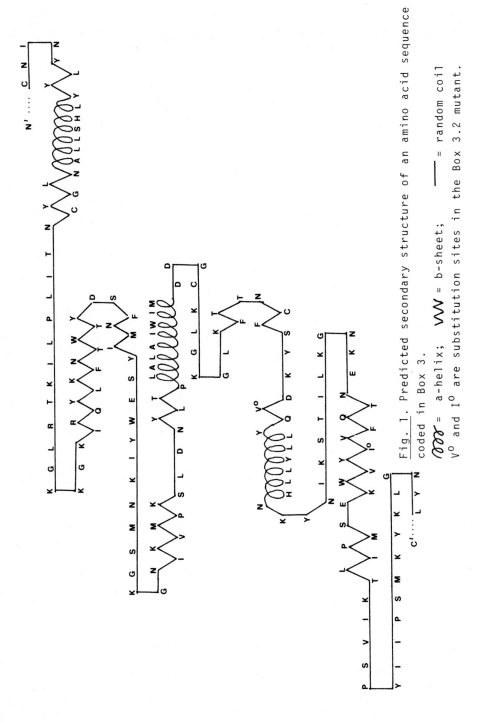

Fig. 1. Predicted secondary structure of an amino acid sequence coded in Box 3.

(helix symbol) = a-helix; (zigzag symbol) = b-sheet; —— = random coil

V^{o} and I^{o} are substitution sites in the Box 3.2 mutant.

The Organization and Expression of the Mitochondrial Genome
A.M. Kroon and C. Saccone, editors

YEAST MITOCHONDRIAL CYTOCHROME OXIDASE GENES

Alexander Tzagoloff, Susan G. Bonitz, Gloria Coruzzi and Barbara E. Thalenfeld
Department of Biological Sciences, Columbia University, New York, N.Y. 10027
and
Giuseppe Macino
Istituto di Fisiologia Generale, Università di Roma, Roma, Italy

INTRODUCTION

Mutations in yeast mitochondrial DNA (mtDNA) have been shown to cause specific lesions in cytochrome oxidase (1-3). Such mutations map in three separate loci termed oxi1, oxi2 and oxi3 (4). The oxi loci are widely distributed on the genome, in each case being separated by other genes with functions unrelated to cytochrome oxidase. These observations coupled with the finding that the oxi loci behave as separate complementation groups (5) indicate the presence of three separate cytochrome oxidase genes in yeast mtDNA.

We have recently completed the DNA sequences of each oxi locus. The nucleotide sequences have enabled us to identify the cytochrome oxidase genes and to deduce the amino acid sequences of the their respective products. In this paper we review evidence that oxi1 codes for subunit 2, oxi2 for subunit 3 and oxi3 for subunit 1 of the enzyme.

ENRICHMENT FOR THE CYTOCHROME OXIDASE GENES

To obtain sufficient amounts of mtDNA to carry out the sequencing studies, we selected ρ^- clones with genetic markers confined to either oxi1, oxi2 or oxi3. The clones were isolated from the respiratory competent haploid strain of Saccharomyces cerevisiae D273-10B. A partial restriction map of the D273-10B genome had already been derived by Morimoto and Rabinowitz (6) and a number of other genes had been sequenced in this strain (7-10).

Several criteria were used in the choice of the ρ^- clones for the sequence analysis: 1) stability of the genotype, 2) mitochondrial genomes with tandemly repeated segments of mtDNA, 3) absence of internal deletions or rearrangements, and 4) a sufficiently small unit genome size for restriction mapping and DNA sequencing; the last requirement could generally be fulfiled if the unit length did not exceed 6 kb.

OXI1 CODES FOR SUBUNIT 2 OF CYTOCHROME OXIDASE

The oxi1 gene was sequenced in a segment of mtDNA retained in the ρ^- clone DS200/A1 (11). This clone was derived from D273-10B/A21 by two consecutive mutageneses with ethidium bromide. DS200/A21 was verified to contain only oxi1 markers and to have a genome consisting of a 4.5 kb segment of mtDNA in a tandem configuration. The restriction map and sequence of the DNA indicated that the genome of this clone had resulted from deletions at 10.5 and 17 map units. The retained segment therefore, corresponded to a region of the wild type mtDNA where the oxi1 locus had previously been mapped (12,13).

The nucleotide sequence of the DS200/A1 segment has revealed the presence of four regions with possible coding functions (Fig. 1). Two of the genes code for $tRNA^{Asn}$ and $tRNA^{Met}$. The third region is a 756 nucleotide long reading frame that has been identified as the oxil gene. The protein encoded by the oxil gene has a molecular weight of 28,480 and an amino acid sequence averaging 50% homology with subunit 2 of bovine cytochrome oxidase (Fig. 2). Since the alignment of the yeast and bovine sequences does not require any significant deletions in the DNA sequence, the gene appears to be co-linear with the protein sequence. This is consistent with earlier observations on the tight genetic linkage of oxil markers (15). The DNA sequence of the oxil gene has been independently obtained by Fox (16) in a cloned restriction fragment of D273-10B mtDNA. The reported sequence of the cloned fragment is identical to that found in the ρ^- clone, demonstrating the validity of using both types of DNAs for sequence analysis.

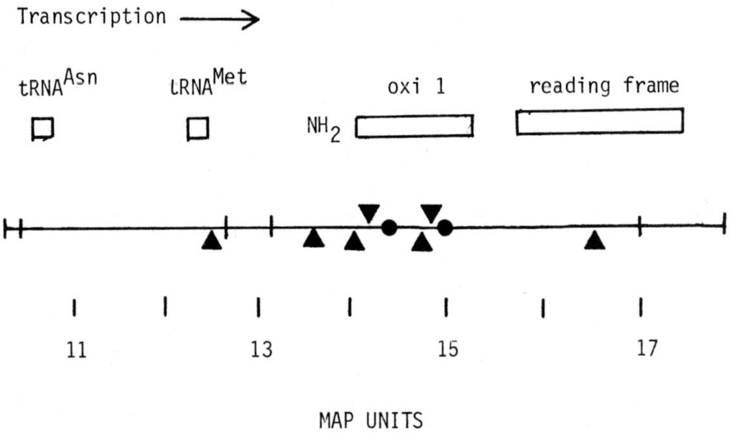

Fig. 1. Location of the tRNA and subunit 2 genes on the physical map of DS200/A1. The following restriction sites have been marked: HinfI (▲), PvuII (●) and HpaII (|). Each map unit is equivalent to 700 base pairs.

fMLDLLRLQLTTFIMNDVPTPYACYFQDSATPNQEGILELHDNIMFYLLVILGLVSWMLYTIVITYSKNPIAYKYIKHG
QTIEVIWTIFPAVILLIIAFPSFILLYLCDEVISPAITIKAIGYQWYWKYEYSDFINDSGETVEFESYVIPDGLLEEG
QLRLLDTDTSIVVPVDTHIRFVVTAADVIHDFAIPSLGIKVDATPGRLNQVSALIQREGVFYGACSELCGTGHANMPI
KIEAVSLPKFLEWLNEQ

Fig. 2. Amino acid sequence of yeast subunit 2 of cytochrome oxidase. The amino acid residues homologous with the bovine protein are underlined. The sequence of the bovine subunit 2 has been determined by Steffens and Buse (14).

The yeast protein is 16 residues longer than bovine subunit 2. There is evidence that both subunit 2 of yeast (17) and *N. crassa* (18) are synthesized as precursors with an amino terminal extension of 15 amino acids. This does not appear to be true of the bovine protein.

One of the more novel features of the mitochondrial genetic system to have emerged from the DNA sequences of mitochondrial genes is the use of the UGA terminator as a tryptophan codon (16,19). Both the yeast and human subunit 2 genes have five in-frame UGA codons (19,20). Four of the five codons occur at positions where the corresponding residues in the bovine protein have been shown to be tryptophanes, suggesting that they function as tryptophan codons rather than terminators. The use of UGA as a tryptophan codon is also supported by recent findings that the tryptophan tRNAs of *N. crassa* (21), human (22) and yeast (23) mitochondria have a 5'-UCA-3' anticodon capable of reading both UGG and UGA.

The third potential coding region in DS200/A1 is an 1164 nucleotide long reading frame with an AUG initiation codon several hundred base pairs downstream from the end of the subunit 2 gene. We are not able to state whether this sequence codes for a protein. The amino acid sequence dictated by this frame suggests a basic protein with a high content of lysine, asparagine and tyrosine. Northern blots of total mitochondrial RNA probed with single stranded DNA fragments complementary to this sequence have failed to reveal any transcripts originating from the region of this reading frame.

OXI2 CODES FOR SUBUNIT 3 OF CYTOCHROME OXIDASE

The *oxi2* locus is also defined by a cluster of closely linked alleles controlling the synthesis of cytochrome oxidase. Mutations in *oxi2* have been mapped between 19 and 21 units on the wild type genome of D273-10B. This region of mtDNA was obtained from the ρ^- clone DS40 selected for the retention of *oxi2* markers (24). The DS40 genome was found to have a repeat length of 4.1 kb extending from 18.6 to 24.3 map units (Fig. 3).

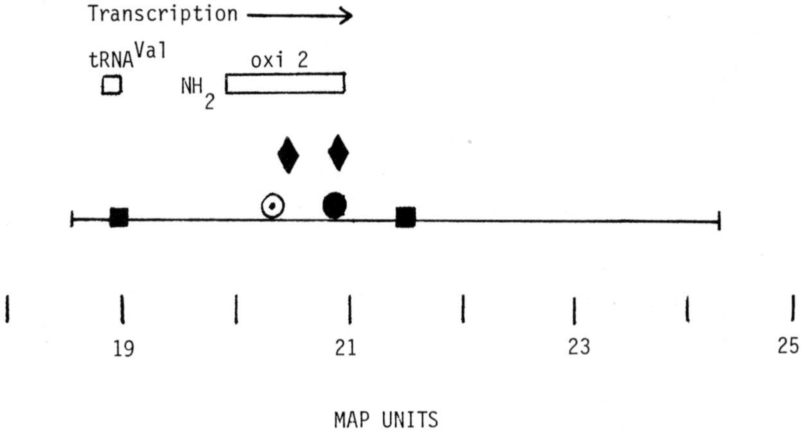

Fig. 3. Location of the valine tRNA and *oxi2* gene on the physical map of DS40. Only the restriction sites for XbaI (■), HindIII (♦), HincII (⊙) and PvuII (●) are shown. The physical limits of the DS40 mtDNA segment is indicated by the line. Each map unit corresponds to 700 base pairs.

The entire mtDNA segment of DS40 has been sequenced and two genes identified. One of the genes, located at 19 map units is a short sequence coding for the $tRNA^{Val}$. The other gene is 810 nucleotides long spanning the region from 19.9 to 21 map units. This sequence has a register with an open reading frame capable of coding for a protein with 267 amino acid residues (24).

Two lines of evidence indicate that the _oxi2_ locus codes for subunit 3 of yeast cytochrome oxidase. The protein encoded in the _oxi2_ gene has an amino acid composition similar to that of _bona fide_ subunit 3 (24). Although the sequence of the yeast protein is not known, some partial sequence data have been obtained on the bovine subunit 3 (Buse and Steffens, personal communication). The deduced sequence of the yeast protein is sufficiently homologous to subunit 3 of bovine cytochrome oxidase to suggest an identity of the two proteins.

The amino acid sequence of subunit 3 of yeast cytochrome oxidase is presented in Fig. 4. It is a hydrophobic protein with a molecular weight of 30,340. This value is 30% higher than estimated by SDS-gel electrophoresis (25,26). Although this discrepancy is most likely due to anomalous behavior of hydrophobic membrane proteins on SDS gels, it is not completely excluded that the _oxi2_ gene might have a short intervening sequence.

fMTHLERSRHQQHPFHMVMPSPWPIVVSFALLSLALSTALTMHGYIGNMNMVYLALFVLLTSSILWFRDIVAEATYLGD
HTIAVRKGINLGFLMFVLSEVLIFAGLFWAYFHSAMSPDVTLGACWPPVGIEAVQPTELPLLNTIILLSSGATVTYSH
HALIAGNRNKALSGLLITFWLIVIFVTCQYIEYTNAAFTISDGVYGSVFYAGTGLHFLHMVMLAAMLGVNYWRMRNYH
LTAGHHVGYETTIIYTHVLDVIWLFLYVTFYWWGV

Fig. 4. Amino acid sequence of yeast subunit 3.

OXI3 CODES FOR SUBUNIT 1 OF CYTOCHROME OXIDASE

The _oxi3_ locus of the yeast mitochondrial genome is a complex region located between the _oli2_ and _par_ resistance markers (4). This region has been estimated to occupy a stretch equivalent to 10-12 kb of DNA (27,28). A large number of of genetically unlinked mutations have been assigned to the _oxi3_ locus. These are either point mutations or deletions. Almost all _oxi3_ deficient mutants lack subunit 1 of cytochrome oxidase (1,4). This has been interpreted to indicate that the _oxi3_ locus codes for subunit 1. The following observations have led to the idea that the _oxi3_ locus is a mosaic gene with long intervening sequences.

1. Subunit 1 of yeast cytochrome oxidase migrates with an apparent molecular weight of 40 kd on SDS-polycarylamide gels (25,26). A co-linear gene should therefore not exceed 1,200 nucleotides. _Oxi3_ mutations, however, scatter over a length of 10 kb (27).
2. Genetically unlinked mutations in _oxi3_ form a single complementation group suggesting a single transcriptional unit coding for one protein (5).
3. The _oxi3_ region gives rise to different size transcripts indicative of a complex pattern of processing involving the excision of stable RNA intron sequences (29,30).
4. Restriction mapping of mtDNA has shown the existence of strain-

specific insertions and deletions in the oxi3 region (31, Hensgens et al, this volume).

In order to sequence the subunit 1 gene, we have isolated the ρ^- clone DS6 with all the oxi3 markers of the yeast genome. DS6 had a mtDNA segment which included the wild type sequence from approximately 35 to 61 map units. The unit length of the DNA was 16.5 kb. Restriction analysis of the mtDNA confirmed that the segment represented a contiguous wild type sequence with all the appropriate restriction sites previously mapped in the oxi3 region of *S. cerevisiae* D273-10B (6). In view of the molecular complexity of the mtDNA, a series of simpler clones were derived from DS6 by ethidium bromide mutagenesis. These were selected for differential retention of various oxi3 markers (Table I). Six clones (DS6/A400, A401, A402, A407, A464 and A422) were found to have mtDNA segments covering the region of DS6 from 41 to 58 map units with sufficient overlap to account for the entire sequence. The mitochondrial genomes of the clones had unit lengths ranging from 2.3 to 6.4 kb. These mtDNAs were used to construct a detailed restriction map of the oxi3 locus and to sequence the gene.

Table I
Genotypes of ρ^- Clones

	Oxi3 markers										
Clone	M8-227	M15-190	M12-193	M3-9	M10-63	M5-121	M11-224	M5-85	M15-233	M6-41	Unit length of mtDNA (kb)
DS6	+	+	+	+	+	+	+	+	+	+	16.5
A401	+	+	+	+	-	-	-	-	-	-	6.4
A400	-	+	+	-	-	-	-	-	-	-	2.6
A402	-	-	+	+	-	-	-	-	-	-	3.1
A462	-	-	-	-	+	+	-	-	-	-	2.3
A407	-	-	-	-	-	-	+	+	+	+	2.6
A422	-	-	-	-	+	+	+	+	+	+	5.2

The retention or loss of the oxi3 markers is indicated by plus and minus signs, respectively.

With the exception of DS6/A462 which was found to have a small internal deletion and rearrangement of its genome, the restriction maps of the other ρ^- mtDNAs indicated that they were coherent segments originating from different parts of the DS6 genome. A restriction map of each mtDNA was constructed by analysis of the products formed with single and combinations of various restriction endonucleases. This information was used to construct the overall physical map of DS6. The placement of restriction sites for some 17 enzymes facilitated the sequence analysis of the region.

Most of the DNA sequence of the oxi3 gene was determined from the sequence analysis of the less complex clones. One region of the DNA, however, was sequenced in a preparative restriction fragment obtained from the DS6 mtDNA.

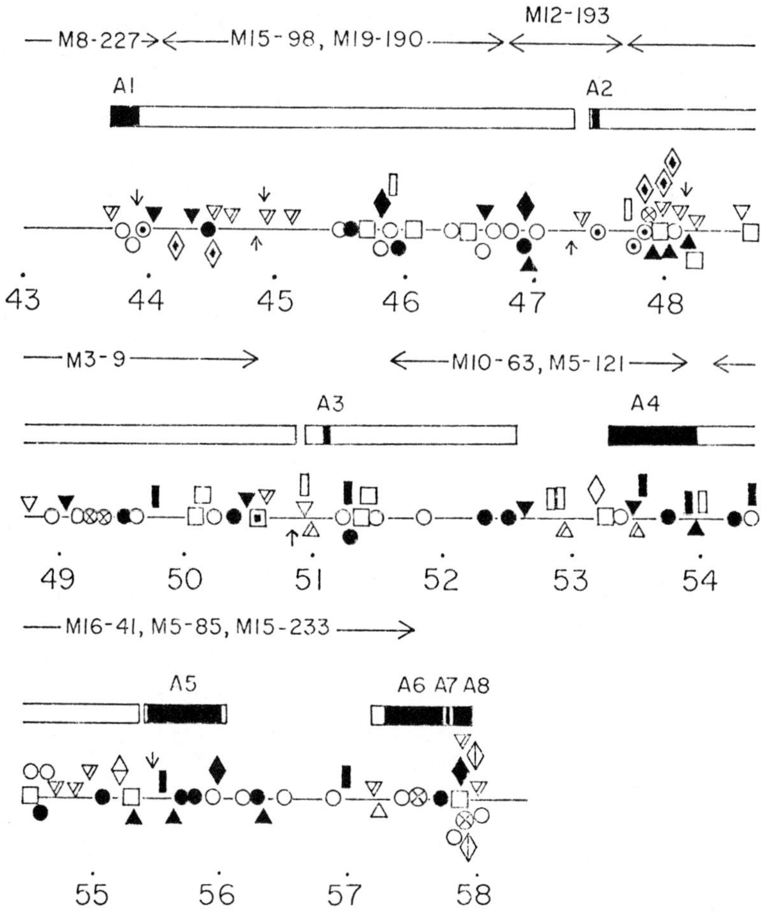

Fig. 5. Restriction map and organization of the oxi3 gene. The location of the restriction sites are shown between 43 and 58 map units. The symbols used to denote the restriction sites are: HpaII (△), HaeIII (▯), HphI (▮), RsaI (◬), HinfI (▲), HincII (◈), MboI (●), MboII (□), AluI (○), TaqI (◆), HhaI (⊙), EcoRI (⊗), EcoRII (↓), BglII (▣), HindIII (◇), BamHI (⬘), and PvuII (⦶). The solid bars indicate the exon reading frames and the open bars the open reading frames in the introns. The eight exons of the subunit 1 gene are designated by the letters A1- A8. The locations of the different oxi3 mutations are indicated by the arrows. The physical limits of DNA within which the mutations have been mapped is based on the deletion end-points of the mtDNA segments retained in the ρ^- clones. Each map unit corresponds to 700 base pairs.

A continuous sequence has been obtained from 43 to 58 map units. This sequence contains six long open reading frames capable of coding for proteins in the range of 30-80 kd. In addition the sequence has some A+T-rich stretches of nucleotides and several short G+C clusters with HpaII and HaeIII sites. There is no information at present about the primary structure of the yeast subunit 1. To identify the exon coding regions of the oxi3 locus, we relied on the homology with the sequence of human subunit 1 provided to us by Drs. Bart Barrell and Fred Sanger. The sequence of the human subunit 1 was deduced from the nucleotide sequence of the gene in human mitochondrial DNA. The homology of the two proteins allowed us to identify 8 coding regions in the oxi3 locus. These were of variable lengths and in most cases exhibited 50% or more homology to the human subunit 1. The first exon (A1) starts with an AUG initiation codon located at approximately 43.6 map units defining the amino terminal end of the gene. The AUG initiator is preceded by a long A+T-rich sequence. The next two exons (A2 and A3) are very short, each coding for only 11-13 amino acids. Most of the protein is encoded in exons A4 and A5 (about 268 amino acids). This middle region of the protein is almost 70% homologous to the human protein. The carboxyl terminal end of the yeast subunit 1 is located in the last coding region (57-58 map units). The homology between the two proteins is rather poor and it is difficult to ascertain from the protein sequences alone whether there are two or three exons in this reading frame. The best alignment of the yeast and human proteins suggests that there are probably three exons (A6, A7, A8) with two short intervening sequences. The entire subunit 1 gene is 9976 nucleotides long. The exon sequences, however, account for only 1530 nucleotides.

◄——— exon A1 ———

fM V Q R W L Y S T N A K D I A V L Y F M L A I F S G M A G T A M S L I I R L E

———► ◄— exon A2 —► ◄— exon A3 —►

L A A P G S Q Y L H G N S Q L F L/A V V G H A V L M I F C/L L V M P A L I G G F G

◄———

N Y L L P L I I G A T D T A F P R I N N I A F W V L P M G L V C L V T S T L V

E S G A G T G W T V Y P P L S S I Q A H S G P S V D L A I F A L H L T S I S S

——— exon A4 ———

L L G A I N F I V T T L N M R T N G M T M H K L P L F V W S I F I T A F L L L

L S L P V L S A G I T M L L L D R N F N T S F F E V A G G G D P I L Y E H L F

———► ◄———

W F F G H P E V Y I L I I P G F G I I S H V V S T Y S K K P V F G E I S M V Y

——— exon A5 ———

A M A S I G L L G F L V W S H H M Y I V G L D A D T R A Y F T S A T M I I A I

P T G I K I F S W L A T I Y G G S I R L A T P M L Y A I A F L F L F T M G G L

T G V A L A N A S L D V A F H D T Y Y V V G H F Y Y V L S M G A I F S L F A G

——— exon A6 ———

Y Y Y W S P Q I L G L N Y N E K L A Q I Q F W L I F I G A N V I F F P M H F L

G I N G M P R R I P D Y P D A F A G W N Y V A S I G S F I A T L S L F L F I Y

——— exon A7 ——— exon A8 ———

I L Y D Q L V N N K S V I Y A K A P S S S I E P L L T S P P A V H S F N T P A

V Q S

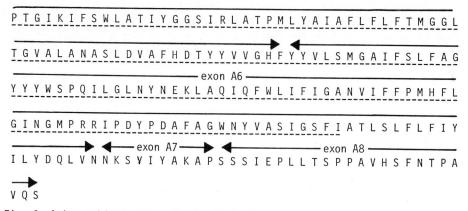

Fig. 6. Amino acid sequence of subunit 1 of yeast cytochrome oxidase. The regions of 50% or higher homology with human subunit 1 are indicated by the dashed line under the sequence. The boundaries of the 8 exons are shown by the arrows.

The amino acid sequence of the subunit 1 of yeast cytochrome oxidase derived from the gene sequence is presented in Fig. 6. The protein is 510 residues long with a very hydrophobic composition. The molecular weight calculated from the amino acid composition is 56 kd. This value is 30% higher than measured from SDS-gels (25,26). We believe that the SDS-gels underestimate the true size of this protein.

INTRON SEQUENCES OF THE OXI3 GENE

As mentioned above the oxi3 gene has six open reading frames. All the exons of subunit 1 are part of these longer reading frames (see Fig. 5). The first four reading frames in particular are extremely long, the exons representing only a small part of potentially longer coding sequences. The function of the intron reading frames is not clear at present. It has been proposed that such intron sequences may code for proteins involved in the splicing of the primary transcipts (32,33). However, direct evidence for the existence of mitochondrially encoded RNA processing enzymes is still lacking. The intron coding sequences differ from the exons in several respects. 1) They utilize codons that are not present in exons or other mitochondrial genes (e.g. CGN codons for arginine, UGG codon for tryptophan). 2) The intron reading frames are very rich in codons for lysine, asparagine and tyrosine. In this respect they resemble the open reading frame downstream from the oxil gene.

It is also of interest that there is extensive DNA and amino acid sequence homology between the first two introns of the oxi3 gene and the first intron of of the cytochrome b (9) and fourth intron of the oxi3 genes. The sequence homology extends to the exon-intron boundaries as well, suggesting common splicing mechanisms determined by specific sequences around the exon junctions.

PHYSICAL LOCATION AND TRANSCRIPTION OF THE CYTOCHROME OXIDASE GENES

The mtDNA segments of the ρ^- clones used to sequence the three genes of yeast cytochrome oxidase have been oriented and precisely located on the wild type map of D273-10B. This was possible due to the presence in the ρ^- genomes of restriction sites that had been mapped on the mtDNA of D273-10B (6). The locations and directions of transcription of the genes is shown in Fig. 7. All three genes are transcribed from the same strand of DNA, the direction of transcription being clockwise according to the convention of Fig. 7.

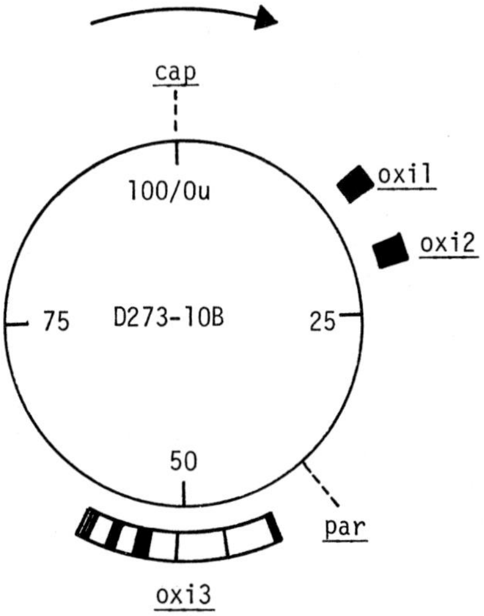

Fig. 7. Organization of cytochrome oxidase genes in the wild type mtDNA of D273-10B. The coding regions of the three genes are depicted by the solid bars. The map units are shown on the inner circle. The cap and par resistance loci provide the orientation of the map. The direction of transcription is indicated by the arrow.

ACKNOWLEGMENTS

We would like to express our gratitude to Dr. Bart Barrell and Dr. Fred Sanger for providing us with the sequence of the subunit 1 of human cytochrome oxidase prior to publication. This research was supported by NIH Research Grant GM25250 and a National Research Servic Award GM07556 to B.E.T.

REFERENCES

1. Tzagoloff, A., Akai, A., Needleman, R.B. and Zulch, G. (1975) J. Biol. Chem. 250, 8236-8242.
2. Groot Obbink, D.J., Hall, R.M., Linnane, A.W., Lukins, H.B., Monk, B.C., Spithill, T.W. and Trembath, M.K. (1976) in "The genetic function of mitochondrial DNA" (C. Saccone and A.M. Kroon, eds.) North Holland Publishing Co. Amsterdam, pp. 163-173.
3. Carignani, G., Dujardin, G. and Slonimski, P.P. (1979) Mol. Gen. Genet. 167, 301-308.
4. Slonimski, P.P. and Tzagoloff, A. (1976) Eur. J. Biochem. 61, 27-41.
5. Foury, F. and Tzagoloff, A. (1978) J. Biol. Chem. 253, 3792-2797.
6. Morimoto, R. and Rabinowitz, M. (1979) Mol. Gen. Genet. 170, 25-48.
7. Macino, G. and Tzagoloff, A. (1979) J. Biol. Chem. 254, 4617-4623.

8. Macino, G. and Tzagoloff, A. (1980) Cell 20, 507-517.
9. Nobrega, F. and Tzagoloff, A. (1980) J. Biol. Chem. in press.
10. Miller, D.L., Martin, N.C., Pham, H.D. and Donelson, J.E. (1979) J. Biol. Chem. 254, 11735-11740.
11. Coruzzi, G. and Tzagoloff, A. (1979) J. Biol. Chem. 254, 9324-9330.
12. Lewin, A., Morimoto, R., Rabinowitz, M. and Fukuhara, H. (1978) Mol. Gen. Genet. 163, 257-275.
13. Borst, P. and Grivell, L.A. (1978) Cell 15, 705-723.
14. Steffens, G.S. and Buse, G. (1979) Hoppe-Seyler's Z. Physiol. Chem. 360, 613-619.
15. Trembath, M.K., Macino, G. and Tzagoloff, A. (1977) Mol. Gen. Genet. 158, 35-45.
16. Fox, T.D, (1979) Proc. Natl. Acad. Sci. U.S.A. 76, 6534-6538.
17. Sevarino, K.A. and Poyton, R.O. (1980) Proc. Natl. Acad. Sci. U.S.A. 77, 142-146.
18. Werner, S. and Machleidt, W. (1980) in "Organization and expression of the mitochondrial genome" (C. Saccone and A.M. Kroon, eds.) North-Holland Publishing Co., Amsterdam (this volume).
19. Macino, G., Coruzzi, G., Nobrega, F.G., Li, M. and Tzagoloff, A. (1979) Proc. Natl. Acad. Sci. U.S.A. 76, 3784-3785.
20. Barrell, B.G., Bankier, A.T. and Drouin, J. (1979) Nature 282, 189-194.
21. Heckman, J.E., Sarnoff, J., Alzner-Deweerd, B., Yin, S. and RajBhandary, U.L. (1980) Proc. Natl. Acad. Sci. U.S.A. in press.
22. Barrell, B.G., Anderson, S., Bankier, A.T., deBruijn, M.H.L., Chen, E., Coulson, A.R., Drouin, J., Eperon, I.C., Nierlich, D.P., Roe, B.A., Sanger, F., Schreier, P.H., Smith, A.J.H., Staden, R. and Young, I.G. (1980) Proc. Natl. Acad. Sci. U.S.A. in press.
23. Silbler, A.-P., Bordonne, R., Dirheimer, G. and Martin, R. (1980) C.R. Acad. Sc. Paris t.290, 695-698.
24. Thalenfeld, B.E. and Tzagoloff, A. (1980) J. Biol. Chem. in press.
25. Mason, T.L., Poyton, R.O., Wharton, D.C. and Schatz, G. (1973) J. Biol. Chem. 248, 1346-1354.
26. Rubin, M.S. and Tzagoloff, A. (1973) J. Biol. Chem. 248, 4269-4274.
27. Morimoto, R. and Rabinowitz, M. (1979) Mol. Gen. Genet. 170, 11-23.
28. Van Ommen, G.-J.B., Groot, G.S.P. and Grivell, L. A. (1979) Cell 18, 511-523.
29. Van Ommen, G.-J.B., Boer, P.H., Groot, G.S.P., DeHaan, M., Roosendaal, E., Grivell, L.A., Haid, A. and Schweyen, R.J. (1980) Cell 20, 173-183.
30. Morimoto, R., Locker, J., Synenki, R.M. and Rabinowitz, M. (1979) J. Biol. Chem. 254, 12461-12470.
31. Sanders, J.P.M., Heyting, C., Verbeet, M. Ph. Meijlink, F.C.P.W. and Borst, P. (1977) Mol. Gen. Genet. 157, 239-261.
32. Church, G.M. and Gilbert, W. (1980) in "Mobilization and reassembly of genetic information" (D.R. Joseph, J. Schultz, W.A. Scott and R. Werner, eds.) Academic Press, N.Y., in press.
33. Slonimski, P.P. (1980) C.R. Acad. Sc. Paris, t.290, 1-4.

The Organization and Expression of the Mitochondrial Genome
A.M. Kroon and C. Saccone, editors

TRANSCRIPTS OF THE OXI-1 LOCUS ARE ASYMMETRIC AND MAY BE SPLICED

THOMAS D. FOX and PAULA BOERNER
Biocenter, University of Basel, CH-4056 Basel, Switzerland

On the basis of DNA sequence analysis[1,2] the oxi-1 locus appears to contain a simple uninterrupted structural gene encoding cytochrome c oxidase subunit II. Transcription of this gene has not yet been carefully investigated, although an 11S RNA species (800 to 1000 nucleotides in length) has been detected which probably corresponds to the mRNA for subunit II[3]. To exploit this simple genetic locus for the study of mitochondrial transcription, we have begun to investigate in detail the mitochondrial RNAs copied from this region of the chromosome. We report here the results of some experiments in which RNA blotted from agarose gels to DBM paper[4] was hybridized to specific DNA probes from the oxi-1 locus and flanking regions. Surprisingly, the results suggest the possibility that this mRNA may be spliced outside of the coding region.

To identify oxi-1 specific transcripts, mitochondrial nucleic acids were blotted from a denaturing (MeHgOH) agarose gel to DBM paper and hybridized to radioactively labeled separated strands of a 105bp HinfI DNA fragment, known to code the N-terminal portion of the protein[1,5] (Fig. 1). The transcribed strand (a) hybridized strongly to an 11S RNA as well as to several minor species ranging in size up to 20S, and to the mtDNA. 11S and 20S transcripts had been previously mapped to this general region[6]. We have observed that the relative amounts of the other minor transcripts varies depending on the yeast strain used. All of the transcripts detected are indeed copied from the oxi-1 region, and not simply cross hybridizing, since they are not detected in the RNA from the oxi-1 deletion mutant M8-171[5] (not shown). The non-transcribed strand (Fig. 1b) hybridized only to the mtDNA, indicating that "anti-message" RNA either is not made or is rapidly degraded (or is present, but electrophoreses with the DNA).

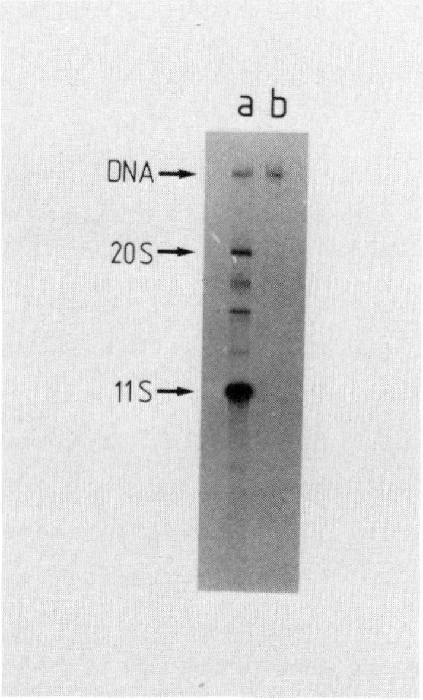

Fig. 1. Total mitochondrial nucleic acids from strain D273-10B were electrophoresed on a 2% agarose gel containing 6 mM methylmercuric hydroxide, and blotted to DBM paper. The paper was cut into strips and hybridized to (a) the transcribed strand (as determined by DNA sequencing) of the 105bp HinfI fragment from the oxi-1 locus (indicated by * in the map of Fig. 2), labeled at the 3'end; (b) the non-transcribed strand of the same fragment.

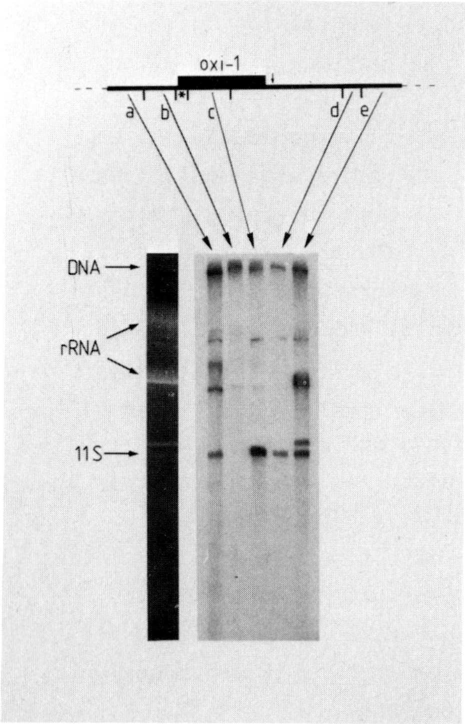

Fig. 2. The top line of the figure represents a 2.5kb HaeIII fragment of D273-10B mtDNA cloned in pBR322. (This fragment is virtually identical to the 2.4kb HpaII fragment described in ref. 5.) Sequences coding structural information for subunit II are indicated by the thickened portion labeled "oxi-1" (753bp in length). The direction of transcription is from left to right. The small arrow next to the coding region indicates the position of the 3' "end" of the mRNA as determined by S_1 nuclease digestion of RNA-DNA heteroduplexes. Restriction fragments labeled a,b,c,d,e, were isolated, radioactively labeled and hybridized to DBM paper strips identical to those of Fig. 1, as indicated. The broken lines represent bacterial DNA and poly-G:C present in probes a and e. A photograph of the ethidium bromide stained gel is shown on the left.

The 11S transcript is the most abundant RNA from this region and hybridizes to all DNA fragments that carry oxi-1 structural gene information (for example, see Fig. 2 c). Presumably it is the mRNA for cytochrome oxidase subunit II. To localize the ends of this RNA, radioactively labeled restriction fragments from around the gene were hybridized to mitochondrial RNA fixed to DBM strips (Fig. 2). Interestingly, the HinfI fragment immediately preceding the coding region (labeled b in the map of Fig. 2) hybridized very weakly if at all to the 11S RNA. However, a restriction fragment further upstream (a) did hybridize with an 11S transcript. One explanation for this result would be that there are two transcripts of very similar size, one hybridizing to restriction fragment (a), the other corresponding to the subunit II mRNA. Another explanation would be that sequences from fragment (a) are spliced onto the mRNA, removing sequences from fragment (b) in the process. In choosing between these alternatives it is relevant to note that the 20S transcript behaves identically to the 11S with respect to hybridization to these restriction fragments.

At the 3' end of the coding region we have mapped the point of divergence between RNA and DNA sequence (or the 3' end of the RNA) to a position about 75 nucleotides beyond translation termination (indicated by the small arrow in the map of Fig. 2), by S_1 nuclease digestion of RNA-DNA heteroduplexes (data not shown). Nevertheless, two restriction fragments further downstream hybridize to 11S transcripts (Fig. 2, d & e), again suggesting the possibility that distal sequences may have been spliced onto the major oxi-1 RNA. Here too however, an independent transcript of similar size has not been ruled out. (Another possibility, that sequences present in the mRNA are repeated outside the gene, has been ruled out by DNA-DNA hybridization control experiments.)

The most distal DNA fragment examined here (Fig. 2 e) also hybridizes to two new RNAs of approximately 12S and 15S, which have not been previously reported. While the 12S species has been detected in all strains examined so far, hybridization at the 15S position (which may represent the small rRNA) was not found with RNA from a strain iso-mitochondrial to 777-3A (not shown).

There is no evidence yet to distinguish whether the higher molecular weight oxi-1 specific RNAs are precursors of the putative mRNA, or simply longer independent transcripts of the region.

We have also compared the approximate levels of the oxi-1 mRNA in cells growing on lactate, galactose and glucose. Mitochondria were isolated from log-phase cells, and their nucleic acids were extracted. Following gel electrophoresis and blotting to paper, the DNA and RNA were hybridized to a labeled oxi-1 specific probe. The relative ratio of oxi-1 mRNA to mtDNA was then estimated by comparing the intensity of hybridization in the 11S band and the mtDNA band in a single track, for each of the growth conditions. No significant difference (greater than a factor of 2) in the relative ratio of oxi-1 mRNA to mtDNA was observed by this technique for these steady-state growth conditions.

ACKNOWLEDGEMENTS

We thank T. Catin for excellent technical assistance. P.B. was supported by a grant from the Sandoz Stiftung zur Förderung der Medizinisch-Biologischen Wissenschaften. This work was supported by a grant from the Schweizerischer Nationalfonds (3.172-1.77).

REFERENCES

1. Fox, T.D. (1979) Proc. Nat. Acad. Sci. USA, 76, 6534.
2. Coruzzi, G. and Tzagoloff, A. (1979) J. Biol. Chem., 254, 9324.
3. Moorman, A.F.M., van Ommen, G.-J.B. and Grivell, L.A. (1978) Molec. Gen. Genet., 160, 13.
4. Alwine, J.C., Kemp, D.J. and Stark, G.R. (1977) Proc. Nat. Acad. Sci. USA, 74, 5350.
5. Fox, T.D. (1979) J. Mol. Biol., 130, 63.
6. van Ommen, G.-J.B., Groot, G.S.P. and Grivell, L.A. (1979) Cell, 18, 511.

© 1980 Elsevier/North-Holland Biomedical Press
The Organization and Expression of the Mitochondrial Genome
A.M. Kroon and C. Saccone, editors

THE SPECIFICATION OF VAR1 POLYPEPTIDE BY THE *VAR1* DETERMINANT

RONALD A. BUTOW, IDA C. LOPEZ, HUI-PING CHANG and FRANCES FARRELLY
The Department of Biochemistry, The University of Texas Health Science Center at Dallas, Dallas, Texas 75235 USA

INTRODUCTION

Over the past few years, our laboratory has been studying a yeast mitochondrial translation product with novel properties, called var1 polypeptide[1-3]. Unlike other mitochondrial translation products which are membrane proteins and are associated with the enzymatic machinery of mitochondrial oxidative phosphorylation, var1 polypeptide is water soluble and is associated with the small mitochondrial ribosomal subunit [4,5]; it is an integral ribosomal protein and behaves on low pH gels as a basic protein not unlike other ribosomal proteins[6,7]. In these respects var1 polypeptide is quite similar to a translation product of *Neurospora* mitochondria which is also associated with the small mitochondrial ribosomal subunit[8]. It is not clear, however, whether other eukaryotes have a mitochondrial translation product like var1.

The most unusual property of var1 polypeptide is its polymorphism detected as mobility differences on SDS-polyacrylamide gels. We have found, or have been able to generate in crosses, at least 15 different var1 species ranging in apparent molecular weight from 40,000 to 44,000[9]. This unusual property raises a number of intriguing questions: What is the molecular basis for the polymorphism at the gene and protein level? Why has the yeast mitochondrial genome retained a mechanism for the fascile generation of polymorphic forms of var1 polypeptide? Are there any metabolic consequences of different forms of the protein? Underlying all of these questions is the expectation that an understanding of var1 will lead to some general principles which would help to clarify the relationship between gene organization and expression. In this connection, a number of facts concerning var1 and its polymorphism have been established. These include the following:

1. The genetic determinant (the *var1* determinant) controlling different polymorphic forms of var1 polypeptide, has been located on mtDNA and physically mapped to the 10th fragment of a Hinc II digest of wild-type mtDNA[10,11].

2. The *var1* determinant specifies a particular form of var1 polypeptide as a result of various combinations within the determinant, of two DNA segments totalling about 100 bp designated *a* and *b*[3,9]. These "inserts" recombine with characteristic frequencies by asymmetric gene conversion in a manner similar to the recombination of ω, a 1 kb intervening sequence located within the 21S rRNA cistron of some yeast strains[12]. Assuming unique positioning, the reassortment of two inserts can only specify four different forms of var1. However, the large number of polymorphic forms of the protein turns out to be due to recombination of *portions* of the *b* insert so that numerous var1 forms can be generated. All var1 forms can unambigously be assigned various genotypes such as a^+b^+, a^+b^-, $a^-b^+_p$ etc. where the notation b^+_p refers to a *var1* determinant which contains a portion of the *b* insert[9].
3. The *var1* determinant region is genetically active and a number of mutants have been isolated in that region giving apparent *syn*⁻ and *mit*⁻ phenotypes[13]. Some of the mutants show complex phenotypes dependent upon the nuclear background.
4. From DNA sequencing data[14], at least 90% of the *var1* determinant region consists of stretches of dA+T with most of the dG+C's located within GC "clusters"[15]. Hence, it is unlikely that the *var1* determinant contains the bulk of the var1 structural gene.

To understand how the *var1* determinant acts at the molecular level to specify the apparent molecular weight of var1 polypeptide, it is essential to know the relationship of the determinant region and the structural gene sequence for the protein. In this report, we summarize some of the approaches we have taken and the results obtained which bear on the resolution of these important questions. In particular, we have carried out experiments to examine, 1) control by *var1* in *trans*, 2) the transcripts associated with *var1*, 3) primary structural differences between different forms of var1 polypeptide and 4) methods for analyzing intermolecular interactions which may be useful not only in understanding the control of var1 forms, but also other gene interactions along the mitochondrial genome.

RESULTS AND DISCUSSION

The *Var1* Determinant

Genetic and physical mapping studies have succeeded in locating the *var1* determinant to Hinc II fragment 10 located between the *ery* and *oli-1* markers (Fig. 1)[2,10,11]. These mapping studies, which included the use of defined petites in transmission (recombination) and zygotic gene rescue[16] analysis, were based on the mobility polymorphism of var1 polypeptide determined on SDS-polyacrylamide gels. Thus, the *var1* determinant region was defined as that locus which specifies the apparent molecular weight of var1 polypeptide.

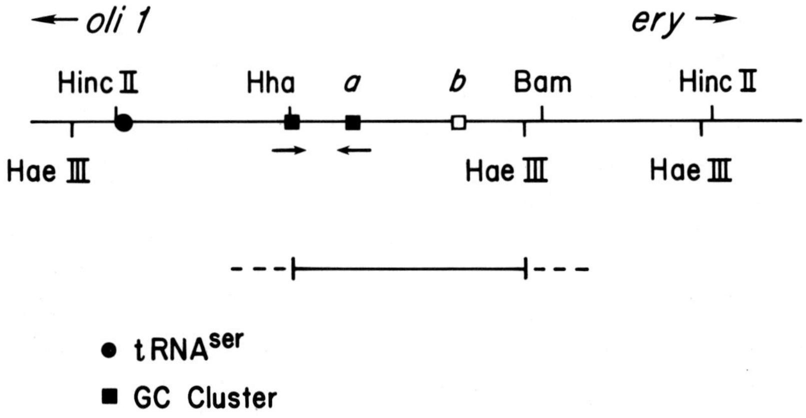

Fig. 1. The *var1* determinant region. Shown is wild-type Hinc II band 10 and a portion of the flanking sequences. The sites labeled *a* and *b* represent the approximate location of the inserts within the *var1* determinant (indicated by the lower line) which correlate with different apparent molecular weight forms of var1 polypeptide.

Recently, we have made use of additional discriminating petites which have sharpened our estimates of the location of the *var1* determinant within Hinc II band 10[10]. This is shown in Fig. 1 as the region extending from the left of the Hha I site rightwards to the first Hae III site toward *ery*. Importantly, this DNA segment encompasses the region containing the *a* and *b* inserts[3], which in various combinations specify different apparent molecular weight forms of var1 polypeptide[3,9]. From fine structure restriction endonuclease mapping, it appears that the *a* insert is a complementary inverted repeat of the 40-50 bp GC cluster around the Hha I site in Hinc II band 10 (Fig. 1); the *b* insert has not yet been defined except for its approximate location within the 1kb Hha-Hae III fragment. The DNA sequence of this 1kb segment has been reported by Tzagoloff *et al*[14] for strain D273-10B, (*var1* $a^{-}b^{+}_{p}$) and preliminary confirmation of a large part of that sequence in a *var1* a^-b^- strain has been obtained by Grossman and coworkers [17]. Because of the extremely high dA+T content of this DNA, it is unlikely that the bulk of the *var1* structural gene is located within the *var1* determinant region.

To understand the molecular basis for the control of var1 polymorphism by the *var1* determinant, we have modified the zygotic gene rescue technique[16] to allow one to look specifically at the mitochondrial translation products produced in zygotes. The procedure is based on the fact that a double auxotrophic yeast strain requiring both methionine and cysteine for growth is unable to incorporate $^{35}SO_4^=$ into protein. If, however, this $SO_4^=$-negative ρ^+ strain is mated to a *met*$^+$ *cys*$^+$ petite, and *in vivo* labeling is carried out with $^{35}SO_4^=$ in the presence of cycloheximide, only the zygote can incorporate label into mitochondrial translation products. This is shown in Fig. 2 which compares the incorporation of $^{35}SO_4^=$ and ^{35}S-methionine into mitochondria isolated at various times from a mating reaction between a *met*$^+$ *cys*$^+$ petite and a ρ^+, $SO_4^=$-negative tester. ^{35}[S]-methionine incorporation into mitochondria is relatively unchanged during the mating, while $^{35}SO_4^=$ incorporation increases dramatically between two and five hours; the only cell type to incorporate $^{35}SO_4^=$ is the

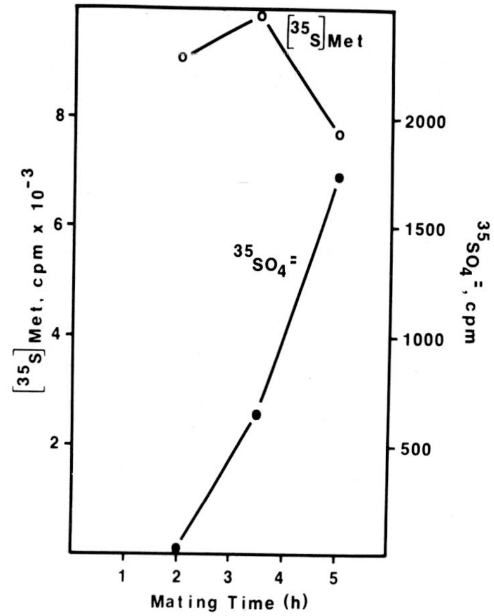

Fig. 2. Kinetics of $^{35}SO_4^=$ and ^{35}S-methionine incorporation into mitochondria. A ρ^+ *met*$^-$ *cys*$^-$ strains was crossed into a *met*$^+$ *cys*$^+$ ρ^-. At various times during mating, cells were labeled in the presence of 600 µg/ml cycloheximide with either $^{35}SO_4^=$ or ^{35}S-methionine. Mitochondria were then prepared and their ^{35}S-radioactive content determined.

zygote. We have taken advantage of this preferential labeling to extend our gene rescue analysis of petites containing the *var1* determinant region. When mitochondrial translation products are analyzed on gels, we find that even as early as 3-3.5 hrs mating, the predominant var1 species made in zygotes is the form specified by the petite *var1* allele; at 5 hrs mating as much as 75% of the total var1 synthesized is the petite-specified form[18]. Considering that cytoplasmic mixing in zygotes is a slow process and not all mitochondria in the zygote fuse[19], these results are particularly striking and strongly suggest that the petite *var1* determinant allele acts in *trans* and effectively competes out the ρ^+ *var1* determinant. We believe this conclusion has important implications in the mechanism by which the *var1* determinant specifies the form of var1 polypeptide.

We should also point out that the procedure of selective labeling of mitochondrial translation products in zygotes should be of some general utility especially if complementing or *trans* interactions between mitochondrial genomes occur which do not result in the restoration of some readily assayed biochemical activity like respiration or the appearance of a cytochrome spectrum.

Var1 Polypeptide

Previous data showed that the electrophoretic mobility of different forms of var1 polypeptide appeared "normal" on SDS-polyacrylamide gels[4]. Peptide mapping studies also showed extensive homologies, although a few fragments could be detected which were unique to a particular var1 form[4]. We have extended our analysis of different var1 species and the following preliminary conclusions can be drawn:

1. Different var1 species all contain a blocked sulfur amino acid at the N-terminus. We presume this to be N-formyl methionine. If so, this would indicate that different forms of var1 polypeptide are not derived from each other as a result of N-terminal post-translational processing.
2. Different var1 forms do not show the same distribution of sulfur-containing amino acids within the first 20 residues at the N-terminus. We do not know if other regions of difference exist elsewhere in the molecules.
3. We cannot detect any major net charge differences between different forms of var1 polypeptide. These preliminary findings provide some insight into the molecular basis for var1 polymorphism. Further chemical characterizations of purified var1 species are in progress.

Transcriptional Mapping

Van Ommen *et al*[20] have carried out a detailed analysis of the transcriptional origin of mitochondrial RNAs in yeast. Transcription of a number of regions of the mitochondrial genome appears to be complex in that multiple overlapping RNA species can be identified. From the analysis of the hybridization of *in vivo*-labeled mitochondrial RNA and cRNA probes, these investigators identified two overlapping RNA species of 17.5S and 19S showing sequence homology to a 2.3 kb fragment encompassing the *var1* determinant region.

We have examined transcripts showing sequence homology to the *var1* region in a number of wild-type and petite strains with different alleles of the *var1* determinant. A summary of our transcriptional mapping studies is shown in Fig. 3 and summarized below:

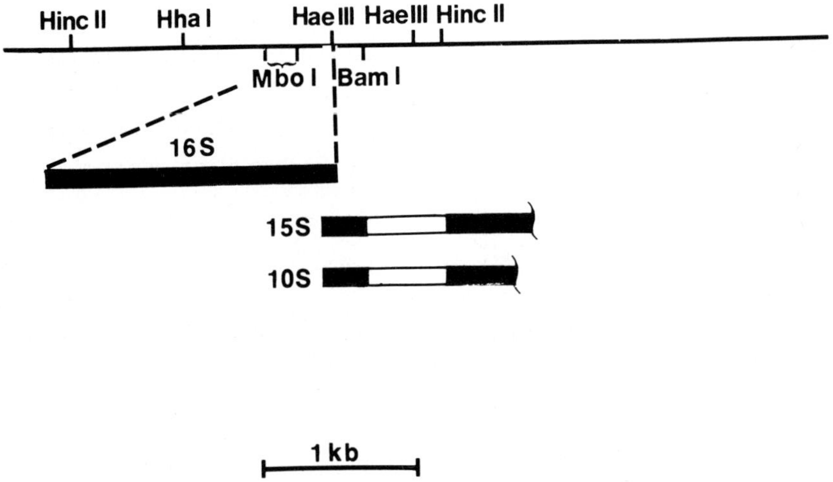

Fig. 3. Transcriptional mapping of the *var1* region. Major mitochondrial RNA species showing sequence homology to segments of the *var1* determinant region and flanking sequences are indicated by the solid bars.

1. A 16S RNA species 1900-2000 nucleotides long shows strong sequence homology to a 250 kb MbO I-Hae III fragment within the *var1* determinant; no other DNA fragment within that region or flanking sequences show strong homology to this species. The 16S RNA varies in size in a strain-dependent manner as does varl polypeptide. For example, it is about 100-200 nucleotides longer in strains with a 44,000 molecular weight form of varl than in a strain with a 40,000 molecular weight form of the protein. This RNA species may be the same as the 17.5S RNA identified by Van Ommen *et al*[20].

2. We find a minor RNA species about 2300-2400 nucleotides long which shows sequence homology to the *var1* region but a fine structure map of that transcript has not yet been obtained.
3. A major RNA species which is indistinguishable from the 15S rRNA and appears to be present in at least the same mass amount as the 15S rRNA, has sequence homology to a portion of Hinc II-bands 10 and 7. Overlapping this 15S RNA is a 10S species which is probably a derivative of the 15S. The 10S RNA also hybridizes to regions of the mitochondrial genome containing the 15S rRNA gene. From these results we tentatively conclude that a portion of Hinc II-bands 7 and 10 contain sequences homologous to the small mitochondrial rRNA and to a 10S derivative of that RNA.

Analysis of Mitochondrial RNAs by Two-Dimensional Gel Electrophoresis

Complementation studies on mutants within the *cob-box* locus suggest that intra- and intermolecular RNA interactions may be important in the regulation of mitochondrial gene expression through control of RNA processing[21]. *Trans* control of var1 polypeptide through the *var1* determinant locus raises the possibility that var1 polymorphism may also involve some types of RNA interactions.

Recently, a small nuclear RNA species present in higher eukaryotes has been shown to have a striking complementary sequence homology to a consensus sequence across the splice junction of mRNAs[22,23]. It has been speculated that these and other types of RNAs could serve as guides to direct RNA splicing[24].

As an approach to detect and study putative interactions among transcripts of the yeast mitochondrial genome in wild-type and in mutant cells, and especially in zygotes where different mitochondrial genomes are placed in *trans* configuration, we have been developing a two-dimensional RNA gel electrophoresis procedure. Briefly, "native" RNAs are electrophoresed on a first-dimension agarose gel, then denatured *in situ* with formamide, glyoxal, etc. and then run in a second dimension on agarose-urea slab gels. The RNA is then blotted onto DBM paper[25] and the immobilized RNAs detected with the appropriate P^{32}-labeled DNA probes from defined segments of the mitochondrial genome. Figure 4 shows an example of a two-dimensional gel of RNA from a wild-type strain visualized by ethidium bromide staining and with a DNA probe from the *var1* determinant region. The major mitochondrial RNA species are well resolved in this system. By hybridization with the *var1* probe, the predominant RNA species (16S, 15S, and 10S) with sequence homology to this region can be seen. A comparison of the ethidium bromide stained gel and the autoradiogram show that the 15S RNA hy-

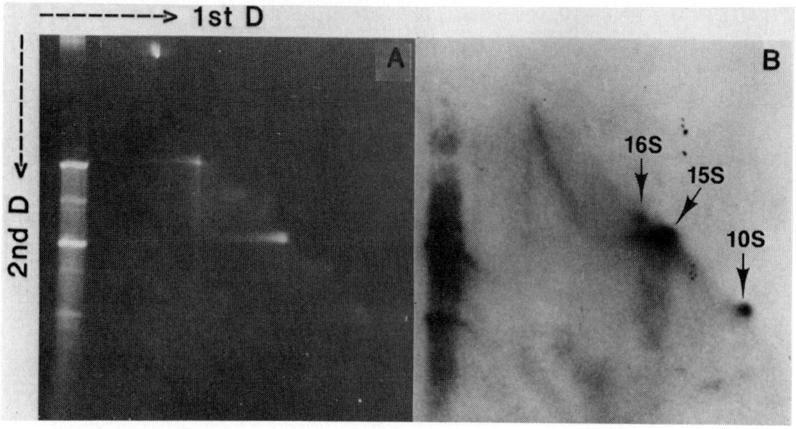

Fig. 4. Electrophoresis of mitochondrial RNAs on two-dimensional gels. Approximately 2 µg of RNA in a volume of 10 µl was loaded on a cylindrical gel (0.25 x 10 cm) composed of 0.5% agarose, 1.44% acrylamide and 0.16% bisacrylamide and electrophoresed at room temperature at 100 v. The gel was then soaked in 75% formamide at room temperature for 10 min followed by 42°C for 10 min, and loaded on a 1.5% agarose 6 M urea slab gel (13 x 13 x 0.3 cm) for the second dimension. An equal amount of RNA was treated identically and run in a separate slot on the same gel. The agarose-urea gel was electrophoresed at room temperature at 25V overnight. After electrophoresis the slab gel was then blotted on to DBM paper[25] for hybridization with a *var1* DNA probe.

bridizing to the probe is not resolved from the 15S rRNA. A detailed analysis of the types of results which could be obtained will be presented elsewhere. However, an obvious possibility with important implications would be RNAs which fall below the diagonal in the second dimension, and particularly, those which appear in vertical arrays. These species would be good candidates for specific intermolecular complexes.

CONCLUDING REMARKS

From our analysis of discriminating petites which separate *var1* recombination from gene rescue in ρ^- x ρ^+ crosses, we can now locate the *var1* determinant region to a DNA segment within Hinc II band 10 extending from the left of an Hha I site to the first Hae III site towards *ery*. We have identified a number of RNA species which contain sequences homologous to this region. One transcript, estimated to be 1900-2000 nucleotides long, varies in size in a strain dependent manner as does var1 polypeptide. This transcript shows strong sequence homology only to a small portion ($\sim$ 250 bp) of the *var1* determinant region and to no other flanking sequences; thus, the bulk of this RNA species must be encoded elsewhere, and perhaps spliced into a *var1* transcript. The 16S "variable" species may be derived through processing from a 19S transcript present in minor amounts in wild-type cells.

Since some part or all of the DNA segment encompassing the *var1* determinant is required to specify var1 polypeptide in *trans* it seems reasonable to consider that the 16S RNA is either the molecular species which determines the particular form of var1 that is expressed (i.e. is a regulatory RNA) or, alternatively, is actually the mRNA for the protein. In either case, it would appear that a transcript of the *var1* region is spliced, perhaps through some intermolecular reaction to another RNA species. The latter could be located elsewhere on the mitochondrial genome or possibly even be of nuclear origin. There is also the possibility that the 16S RNA may not contain any transcript of the *var1* region but rather, have only some sequence homologous to it. Further characterization of the 16S RNA should help to decide among these possibilities. It is of interest to note that 5' splicing to give functional mRNAs has been well documented, for example, in the processing of adenovirus mRNAs[26]. Moreover, apparent intermolecular splicing between most transcripts and influenza virus RNAs seems to occur to generate functional mRNAs[27]. Thus, both intra- and intermolecular RNA recombinations are used to generate functional mRNAs.

Two other RNA species, a 15S and 10S appearing as overlapping transcripts, also contain sequences homologous to a small portion of the *var1* determinant region and to sequences near the *var1* determinant towards *ery* extending into Hinc II band 7. Under a variety of denaturing conditions and using either one or two dimensional gel electrophoresis, we have not been able to distinguish the 15S RNA species from the 15S rRNA. We have also observed hybridization of the 15S and 10S RNA to other regions of the mitochondrial genome including sequences containing the 15S rRNA. Possibly, the sequences distributed along the mitochondrial genome with homology to the 15S rRNA (and the 10S species) might represent common recognition sequences for ribosome binding and initiation of protein synthesis, as seem to be the case in prokaryotic protein synthesis[28].

Acknowledgements

This work was supported by Grants from the USPHS GM 22525, GM 26546 and Grant I-642 from The Robert A. Welch Foundation.

REFERENCES

1. Douglas, M.G. and Butow, R.A. (1976) Proc. Natl. Acad. Sci. USA 73, 1083-1086.

2. Perlman, P.S., Douglas, M.G., Strausberg, R.L. and Butow, R.A. (1977) J. Molec. Biol. 115, 675-694.
3. Strausberg, R.L., Vincent, R.D., Perlman, P.S. and Butow, R. A. (1978) Nature 226, 577-583.
4. Terpstra, P., Zanders, E. and Butow, R.A. (1979) J. Biol. Chem. 254, 12653-12661.
5. Groot, G.S.P., Mason, T.L. and Van-Harten-Loosbrook, N. (1979) Molec. Gen. Genet. 174, 339-342.
6. Gorenstein, C. and Warner, J.R. (1976) Proc. Natl. Acad. Sci. USA 73, 1547-1551.
7. Collatz, E., Lin, A., Stoffler, G., Tsurugi, K. and Wool, I. (1976) J. Biol. Chem. 251, 1808-1816.
8. Lambowitz, A.M., Chua, N.H., and Luck, D.J.L. (1976) J. Mol. Biol. 107, 223-253.
9. Strausberg, R.L. and Butow, R.A., Submitted for publication.
10. Vincent, R.D., Perlman, P.S., Strausberg, R.L. and Butow, R.A. Current Genetics, In press.
11. Butow, R.A., Vincent, R.D., Strausberg, R.L., Zanders, E. and Perlman, P.S. (1977) in Mitochondria 1977: Genetics and Biogenesis of Mitochondria (Bandlow, W., Schweyen, R.J., Wolf, K. and Kaudewitz, F. eds) Walter de Gruyter, Berlin pp. 317-335.
12. Bos, J.L., Heyting, C., Borst, P., Kornberg, A.C. and Van Bruggen, E.F. J. (1978) Nature 275, 336-338.
13. Zassenhaus, P. and Perlman, P.S. Unpublished observations.
14. Tzagoloff, A., Macino, G., Nobrega, M.P., and Li, M. (1979) in Extrachromosomal DNA: ICN-UCLA Symp. on Molecular and Cellular Biology (Cummings, D., Borst, P. Dawid, I., Weissman, S. and Fox, C.F., eds.) Academic Press, New York pp. 339-355.
15. Prunell, A. and Bernardi, G. (1977) J. Mol. Biol. 110, 53-74.
16. Strausberg, R.L. and Butow, R.A. (1977) Proc. Natl. Acad. Sci. USA 74, 2715-2719.
17. Grossman, L. and Hudspeth, M.I., Unpublished results.
18. Lopez, I.C., Unpublished results.
19. Strausberg, R.L. and Perlman, P.S. (1978) Molec. Gen. Genet. 163, 131-144.
20. Van Ommen, G.J.B., Groot, G.S.P. and Grivell, L.A. Cell (1979) 18, 511-523.
21. Church, G.M., Slonimski, P. and Gilbert, W. (1979) Cell 18, 1209.
22. Lerner, M.L., Boyle, J.A., Mount, S.M., Wolin, S.L., and Steitz, J.A (1980) Nature 283, 220-224.

23. Rogers, J. and Wall, R. (1980) Proc. Natl. Acad. Sci. USA 77, 1877-1879.
24. Murry, V. and Holliday, R. (1979) FEBS Letters 106, 5-7.
25. Alwine, J.C., Kemp, D.J. and Stark, G.P. (1977) Proc. Natl. Acad. Sci. USA 74, 5350.
26. Berget, S.M., Moore, C. and Sharp, P.A. (1977) 74, 3171-3175.
27. Krug, R.M., Brom, B.A. and Bouloy, M. (1979) Cell 18, 329-334.
28, Shine, J. and Dalgarno, L. (1975) Nature 254, 34-38.

© 1980 Elsevier/North-Holland Biomedical Press
The Organization and Expression of the Mitochondrial Genome
A.M. Kroon and C. Saccone, editors

THE NUCLEOTIDE SEQUENCE OF THE tsm8-REGION ON YEAST MITOCHONDRIAL DNA

WOLFHARD BANDLOW[1,2], ULRIKE BAUMANN[1] and PETER SCHNITTCHEN[1]

1) Genetisches Institut der Universität München, Munich, Germany
2) Institut de Biologie Moléculaire, Université Paris VII, Paris, France

INTRODUCTION

The mitochondrial genome of *Saccharomyces cerevisiae* codes for at least two ribosomal RNAs, about 25 tRNAs and 8 major polypeptide species (1). In addition, mutational sites have been described which clearly map outside genetically characterized loci of known genes. Among these are the loci P (conferring resistance to paromomycin of the small ribosomal subunit,2),tsm8 (causing partial shut down of mitochondrial gene expression after a temperature shift, 3) and some other less defined mutations, 4). In none of the cases a gene product has been described.

RESULTS AND DISCUSSION

In the present study a variety of mit^- markers mapping in the C to oxi1 segment of mitochondrial DNA was used to select and characterize rho^- clones which have retained the marker tsm8 but have lost most of the flanking markers (Fig. 1A). One of these strains, PM 1C6573, containing the markers tsm8 and 982ts (5) has been restriction mapped (Figs. 1B and 2) and partially sequenced. From a comparison of the restriction maps of PM 1C6573 *tsm8*$^+$ *982ts*$^+$and BS 52 *tsm8*$^+$ *982ts*$^+$ (6) with PM G331 *tsm8*° *982ts*$^+$ it is evident that the DNA sequence of the latter is completely contained in that of the former two (Fig. 1B). Since the junction fragment of BS 52 is 550 bp long and the AluI5ooHaeIII fragment of PM 1C6573 contains mainly AT-rich sequences whith the exception of the first 55 bp to the right of the Alu site, the locus tsm8 is localized in a segment extending from about 55o bp to the left of the AluI / MboII cluster to 55 bp to its right. The nucleotide sequence of this segment is given in Fig. 3.

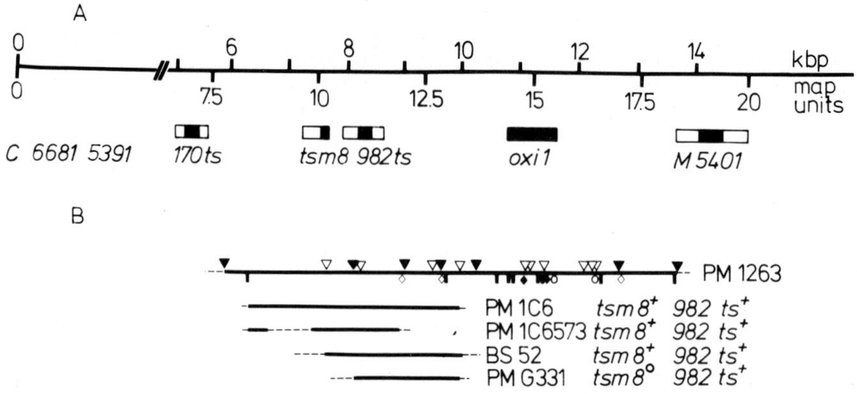

Fig. 1A. Selection of rho⁻ petite clones carrying genetic markers of the C to oxi1 region. Strain SM 1313-1C was used as the parent. Rho⁻ petites were induced with ethidium bromide. 8 oxi1 marker alleles have been tested.
Fig. 1B. Comparison of restriction maps of PM 1C6573 with PM H1263, BS 52 and PM G331. The restr-ction data of strain BS 52 are taken from ref. 6. Restriction sites for ▽ AluI, ▼ HaeIII, | Hin fI, ○ MboI, ◆ PvuII, ◇ TacI.

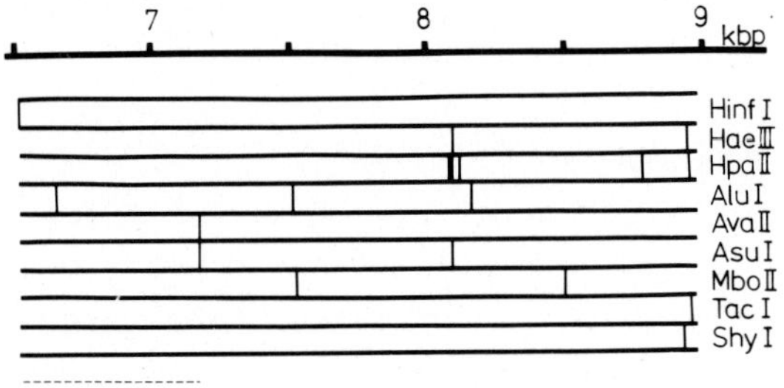

Fig. 2. Fine structure restriction map of strain PM 1C6573. The strain was isolated from a rho⁻ petite, PM 1C6, after ethidium bromide treatment. 23 restriction nucleases have been used, some of which do not cut. ---- junction fragment.

In its central section, to the left and right of the AluI / MboII cluster (position 34, in italics) it exhibts a GC-content of 42% which is higher than in all mitochondrial genes sequenced so far (*e.g.* 8-11), but only about 90 bp long.

```
-393                                              -350
TTCGAATTCTATATATATAATATTATATAATATAATATATATA ATATTAGGTGTTAGTTAATAAT

                                         -300             A
ATTGCTTATTATATATAAATAAGATATG CTATAGCTCCTTCGGGAGGTCCAGACCCCCCTCCCCT
                                                  Ava II
           -25o                                                        -200
GCGGGAGGGGAGC GCTTTTATTAAAAATATTATAATTAAATAATAATATAAATAATTTATAAT A

                                                              -15o
TAACAATATATACTTATAAATAATATTAACTTATATAATTATAACTATA ATTATTATTAATAAAT

                                          -100
CCGTGGTAAGGACTTTATATTTATATTATATATA TATGGATATATAAACTTTAAATAGGATAAAT

                          -50
ATATTATAATAATTATATAAA TATATATATATTATATTAGGATAATAATATATATATATATTAAT

        1                                                           50
ATATA GGAGGGGATTTTCAATGTTGGTGGTTGAGTTGAGCTGTTAAACTCAATTGA CCTAGGTC
                                      Alu I
                                       1oo
TTCCTAGGTTCCATTCCTATTCCCTTCCTAAATAATTTATTA TTAATTATATATTATTATAATCC
Mbo II                                       ochre
                                 150
ATTAGAATTAATATAATCCATAGATAA TTAATTTAATACACAATTTAATATATAGATTATATATA

          2oo                                                        250
TATATACCTTAT TAAAAAAAAATTATATAATAATTATATTAATATATTTATATATATAAATT AA
```

Fig.3. Partial nucleotide sequence of strain PM 1C6573. The sequence was determined from fragments resulting from digestions with Alu I, Mbo II and from a combination of Ava II and Hae III. 5′-labelled ends were separated by secondary cleavage and electrophoresis on low melting agarose. Partial base specific chemical cleavage was performed according to (7). The sequence of the strand assumed to be non-coding is given in 5′ to 3′ direction. The GC-rich stretch under consideration extends from position 1 to position 86.

Three reasons for the shortness of this sequence may be considered:

i) it codes for a tRNA. Supporting evidence for this possibility is given by the fact that the stretch is rich in inversely repeated complementary sequences which permit base pairing in the form of a clover leaf structure. The resulting tRNA is 89 bases long and has in its anticodon loop exceptionally 8 unpaired bases with GUU as the anticodon for asparagin (position 47 in Fig. 3). AsN-tRNA maps close to tsm8 (5).

ii) the GC-rich sequence is part of a gene coding for a polypeptide, but the junction of PM 1C6573 coincides with the left rim of the

GC-rich stretch. Although the junction lies within the Alu 850 Alu fragment, it appears to be left of the isolated Ava II site,since this site is present at the same place both in the wild type and in strain PM H1263 (see Fig. 2).

iii) the AT-rich stretch to the left of the GC-rich sequence is part of an intron within a polypeptide coding gene which is excized from the primary transcript during processing events. Although intron/exon junctions have not yet been defined on mitochondrial DNA, introns apparently do not consist of only AT. The open reading frame found when taking the TAA triplett at position 91 (Fig. 3) as the translational stop codon could support either of the latter two possibilities. (The other five frames contain internal stops). Transcript studies combined with the identification of the mutational alteration in *tsm8*$^-$ are underway and should permit the decision between the three alternatives and thus clarify the question whether the tsm8-mutation lies within AsN-tRNA.

ACKNOWLEDGEMENTS

W.B. thanks C. Gaillard for introduction into nucleotide sequencing and G. Bernardi for hospitality during an EMBO fellowship. This work was further supported by the Deutsche Forschungsgemeinschaft (grant Ba 415/15).

REFERENCES

1. Borst, P. and Grivell, L.A. (1978) Cell 15, 705-723.
2. Kutzleb, R., Schweyen, R.J. and Kaudewitz, F. (1973) Molec.Gen. Genet. 125, 91-98.
3. Bandlow, W. Metzke, R., Klein, A., Kotzias, K., Doxiadis, I., Bechmann, H., Schweyen, R.J. and Kaudewitz, F. (1977) Europ. J. Biochem. 76, 372-382.
4. Mahler, H.R., Hanson, D., Miller, D., Lin, C.C., Alexander, N.J., Vincent, R.D. and Perlman, P.S. (1978) Biochemistry and Genetics of Yeasts (Bacila, M., Horecker, B.L. and Stoppani, A.O.M., eds.) Academic Press, New York, pp. 513-547.
5. Coletti, E., Frontali, L., Palleschi, C., Wesolowski, M. and Fukuhara, H. (1979) Molec. Gen. Genet. 175, 1-4.
6. Wesolowski, M. (1979) Thesis, Université de Paris Sud, Orsay.
7. Maxam, A.M. and Gilbert, W. (1977) Proc. Natl. Acad. Sci. 74, 560-564.
8. Li, M. and Tzagoloff, A. (1979) Cell 18, 47-53.
9. Hensgens, L.A.M., Grivell, L.A., Borst, P. and Bos, J.L. (1979) Proc. Natl. Acad. Sci. 76, 1663-1667.
10. Fox, T.D. (1979) Proc. Natl. Acad. Sci. 76, 6534-6538.
11. Tzagoloff, A., Nobrega, M., Akai, A. and Macino, G. (1980) Current Genet., in the press.

© 1980 Elsevier/North-Holland Biomedical Press
The Organization and Expression of the Mitochondrial Genome
A.M. Kroon and C. Saccone, editors

FURTHER CHARACTERIZATION OF RAT-LIVER MITOCHONDRIAL DNA

C. SACCONE[1], P. CANTATORE[2], G. PEPE[1], M. HOLTROP[3], R. GALLERANI[1], C. QUAGLIARIELLO[2], G. GADALETA[1] and A.M. KROON[3].
[1]Istituto di Chimica Biologica, Università di Bari, Via Amendola 165/A, 70126 Bari, Italy; [2]Dipartimento di Biologia Cellulare, Università della Calabria, Arcavacata di Rende, Cosenza, Italy and [3]Laboratory of Physiological Chemistry, State University, Bloemsingel 10, 9712 Kz, Groningen, The Netherlands.

INTRODUCTION

In previous publications of our groups[1-4] we have presented various results of experiments on the gene organization of rat-liver mitochondrial (mt) DNA. The physical map of the rat-liver mtDNA was constructed using about twelve different restriction enzymes. Mitochondrial genes were localized by hybridization of labeled ribosomal and transfer RNA species to stripfilters of DNA fragments. Total mtRNA, labeled _in vitro_ with ^{125}I-iodine was also hybridized to DNA fragments in the presence of excess non-labeled mt-rRNAs and mt-tRNAs. From these experiments the location of potential messenger RNAs was provisionally inferred. The results obtained were, on the whole, in good agreement with the data from other authors, who studied the gene organization of mtDNA of animal cells. The preliminary studies allowed us to pick up the main features of the organization of the rat-liver mitochondrial genome, which can be summarized as follows: 1. The ribosomal RNA genes are adjacent and separated by a region of only 200 basepairs, containing a tRNA gene; 2. the genes for the tRNAs are scattered over the genome; 3. the messenger RNA region is rather condensed, leaving the area of the rRNAs and the D-loop and the regions directly flanking this area, free of information for messenger RNAs. In this paper we will describe the results of a more detailed examination of parts of the mtDNA of rat liver.

MATERIALS AND METHODS

Cloned fragments of rat-liver mtDNA were prepared by digestion of the respective recombinant plasmids with Eco R1. The plasmid

DNAs were prepared from E. coli cells using the following procedure. The cells were incubated in the cold in presence of 2 mg/ml lysozyme, 65 mM EDTA and were lysed at a final concentration of Triton X-100 0.05%. The suspension was centrifuged 1 h - 35,000 rpm in 50 Ti Spinco Rotor at 0°C; the supernatant was extracted once with an equal volume of phenol-chloroform (1:1 v/v). The nucleic acids were precipitated with ethanol and collected by centrifugation after 16 h standing in the cold. Contaminating RNA was digested with the aid of T1-and pancreatic ribonuclease. After a second phenol/chloroform extraction the DNA was again precipitated with ethanol. The resulting precipitate was dissolved in TE buffer (10 mM Tris-HCl, 0.1 mM EDTA, pH 7). Plasmid DNA was then isolated by CsCl-ethidiumbromide centrifugation (60 h; 40,000 rpm in 50 Ti Spinco rotor at 23°C). The plasmid DNA collect after the run was extensively dialyzed against TE buffer and concentrated by ethanol precipitation. Restriction endonuclease digestions were carried out by using the standard conditions for each enzyme. Restriction enzyme fragments were separated either on polyacrylamide gels or agarose gels, depending on their size. The fragments were eluted from the gels by buffer extraction and concentrated by ethanol precipitation. All further methods have been described in detail previously[1-4].

RESULTS AND DISCUSSION

The heterogeneity of rat-liver mtDNA

One single rat has only one type of mtDNA; the mtDNA in the various organs and tissues is the same. However, within a strain of rats more than one type of mtDNA may exist. In total 4 types of rat mtDNA have been observed, types A-D[5]. In our Wistar strain the types A and B occur. We have shown the strictly maternal and cytoplasmic inheritance of the mtDNA-character[6]. We furthermore analysed the type differences of a small fragment of the mtDNA. It appears that the differences which were originally observed as variations in the restriction fragment patterns are based an point mutations[4]. There is no indication so far for type differences based on either deletions and insertions or on nucleotide modifications[7]. The sequence divergence between types A and B is in the order of 2% or less.

An interesting observation was that for the small Hind III fragment F, that was completely sequenced, one of the six reading frames could be decoded in a way that the three mutations of type A as compared to type B are translationally counterbalanced by the degenerancy of the genetic code. In this frame a carboxylterminus was indicated. However, in view of the later observations that the UGA triplet is coding for tryptophane in mitochondria[8], we have to assume that the frame reads through the complete fragment. Base-sequence analysis of the neighbouring fragments will reveal if the sequence is really part of a gene coding for a mitochondrial protein.

The cloning of rat-liver mtDNA

Analysis of the Hind III fragment F was relatively easy since the fragment had a suitable length and could be easily obtained from total rat-liver mtDNA after digestion with one enzyme only. To obtain fragments of other parts of the mtDNA for purpose of base-sequence analysis multiple digestions and reisolations are necessary and the inherent losses of material during these steps were prohibitive for using experimental animals as the starting material. To get a more detailed insight into the genetic function of mtDNA we have, therefore, undertaken experiments with Eco R1 restriction fragments, cloned in plasmids. Recombination of the fragments with the plasmid PSF2124 ($ap^{+}col^{+}$), transfection into E. coli W5445 and initial cloning experiments were performed by Dr. Janine Doly from the group of Dr. G. Bernardi in Paris. With this host/vector system stable transformants were obtained for the Eco R1 fragments A,B,C,D and E. We started our studies with the further characterization and sequence analysis of the fragments D,B and E.

The Eco R1 fragment D

The Eco R1 fragment D is 1950 basepairs (bp) long and it contains the information for the large ribosomal RNA, that has a sedimentation coefficient of 16S. Fig. 1 shows the restriction fragment map of Eco D with sites for the enzymes Hind III, Hpa II, Hha I, Hae III, Hinf I, Taq I and Sau 3a. Hind III produces 2 fragments of 1500 and 450 bp respectively, in a complete double

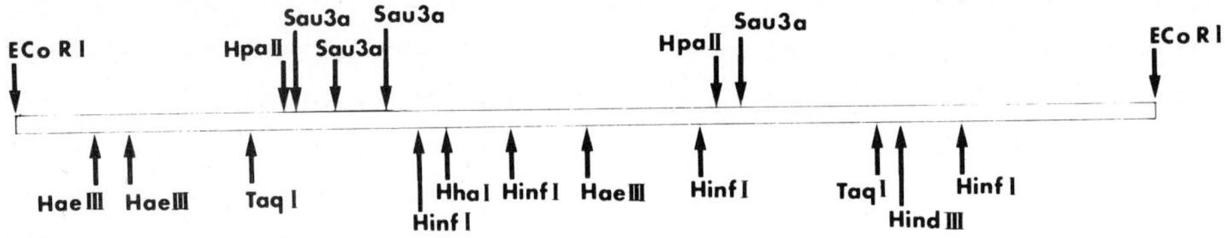

Fig. 1. Fine map of Eco R1 fragment D. At the left is the Eco R1 cleavage site between the Eco R1 fragments D and E, at the right the cleavage site between the Eco R1 fragments D and A. 1 cm corresponds to 115 bp. The fragment was isolated from the recombinant plasmid.

digest of the rat-liver mtDNA previously designated EH_3 and EH_6[1]. The strands of these two fragments were separated on agarose gel, transferred to stripfilters and hybridized to labeled rRNA. As

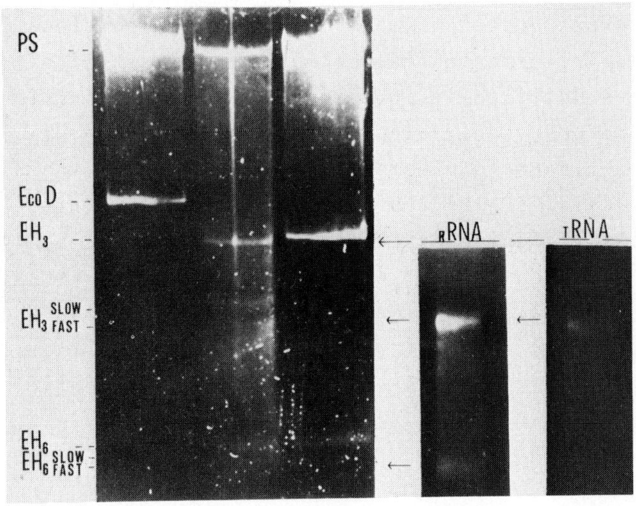

Fig. 2. Hybridization of mt-rRNA and mt-tRNA with separated strands of the Hind III fragments of Eco D (EH_3 and EH_6). At the left is shown a fluorograph of the strands separated on a 1.4% agarose gel after alkaline denaturation. At the right the autoradiograms of the stripfilter hybridizations are shown.

shown in Fig. 2 the rRNA exclusively hybridized to the fast (heavy) strand of both fragments. The ribosomal RNA gene encompasses the Hind III restriction site, its 5' end lying very close to the cleavage site between the Eco Rl fragments A and B. Fig. 2 also shows that the heavy strand of EH_3 contains information for one or two tRNAs[2,3].

With the aid of the enzymes indicated in Fig. 1 we were able to split Eco D into well-defined fragments sufficiently small to determine the nucleotide sequence, using the method of Maxam and Gilbert[9]. The fragments were labeled at either the 3' or the 5' end. The sequence is almost completely known; a few minor internal regions are still lacking. However, some interesting features are obvious.

Comparing the sequence with that of the 5' end of the 23S ribosomal RNA gene of E. coli brings to light two short stretches with homologous base sequence. Due to this homology we were able to locate the 5' end of the 16S rRNA rather precisely. In accordance with our hybridization data the 16S rRNA gene starts very close to the cleavage site between Eco D and Eco A. The homology between the E. coli and rat mitochondrial large RNA genes regards two five-bp blocks, which are shown in Fig. 3. It is interesting

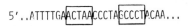

17 S RIBOSOMAL RNA (HAMSTER CELL MITOCHONDRIA)	5'pACUAACUCUAGCCCUCAUUU
16 S RIBOSOMAL RNA GENE (RAT LIVER MITOCHONDRIA)	5'..ATTTTGAACTAACCCTAGCCCTACAA...
23 S RIBOSOMAL RNA GENE (E. COLI CELLS)	5'..GGUUAAGCGACTAACCGTACACGGTGGATGCCCTGGCA..

Fig.3.A comparison of the 5'end of the rat-liver mitochondrial 16S rRNA gene with the large ribosomal RNA sequences from other sources.

to note that the same homology has been observed for the E. coli and hamster mitochondrial rRNAs (Dubin and Baer, these proceedings). We have recently reported base sequence homology between the large ribosomal rRNA genes of mitochondria of Neurospora crassa and rat liver[10]. The differences in length of the 16S rRNA from rat liver mitochondrial ribosomes and the large ribosomal RNAs of E. coli and Neurospora mitochondria is considerable. Further sequence analysis will show up other regions with a high degree of conservation. It is tempting to speculate that the conserved sequences may have something to do with the conformation of the ribosomal peptidyltransferase, which shows similar sensitivity to inhibition to e.g. chloramphenicol in all mitochondrial and bacterial ribosomes studied so far.

From electronmicrographs of hybrids between 16S rRNA and separated strands of Eco D the length of the 16S rRNA was calculated to be about 1500 nucleotides. Firm indications for the presence of intervening sequences within the 16S rRNA gene were not obtained.

The Eco R1 fragment B

The Eco R1 fragment B was chosen for further analysis because it should contain information for one or more polypeptides synthesized within the mitochondria and for 5 or 6 tRNAs.

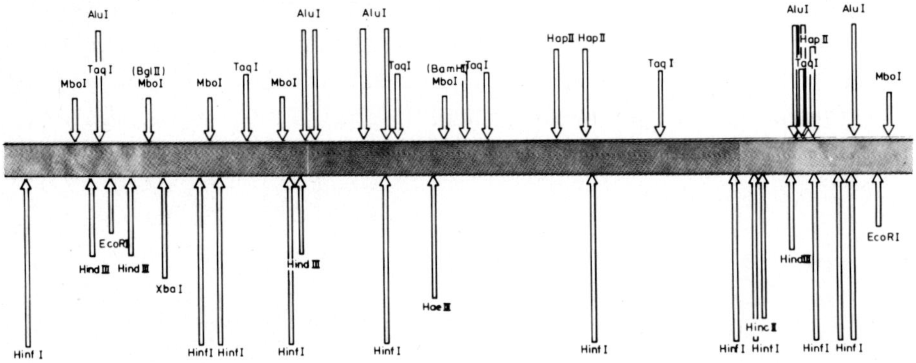

Fig. 4. Fine map of Eco R1 fragment B. At the left the Eco R1 cleavage site between Eco B and Eco A, at the right the site between Eco B and Eco C. 1 cm corresponds to 255 bp. For Hae III, Hinf I and Alu I not all sites are indicated.

Fig. 4 shows the restriction fragment map of Eco B with sites for the enzymes Hind III, Hinc II, Hinf I, Hap II, Taq I, Mbo I, Xba I, Hae III and Alu I. The Mbo I sites sensitive to Bgl II and Bam HI are also indicated. With respect to Mbo I it should be mentioned that plasmid DNA is methylated rendering the Eco B fragment obtained from the plasmid resistant to digestion with Mbo I. Instead Sau 3A was used. For Hae III, Hinf I and Alu I not all restriction sites have been localized. Base sequence analysis has as yet insufficiently progressed to draw conclusions about the genetic function of the fragment. By analysis of the fragments obtained by digestion of Eco B with Hinc II and labeled in the Eco R1 site, it could be shown that the Taq I restriction site is 640 bp away from the Eco R1 site between Eco A and Eco B. This means that the preliminary orientation of the sequenced fragment Hind III F was correct. The two read-through frames of this fragment, therefore, start in Eco B. From the further analysis it is clear that only the reading frame in which the differences of the types A and B are not expressed as differences in the amino acid sequence has a AUG-start codon in the Eco B region sequenced, whereas in the other frame neither a stopcodon or a startcodon has been detected yet (Fig. 5). The direction of transcription is from

```
Eco R1-C ....... 5' G T G T[T A A]T G G G C G T A T[T A G]C C T C A G T C T[A T G]

C T G C G C G T C A T C G C C A C[A T G]T G T[A T G]C C G T C A C A C G C C A T T
                                  MET - CYS - MET - PRO - SER - HIS - ALA - ILE -

A[T A A]T C G A C A C C A T A A G C T T C A C G G G A G C C A C A A T A C[T A A][T A
 ILE - ILE - ASP - THR - ILE - SER - PHE - THR - GLY - ALA - THR - ILE - LEU - ILE
                                        ●
A]T C G C C C A C G G C T[T A A]C C T C A T C A C T C T T A T T C T G C C[T A G]C A
 ILE - ALA - HIS - GLY - LEU - THR - SER - SER - LEU - LEU - PHE - CYS - LEU - ALA
                                        ●
A A C A C C A A C T A C G A A C G A A T T C A C A G C C G A A C T A[T A A]T T A[T A
 ASN - THR - ASN - TYR - GLU - ARG - ILE - HIS - SER - ARG - THR - ILE - ILE - ILE
  ●
G]C T C G A G G A T T A C A A A[T A A]T C T T T C C A T T G A[T A G]C A A C A T G A
 ALA - ARG - GLY - LEU - GLN - ILE - ILE - PHE - PRO - LEU - ILE - ALA - THR - TRP

T G A C T A T[T A G]C A A G C T T  3'  ............... Eco R1-A
 TRP - LEU - LEU - ALA - SER -
```

Fig. 5. Basesequence of a small part of the junction between the Eco R1 fragments B and A. The sequence shows the N-terminus of an as yet unidentified reading frame. All stopcodons indicated lay in the same reading frame. In the third reading frame no start- or stop- codons were detected. The grey dots show the positions at which type B mtDNA differs from type A mtDNA by a point mutation.

right to left. As outlined for Eco D this is the direction of transcription of the heavy strand. Since the sequences of this potential gene in the Eco A fragment are still lacking, the molecular weight and possible identity of the product encoded remain unknown.

The Eco R1 fragment E

Sequencing studies with Eco R1 fragment E have been also undertaken. The fragment is 650 bp long and should contain one or two tRNA genes. The restriction fine map is given in Fig. 6. We know

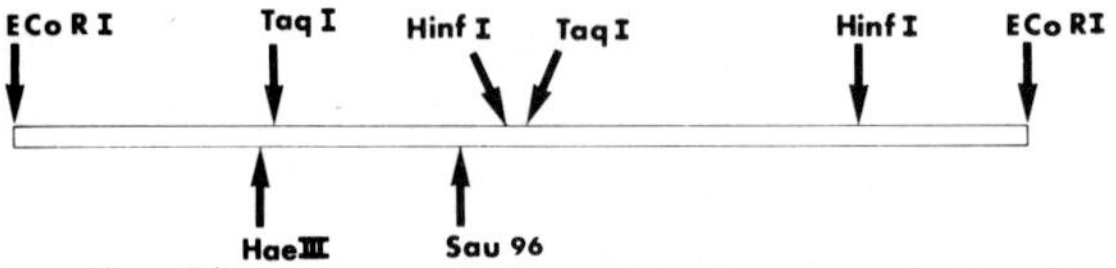

Fig. 6. Fine map of Eco RI fragment E. At the left the Eco R1 site between the fragments Eco E and Eco H, at the right between the fragments Eco E and Eco D. 1 cm corresponds to 65 bp.

the sequence of about 200 bp, mainly close to Eco R1 H side. This short sequence shows neither a clear reading frame nor an indication for a complete tRNA gene.

CONCLUSION

Although far from complete, our data on fine mapping and sequence analysis, confirm and extend similar data by other authors and with other organisms. A further sequence analysis to completion of the whole mtDNA will be necessary to obtain insight into the function of the mitochondrial genome. The rat-liver system is specially suitable for this purpose, because much work has already been done on the characterization of the enzyme complexes containing the mitochondrial translation products.

ACKNOWLEDGEMENTS

These studies were supported by grants to C.S. from the Consiglio Nazionale delle Ricerche (CNR), Italy and to AMK from the Netherlands Foundation for Chemical Research (SON) with financial aid from the Netherlands Organization for the Advancement of Pure Research (ZWO). The coöperation between the laboratories is highly facilitated by NATO Research Grant No 1484. The authors wish to thank Mrs Marisa Badini for preparing the manuscript.

REFERENCES

1. Kroon, A.M., Pepe, G., Bakker, H., Holtrop, M., Bollen, J.E., Van Bruggen, E.F.J., Cantatore, P., Terpstra, P. and Saccone, C. (1977) Biochim. Biophys. Acta, 478,128-145.
2. Saccone, C., Pepe, G., Bakker, H., Greco, M., De Giorgi, C. and

Kroon, A.M. (1979) in: Macromolecules in the functioning cell (Salvatore, F., Marino, C. and Volpe, P., eds), pp. 31-47, Plenum, New York.

3. Greco, M., Pepe, G., Bakker, H., Kroon, A.M. and Saccone, C. (1979) Biochem. Biophys. Res. Commun., 88, 199-207.

4. De Vos, W.M., Bakker, H., Saccone, C. and Kroon, A.M. (1980) Biochim. Biophys. Acta, 607, 1-9.

5. Hayashi, J., Yonekawa, H., Gotoh, O., Tagashira, Y., Moriwaki, K. and Yosida, T.H. (1979) Biochim. Biophys. Acta, 564,202-221.

6. Kroon, A.M., De Vos, W.M. and Bakker, H. (1978) Biochim. Biophys. Acta, 519, 269-273.

7. Groot, G.S.P. and Kroon, A.M. (1979) Biochim. Biophys. Acta, 564, 355-357.

8. Barrell, B.G., Bankier, A.T. and Drouin, J. (1979) Nature, 282, 189-194.

9. Maxam, A.M. and Gilbert, W. (1977) Proc. Natl. Acad. Sci. USA, 74, 560-564.

10. Kroon, A.M., De Vries, H. and Saccone, C. (1980) in: Endosymbiosis and Cell Research (Schwemmler, W. and Hchenk, H., eds) W. de Gruyter, Berlin, in press.

© 1980 Elsevier/North-Holland Biomedical Press
The Organization and Expression of the Mitochondrial Genome
A.M. Kroon and C. Saccone, editors

NUCLEOTIDE SEQUENCES OF THE CLONED EcoA FRAGMENT OF RAT MITOCHONDRIAL DNA

M. KOBAYASHI, K. YAGINUMA, T. SEKI AND K. KOIKE
Cancer Institute, Japanese Foundation for Cancer Research, Kami-Ikebukuro
Toshima-ku, Tokyo 170 (Japan)

INTRODUCTION

Due to gene cloning and DNA sequencing, our knowledge of the structure of the mammalian mitochondrial genome has advanced rapidly since a few reports first indicated that this circular DNA of approximately 16 kilobase pairs can autonomously replicate and codes for ribosomal, transfer and polyA-containing messenger RNAs[1,2]. Restriction endonuclease EcoRI cleaves rat mitochondrial DNA into seven distinct fragments (EcoA - EcoG) and all of these fragments were cloned by using phage λ as a vector. One of the recombinant phages, λgtWES: mtA102, contains the largest fragment EcoA of about 6.3 kilobase pairs[3,4].

It has been suggested by RNA transcript mapping that the EcoA DNA in the genome of rat mitochondria contains the structural genes coding for 12S rRNA, several 4S RNAs and some polyA-containing mRNAs[5]. Cleavage maps of the EcoA

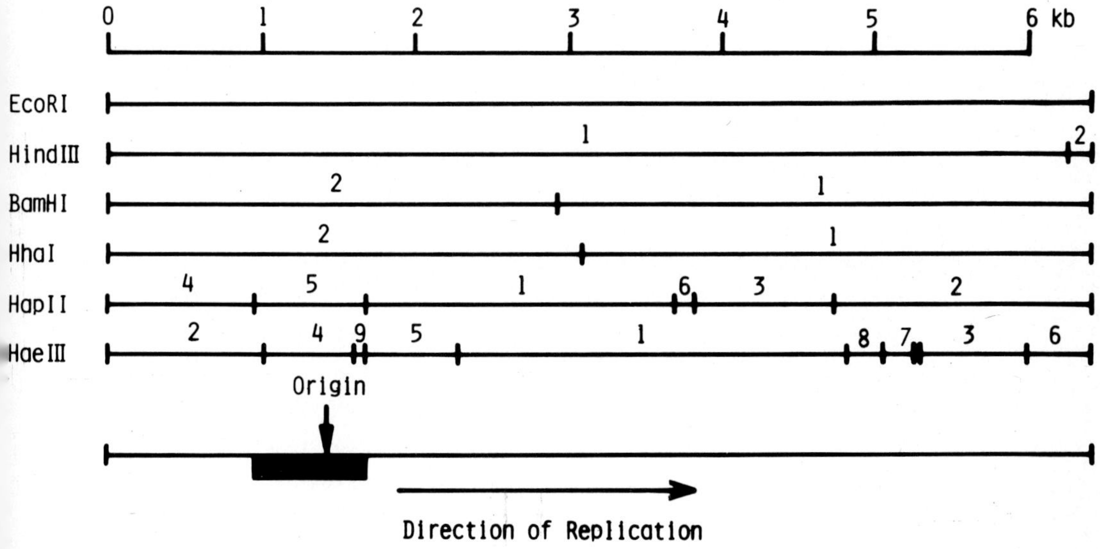

Fig. 1. Cleavage maps of the EcoA fragment[4,6].

fragment were made for restriction endonucleases, HindIII, BamH1, HhaI, HpaII and HaeIII[4,6]. Origin of DNA replication lies in the HpaEcoA5 fragment produced by HpaII digestion of the EcoA fragment, as shown in Fig. 1. The total nucleotide sequence of the HpaEcoA5 DNA has been reported previously[6].

Since we have been sequencing the other HpaII fragments, here we describe these nucleotide sequences. In the sequence of EcoA we identified several putative tRNA genes and 12S rRNA gene in the population of the HpaII fragments. Transfer RNA sequences are of interest because they are expected to contain the eukaryotic or prokaryotic promoters and termination signals, as well as, perhaps, other kinds of regulatory sequences which influence the processing of mitochondrial transcripts. Recent suggestion that a $tRNA^{Asp}$ gene can be transcribed together with cytochrome oxidase II gene and processed correctly[7] accessed us to identify such transcriptionally important sequences. Organization of these genes relative to the replication origin is also an interesting subject from an evolutional point of view.

MATERIALS AND METHODS

Purification of cloned mitochondrial DNA fragment. The clone used in the present experiments is one of the recombinant phages, λgtWES:mtA102, which carries the largest fragment EcoA of rat mitochondrial DNA. All subsequent growth and purification of recombinant phage were carried out according to the previous method[3]. The EcoA DNA was excised from the purified recombinant phage DNA by digestion with restriction endonuclease EcoRI and separated by electrophoresis on 0.7 % agarose gel.

Intact mitochondrial DNA was extracted from rat liver by the SDS-phenol method and closed-circular DNA was purified by two cycles of buoyant density centrifugation in CsCl-ethidium bromide, as described previously[8].

Restriction endonuclease digestion and fragment purification. DNA samples were digested with restriction endonuclease, HindIII, BamHI, HhaI, HpaII, HaeIII, AluI, HinfI, AvaII, MboI or XbaI as described earlier[4,9]. The DNA fragments were separated by electrophoresis on agarose or polyacrylamide gel. After the run, the DNA bands were stained by 2 μg/ml ethidium bromide. DNA was extracted from the gel by homogenization and incubation with 0.5 M ammonium acetate/0.1 % SDS/0.01 M magnesium acetate/0.1 mM EDTA for 24 h at room temperature.

Sequencing procedure. 5'-End labeling of the DNA fragment was performed with T4 polynucleotide kinase and [γ-^{32}P]ATP. Strand separation of 5'[^{32}P]labeled double-stranded DNA was performed, and the resulting DNA carrying labeled phosphate at one end was subjected to sequence analysis as described by Maxam and Gilbert[10] with minor modifications[6].

Gel electrophoresis and autoradiography. Agarose gels of 0.7 - 2 % (w/v) were prepared according to Sharp et al[11]. Polyacrylamide gels of 5, 8 or 10 % were prepared as described by Maxam and Gilbert[10]. Sequencing gels (0.5 mm thick) consisted of 20 or 10 % polyacrylamide and 7 M urea. Autoradiography was performed with Kodak X-Omat XRP-1 in combination with Dupont Cronex Lightning plus intensifying screen.

RESULTS AND DISCUSSION

Nucleotide sequence of HapEcoA5 DNA. The origin-containing HpaEcoA5 DNA excised from EcoA was further cleaved into smaller fragments by restriction endonucleases and the products were subjected to sequence analysis. The total nucleotide sequence of 717 base pairs for HpaEcoA5 thus determined is shown in Fig. 2.

As it has been indicated that the initial event in replication of rat mitochondrial DNA is the synthesis of a short heavy-strand initiation fragment[9], experiments were performed to find out the exact location of replication origin by analyzing the 5'-end portion of the short initiation fragment made in vivo. The electrophoretic separation of the short fragment labeled at the 5'-end revealed that there are at least two different species in size. A prominent species migrated at a position corresponding to a size of about 680 nucleotides and a minor one is corresponding to a size of about 630 nucleotides. The sequencing work indicated that the 5'-end regions of both fragments carry almost the same sequence, although their 5'-end are heterogeneous and about 60 % of both major and minor fragments contain ribonucleotide(s) at their 5'-end. By comparing with the sequence of the initiation fragment, one of the start point of the deoxyribonucleotide polymerization for the major initiation fragment was assigned on the sequence at the 425th base on the heavy strand[6].

Possible secondary structures around replication origin. Examination of the nucleotide sequence of HpaEcoA5 reveals several regions carrying two-fold rotational symmetry and palindrome structures. These structures between nucleotide residues 331 and 560 are shown in Fig. 3 by the heavy strand sequence. In the region upstream from the DNA start point (nucleotide 425), two palindrome structures composed of only AT residues are observed. The ATAAAATA sequence immediately precedes the DNA start point and the other is a structure between nucleotide residues 386 and 402. On the other hand, in the region downstream from the DNA start point, there are four structures of two-fold rotational symmetry and palindrome. Two adjacent hairpin structures are immediately

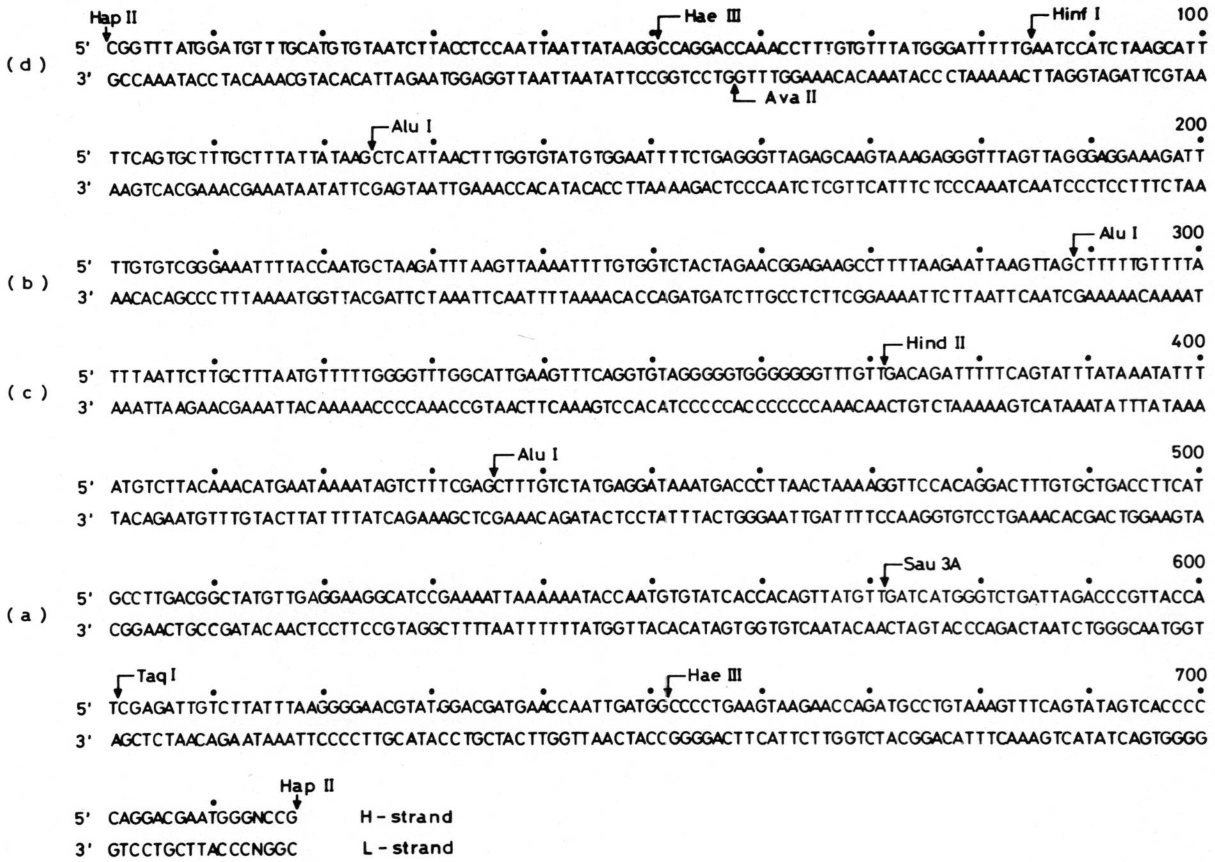

Fig. 2. The total nucleotide sequence of the HapEcoA5 DNA[6].

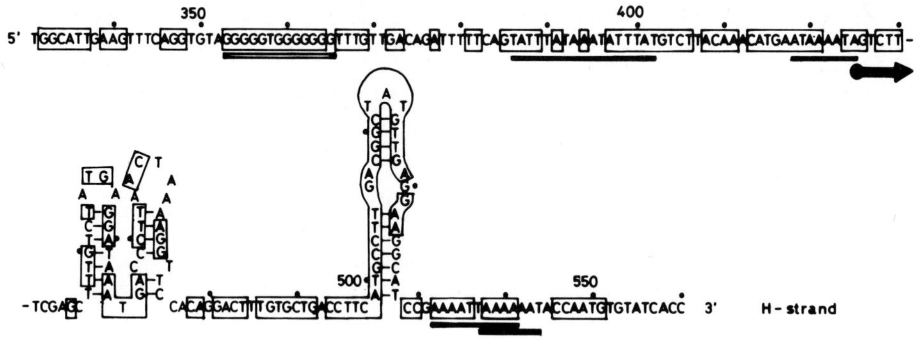

Fig. 3. Possible secondary structures of the region around the replication origin [6]. The start point for DNA synthesis is at the nucleotide residue 425 and is shown by a dot with an arrow, indicating the direction of elongation. Palindrome structures are underlined by solid bars. G culster is underlined by open bar.

beyond the DNA start point (nucleotides 437-475). The third hairpin structure is constructed at the region between nucleotide residues 499 and 528. This hairpin is followed by an AT sequence of 14 residues which contains two palindrome structures overlapping each other (nucleotides 532-541 and 537-544).

As described above, there is a palindrome structure composed of 17 AT base-pairs in the region 23 nucleotides away from the DNA start point. This region includes an interesting sequence of TATAAATA. It is possible to speculate that this region will serve as a recognition site for the RNA polymerase to initiate transcription. Another remarkable feature is the presence of a 13-base G cluster at the nucleotide residues 353 to 365. It is interesting to note that the identical sequence exists at the region upstream from the replication origin in HeLa cell and mouse mitochondrial DNA[12,13]. This might be a signal for the unidirectional mode of the mitochondrial DNA replication.

Location of tRNA^{Phe} gene. Hybridization experiments have suggested that the structural genes for tRNA and 12S rRNA exist in the region far upstream from the DNA start point[6]. One tRNA gene has been mapped in the region near the 5'-end of the 12S rRNA coding sequence on the heavy strand[1,5]. Knowing the approximate position of the 4S RNA coding sequence, we analysed in detail the region far upstream in HpaEcoA5 by looking for a clover-leaf structure. In fact, we were able to locate one tRNA gene, which is coded on the light strand, that is, the tRNA product hybridizes to the heavy strand. Between nucleotide residues 29 and 150 the tRNA gene is shown in the clover-leaf structure in Fig. 4.

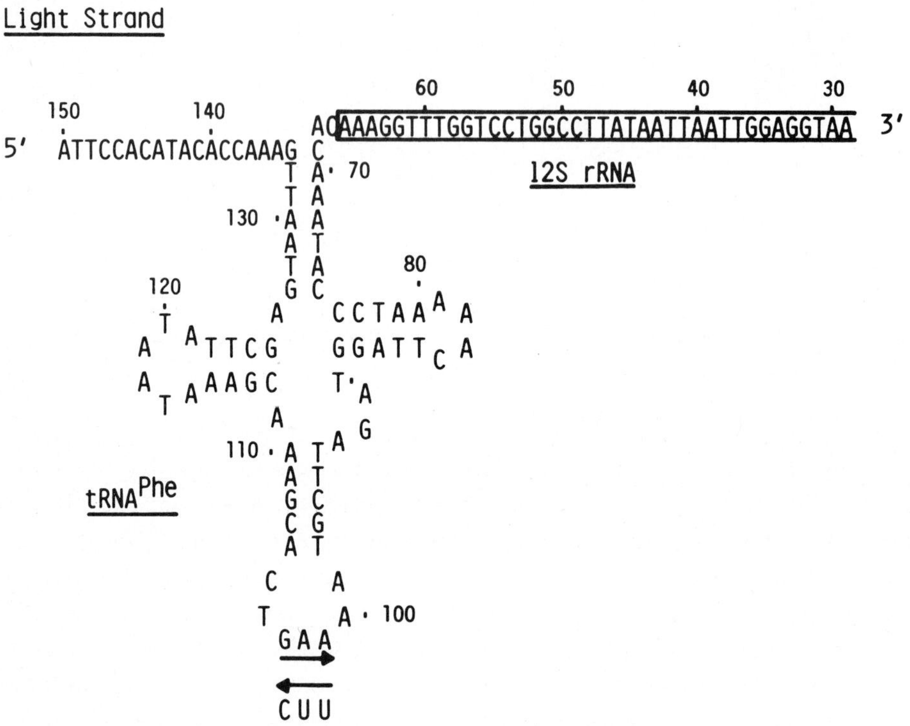

Fig. 4. The DNA sequence of the $tRNA^{Phe}$ gene arranged in the clover-leaf structure.

This tRNA has an anticodon corresponding to phenylalanine. The tRNA has an overall structures similar to that of standard prokaryotic tRNA, but it exhibits some distinctive features. The 5'-end of the tRNA coding sequence corresponds to the residue 133 on the heavy strand, and the 3'-end to the 68th residue. If this tRNA has a CCA sequence at the 3'-end, then this is not encoded in the DNA of this region. There is no intervening sequence in the $tRNA^{Phe}$. The sequence in the right-hand loop of this tRNA significantly differ from the standard prokaryotic tRNA pattern.

The in vitro 5'-end labeled 12S rRNA was characterized in its 5' termini by a complete alkali digestion and fractionation on PEI thin layer plates. The result obtained indicated that the 5'-end of 12S rRNA was pAp. Sequencing of the 5'-end of 12S rRNA was performed according to the method of Donis-Keller et al[14] using PEI thin layer plates. There may exist two species of 12S rRNA that differ in molecule length by at least one nucleotide at the 5'-end. The 5'-end of the major 12S rRNA corresponds to a residue in the heavy strand at the 66th base. Thus, it is possible to speculate that 12S rRNA gene sequence is adjacent to the 3'-end of the $tRNA^{Phe}$. A similar contiguous nature of the two genes in human mitochondrial DNA has recently been reported by Crews and Attardi[12]. Although the significance of such head and tail organization of the 12S rRNA and the $tRNA^{Phe}$ in rat mitochondrial DNA is uncertain, this raises an interesting discussion concerning the mechanism of synthesis and processing of mitochondrial rRNA species. It is conceivable that the $tRNA^{Phe}$ is processed from the same precursor by a endonucleolytic cleavage or is derived from an independent transcript.

Nucleotide sequence of HpaEcoA2 DNA. The HindIII site-containing HpaEcoA2 DNA, which is at the right side end of EcoA, was further cleaved into smaller fragments by restriction endonuclease, HaeIII, HinfI or AluI and the resulting products were subjected to electrophoresis. Fine cleavage maps of the HpaEcoA2 DNA thus constructed are shown in Fig. 5. Present sequencing study revealed a very small HaeIII fragment (24 base pairs) between HaeEcoA7 and HaeEcoA3 fragments which was missed in the previous map[4].

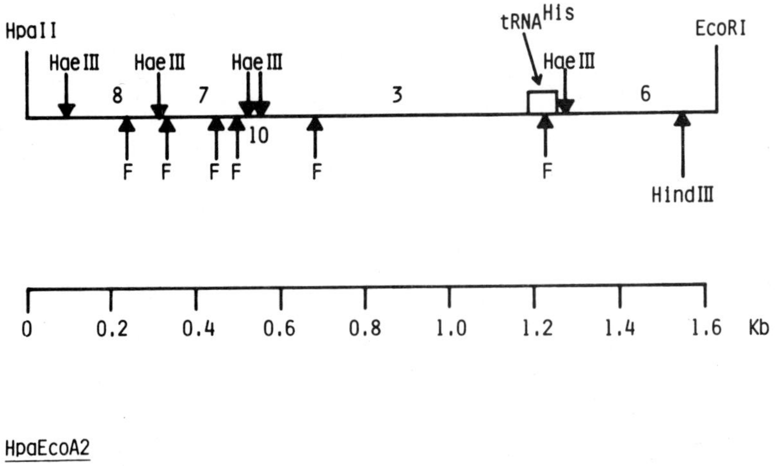

Fig. 5. Cleavage maps of the HpaEcoA2 fragment. Position of the $tRNA^{His}$ gene is shown in a box. F: HinfI cleavage site

Based on the present cleavage maps, the nucleotide sequence of the HpaEcoA2 DNA was determined.

Location of tRNA^{His}. As mentioned above, an RNA transcript map has been constructed on the basis of hybridization of the various 4S RNA species with separate restriction fragments. It suggested that at least two regions in the HpaEcoA2 DNA are hybridized to the 4S RNA. Knowing the approximate positions of these 4S RNA coding sequences, we analyzed our nucleotide sequences in detail by looking for a clover-leaf structure. Here we were able to locate one tRNA gene, which is coded on the light strand. In the sequence between nucleotide residues 374 and 441 the tRNA gene, in its characteristic clover-leaf structure, is shown in Fig. 6. This tRNA has an anticodon corresponding to histidine.

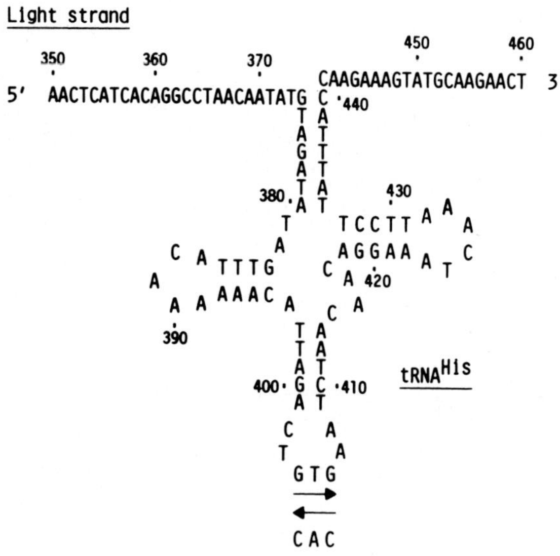

Fig. 6. The DNA sequence of the tRNA^{His} gene arranged in the clover-leaf structure.

The 5'-end of tRNA coding sequence probably corresponds to the residue 374 on the heavy strand, and the 3'-end to the residue 441. In this tRNA^{His} CCA is not found at the 3'-end as in tRNA^{Phe}, that is, there is no coding sequence for CCA in the DNA of this region. There is no intervening sequence in the tRNA^{His}.

The sequence in the left-hand loop of this tRNA significantly differs from the corresponding sequence found in the standard prokaryotic tRNA. The sequence downstream from this tRNA region can also be made into a clover-leaf structure, however, it widely deviates from the standard tRNA pattern.

As mentioned before, the contiguous nature of the mitochondrial genes raises an interesting possibility in processing the primary transcript. If a large size, but not in full length, transcript of the heavy strand is processed to give the tRNAs and the mRNAs, then a single cleavage between two genes would leave no 5'-leader sequence on the mRNA. Although we have no knowledge for the adjacent gene, there is no ATG codon immediately adjacent to the 3'-end of $tRNA^{His}$. There exists a short region (8 base pairs) between the 3'-end of $tRNA^{His}$ and the first ATG codon. Data suggest that the tRNA gene is not always contiguous to the adjacent mRNA. This 8-base sequence, AAGAAAGT, is presumed to be involved in leader function preceding the initiation codon. If this is the signal for some recognition of the adjacent mRNA, the $tRNA^{His}$ could be processed from the large transcript before the translation of the mRNA, although no evidence has yet been found for the ribosome recognition site. In the upstream from the $tRNA^{His}$, there is no 9-base repeat structure, as reported in the case of human $tRNA^{Lys}$ [7]. Details of the nucleotide sequences of the HpaEcoA2 or the other HpaEcoA fragments and the locations of other mitochondrial tRNA genes will be published elsewhere.

REFERENCES

1. Wu, M. et al. (1972) J. Mol. Biol., 71, 81-93.
2. Hirsch, M. et al. (1974) Cell, 1, 31-35.
3. Kobayashi, M. and Koike, K. (1979) Gene, 6, 123-136.
4. Koike, K. et al. (1979) Cold Spring Harb. Symp. Quant. Biol., 43, 193-201.
5. Greco, M. et al. (1979) Biochem. Biophys. Res. Commun., 88, 199-207.
6. Sekiya, T. et al. (1980) Gene, in press.
7. Barrell, B. G. et al. (1979) Nature, 282, 189-194.
8. Koike, K. et al. (1974) Biochim. Biophys. Acta, 361, 144-154.
9. Koike, K. and Kobayashi, M. (1977) Gene, 2, 299-316.
10. Maxam, A. M. and Gilbert, W. (1977) Proc. Natl. Acad. Sci. USA, 74, 560-564.
11. Sharp, P. A. et al. (1973) Biochemistry, 12, 3055-3063.
12. Crews, S. et al. (1979) Nature, 277, 192-198.
13. Gillum, A. M. and Clayton, D. A. (1979) J. Mol. Biol., 135, 353-368.
14. Donis-Keller, H. et al. (1977) Nucl. Acid Res., 4, 2527-2538.

The Organization and Expression of the Mitochondrial Genome
A.M. Kroon and C. Saccone, editors

SEQUENCE & STRUCTURE ANALYSIS OF MITOCHONDRIAL RIBOSOMAL RNA FROM HAMSTER CELLS

DONALD T. DUBIN and RICHARD J. BAER
Department of Microbiology, CMDNJ-Rutgers Medical School, Piscataway, N.J. 08854 (U.S.A.)

INTRODUCTION

Hamster cell mit* ribosomes contain unusually small ribosomal RNA's that we have designated "17S" and "13S" RNA. Their behavior on acrylamide gel electrophoresis[1] and density gradient sedimentation[1,2] indicates that they resemble homologous RNA's from other animal mitochondria in being roughly 1600 and 1000 nucleotides long. 17S and 13S RNA are also distinctive in their modification status, and in this regard, too, they appear representative of animal mit rRNA as a class[3]. We summarize here the results of our recent studies on modified, and terminal, sequences of 17S and 13S RNA.

Both animal[4] and fungal[5] mit ribosomes lack conventional 5S rRNA. However, we have found in hamster mitochondria an RNA species, "$3S_E$" RNA, that is a candidate for a mitochondrial 5S RNA-equivalent on the basis of its size and abundance relative to 17S and 13S RNA[2,6], and its absence of modified nucleotides[7]. This RNA has now been sequenced and its secondary structure assessed.

METHODS

Mit RNA was prepared from cultured BHK-21 cells, 17S and 13S RNA were purified by density gradient sedimentation, and $3S_E$ RNA was purified by sequential "warm" and "cool" gel electrophoresis of 4S fractions, all as previously described[6,8]. Classical sequencing procedures followed the approaches of Sanger and colleagues; details appear in the legend to Table 1. 5'- and 3'- end-labeling was performed, on samples containing 10 to 20 pmoles of termini, using 5'-^{32}P-pCp and T4 RNA ligase[9] or γ-^{32}P-ATP and T4 polynucleotide kinase[10]. Terminally labeled RNA was repurified by acrylamide gel electrophoresis. Such RNA was subjected to mobility shift analysis[11] after partial hydrolysis with acid[12] or formamide[13], and ladder analysis[14] after partial chemical hydrolysis[15] or partial hydrolysis with ribonuclease T1, U2, A, Phy I, or micrococcal nuclease in the presence of calcium (following refs. 12,13,16). Structural analyses were performed with nuclease S1 (ref. 17) or T1 (ref. 18).

*Abbreviations. Mit, mitochondrial; Cyt, cytoplasmic; LSU, large ribosomal subunit; SSU, small ribosomal subunit.

RESULTS & DISCUSSION

Methylated Oligonucleotides of 17S and 13S RNA.

Our findings on the methylated sequences of 17S and 13S RNA were of interest in themselves, and also contributed to interpretations of ladder patterns. Information on T_1-released methylated oligonucleotides has appeared[3]. Table 1 summarizes our current knowledge, based on combining oligonucleotide fingerprint data from ribonucleases A, and T_1, digests. The suspected similarity between 17S RNA oligonucleotide 1 and a similarly modified sequence of 28S RNA[23] was confirmed. Fig. 1 demonstrates that the 17S and 28S RNA oligonucleotides in question can be drawn as homologous "hairpins"; furthermore, there is a sequence in the 3'-portion of coli 23S RNA[24] that yields a hairpin with a similar loop and a 5 base-pair stem, also shown in Fig. 1. [23S RNA contains a single UmG residue (HsuChen & Dubin, unpublished) which has not been located within the molecule; we predict that it will be found in the loop as shown in Fig. 1.] These results foreshadow a recurring theme: the relative conservation of fam-

TABLE 1

MODIFIED OLIGONUCLEOTIDES OF HAMSTER MITOCHONDRIAL R-RNA

17S RNA	13S RNA
1. GUU·Um·Gm·UUCAACG	1. YGG·m^6_2A·m^6_2A·AGU
2. YG(G,Gm·G)U	2. YG·m^4C(C,m^5C)G
	3. YGGGA·m^5U·UAG

For standard fingerprints RNA was prepared (as described in Methods) from cells labeled with 100 mci of ^{32}Pi and 15 mci of [methyl-^{3}H]methionine. Samples were digested with 0.05 μg of ribonuclease T_1, or 0.075 μg of ribonuclease A, per μg of RNA, for 30 min. at 37° in 2-3 μl of 0.01 M Tris·HCL containing 1mM EDTA[19]. Fingerprints were generated using cellulose acetate electrophoresis and homochromatography[20], and methylated oligonucleotides were located by eluting spots and determining their ^{3}H contents. To establish the sequence of 17S RNA oligonucleotide 1, we prepared post-labeled digests using T1-plus-phosphatase followed by polynucleotide kinase and γ-^{32}P-ATP[21]. Secondary and tertiary analyses were performed using the approaches of Sanger and colleagues[19] and modifications thereof (refs. 11,22 and Baer & Dubin, in preparation).

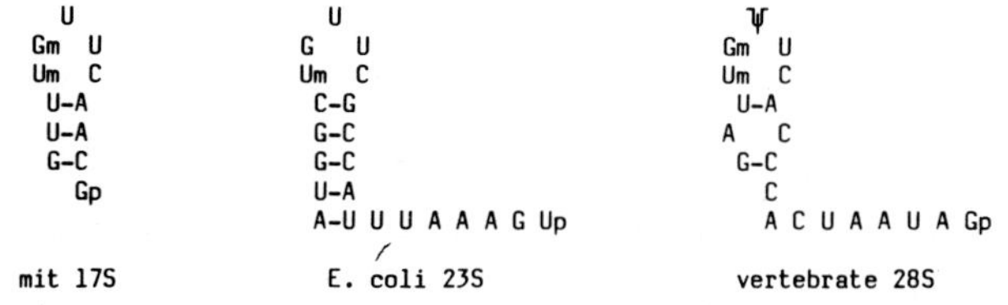

Fig 1. Hairpins involving "UmG(m)" Oligonucleotides of LSU RNA.

ilies of single-stranded regions rich in modified residues; and of duplex regions, with little conservation of primary sequence within them.

We have not yet located the 17S RNA methylated oligonucleotides in the molecule as a whole, but two of the three methylated oligonucleotides of 13S RNA have been located, in the 3'-portion (v.i.).

5'-End Groups of Mitochondrial RNA.

Preliminary to end-labeling studies, native 5'-termini were investigated on samples heavily labeled *in vivo* with ^{32}P. After digestion with T2 ribonuclease all RNA species examined yielded pNp termini in amounts commensurate with their chain lengths. As summarized in Table 2, $3S_E$ and 17S RNA each yielded a single predominant end-group, pGp and pAp respectively. 13S RNA unexpectedly yielded two end-groups, pUp and pAp, in approx. equal amounts. We list for comparison the two main centrifugal fractions, $12S_E$ and $20S_E$ RNA, that we find in hamster mit poly A (+) RNA[25]. $12S_E$ RNA, which sediments in the region of 13S RNA and runs as several discrete bands in denaturing acrylamide gels, yielded mainly pAp. Paradoxically, $20S_E$ RNA, which sediments slightly slower than 17S RNA and runs as a single discrete band in such gels, resembled 13S RNA in yielding two termini in equal amounts.

TABLE 2

5'-TERMINI OF *IN VIVO* LABELED MITOCHONDRIAL RNA

	pGp	pUp	pAp	pCp
		% of Total pNp		
$3S_E$ RNA	89	<3	<3	<3
13S RNA	6	44	41	9
17S RNA	4	4	89	4
$12S_E$ RNA	16	7	75	1
$20S_E$ RNA	46	8	46	1

RNA was purified from a culture heavily labeled *in vivo* with ^{32}P as for Table 1; $12S_E$ and $20S_E$ RNA fractions were purified by centrifugation in parallel with 13S and 17S RNA. Aliquots were hydrolyzed with T2 ribonuclease followed by DEAE column chromatography, and "-4" peaks were then characterized by DEAE paper electrophoresis[6].

As illustrated in Table 3, 5'-terminally post-labeled $3S_E$, 13S and 17S RNA yielded end-groups in accord with the *in vivo* results. Especially noteworthy is the fact that 13S RNA again appeared heterogeneous, yielding 40 to 60% each (in different runs) of pA and pU after P1 digestion. The end group of 18S RNA, pU, was as expected[26].

TABLE 3

5'-TERMINI OF *IN VITRO* LABELED RNA

	pG	pU	pA	pC
	% of Total pN			
$3S_E$ RNA	81	4	12	3
13S RNA	<1	52	46	1
17S RNA	2	6	90	3
18S RNA	2	85	9	5

Samples were labeled at 5'-Termini as described in the Text and aliquots were digested with nuclease P1 (0.5 mg/ml, 30 min. at 37°) and fractionated by paper electrophoresis at pH 3.5

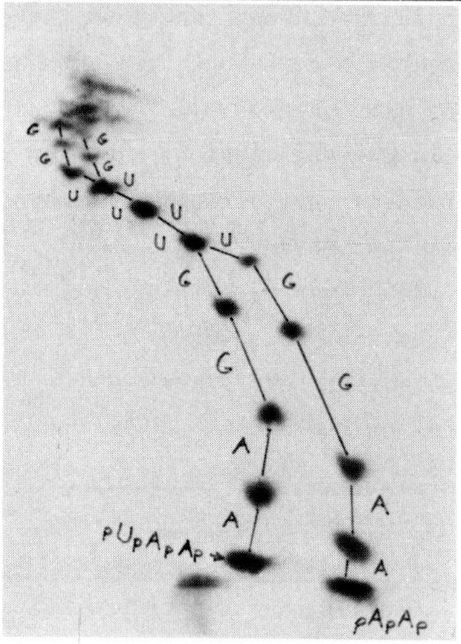

Fig 2. Mobility Shift Analysis of 5'-Labeled 13S RNA. A sample of 5' -labeled RNA was subjected to partial acid hydrolysis[12] followed by mobility shift analysis[13]. First dimension (electrophoresis at pH 3.5), right to left; second dimension (homochromatography), down.

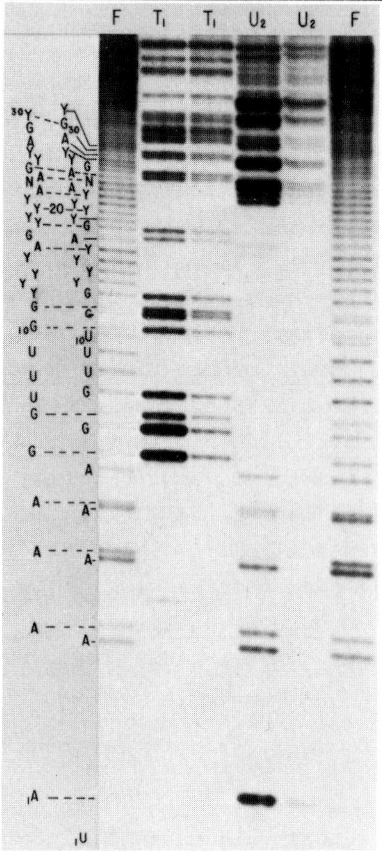

Fig 3. Ladder Analysis of 5'-Labeled 13S RNA. RNA was subjected to partial digestion with formamide ("F"), ribonuclease T1, or ribonuclease U2, followed by electrophoresis through a 20% acrylamide gel, as described in Methods. We have indicated our reading of the patterns assuming nested sequences, the shorter one on the left and the longer on the right.

5'-Terminal Sequence of 13S RNA.

Despite its heterogeneity we were able to obtain considerable sequence information on 5'-labeled 13S RNA. Indeed, the terminal heterogeneity provided some pleasing patterns. Mobility shift experiments registered the presence of two "nested" sequences as what we shall refer to as left and right tracks (Fig. 2). The left track was longer by one residue (demonstrated in shorter runs than that of Fig. 2), but otherwise differed from the right by the equivalent of a "U-shift." The terminal nucleotides were shown[6] to be pUp for the left and pAp for the right track. The four left-track and five right-track nucleotides were eluted from the plate of Fig. 2 (in which several oligonucleotides ran into the wick). P1 digestion yielded only pU from the left track and pA from the right track oligonucleotides. The next three spots were presumed to be common to the two sequences due to the compensating effect of the $G_6 \rightarrow U_7$ shift of the right track. Indeed, on P1 digestion these spots all yielded pU and pA in approx. equal amounts. These and similar results showed that the 5'-terminal sequences of 13S RNA are (pU)AAAAGGUUUGG, where about half the molecules have, and half lack, the terminal pU. Ladder gels of such RNA samples after partial enzymatic digestion also proved surprisingly readable (assuming continued nesting), as illustrated by Fig. 3. Fig. 4 presents our current sequence inferences through residue 75 on the basis of such studies, compared with sequences for the homologous HeLa mit ("12S") RNA (based largely on mit DNA sequence data)[27] and *E. coli* 16S RNA[28]. The two mit species are similar in being terminally heterogeneous, although in HeLa the variable residue is pA and occurs on only a minor fraction of molecules[27]. There are two regions of close sequence correspondence in the two mit species, 2-20 and 27-30 in Fig. 4; and two of likely correspondence (given the ambiguity in our pyrimidine identifications), 42-52 and 64-71. The other regions are surprisingly dissimilar. Two of the regions homologous in the mit species also bear some homology to regions in similar locations of *E. coli* 16S RNA. In addition, a third sequence of hamster 13S RNA (36-40) has a possible counterpart in 16S RNA.

We have also sequenced hamster cyt 18S RNA approx. 100 nucleotides in from

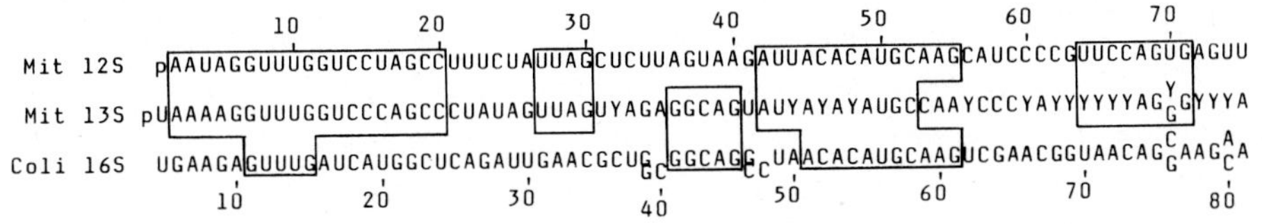

Fig. 4. 5'-Terminal Sequences of SSU RNA's. The numbers at the top refer to the longer of the two 13S RNA sequences.

the 5'-end (data not shown) and found only one apparent region of homology to the mit or coli sequences of Fig. 4. Residues 24-31 of 18S RNA were CAUAm•UGCU which resembles mit 46-52 or 53, and coli 52-58. Rat 18S RNA has been reported to contain CACAm•UG(C,U,U) in this region[29], possibly providing closer homology to the HeLa mit and coli species.

5'-Terminal Sequence of 17S RNA.

The patterns obtained from sequence analysis of 5'-labeled 17S RNA were more conventional than those from 13S RNA, reflecting absence of terminal heterogeneity. Our current inferences on the 5'-terminal 75 residues, obtained from mobility shift and ladder analysis (data not shown), appear in Fig. 5, together with comparative data on the homologous HeLa mit ("16S") RNA[27] and *E. coli* 23S RNA[24]. The most striking facet of the 17S RNA sequence is the presence of only a single G through residue 67. In addition, there was a stretch of correspondence, from positions 2 through 15, with HeLa 16S RNA (again making "favorable" assumptions for "Y's"), but rather little correspondence thereafter. There were also within the 5'-terminal 15 residues of 17S RNA two stretches of 5-nucleotide correspondence to regions near the 5'-end of coli 23S RNA.

```
                 10        20        30        40        50        60        70
mit 16S    pGCUAAAYYYAGCCYYAAACCYACUYYACCCYA
mit 17S    pACUAACUCUAGCCCUCAUUUUUUCAAUCUAUAAAAUUUUAACCUACAAACUAAAACAUUCACUAAAAGAAGUAUC
coli 23S   GACUAAG---UGCCCUGGCAGUCAGAGGCGAUGAAGGACGUGCUAAUCUGCGAUAAGCGUCGGUAAGGUGAUAUGA
           10         30       40        50        60        70        80        90
```

Fig. 5. 5'-Terminal Sequences of LSU RNA. The 3 dashes in the 23S RNA sequence represent a stretch of 13 residues with no apparent homology to mit RNA.

Sequence and Structure of $3S_E$ RNA.

The primary sequence of $3S_E$ RNA was established by means of mobility shift and ladder analyses to be:

```
  60        50        40        30        20        10
pGGAGAAUGUAUGCAAGAGCUGCUAACUCCUGCUACCAUGUAUAAUAACAUGGCUUUCUUACCA(OH)
```

The underlined stretches can form the stem of 2 hairpins whose existence was supported by S1 nuclease analysis; in addition, a modest duplex stem can be drawn involving residues 6-10 and 58-62. The resulting configuration contains the fundamental elements of the "universal" structure proposed for prokaryotic 5S RNA[30], despite the apparent absence of primary sequence homology between $3S_E$ RNA and conventional 5S RNA.

3'-Terminal Sequence and Structure Analysis of 13S RNA.

Our most extensive studies have involved 3'-terminally labeled 13S RNA. Methodology was essentially as in the earlier sections, and the data will be presented in detail elsewhere. We present here our inferred primary structure 220 residues "in" and some secondary structural features. Fig. 6 compares 13S RNA with coli 16S RNA[28] and *B. mori* 18S RNA[31], the longest prokaryotic and eukaryotic cyt SSU RNA 3'-terminal sequences published. Our alignments were guided by earlier work on conserved regions of SSU RNA. Sections 1 and 2 correspond to regions of *E. coli* 16S RNA containing T1 oligonucleotides that are highly conserved among bacterial SSU RNA's[32] and between coli 16S and *B. mori* 18S RNA[31]. Approx 100 16S or 18S RNA residues, and 65 13S RNA residues, fall between these regions, and there indeed appears to be relatively little homology here. The longest stretches of apparent primary sequence conservation occur in regions of putative single-strandedness (v.i.) that in addition tend to contain methylated nucleotides: namely 1c and 2a. It is notable that these regions are also rich in kethoxal-sensitive G residues that are "protected" by association of 50S and 30S units[33,34]. These correlations support the idea that some primary subsequences in the 3'-region of SSU RNA play universal roles in the association-dissociation reaction of ribosomes (perhaps via interaction with 5S

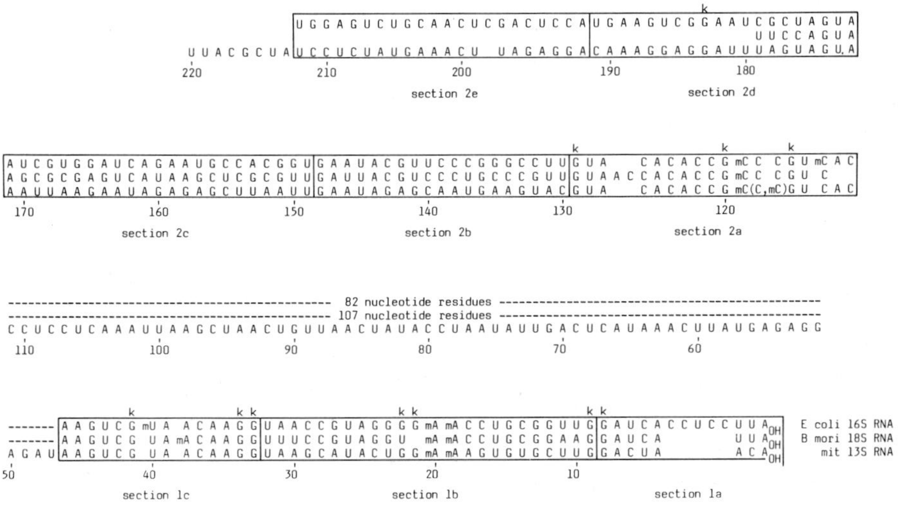

Fig. 6. 3'-Terminal Sequences of SSU RNA. The "mA's" in section 1b are m_2^6A; that in 1c of 18S is presumed to be m^6A (ref. 29). The "mC's" of 13S RNA section 2a are those of oligonucleotide 2 (Table 1); that of 18S RNA is Cm; and those of 16S RNA are m^4Cm and m^5C, respectively[28]. "k" indicates kethoxal sensitive residues of 16S RNA[33,34].

or LSU RNA[33-35], and that such roles may be fine-tuned by post-transcriptional modification.

Another noteworthy feature of the primary sequences involves the extreme terminal region (section 1a). This region has in *E. coli* been shown to play a role in mRNA-ribosome binding via base-pairing with the messenger molecules (see 36,37). Such a "Shine-Dalgarno" mechanism is probably not operative in eukaryotic cyt systems[38] and indeed the crucial CCUCC is missing from an otherwise homologous stretch in 18S RNA (Fig. 6). Mit 13S RNA resembles 18S RNA in this regard, and indeed we can find no appropriate regions in the sequences published for mit RNA's that could mediate Shine-Dalgarno interactions with 16S <u>or</u> 13S RNA sequences. Mit mRNA also lacks "caps"[39,40]; does it employ novel mechanism(s) in binding to ribosomes?

Our secondary structure inferences resulted from a search for possible stable "hairpins" common to each of the aligned sequences. Three were found (Fig. 7), and their reality was supported by experiments employing nucleases S1 and T1 as structural probes (Baer & Dubin, in preparation). We indicate in Fig. 6 the primary sequences corresponding to hairpins I, II and III: namely 1b, 2c and 2e respectively. The putative double-stranded stems show less primary sequence conservation than do single-stranded stretches within the conserved

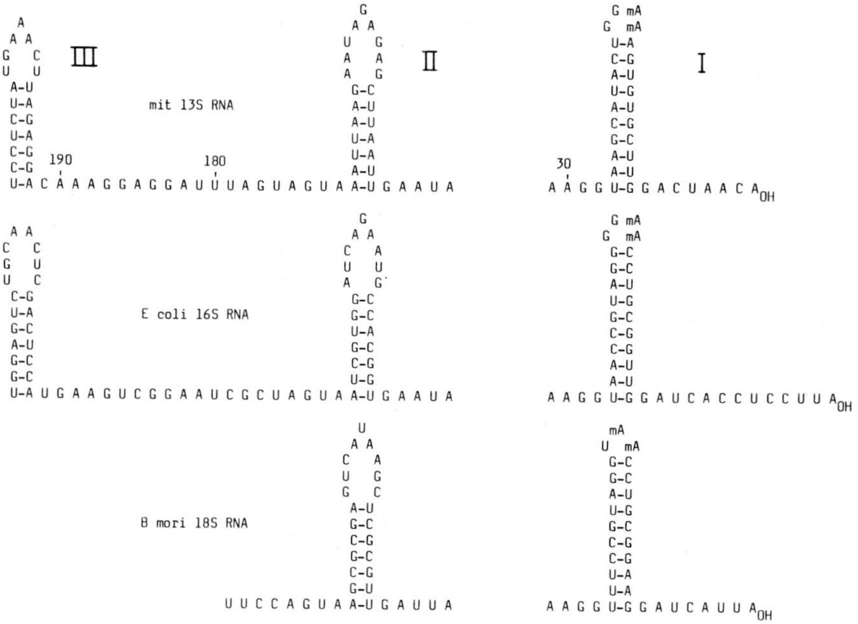

Fig. 7. Proposed Conserved Hairpins of SSU RNA.

regions. This supports the evolutionary conservation and likely functional importance of the hairpins as hairpins; presumably "compensating" mutations in stems are well tolerated.

Did Mitochondria Arise Endosymbiotically?

Our results suggest a somewhat closer similarity between mit and *E. coli*, than between mit and cyt, rRNA sequences, and can thus be scored as favoring the endosymbiotic theory of the origin of mitochondria. In this connection we note that the stem of 13S RNA hairpin I constitutes a good example of preservation of secondary structure in the face of divergence of primary sequence: 4 of 10 base pairs differ from 16S RNA, and 6 of 10 from 18S RNA. The divergence from a homologous structure in maize chloroplast mit 16S RNA[41] is only 3 of 10; is this a hint that the lineages of animal mitochondria and of some chloroplasts (cf ref. 42) may be related?

ACKNOWLEDGEMENTS

This work was supported by N.I.H. Grants GM-14957 and M.S.R.P. 27-9858, U.S.P.H.S. R.J.B. is a Predoctoral Trainee under Institutional National Research Service Award CA-09069, U.S.P.H.S. We thank Dr. G. Cleaves for his assistance with computer resources, Dr. R. Wurst for a generous gift of nuclease S1, and Mrs. K. Timko and Miss T. Azzolina for their expert technical help.

REFERENCES

1. Montenecourt, B.S., Langsam, M.E. & Dubin, D.T. (1970) J. Cell Biology 46, 245-251.
2. Dubin, D.T. & Friend, D.A. (1972) J. Mol. Biol. 71, 163-175.
3. Dubin, D.T., Baer, R.J., Davenport, L.E., Taylor, R.H. & Timko, K.D. (1979) in "Transmethylation," Usdin E., Borchardt, R.T. & Creveling, C.R., Eds., Elsevier/North Holland, N.Y. pp. 389-398.
4. Dubin, D.T. & Montenecourt, B.S. (1970) J. Mol. Biol. 48, 279-295.
5. Lizardi, P.M. & Luck, D.J.L. (1971) Nature New Biol. 229, 140-142.
6. Dubin, D.T. & Taylor, R.H. (1980) FEBS Letters, 109, 223-227.
7. Dubin, D.T., Jones, T.H. & Cleaves, G.R. (1974) Biochem. Biophys. Res. Commun. 56, 401-406.
8. Dubin, D.T. (1974) J. Mol. Biol. 84, 257-273.
9. England, T.E. & Uhlenbeck, O.C. (1978) Nature 271, 560-561.
10. Efstratiadis, A., Vournakis, J.M., Donis-Keller, H., Chaconas, G., Dougall, T.K. & Kafatos, F.C. (1977) Nuc. Acids Res. 4, 4165-4174.
11. Silberklang, M., Gillum, A.M. & RajBhandary, U.L. (1977) Nuc. Acids Res. 4, 4091-4108.
12. Krupp, G. & Gross, H.J. (1979) Nuc. Acids Res. 6, 3481-3490.

13. Simoncsits, A., Brownlee, G.G., Brown, R.S., Rubin, J.R. & Guilley, H. (1977) Nature 269, 833-836.

14. Sanger, F. & Coulson, A.R. (1978) FEBS Letters 87, 107-110.

15. Peattie, D.A. (1979) Proc. Nat. Acad. Sci., USA 76, 1760-1764.

16. Donis-Keller, H., Maxam, A.M. & Gilbert, W. (1977) Nuc. Acids Res. 4, 2527-2538.

17. Wurst, R.M., Vournakis, J.N. & Maxam, A.M. (1978) Biochemistry 17, 4493-4499.

18. Pavlakis, G.M., Lockard, R.E., Vamvakopoulos, N., Rieser, L., RajBhandary, U.L. & Vournakis, J.N. (1980) Cell 10, 91-102.

19. Brownlee, G. (1972) "Determination of Sequences in RNA." American Elsevier, New York.

20. Volckaert, G., Min Jou, W., & Fiers, W. (1976) Analyt. Biochem. 72, 443-446.

21. Frisby, D. (1977) Nuc. Acids Res. 4, 2975-2996.

22. Volckaert, G. & Fiers, W. (1977) Analyt. Biochem. 83, 228-239.

23. Eladari, M-E, Hampe, A. & Galibert, F. (1977) Nuc. Acids Res. 4, 1759-1767.

24. Brosius, J., Dull, T.J. & Noller, H.F. (1980) Proc. Nat. Acad. Sci., USA 77, 201-204.

25. Cleaves, G.R., Jones, T. & Dubin, D.T. (1976) Arch. Biochem. Biophys. 175, 303-311.

26. Eladari, M-E., & Galibert, F. (1975) Eur. J. Biochem. 55, 247-255.

27. Crews, S. & Attardi, G. (1980) Cell 19, 775-784.

28. Carbon, P., Ehresmann, B. & Ebel, J.P. (1979) Eur. J. Biochem. 100, 399-410.

29. Choi, Y.C. & Malinowski, H. (1980) Fed. Proc. 38, 2200.

30. Fox, G.E. & Woese, C.R. (1975) Nature 225, 505-507.

31. Samols, D.R., Hagenbüchle, O. & Gage, L.P. (1979) Nuc. Acids Res. 7, 1109-1119.

32. Woese, C.R., Fox, G.E., Zablen, L., Uchida, T., Bonen, L., Pechman, K., Lewis, B.J. & Stahl, D. (1975) Nature 254, 83-86.

33. Chapman, N.M. & Noller, H.F. (1979) J. Mol. Biol. 109, 131-149.

34. Herr, W., Chapman, N.M. & Noller, H.F. (1979) J. Mol. Biol. 130, 433-449.

35. Azad, A.A. (1979) Nuc. Acids Res. 7, 1913-1929.

36. Shine, J. & Dalgarno, L. (1974) Proc. Nat. Acad. Sci., USA 71, 1342-1346.

37. Steitz, J.A. & Jakes, K. (1975) Proc. Nat. Acad. Sci., USA 72, 4734-4738.

38. De Wachter, R. (1979) Nuc. Acids Res. 7, 2045-2054.

39. Taylor, R.H., and Dubin, D.T. (1975) J. Cell Biol. 67, 428a.

40. Grohmann, K., Amalric, F., Crews, S. & Attardi, G. (1978) Nuc. Acids Res. 5, 637-651.

41. Schwarz, A., & Kossel, H. (1979) Nature 279, 520-522.

42. Woese, C.R. (1977) J. Molec. Evol. 10, 93-96.

The Organization and Expression of the Mitochondrial Genome
A.M. Kroon and C. Saccone, editors

THE ADENINE AND THYMINE-RICH REGION OF *DROSOPHILA* MITOCHONDRIAL DNA MOLECULES.

DAVID R. WOLSTENHOLME, CHRISTIANE M.-R. FAURON AND JUDY M. GODDARD
Department of Biology, University of Utah, Salt Lake City, Utah 84112 (U.S.A.)

INTRODUCTION

The circular mitochondrial DNA (mtDNA) molecules of species of the genus *Drosophila* are peculiar in that unlike other metazoan mtDNAs they contain a single region which is exceptionally rich in adenine and thymine. It was found independently by three different groups[1-3] that a portion of *Drosophila melanogaster* mtDNA denatured at a lower temperature than the remainder, and therefore appeared to be rich in adenine and thymine (A+T). Using electron microscope denaturation mapping, Peacock et al.[3] showed that the majority of the A+T-rich DNA was confined to a limited region of each molecule. Fauron and Wolstenholme[4] and Klukas and Dawid[5] using heat and alkali denaturation mapping respectively demonstrated that in fact there is a single A+T-rich region of constant size (approximately 5.1 kb) in *D. melanogaster* mtDNA molecules. The former workers[4] further showed that the mtDNA molecules from other *Drosophila* species also contain a region which denatures under the same conditions as the A+T-rich region of mtDNA of *D. melanogaster*. In *D. virilis* the A+T-rich region was about 1.0 kb. In different species of the melanogaster group the A+T-rich region varied in size from 1.0 kb in *D. ananassae* to 5.1 kb in *D. melanogaster*, and these size differences appeared to account for differences in size of the mtDNA molecules (15.9 kb to 19.5 kb) which contained them. However, the distribution of sizes of the A+T-rich regions did not follow a simple taxonomic pattern.

We present here our recent studies which pertain to the sequence organization of the A+T-rich region of *Drosophila* mtDNA molecules.

MATERIALS AND METHODS

The origin of each of the different species, strains, and lines, of *Drosophila* used in these investigations, and details of the methods employed have been given elsewhere[4, 6-10]. *Alu*I restriction endonuclease was obtained from New England Biolabs and digestion conditions were those recommended by the supplier.

RESULTS

Using agarose gel electrophoresis and electron microscopy we mapped on the mtDNA molecules of *D. melanogaster*, *D. simulans*, *D. mauritiana*, *D. yakuba*, *D. takahashii* (all members of the melanogaster group of the subgenus Sophophora) and *D. virilis* (subgenus Drosophila), the sites sensitive to cleavage by the restriction enzymes *Eco*RI and *Hin*dIII, relative to the A+T-rich region[6]. Comparisons of these maps (Fig. 1) revealed that all molecules have three *Eco*RI sites and two *Hin*dIII sites in common, and that species of the melanogaster group have one further *Hin*dIII site in common. Further, with the exception of the two *Eco*RI sites which are unique to *D. yakuba* mtDNA, all *Eco*RI and *Hin*dIII sites are found in mtDNA molecules of more than one species. The most interesting observation is that the A+T-rich regions, which vary in size among the different species examined, occupy an homologous position on the molecules.

From a sample of embryo mtDNA from *D. melanogaster* (Ore.), a fraction comprising > 99% circular molecules (> 90% covalently closed) was obtained by cesium chloride-ethidium bromide buoyant density equilibrium centrifugation, and digested with the restriction enzyme *Alu*I. At least 13 fragments were apparent following electrophoresis of this digest (Fig. 2). One fragment (≃ 5.5 kb) was more than seven times larger than the remaining fragments (all < 0.75 kb). Electron microscopy was used to confirm the size of the largest fragment (5.5 ± 0.2 kb, n = 30). *Alu*I-digested *D. melanogaster* mtDNA was next heated at 40°C for 10 min in 0.05 M sodium phosphate (pH 7.8) and 10% formaldehyde (adequate conditions for total denaturation of the A+T-rich region of this mtDNA[4]) and prepared for electron microscopy. All of the fragments of the largest size class contained a region of denaturation of 5.2 ± 0.3 kb (n = 30) bounded on each side by undenatured regions of 0.20 ± 0.01 and 0.35 ± 0.02 kb respectively (Fig. 3). As the A+T-rich region in undigested circular molecules was found to be approximately 5.1 kb (Fig. 1), these data indicate that *Alu*I does not cleave the A+T-rich region of *D. melanogaster* mtDNA molecules and that this region accounts for approximately 90.5% of the largest *Alu*I-produced fragment.

A DNA sample, determined by electron microscopy to comprise > 96% intact A+T-rich region-containing *Alu*I fragments was obtained by sucrose (5-20%) velocity sedimentation of an *Alu*I digest of *D. melanogaster* mtDNA. The buoyant density of the DNA in this sample was 1.666 g/cm^3 (Fig. 4).

The buoyant density of whole double-stranded mtDNA molecules from *D. melanogaster* was 1.681 g/cm^3 (Fig. 4). From a consideration of the distribution of DNA in a buoyant density gradient of *Hin*dIII-digested *D. melanogaster* mtDNA (Fig. 4), it appeared that the A+T-rich region-containing fragment, of which the A+T-rich region accounts for 83.7% (Fig. 1), was 1.668 g/cm^3, and the mean

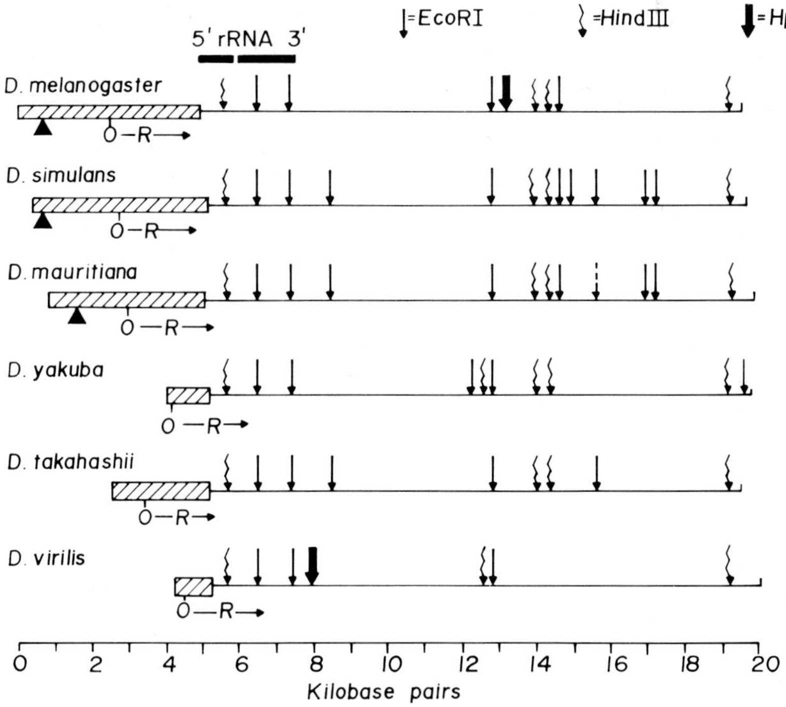

Fig. 1. Maps of the mitochondrial genomes of the six *Drosophila* species indicated, showing the relative positions of the A+T-rich regions (hatched areas) and the sites at which *Eco*RI and *Hin*dIII cleave. *Hpa*I sensitive sites have been mapped only for *D. melanogaster* and *D. virilis*. The genomes have been oriented so as to maximize the coincidence of enzyme sensitive sites and then aligned by the common *Eco*RI site nearest the A+T-rich region. This site defines the right end of the A+T-rich region in the figure and all molecules are linearized at what is then the left end of the A+T-rich region. The origin (O) and direction of replication (R) are shown for each genome. The position and polarity of the rRNA molecules transcribed from *D. melanogaster* mtDNA are from Klukas and Dawid[5]. The sizes of A+T-rich regions of mtDNA molecules of different lines and strains of *D. melanogaster, D. simulans* and *D. mauritiana* vary by the amounts shown to the left of the triangles.

buoyant density of the other three fragments was 1.685 g/cm^3. If the GC content of the A+T-rich region-containing *Alu*I and *Hin*dIII fragments, calculated from their respective buoyant densities, is corrected for the GC content of the segments they contain which lie outside the A+T-rich region (assuming this value to be equal to the mean GC content of the remainder of the molecule; 24.5%), values of 4.1 and 5.0% GC respectively are obtained for the A+T-rich region. These compare with values of 7.6% GC[2], 4.2% GC[1], and 0-5% GC[11]

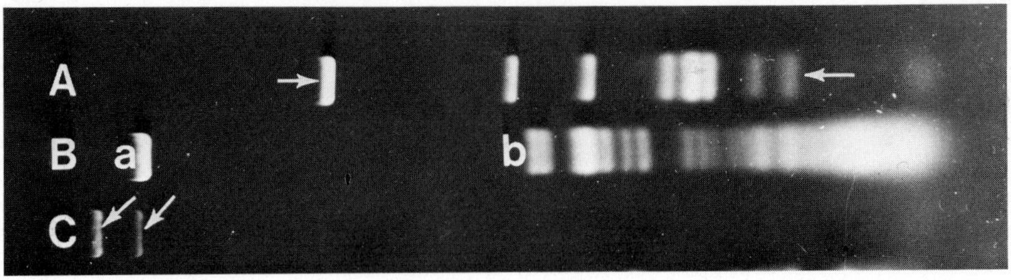

Fig. 2. Fluorescence photograph showing the distribution of ethidium bromide-stained bands in a 1.5% agarose slab gel after electrophoresis. The center lane (B) contains the *Alu*I digestion products of *D. melanogaster* mtDNA. The left-most band (a) comprises the A+T-rich region-containing fragment which is 5.5 kb, 7.9 times the size of the second largest fragment (band b, 0.7 kb). The standard DNAs in lanes A and C are, respectively, *Hae*III digestion products of SV40 DNA (arrows indicate fragments of 1.7 and 0.18 kb), and the *Eco*RI digestion products of *D. melanogaster* mtDNA (arrows indicate fragments of 11.5, 5.4 kb).

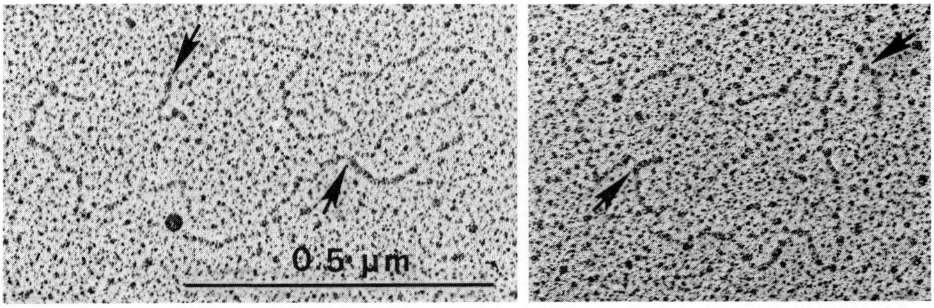

Fig. 3. Electron micrographs of fragments resulting from *Alu*I digestion of *D. melanogaster* mtDNA, followed by heating at 40°C for 10 min in 0.05 M sodium phosphate (pH 7.8) and 10% formaldehyde. The limits of the denatured region of each fragment, which compares in size to the A+T-rich region, are shown by arrows.

calculated from the results of thermal melting studies.

When a sample of A+T-rich region-containing *Alu*I fragments was centrifuged to equilibrium in alkaline cesium chloride, two peaks in a single broad band were observed at ρ = 1.724 g/cm^3 and ρ = 1.727 g/cm^3 (Fig. 4). This suggests that little if any base bias exists between the two complementary strands of the A+T-rich region of *D. melanogaster* mtDNA. Further, as the

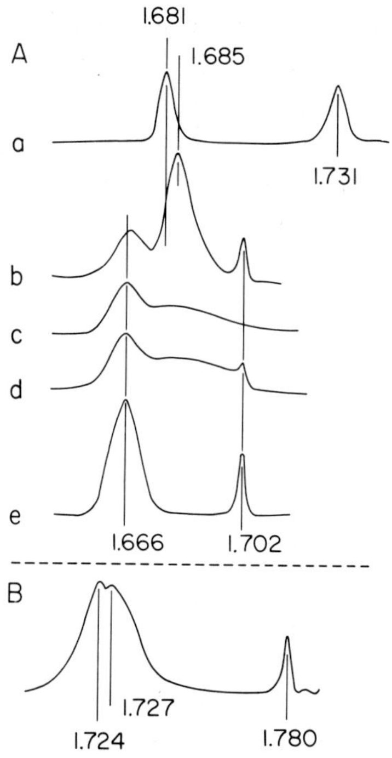

Fig. 4. Photoelectric scans using 260 nm illumination of cesium chloride equilibrium buoyant density gradients of mtDNA from *D. melanogaster*. A. Neutral cesium chloride gradients. a, whole, native mtDNA (ρ = 1.681 g/cm^3). The reference band (ρ = 1.731 g/cm^3) is native DNA of *Micrococcus lysodeikticus*. b, mtDNA digested to completion with *Hin*dIII. c, and d, mtDNA digested to completion with *Alu*I. e, the largest (A+T-rich region-containing) *Alu*I fragment isolated by sucrose gradient sedimentation. The reference band (ρ = 1.702 g/cm^3) in b, d and e is native kinetoplast DNA of *Crithidia acanthocephali*[10]. B. Alkaline cesium chloride gradient of the largest (A+T-rich region-containing) *Alu*I fragment. The reference band (1.780 g/cm^3) represents native kinetoplast DNA of *C. acanthocephali*[10].

nuclear A+T-rich satellite DNA (ρ = 1.672 g/cm^3 in neutral cesium chloride) of *D. melanogaster* separates into two distinct components of p = 1.713 g/cm^3 and ρ = 1.745 g/cm^3 in alkaline cesium chloride[3], this DNA, and the A+T-rich region of mtDNA molecules are clearly indicated to differ in nucleotide sequence.

The distribution of sizes among A+T-rich regions of mtDNA molecules of eight *Drosophila* species suggested to us[4] that they represent integral multiples of a 0.5 kb unit. In view of this, we have attempted to test in the following way, the possibility that the A+T-rich region of *D. melanogaster* mtDNA comprises a tandemly repeated sequence. A+T-rich region-containing *Alu*I fragments of *D. melanogaster* mtDNA ($\simeq$ 4 μg/ml), obtained as described above, were denatured by dialysis against 95% formamide for 1 hour. A sample of this DNA was removed and, using electron microscopy, it was ascertained that strand separation was complete, and that > 90% of the single strands were the length of the original duplex. The remainder of the DNA was renatured by dialysis against 35% formamide for 1 hr, and for 2 hr, and the products were examined in the electron microscope following preparation by the formamide protein monolayer technique.

If the A+T-rich region comprises tandemly repeated sequences, then under conditions where only a fraction of the DNA had been allowed to renature, molecular forms could be expected, interpretable as resulting from the pairing of one segment of an A+T-rich region with a different segment of a complementary A+T-rich region. Further, such forms should circularize upon further annealing. Regardless of whether the A+T-rich region comprises tandemly repeated sequences, perfect double-stranded molecules the length of the original double-stranded fragment would be expected among the renaturation products. It was found that in the preparation annealed for 1 hr approximately 50% renaturation of the DNA had occured. Apparently totally double-stranded linear molecules (5.5 ± 0.2 kb, n = 50) the size of the original *Alu*I restriction fragment were abundant in this sample, and were the predominate molecular form in the sample annealed for 2 hr. Most important, molecular forms expected only if the A+T-rich region contained tandemly repeated sequences were not observed. A similar result was obtained when the experiment was repeated using the A+T-rich region-containing *Eco*RI fragments of *D. melanogaster* mtDNA.

In electron microscope preparations of renaturation products of *Alu*I fragments, we failed to detect double-stranded molecules containing internally unpaired regions, or single-stranded loops such as might be expected if extensive differences (> 50 base pairs[12]) occurred between nucleotide sequences in any region of different molecules in the samples used.

In an attempt to gain information on the relationship of nucleotide sequences in the A+T-rich regions of mtDNA molecules of *D. melanogaster*, *D. simulans* (Cal.), *D. mauritiana* (line I, see below), *D. yakuba*, *D. takahashii*, and *D. virilis*, we constructed a series of heteroduplexes[6]. This was accomplished by mixing nicked circular mtDNA molecules of one species with a five-fold excess of fragments obtained by *Eco*RI digestion of circular mtDNA molecules of a second species, denaturing the molecules by dialysis against 95% formamide, and then renaturing by dialysis against 35% formamide (highly permissive for base pairing). In samples in which approximately 50% of the DNA was renatured (ascertained by electron microscope examination of samples) circular molecules containing double-stranded segments were located and photographed, and then analyzed in regard to the lengths of the component double-stranded and single-stranded segments. The results of this study were quite straightforward. Complete pairing of molecules outside the A+T-rich region was found in all heteroduplexes examined. However, in contrast, the A+T-rich regions of mtDNA molecules failed to pair in the following heteroduplex combinations; *D. melanogaster:D. yakuba*; *D. melanogaster:D. takahashii*; *D. melanogaster:D.*

virilis; D. takahashii:D. yakuba; and *D. virilis:D. yakuba*. The results of control experiments ruled out the possibility that in these heteroduplexes, failure of pairing of A+T-rich regions resulted from insufficient annealing conditions, or from the circularity of one molecule in some way preventing pairing of the A+T-rich DNA. In heteroduplexes formed between *D. melanogaster* and *D. simulans* mtDNA molecules and between *D. melanogaster* and *D. mauritiana* mtDNA molecules, up to 35% of the A+T-rich regions appeared double-stranded. Only in heteroduplexes between *D. simulans* and *D. mauritiana* did the A+T-rich regions appear to be totally double-stranded. These data clearly indicate that much more extensive divergence of sequences has occurred in A+T-rich regions than in other regions of *Drosophila* mtDNA molecules.

From the results of experiments using gel electrophoresis and electron microscopy, we determined that the mtDNA molecules of some flies taken from a stock of *D. mauritiana* (originally collected from Mauritius Island by Dr. L. Tsacas) contained 7 *Eco*RI sites and an A+T-rich region of 3.8 kb. These flies were designated line II. The mtDNA molecules of other flies taken from the same stock of *D. mauritiana* contained the same 7 *Eco*RI sites plus one extra *Eco*RI site, and an A+T-rich region of 4.6 kb (line I). We next constructed heteroduplexes between *Eco*RI fragments of *D. mauritiana* line II mtDNA and either circular molecules or *Eco*RI fragments of *D. mauritiana* line I mtDNA, in the manner described above. In each heteroduplex which included fragments containing A+T-rich regions of both types of molecule, two or three regions of strand separation were observed in a centrally located 1.0 kb portion of the A+T-rich region. The two strands in each of these regions were of unequal length. It was noted that the sum of the differences in length (1.2 to 2.0 kb in different heteroduplexes) was greater than the overall difference in length (approximately 0.8 kb) of the A+T-rich regions of line I and line II mtDNA molecules. This suggests that within at least one portion of the A+T-rich region the overall length of the nucleotide sequence of line II mtDNA is greater than the length of the corresponding nucleotide sequence of line I mtDNA.

In contrast to the complete pairing of A+T-rich regions found in heteroduplexes of mtDNA molecules of *D. mauritiana* line I and *D. simulans* Cal. (see above), we observed segments of strand separation in A+T-rich regions of heteroduplexes of mtDNA molecules of *D. mauritiana* line II and *D. simulans* Cal. These segments of strand separation compared in number, size and position to the segments of strand separation found in heteroduplexes of *D. mauritiana* line I and II mtDNAs.

We crossed *D. mauritiana* line I and line II flies in the four possible combinations and compared the numbers of F_1 and F_2 progeny. The results failed to provide evidence of infertility between *D. mauritiana* flies containing different kinds of mtDNA molecules.

We used the F_1 progeny of the crosses mentioned above to determine the mode of inheritance of *Drosophila* mtDNA. MtDNA was extracted from F_1 females and analyzed by agarose gel electrophoresis following *Eco*RI digestion. Comparisons of band patterns obtained clearly indicated that the mtDNA of progeny was derived mainly if not completely from the female parent.

The A+T-rich regions of mtDNA molecules from two strains of *D. melanogaster*, Nagasaki (Japan) and L-M. (USSR) were found to be 0.8 kb and 0.4 kb respectively smaller than the A+T-rich region of *D. melanogaster* (Oregon-R-Utah (Ore.)), the strain which we have used previously. We constructed heteroduplexes first between *Eco*RI or *Hin*dIII fragments of *D. melanogaster* Ore. and nicked circular molecules of *D. melanogaster* Nagasaki mtDNAs, and then in the two reciprocal combinations of restriction fragments and circular molecules. In all of the heteroduplexes examined the A+T-rich regions appeared completely double-stranded. This suggests that the overall difference in length of the A+T-rich regions of the two strains of *D. melanogaster* studied result from many small differences scattered throughout the A+T-rich regions.

Differences in length of 0.3 kb were also found between A+T-rich regions of mtDNA molecules of two different strains of *D. simulans*, one from California (Cal.) and the other from Peru. Examination of heteroduplexes again failed to reveal regions of strand separation within the A+T-rich regions.

Using the relative locations of the *Eco*RI sites and the A+T-rich region we determined that in all of the species mentioned above (Fig. 1), replication originates within the A+T-rich region and proceeds unidirectionally around the molecule towards the nearest common *Eco*RI site[9, 13]. In mtDNA molecules of *D. melanogaster*, *D. simulans*, and *D. mauritiana* the replication origin is located near the center of the A+T-rich region, while in molecules of *D. yakuba*, *D. takahashii*, and *D. virilis* it lies close to that end of the A+T-rich region which is distal to the nearest common *Eco*RI site (Fig. 1).

DISCUSSION

It is clear from the data presented that much greater sequence differences occur in the A+T-rich regions than in other regions of mtDNA molecules, not only from different *Drosophila* species, but also from different female lines of the same species. Defining all of the actual differences in the nucleotides between any two A+T-rich regions is not a straightforward task, particularly where the larger A+T-rich regions are concerned, as restriction enzyme cleavage sites are either rare or absent from these regions. Also, to date we have failed in our attempts to clone in bacteria, *Drosophila* mtDNA segments containing an A+T-rich region.

Our data indicate that mtDNA of *Drosophila*, like mtDNA of a number of vertebrates[14-16] is inherited mainly if not completely from the female parent. At this time the possibility of a paternal contribution to the mtDNA of progeny of any of the metazoan species examined cannot be ruled out. In *Drosophila* the nebenkern, the sperm mid-piece component derived by fusion of mitochondria in spermatids, actually enters the ovum[17]. It is not known whether this structure contains mtDNA. If it does then the amount of this mtDNA, relative to the amount of mtDNA contained in the egg (10^5 circular molecules[11]) is likely to be too low to be detected among progeny mtDNA by the methods we have used to date.

The A+T-rich region of the mtDNA of all of the species we have studied contains the origin of replication. However, the function of the remainder of this region remains unknown. As the guanine + cytosine content is indicated to be very low, it seems unlikely that the A+T-rich region contains sequences coding for proteins. This argument is supported by two reported failures to detect RNA transcripts which map in the A+T-rich region[18, 19].

Our observations on sequence relationships of the A+T-rich regions of mtDNAs of different species, and of different lines or strains of the same species, are consistent with the notion that mutations in this region which result in either a change of A to T or T to A, or losses and gains of stretches of A and T nucleotides can occur within wide limits without resulting in a selective disadvantage to the organelle containing such molecules. However, any function or model of evolution which might be proposed for the A+T-rich region must take into consideration not only the extensive differences of sequences in this region among *Drosophila*, but also the fact that evidence for heterogeneity in sequences in mtDNA molecules from *Drosophila* of any one female line has not been obtained to date.

ACKNOWLEDGEMENTS

This work was supported by National Institutes of Health Grant Nos. GM-17375 and K4-GM-70104, American Cancer Society No. NP-41B, and National Science Foundation Grant No. BMS-74-21955. We thank L. M. Okun for discussions.

REFERENCES

1. Polan, M. L., Freidman, S., Gall, J. G. and Gehring, W. (1973) J. Cell Biol. 56, 580-589.

2. Bultmann, H., and Laird, C. D. (1973) Biochim. Biophys. Acta 299, 196-209.

3. Peacock, W. J., Brutlag, D., Goldring, E., Appels, R., Hinton, C. and Lindsley, D. C. (1974) Cold Spring Harbor Symp. Quant. Biol., 71, 343-448.

4. Fauron, C. M.-R. and Wolstenholme, D. R. (1976) Proc. Natl. Acad. Sci. USA 73, 3623-3627.

5. Klukas, C. K. and Dawid, I. B. (1976) Cell 9, 615-625.

6. Fauron, C. M.-R. and Wolstenholme, D. R. (1980a) Nucleic Acids Res. 8, 2439-2452.

7. Fauron, C. M.-R., and Wolstenholme, D. R. (1980b) Submitted for publication.

8. Wolstenholme, D. R. and Fauron, C. M.-R. (1976) J. Cell Biol. 71, 434-448.

9. Goddard, J. M. and Wolstenholme, D. R. (1978) Proc. Natl. Acad. Sci. USA 75, 3886-3890.

10. Fouts, D. L., Manning, J. E. and Wolstenholme, D. R. (1975) J. Cell Biol. 67, 378-399.

11. Goldring, E. S. and Peacock, W. J. (1977) J. Cell Biol. 73, 279-286.

12. Davis, R. W., Simon, M. and Davidson, N. (1971) In Methods in Enzymology, eds. Grossman, L. and Moldave, K. (Academic Press, New York), Vol. XXI, pp. 413-428.

13. Goddard, J. M. and Wolstenholme, D. R. (1980) Nucleic Acids Res. 8, 741-757.

14. Dawid, I. B., and Blacker, A. W. (1972) Developmental Biol. 29, 152-161.

15. Hutchison, C. A. III, Newbold, J. E., Potter, S. S. and Edgell, M. H. (1974) Nature 251, 536-538.

16. Buzzo, K., Fouts, D. L. and Wolstenholme, D. R. (1978) Proc. Natl. Acad. Sci. USA 75, 909-913.

17. Perotti, M. E. (1973) J. Ultrastructure Res. 44, 181-198.

18. Bonner, J. J., Berninger, M., and Pardue, M. L. (1978) Cold Spring Harbor Symp. Quant. Biol. 42, 803-814.

19. Battey, J., Rubenstein, J. L. R., and Clayton, D. A. (1979) Proc. 8th Ann. ICN-UCLA Symp. Molecular and Cellular Biology: Extrachromosomal DNA, pp. 427-442, ed. D. Cummings, I. B. Dawid, P. Borst, S. Weissman, and F. Fox. Academic Press, New York.

MITOCHONDRIAL REPLICATION, TRANSCRIPTION AND TRANSLATION

The Organization and Expression of the Mitochondrial Genome
A.M. Kroon and C. Saccone, editors

EXPRESSION OF THE MITOCHONDRIAL GENOME OF YEAST

ANTHONY W. LINNANE, A.M. ASTIN, M.W. BEILHARZ, C.G. BINGHAM, W.M. CHOO, G.S. COBON, S. MARZUKI, P. NAGLEY and H. ROBERTS
Department of Biochemistry, Monash University, Clayton, Victoria, 3168, Australia

INTRODUCTION

In recent years the overall genetic organization and function of the yeast mitochondrial genome has been studied in detail by application of a wide range of genetic and biochemical approaches[1-3]. The genetic fine structure of some of the genes has been determined by analysis of mutants, or by direct determination of the DNA sequence. Discrete mRNA transcripts coded by different regions have been recognized together with longer RNA molecules that represent the immediate precursors of the mature transcripts[4,5]. Other investigations have been concerned with the integration of the products of mitochondrial protein synthesis together with polypeptides synthesized outside the mitochondrion to form multisubunit enzyme complexes of the inner mitochondrial membrane[2,6].

In this paper we describe the approaches we have adopted to address some specific questions concerning the expression of yeast mtDNA. The *oli2* gene has been precisely mapped on mtDNA and its gene product identified as the 20,000 dalton subunit of the mitochondrial ATPase (mtATPase). This subunit is shown to be involved in the coupling of oxidative phosphorylation. The function of the mtATPase can be influenced by interactions between the two mitochondrially synthesized components, namely the 20,000 dalton subunit and the 7,600 dalton proteolipid subunit coded by the *oli1* gene. Our studies on mitochondrial RNA have revealed the existence of a novel double stranded transcript of yeast mtDNA which has the properties of a transcript of the entire mitochondrial genome, and thus could be the precursor to mature fully processed transcripts.

CHARACTERIZATION OF THE STRUCTURAL GENE OF THE 20,000 DALTON MITOCHONDRIALLY SYNTHESIZED SUBUNIT OF THE mtATPase

We have previously reported that some *mit*⁻ mutations in the *oli2* region of the mtDNA result in the loss of synthesis of the 20,000 dalton mtATPase subunit[7]. In some of these mutants, new mitochondrial translation products with apparent molecular weights of less than 20,000 were observed. These new polypeptides co-precipitated with the mtATPase when the enzyme complex was isolated using a specific antiserum against purified oligomycin sensitive ATPase complex, which indicates that the new products either retain some antigenic determinants of an ATPase subunit, or are physically associated with the ATPase complex. These results suggest that the *oli2* mutations are probably in the structural gene of the 20,000 dalton subunit, and result in the premature termination of translation during the synthesis of this subunit.

We have now isolated several more *mit*⁻ mutants carrying mutations in the *oli2* region, and have obtained a partial order of the mutations by petite deletion analysis (Fig. 1). Petite strain G4 rescues five of the *mit*⁻ mutations which cause new mitochondrial translation products to appear. The mutation in strain M27-14 gives rise to the shortest new polypeptide observed (10,000 daltons). The remaining four *mit*⁻ mutations which lie towards the end of the *oli2* region nearest to the cytochrome *b* gene (see legend to Fig. 1) produce polypeptides of between 14,000 and 19,000 daltons. Assuming that these *mit*⁻ mutations

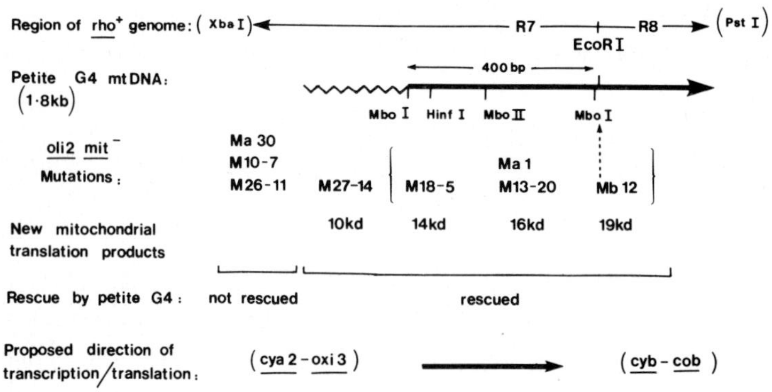

Fig. 1. Physical mapping of the *oli2* region of mtDNA. The top line indicates the position of the *Eco*RI site between the *Eco*RI restriction fragments R7 and R8 which contain sites for *Xba*I and *Pst*I, respectively. Below this is a map of some restriction sites in petite G4. The zig-zag line indicates the uncertainty in mapping the left hand end of this petite. The order of the four *oli2 mit*⁻ mutations in parentheses is based only on the sizes of the new polypeptides produced in each mutant. A number of other mutations also rescued by G4 lie in this group. These include several *mit*⁻ mutations, as well as C58 (Table 1) and *oli23*[14], but genetic analysis has so far not indicated the positions of these loci with respect to the other loci (e.g. Mb12) in this group. The mutation in strain M27-14 is separable from these, as it is rescued by another petite (not shown) which also rescues the left-hand group of mutations (e.g. Ma30) but not the mutations in parentheses. Since the latter petite also rescues some *cya2* mutations, the marker order is deduced to be: *cya2* - Ma30 - M27-14 - Mb12.

cause termination of translation, then the mutations producing the longest polypeptides lie nearest to the C-terminal coding sequence of the *oli2* gene; the probable orientation of the gene is shown in Fig. 1. The mutations in strain Ma30, M10-7 and M26-11, mapped as lying towards the *cya2-oxi3* end of the *oli2* gene (see legend to Fig. 1), cause the disappearance of the 20,000 dalton subunit, but no new mitochondrial translation products were observed.

One of the *oli2* mutations (in Mb12) has been precisely located on the wild-type mitochondrial genome. In comparative studies of restriction enzyme digestion of Mb12 mtDNA and mtDNA of its wild-type parent, mtDNA of Mb12 was found to lack the *Eco*RI restriction site located between the unique *Pst*I site and the nearby *Xba*I site (cf. Fig. 1). The loss of the *Eco*RI site is presumably due to a base change (or insertion, or deletion) within the hexanucleotide recognition site for this enzyme. The *oli2* gene thus overlaps this *Eco*RI site between wild-type *Eco*RI fragments 7 and 8 (cf. ref. 8). This more precise location of the *oli2* gene is consistent with our previous mapping studies[9] and those carried out in other laboratories[10]. Some restriction enzyme sites in petite G4 have been identified in the region of the *Eco*RI site (Fig. 1). These sites are consistent with those predicted by the DNA sequence of a petite that rescues *oli2* mutations (A. Tzagoloff, personal communication), in which a protein of approximately 27,000 daltons is apparently encoded. It is thus important to establish the relationship between this predicted amino-acid sequence and the properties of the 20,000 dalton ATPase subunit.

THE 20,000 DALTON mtATPase SUBUNIT IS ESSENTIAL FOR THE COUPLING OF OXIDATIVE PHOSPHORYLATION

All the *oli2 mit*$^-$ mutants which have previously been reported from this and other laboratories[7,11] contain very low mitochondrial ATPase activity which is not sensitive to inhibition by oligomycin. This activity is attributed to F_1-ATPase which cannot be assembled to the F_0 sector because of the gross alteration in the structure of the 20,000 dalton subunit in the mutant strains. The *mit*$^-$ mutants are thus not very useful in the investigation of the role of the 20,000 dalton subunit in oxidative phosphorylation.

More recently, we have reported[12] the isolation of two *oli2* mutants which, as a result of single mutations, showed a partial *mit*$^-$ phenotype at 28°C, and respiratory deficiency conditional upon growth at low temperatures. Unlike the *mit*$^-$ strains, these *oli2* mutants contained assembled F_0F_1-ATPase as indicated by the ability of oligomycin to completely inhibit the mutant mitochondrial ATPase (although the mutant ATPase is slightly resistant to lower concentrations of the antibiotic). However, the mutant ATPase appears to be unable to synthesize ATP at the normal rate: although the mutants contained a functional respiratory chain, they almost completely lacked the ability to grow by oxidative metabolism when forced to grow with a generation time of 7 h in glucose-limited chemostat cultures. Under these conditions, the mutant mitochondria catalysed ATP-$^{32}P_i$ exchange at only 10% of the rate of the parent strain, and gave P:O ratios of only 0.2 (compared with at least 1.5 in the parent strain), indicating that there is a defect in the coupling mechanism[12].

TABLE 1

RESTORATION OF OXIDATIVE METABOLISM IN STRAIN C58 BY LOWERING THE GROWTH RATE

Cells were grown in glucose-limited chemostat cultures. The calculated generation time is given by ln2/dilution rate. The glucose concentration of the inflowing medium was 55 mM. The outflowing medium contained less than 0.1 mM glucose and less than 20 mM ethanol. Whole cell respiration was measured polarographically at 30°C in 2.5 ml of 50 mM phosphate buffer (pH 7.0). The cell concentration was 1 mg dry wt/ml and the reaction was started by the addition of 10 µl of absolute ethanol. The concentration of the uncoupler, CCCP, was 50 µM.

Strain	Generation time (h)	Growth yield (g cell dry wt per mole glucose)	Cellular Respiration	
			Unstimulated rate (nmol O min^{-1} mg^{-1})	Stimulation by uncoupler (%)
J69-1B	10	135	68.0	150
C58	9	28	31.1	0
C58	13	56	55.8	17
C58	28	73	77.2	50

The coupling defect in the mutant strains is probably due to a small modification of the primary structure of the 20,000 dalton ATPase subunit, because the alteration is not detectable in SDS-polyacrylamide gels of immunoprecipitated ATPase. The altered subunit then either cannot be precisely orientated in the ATPase complex, or is unable to function in the normal way. We have investigated the assembly of a functional ATP synthetase in one of these *oli2* mutants by altering the growth rate in the chemostat cultures (Table 1). When the generation time was allowed to increase from 9 h to 28 h (by lowering the dilution rate) there was a partial recovery of oxidative phosphorylation in the mutant strain, as shown by an increase in the growth yield and by the ability of an uncoupler to stimulate respiration. Thus it is possible that when the mutant cells are forced to grow faster a final step in the assembly of the mtATPase which determines the precise orientation of the altered 20,000 dalton subunit in the membrane sector of the ATPase becomes rate-limiting in the assembly of the coupling mechanism.

INTERACTIONS OF THE ATPase COMPONENTS: A MUTATION IN THE STRUCTURAL GENE OF THE mtATPase PROTEOLIPID RESTORES OXIDATIVE METABOLISM IN AN *oli2* MIT^- STRAIN

We have isolated numerous spontaneous revertants of several *oli2* mit^- strains. The majority of these were phenotypically indistinguishable from the original wild-type strain (rho^+ *oli-s*). However, one revertant (denoted Ma30-11) which was isolated from strain Ma30, could grow normally by oxidative metabolism but was resistant to the inhibitors of oxidative phosphorylation, oligomycin and venturicidin. Genetic recombination analysis, and marker rescue experiments using a petite strain which retained only the *oli1* gene,

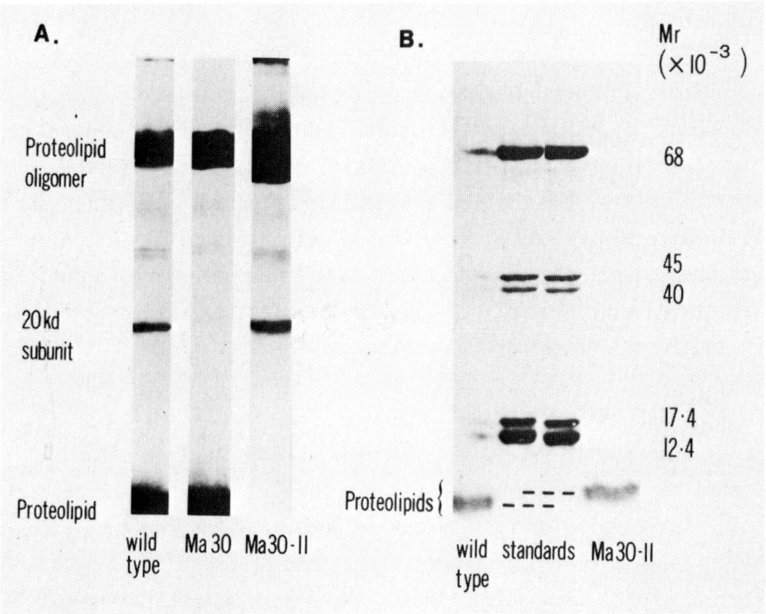

Fig. 2. A. Mitochondrial translation products associated with purified mtATPase. Cells were labelled with [^{35}S]-sulphate in the presence of cycloheximide[14], mitochondria were prepared, and the mtATPase was isolated by immunoprecipitation with a specific antiserum raised against purified F_0F_1-ATPase using a modified method of Kessler[23]. The isolated mtATPase was analysed by polyacrylamide gel electrophoresis in the presence of SDS[14] and mitochondrial translation products were visualized using fluorography. B. Purified proteolipid subunit from mtATPase. The proteolipid was isolated from mitochondria by the method of Sebald et al.[24] The extracted proteolipid was electrophoresed on SDS-polyacrylamide slab gels and was visualized with Coomassie brilliant blue G-250.

indicated that this revertant carries a mitochondrial suppressor mutation which maps in the *oli1* region of the mtDNA.

An analysis of the mitochondrially synthesized subunits of the mtATPase by polyacrylamide gel electrophoresis in the presence of SDS showed that in the suppressed strain the monomeric form of the proteolipid subunit (M_r 7,600) was absent but the oligomeric form of the proteolipid (M_r 50,000) was present (Fig. 2A). Dissociation of the proteolipid oligomer by treatment with strong alkali produced a monomer with a slightly lower mobility than the normal polypeptide in SDS polyacrylamide gels. This observation was confirmed by isolating the proteolipid from the suppressed strain (Fig. 2B). In the suppressed strain, the alteration in the proteolipid was associated with the reappearance of apparently normal amounts of the 20,000 dalton mtATPase subunit (Fig. 2A) whereas mtATPase isolated from the *oli2* mutant Ma30 contained virtually no 20,000 dalton subunit.

It thus appears that in strain Ma30-11 interactions between genetic loci have occurred to produce a functional mtATPase complex. It is evident that the defective 20,000 dalton subunit in the mutant Ma30 is either not properly assembled into the complex or is more susceptible to degradation. In the suppressed strain, therefore, the altered proteolipid encoded by the *oli1* gene could either permit assembly of the altered 20,000 dalton subunit into the ATPase or protect the subunit from degradation. This phenomenon represents one type of interaction between genetic loci, probably at the level of their gene products in the inner mitochondrial membrane.

PLEIOTROPIC MUTATIONS AND RNA PROCESSING

A second type of interaction involving different genetic loci on mtDNA occurs where mutants mapping at one locus not only have a modified gene product of this locus, but also fail to synthesize one or more proteins coded by other regions of mtDNA. These pleiotropic effects have been observed for mutations at the *cyb-cob-box*[13], *var1* [14] and *oli2*[12] loci. Moreover, a number of mutations have been recently recognized in the region of mtDNA between the *ery1* and *var1* loci which codes for no known mitochondrial protein[15]. These novel mutations have various pleiotropic effects on one or more of the major mitochondrial membrane complexes including ATPase, cytochrome oxidase and the cytochrome bc_1 complex.

A unifying scheme has been proposed by Linnane et al.[3] to explain these phenomena. It was suggested that yeast mitochondria contain long precursor RNA molecules that are transcribed from extended stretches of mtDNA covering several different gene coding regions; these possibly represent transcripts of the entire mitochondrial genome. The processing of this RNA into mature monocistronic RNA transcripts would require the specific folding of this RNA to present a substrate for processing enzymes. The processing reactions would include the cleavage of the multicistronic precursor RNA into transcripts of individual genes, as well as the removal by splicing of intervening sequences from transcripts of those genes containing introns[16,17]. It is envisaged that mutations which prevent the correct folding of the precursor RNA could interfere with the processing of transcripts arising from one or more separate

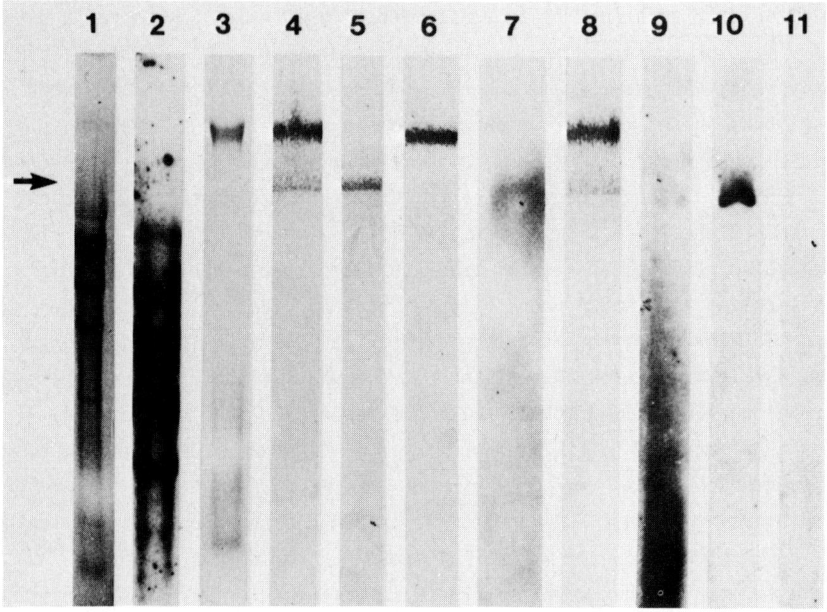

Fig. 3. Agarose gel electrophoresis of mitochondrial RNA. All samples were glyoxylated[25] and run in 1.5% agarose gels in 10 mM sodium phosphate buffer, pH 6.8. Gels were either stained with ethidium bromide (lanes 1 and 3) or transferred to diazotized paper[26] and hybridized to ^{32}P-labelled *rho*$^+$ mtDNA (lanes 2,4-6) or ^{125}I-labelled 21S rRNA (lanes 7-11). All samples were prepared from the *rho*$^+$ strain JM6, which lacks the cytoplasmic dsRNA associated with killer functions in yeast[27]. Lanes 1 and 2, total mitochondrial RNA extracted using a phenol extraction procedure; lanes 3-11, mitochondrial nucleic acids extracted using guanidine hydrochloride; prior to electrophoresis some of the guanidine extracted nucleic acids were treated further as follows: lane 5, treated with DNase; lane 6, treated with RNase; lane 7, Cs_2SO_4 gradient fraction (density 1.58 g/ml); lane 9, guanidine extract treated with proteinase K, phenol, then DNase; lane 10, LiCl soluble fraction treated with DNase; lane 11, LiCl insoluble fraction treated with DNase. The arrow indicates the position of mt-dsRNA.

regions of mtDNA, and thus potentially prevent synthesis of several gene products. Indeed Church et al.[18] have shown that mutations at the *box3* and *box7* loci in the cytochrome *b* gene not only prevent the correct processing of cytochrome *b* mRNA, but also interfere with processing of RNA transcribed from the *oxi3* region specifying subunit I of cytochrome oxidase.

CHARACTERIZATION OF mt-dsRNA: A LONG DOUBLE STRANDED RNA SPECIES IN YEAST MITOCHONDRIA

We have recently obtained evidence for the presence of a long double stranded RNA species (denoted mt-dsRNA) in yeast mitochondria which may be related to the putative long precursor RNA molecules discussed above. Whilst the mt-dsRNA is not sufficiently abundant in mitochondria to be detected in mtRNA isolated by a conventional phenol extraction procedure (Fig. 3, lanes 1 and 2), it is enriched in mitochondrial nucleic acid preparations made using guanidine hydrocloride[19]. The ethidium stained agarose gel profile of the guanidine extract indicated the presence of a slow moving band (Fig. 3, lanc 3), which consisted of DNA (see below). Transfer of this track to diazotized paper and hybridization to a labelled *rho*$^+$ mtDNA probe (Fig. 3, lane 4) clearly indicated the presence of two slowly moving bands complementary to the mtDNA probe. Treatment of the guanidine extract with DNase (lane 5), and RNase (lane 6), demonstrated that the upper band contained mtDNA whilst the lower band consisted of RNA. This latter band (arrowed in Fig. 3) contains the mt-dsRNA as will be seen from results presented below.

The hybridization of this RNA band to *rho*$^+$ mtDNA does not in itself indicate which particular mtDNA sequence is hybridizing. Accordingly, a number of ^{32}P-labelled petite mtDNA probes were prepared which represent discrete sequences drawn from many different regions of the mitochondrial genome, and which collectively cover most of the known genes. The RNA band in the guanidine extract was found to hybridize to all ten petite probes tested (data not shown). The finding that the RNA band contains sequences from all over the mitochondrial genome raises the question as to whether the sequences which hybridize to the various petites are physically in the same RNA molecules, or whether a series of individual long transcripts from each mtDNA region are co-migrating at the limiting mobility of RNA on this agarose gel. Electron microscopy of the RNA present in the guanidine extract supports the contention that there are indeed RNA molecules covering more than one coding region in mtDNA. The modal length of RNA molecules in guanidine extracts rigorously treated with DNase was found to be about 7 μm (approx. 21 kb) (most abundant size range 3-12 μm). Some molecules exceeded 15 μm, with the longest molecule about 22 μm in length which approaches the complete genome size of 25 μm. An important feature of these electron micrographs was that the RNA appeared double stranded. An example of an 8.2 μm molecule spread in the presence of a double stranded circular DNA molecule (pBR322) is shown in Fig. 4.

Three lines of evidence have been pursued to confirm the double stranded nature of this long mitochondrial RNA. First, the guanidine hydrochloride extract was subjected to isopycnic centrifugation in a Cs_2SO_4 density gradient. The gradient conditions were chosen so that the DNA (density 1.42 g/ml) is found at the top of the tube, and single stranded RNA (density 1.64 g/ml) formed a band near the bottom of the tube[20]. One central fraction of the gradient was found to contain mt-dsRNA at a density of 1.58 g/ml which is that expected of double stranded RNA[20]. Second, the putative double stranded nature of mt-dsRNA predicts that it would contain sequences complementary to mature transcripts of mtDNA. The 21S mitochondrial rRNA species is representative of such mature transcripts. Fig. 3 (lane 7) shows the result of hybridizing ^{125}I-labelled 21S-rRNA to the electrophoretically resolved molecules from the 1.58 g/ml region of a Cs_2SO_4 gradient in which the guanidine extract had been separated. This fraction, which consists of purified mt-dsRNA, is seen to contain one region of hybridization to the iodinated 21S rRNA at the

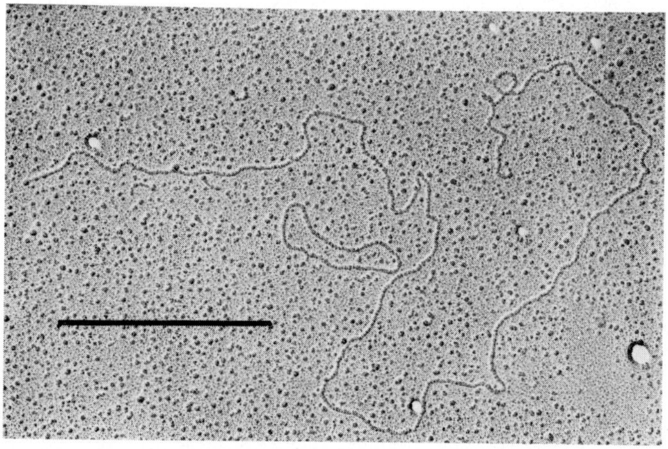

Fig. 4. Electron microscopy of RNA in guanidine hydrochloride extracts of yeast mitochondria, treated with DNase and spread using a formamide technique. The circular molecule is a relaxed circle of pBR322 (1.3 μm) mixed with the mitochondrial RNA after inactivation of the DNase with 10 mM EDTA. The bar indicates 1 μm.

position of mt-dsRNA. The guanidine extract that had not been fractionated in Cs_2SO_4 also showed the mt-dsRNA band hybridizing to iodinated 21S-rRNA (Fig. 3, lane 8) in addition to the expected hybridization due to mtDNA. Treatment of the guanidine extracted mitochondrial nucleic acids with proteinase K, followed by phenol extraction and a subsequent DNase digestion, failed to prevent hybridization of mt-dsRNA to the labelled rRNA probe (Fig. 3, lane 9). Thus, the low mobility on agarose gels of mt-dsRNA cannot be attributed to aggregation of the RNA with proteins. The third line of evidence concerning the double stranded nature of mt-dsRNA is illustrated in lanes 10 and 11 of Fig. 3 where it can be seen that the mt-dsRNA is not precipitated by 2M LiCl; this is a characteristic property of double stranded nucleic acids.

Our estimate of the level of mt-dsRNA in mitochondria is of the order of 0.01% of total mitochondrial RNA. Jakovcic et al.[21] concluded that whilst the majority of transcripts of mtDNA apparently result from asymmetric transcription, their RNA-DNA hybridization data did not exclude the existence of a small proportion of symmetrical transcripts. The levels at which we detect mt-dsRNA in the present work lie well below the maximal levels estimated by Jakovcic et al.[21]

The physical size of mt-dsRNA and its sequence content suggest that symmetrical transcription of extended regions of yeast mtDNA takes place, as occurs on HeLa cell mtDNA[22]. The mt-dsRNA could arise by annealing of separate long transcripts of both complementary strands of mtDNA. The question arises as to whether the yeast mitochondrial genome can be transcribed in its entirety in one transcriptional run through. Current experiments in our laboratory are directed towards an examination of the long RNA strands found in mt-dsRNA as possible precursors of mature transcripts, and the way in which pleiotropic effects of some mutations may be expressed through faulty processing of these RNA species.

REFERENCES

1. Cummings, D.J., Borst, P., Dawid, I.B., Weissman, S.M. and Fox, C.F., eds. (1979) Extrachromosomal DNA, Academic Press, New York, pp.1-564.
2. Tzagoloff, A., Macino, G. and Sebald, W. (1979) Ann. Rev. Biochem. 48, 419-441.
3. Linnane, A.W., Marzuki, S., Nagley, P., Roberts, H., Beilharz, M.W., Choo, W.M., Cobon, G.S., Murphy, M. and Orian, J. (1980) in the Plant Genome (Davies, D.R. and Hopwood, D.A., eds.), The John Innes Charity, Norwich, pp.99-110.
4. Grivell, L.A., Arnberg, A.C., Boer, P.H., Borst, P., Bos, J.L., van Bruggen, E.F.J., Groot, G.S.P., Hecht, N.B., Hensgens, L.A.M., van Ommen, G.J.B. and Tabak, H.F. (1979), see ref. 1, pp.305-324.

5. Levens, D., Edwards, J., Locker, J., Lustig, A., Merten, S., Morimoto, R., Synenki, R. and Rabinowitz, M. (1979), see ref. 1, pp. 287-304.

6. Schatz, G. (1979) FEBS Lett. 103, 201-211.

7. Roberts, H., Choo, W.M., Murphy, M., Marzuki, S., Lukins, H.B. and Linnane, A.W. (1979) FEBS Lett. 108, 501-504.

8. Morimoto, R. and Rabinowitz, M. (1979) Mol. Gen. Genet. 170, 11-23.

9. Choo, K.B., Nagley, P., Lukins, H.B. and Linnane, A.W. (1977) Mol. Gen. Genet. 153, 279-288.

10. Morimoto, R., Merten, S., Lewin, A., Martin, N.C. and Rabinowitz, M. (1978) Mol. Gen. Genet. 163, 241-255.

11. Foury, F. and Tzagoloff, A. (1976) Eur. J. Biochem. 68, 113-119.

12. Murphy, M., Roberts, H., Choo, W.M., Macreadie, I., Marzuki, S., Lukins, H.B. and Linnane, A.W. (1980) Biochim. Biophys. Acta (in press).

13. Claisse, M.L., Spyridakis, A. and Slonimski, P.P. (1977) in Mitochondria 1977. Genetics and Biogenesis of Mitochondria (Bandlow, W., Schweyen, R.J., Wolf, K. and Kaudewitz, F., eds.), de Gruyter, Berlin, pp. 337-344.

14. Murphy, M., Choo, K.B., Macreadie, I., Marzuki, S., Lukins, H.B., Nagley,P. and Linnane, A.W. (1980) Arch. Biochem. Biophys. (in press).

15. Machin, K., Astin, A., Macreadie, I., Polkinghorne, M. and Lukins, H.B. (1980) Proc. Aust. Biochem. Soc. 13, 125.

16. Borst, P. and Grivell, L.A. (1978) Cell 15, 705-723.

17. Bernardi, G. (1978) Nature 276, 558-559.

18. Church, G.M., Slonimski, P.P. and Gilbert, W. (1979) Cell 18, 1209-1215.

19. Beilharz, M.W., Cobon, G.S., Nagley, P. and Linnane, A.W. (1980) submitted for publication.

20. Szybalski, W. (1968) Methods Enzymol. 12B, 330-360.

21. Jakovcic, S., Hendler, F., Halbreich, A. and Rabinowitz, M. (1979) Biochemistry 18, 3200-3205.

22. Murphy, W.I., Attardi, B., Tu, C. and Attardi, G. (1975) J. Mol. Biol. 99, 809-814.

23. Kessler, S.W. (1975) J. Immunol. 115, 1617-1624.

24. Sebald, W., Wachter, E. and Tzagoloff, A. (1979) Eur. J. Biochem. 100, 599-607.

25. McMaster, G.K. and Carmichael, G.G. (1977) Proc. Natl. Acad. Sci. USA, 74, 4835-4838.

26. Alwine, J.C., Kemp, D.J. and Stark, G.R. (1977) Proc. Natl. Acad. Sci. USA, 74, 5350-5354.

27. Vodkin, M. (1977) J. Bacteriol. 132, 346-348.

The Organization and Expression of the Mitochondrial Genome
A.M. Kroon and C. Saccone, editors

TRANSCRIPTION AND PROCESSING OF YEAST MITOCHONDRIAL RNA

DAVID LEVENS, ARTHUR LUSTIG, BARUCH TICHO, RICHARD SYNENKI, SYLVIE MERTEN, THOMAS CHRISTIANSON, JOSEPH LOCKER AND MURRAY RABINOWITZ
Departments of Biochemistry, Biology, Medicine and Pathology, The University of Chicago, Chicago, Illinois (U.S.A.)

INTRODUCTION

The mitochondrial DNA of yeast codes for messages specifying three of seven subunits of cytochrome oxidase, four of nine subunits of ATPase, one subunit of the cytochrome bc_1 complex and a peptide, Var 1, associated with mitochondrial ribosomes[1-3]. Mitochondrial ribosomes contain a 21S and a 14S rRNA which are both encoded by mtDNA. In addition about 25 tRNA's are transcribed from mtDNA. Yeast mitochondria contain a 75kb, circular genome with an A-T content of 82%. While many transcripts associated with mitochondrial gene products have been mapped[4,5], little is known about the regulatory components governing the expression of this genome. Recently our laboratory has been studying the transcription and processing of mitochondrial RNA's. We have used a variety of techniques to locate the sites of transcriptional initiation and to define the intermediates in the processing of the large rRNA. We have also studied the mitochondrial RNA polymerase and have begun to investigate its role in the synthesis of mitochondrial RNA.

Since the mitochondrial RNA polymerase is the product of a nuclear gene and hence is translated on cytoplasmic ribosomes, the expression of this gene is likely to play a regulatory role in the derepression of the mitochondrial genome. The RNA polymerase is a matrix enzyme that must cross both mitochondrial membranes to reach its site of action. In contrast cytochrome oxidase, which is an end product of derepression, is a membrane protein assembled from both sides of the mitochondrial membranes[3]. We have studied the biosynthesis of cytochrome oxidase and have found that the four nuclear coded subunits are translated as distinct and separate polypeptides *in vitro*. The expression of the cytochrome oxidase as well as the mitochondrial RNA polymerase genes is determined by the physiologic status of the yeast cell. Therefore the regulation of the synthesis of mRNA's directing the synthesis of these peptides was investigated.

RESULTS AND DISCUSSION

Both petite and grande strains of yeast extensively transcribe their mitochondrial genomes[5]. Our laboratory has previously shown that at least 70% of one single-strand equivalent of mtDNA is represented in transcripts[6]. Transcription is largely asymmetric and is likely to occur from more than one promoter[6,7]. Mitochondrial genes are separated by regions of DNA with an A-T content of greater than 95%[8]. Although the high A-T regions are transcribed, their information content, if any, remains obscure. At least three mitochondrial genes possess intervening sequences: the 21S rRNA gene, the cytochrome b (COB) gene and the OXI-3 gene[2,9]. These intervening sequences are present in mitochondrial transcripts[10,11]. We have examined the transcription and processing of mitochondrial RNA in petites and have found that, in some cases, petites transcribe and process their RNA's in a manner similar to that observed in the grande[5]. The aggregate molecular weight of all transcripts associated with some regions of the genome exceeds the coding capacity of the DNA. We concluded that multiple species of RNA are present for individual loci, probably representing intermediates in the processing of the primary transcripts.

We have studied the transcription and processing of the 21S rRNA gene in the petite F11[11]. This petite retains a 12 kb piece of the grande genome encoding the 21S rRNA as well as a few tRNA's. The F11 genome has been extensively dissected with restriction enzymes allowing the construction of a fine structure map[12]. Using defined DNA probes we have examined the sequences present in F11 transcripts immobilized on diazobenzyloxymethyl cellulose paper. This technique coupled with electron microscopic analysis of R-loop hybrids of F11 RNA to F11 DNA has allowed us to propose a processing scheme for the maturation of the 21S rRNA. The largest transcript we have detected from this region contains an intervening sequence and a 3' extension not found in the mature rRNA. The intervening sequence and the 3' extension are removed in discrete steps. We have determined the extent of transcription in F11 by hybridizing total F11 RNA labeled with γ-^{32}P-ATP and polynucleotide kinase after nicking with alkali to F11 DNA immobilized on nitrocellulose. This hybridization defined the minimum fraction of the F11 genome transcribed. Other sequences, if present, must be of low abundance and hence are either poorly transcribed or rapidly degraded. None of the techniques used thus far to study the transcription and maturation of the 21S rRNA in F11 have identified the transcription initiation sites nor indicated if the 5' end of the 21S rRNA primary transcript is processed.

To determine the sites of the initiation of transcription, a method was sought that would specifically label the 5' ends of primary transcripts. Labeling with γ-^{32}P-ATP and polynucleotide kinase was considered to be inadequate.

Since yeast mitochondria extensively process their RNA's with cleavages likely to occur at any point along the body of the transcript, polynucleotide kinase could label the 5' terminus of a cleavage product from the 3' end of a primary transcript. Upon hybridization to mtDNA, the label would then map at a site distant from the point of the initiation of transcription. Therefore we exploited a feature accepted as unique to primary transcripts: the possession of a polyphosphate terminus. On the assumption that mitochondrial primary transcripts retain their 5' triphosphate nucleotide, we undertook to specifically label that nucleotide. To accomplish this we purified the enzyme guanylyl transferase from vaccinia virus cores[13,14]. Guanylyl transferase catalyzes the coupling of GTP to a di- or triphosphate terminated polyribonucleotide with the formation of a 5'-5' triphosphate linkage. In this reaction the α-phosphate of the GTP is incorporated into the cap while pyrophosphate is released. In this manner α-^{32}P-GTP specifically labels primary transcripts. This procedure has been used to identify and locate the initiation site of transcription for the 40S rRNA precursor in *Xenopus laevis*[15,16].

In vitro capped grande mitochondrial RNA displayed a small number of transcripts on agarose-urea gel electrophoresis and autoradiography (Fig. 1). The labeling was apparently specific for mitochondrial RNA since no cytoplasmic species were detected even with preparations highly contaminated with the cytoplasmic rRNA's. Paper electrophoresis of the capped RNA following treatments with nuclease P1 and bacterial alkaline phosphatase or alkali, followed by nuclease P1, indicated that the α-^{32}P-GTP was incorporated into a *bona fide* cap structure. The 21S rRNA was strongly labeled indicating that this species retains its 5' triphosphate. The 14S rRNA was not labeled but a transcript of slightly lower electrophoretic mobility was present. While this transcript was not visible in ethidium bromide stained agarose-urea gels of grande mitochondrial RNA, it was prominent in petites retaining this region. We concluded that the small rRNA is derived from a larger precursor following cleavage of the 5' end of the primary transcript. We also mapped transcription initiation sites on the mitochondrial genome by hybridizing capped total grande mitochondrial RNA to mtDNA (Fig. 2). Prior to hybridization, the RNAs were nicked with alkali to insure hybridization of each capped 5' terminus to a single restriction fragment. We located at least five initiation sites in an Eco R1-Hinc II restriction enzyme digest of mtDNA from the grande strain MH41-7B. Intense hybridization was observed to the restriction fragment containing the 5' end of the 21S rRNA. Since the 5' end of the large rRNA is likely to be protected by ribosomal structure from the action of phosphatases and nucleases it is probably the most abundant 5' triphosphate terminus in mitochondrial RNA.

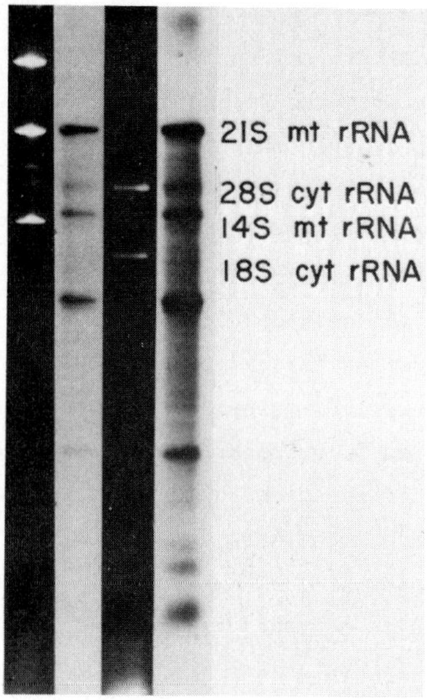

Fig. 1. Agarose-urea gel electrophoresis of capped MH41-7B RNA on 1.5% agarose-6M urea gels. Left to right: unlabeled MH41-7B RNA stained with ethidium bromide, autoradiogram of capped RNA, ethidium bromide stained capped RNA (this preparation was highly contaminated with cytoplasmic RNA), autoradiogram of capped RNA exposed for a longer time.

The fragments containing transcription initiation sites are shown in Figure 2. Finer restriction maps are required to correlate transcription initiation sites with specific genes. The examination of petites which transcribe their DNA's similarly to the grande should also be useful for detailed analysis of individual transcription units.

We have used guanylyl transferase to extend our knowledge of transcription in petite F11. Agarose urea gels of capped total F11 RNA revealed the presence of triphosphate termini on the 21S rRNA and its precursors (Fig. 3). In addition, a transcript of 0.7 kb was observed. Transcription initiation sites were mapped by hybridizing alkali nicked capped F11 RNA to an Hpa II-Hinc II restriction enzyme digest of F11 DNA immobilized on nitrocellulose (Fig. 4). Hybridization to two fragments was detected. Fragments 1 and 7

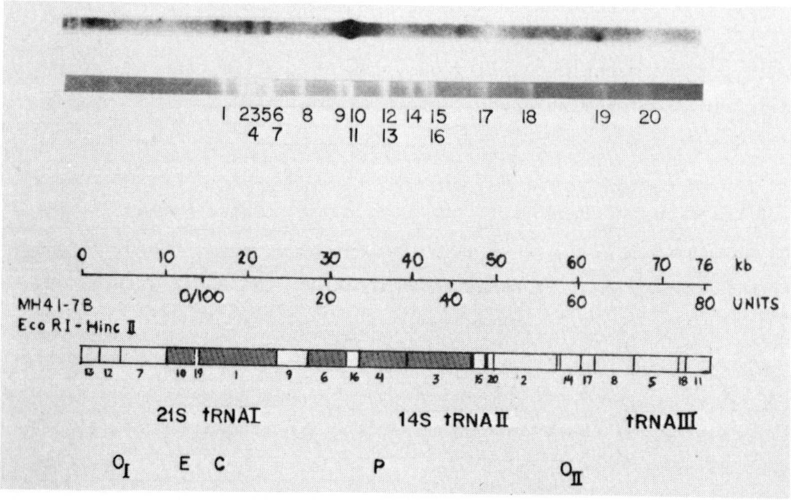

Fig. 2. Hybridization of capped MH41-7B RNA to MH41-7B mtDNA digested with Eco R1 and Hinc II: Total mitochondrial RNA was capped and hybridized to mtDNA immobilized on nitrocellulose (on top). Thick lines on bottom indicate fragments containing transcription initiation sites.

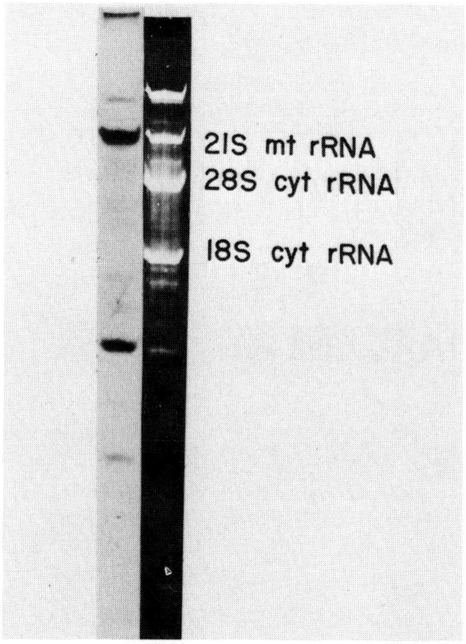

Fig. 3. Agarose-urea gel electrophoresis of capped F11 RNA on 1.5% agarose-6M urea gels. Left: ethidium bromide stain of capped RNA. Right: autoradiogram of capped RNA.

were clearly visualized. When the same restriction enzyme digest was probed with 21S rRNA labeled with γ-^{32}P-ATP and polynucleotide kinase, hybridization to fragment 1 was detected suggesting that this fragment contains the sequence responsible for the initiation of 21S rRNA transcription. The 0.7 kb transcript was eluted from agarose-urea gel and observed to hybridize with fragment 7 of the Hpa II-Hinc II restriction enzyme digest of F11 mtDNA (Fig. 4). The transcription initiation site of the 0.7 kb transcript is far removed from the 21S rRNA gene indicating the presence of multiple initiation sites. Direct sequence analysis of 21S rRNA labeled with either guanylyl transferase and α-^{32}P-GTP or polynucleotide kinase and γ-^{32}P-ATP yielded the same sequence

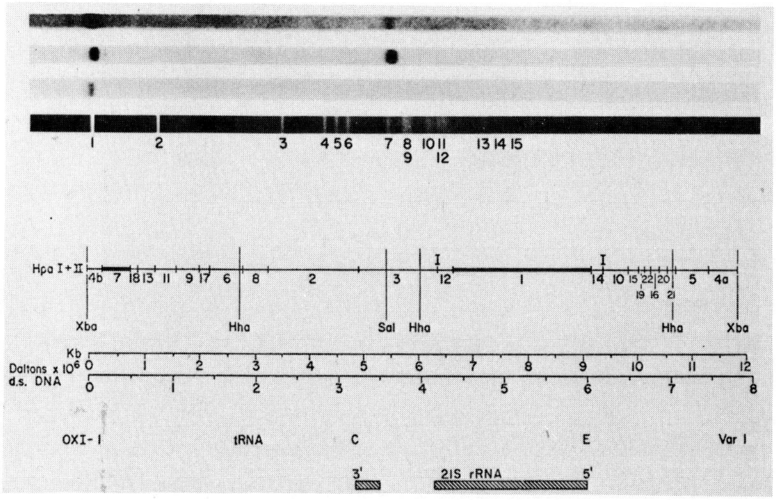

Fig. 4. Hybridization of capped F11 RNA to F11 mtDNA digest with Hpa II and Hinc II. Top to bottom: hybridization of total capped F11 mitochondrial RNA, hybridization of the 0.7 kb, transcript after elution from agarose urea gels, hybridization of RNA eluted from a portion of the agarose-urea gel containing no identifiable transcript indicating the presence of capped 21S rRNA degradation products throughout the gel, DNA-gel, and map indicating transcription initiation sites in F11 genome.

(Fig. 5). Therefore not a single nucleotide is cleaved from the 5' end of the 21S rRNA primary transcript. All other known large rRNA's are processed at their 5' ends. The 21S rRNA is initiated with A. In contrast, analysis of the 0.7 kb transcript by complete digestion with pancreatic ribonuclease suggests that this RNA is initiated with a pyrimidine, a rare occurrence in other organisms.

pppAGUUUUUAGUAGAAUAAUAGA...

Fig. 5. Sequence of the 5' end determined for 21S rRNA labeled with either guanylyl transferase or polynucleotide kinase (after treatment with bacterial alkaline phosphatase). Nucleotides 4-7 are cleaved only weakly by *Bacillus cereus* ribonuclease and hence are identified with greater uncertainty than the surrounding sequence.

The development of a method for identifying the 5' ends of primary transcripts provides a reference point to study the specificity of *in vitro* transcriptional initiation of mtDNA by the mitochondrial RNA polymerase. The enzyme has been purified to greater than 90% homogeneity in our laboratory and has been shown to contain a 45000 MW polypeptide. The RNA polymerase transcribes the DNA of petite F11 in a highly non-random manner Fragment 1 of an Hpa II-Hinc II restriction enzyme digest of F11 mtDNA is heavily transcribed *in vitro* and is the same fragment that contains the site for the initiation of 21S rRNA gene transcription. In addition, fragments 4 and 5 are heavily transcribed *in vitro*. These fragments are adjacent to each other with fragment 4 also neighboring fragment 7, which contains the initiation site for the 0.7 kb nucleotide transcript found in F11. If the initiation sites *in vivo* and *in vitro* are the same and if transcription begins in fragment 7 just at its junction with fragment 4, continuous labeling with α-^{32}P-UTP might not result in significant hybridization to fragment 7. *In vitro* transcription reactions in the presence of γ-^{32}P nucleotide triphosphates should allow us to determine definitively whether the purified RNA polymerase possesses the ability to initiate transcription with biological specificity.

Another aspect of our work involves the control the nuclear genome exerts on individual nuclear-coded polypeptides of structural and regulatory proteins destined for the mitochondria. In particular, we have examined the nature and regulation of the initial translation products of mitochondrial RNA polymerase and the nuclear-coded peptides (IV-VII) of cytochrome oxidase.

RNA isolated from cells grown on a variety of carbon sources was used to direct a reticulocyte cell free ribosomal system. The products were subjected to immunobinding using antibodies prepared against the 45000 MW peptide of mitochondrial RNA polymerase. When RNA from raffinose grown cells was used to stimulate the cell free system a series of three polypeptides of 46000 to 48000 MW was produced which immunoreacted with anti-45000 MW peptide antibodies (Fig. 6). However, only one of these three polypeptides was competed effectively by the addition of unlabeled mitochondrial RNA polymerase. Similarly, ^{125}I-mtRNA polymerase immunoreactive with the anti-45 Kd peptide was competed only by the addition of unlabeled mitochondrial RNA polymerase. In no case is competition accomplished by the addition of *E. coli* RNA polymerase.

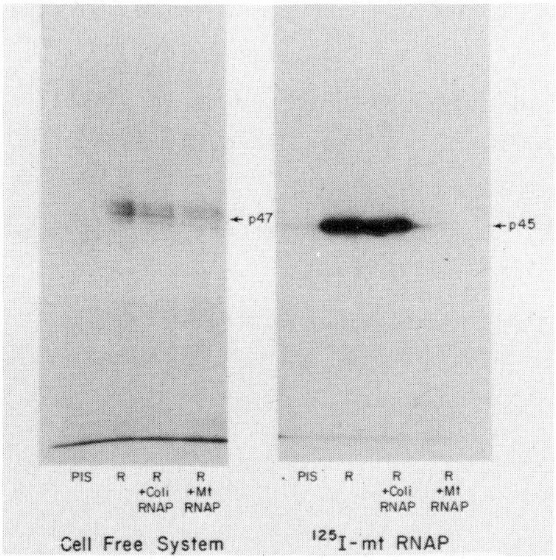

Fig. 6. Left: Gel electrophoresis of cell free products immunobound with pre-immune sera (PIS) or anti-45000 MW peptide (R) in the presence or absence of *E. coli* RNA polymerase and mitochondrial RNA polymerase. Right: Gel electrophoresis of ^{125}I-mtRNA polymerase immunobound as above.

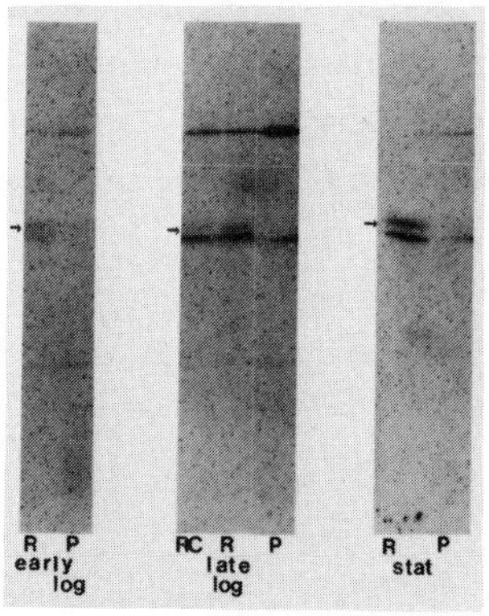

Fig. 7. Gel electrophoresis of *in vitro* products directed by total RNA isolated from early log (left), late log (middle) and stationary glucose-grown cells (right) immunobound with pre-immune sera (P) or anti-45000 MW peptide (R) in the presence (RC) or absence of mitochondrial RNA polymerase.

When RNA isolated from different stages of glucose growth are used to program the cell free system and the products subjected to immunobinding, an induction of the precursors is observed (Fig. 7). During early logarithmic growth very small amounts of the polymerase 'precursor' and the associated peptides are produced. However, late in logarithmic growth, an induction of the presumptive precursor is observed. In stationary phase a burst of 'precursor' synthesis appears to take place. Thus, the synthesis of the presumptive precursor as well as the associated peptides are induced during mitochondrial derepression in late log and stationary phase.

Our data indicate that yeast RNA produces a product of 47000 MW immunologically related to RNA polymerase suggesting the presence of a cytoplasmic precursor. This peptide as well as the associated proteins are regulated by the cell's physiological state.

In addition to studying mitochondrial RNA polymerase we have analyzed the nature and regulation of the nuclear-coded peptides of cytochrome oxidase. Antibodies were raised against holo-cytochrome oxidase and to some of its individual peptides. When antibodies to cytochrome oxidase (COAb) were reacted with a ^{35}S-labeled mitochondrial lysate, all of the peptides of oxidase with the exception of subunit VI, which is not labeled by this procedure, were immunobound. Two independent antibodies raised against subunit V (CPAb, VAb [provided by G. Schatz]) specifically bind this peptide as well as small amounts of higher molecular species (Fig. 8A). Subunit VI is immunobound by anti-holocytochrome oxidase and anti-subunit VI (VIAb, kindly provided by G. Schatz) when ^{125}I-cytochrome oxidase was used as antigen (Fig. 8B).

RNA isolated from cells during a high to low glucose shift was used to stimulate the reticulocyte cell free system. The products were immunobound and fractionated on 15% SDS-polyacrylamide gels[17] (Fig. 8C). Anti-subunit V

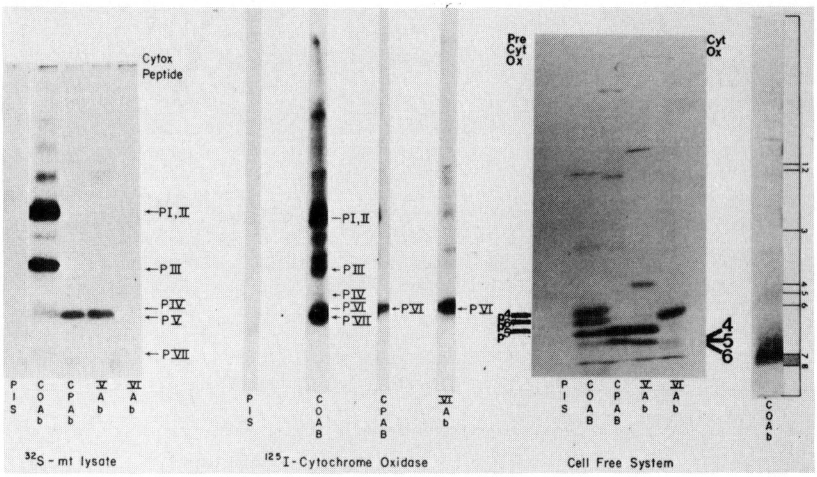

Fig. 8. A. ^{35}S-labeled lysate immunobound with a series of antibodies discussed in text (PIS: pre-immune sera) and fractionated by 15% SDS-PAGE. B. ^{125}I-cytochrome oxidase immunobound with a series of antibodies as described in text and fractionated by 15% SDS-PAGE. C. ^{35}S-*in vitro* products immunobound with identical antibodies as in part A and fractionated by 15% SDS-PAGE. Arrows indicate the position of precursors (left) and products (right). D. ^{35}S-*in vitro* product directed by 8S poly(A) RNA, immunobound with anti-holocytochrome oxidase (COAb) and fractionated by 20% SDS-PAGE. Staining pattern is shown on the right.

(12500 MW) reacts with a peptide at ~15000 MW while antibodies raised to subunit VI (12000 MW) bind a peptide of ~16000 MW *in vitro*. Antibodies prepared against holo-cytochrome oxidase bind these peptides as well as a protein at 17000 MW. A peptide migrating near the front on 15% SDS-gels immunoreactive with cytochrome oxidase antibody is more clearly visualized with 8S-poly(A) RNA is used to direct the cell free system. The *in vitro* product migrates slower than subunit VII on 20% SDS-polyacrylamide gels. Since 8S RNA does not code for any of the other immunoreactive species, this peptide is unlikely to be a degradation product of other proteins and is likely to be related to subunit VII (Fig. 8D). The remaining peptide at 12000 MW is most probably related to subunit IV (14000 MW). Higher molecular species are also present with some antibodies.

RNA fractionated on 5-20% linear sucrose gradients (Fig. 9) was used to program the cell free system and analyzed using antibodies to glyceraldehyde 3-phosphate dehydrogenase, holo-cytochrome oxidase, or subunit V (CPAb). Low molecular weight RNA (8-15S) codes for the smaller products immunobound by these antibodies while the higher molecular weight RNA (~28S) codes for the larger species at 90000 and 80000 MW. Thus, the immunobound products do not represent proteolytic fragments of a larger molecular weight polyprotein. These primary translation products are likely to be precursors for each of the peptides of cytochrome oxidase. The nature of the higher molecular weight products are unknown but do not represent 'polyprotein precursors' since the identities of higher molecular weight products immunobound are highly dependent upon the antibody utilized. A schematic representation of the proposed precursor-product relationship for each of the peptides is shown in Fig. 10.

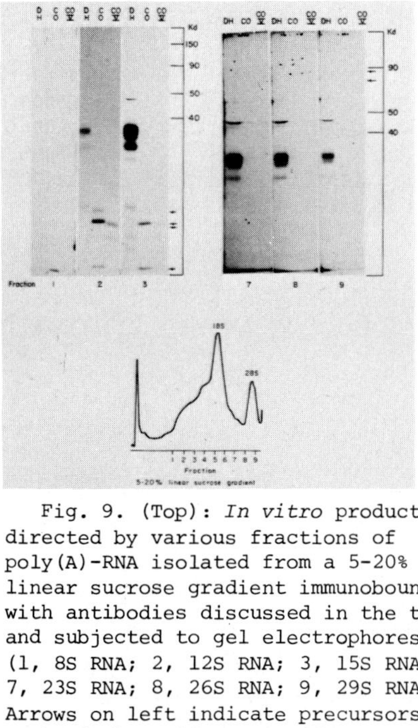

Fig. 9. (Top): *In vitro* products directed by various fractions of poly(A)-RNA isolated from a 5-20% linear sucrose gradient immunobound with antibodies discussed in the text, and subjected to gel electrophoresis (1, 8S RNA; 2, 12S RNA; 3, 15S RNA; 7, 23S RNA; 8, 26S RNA; 9, 29S RNA). Arrows on left indicate precursors for each of the oxidase peptides. Arrows on right indicate positions of higher molecular weight forms. (Bottom): Typical 5-20% linear sucrose gradient.

In order to ascertain whether these peptides are capable of becoming associated with mitochondria, *in vitro* mitochondria were incubated with the cell free product by the method of Schatz[18].

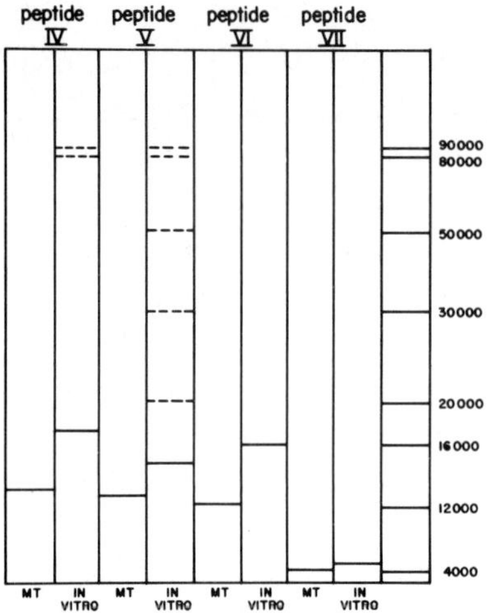

Fig. 10. Schematic representation of the relationship between *in vitro* precursors and mitochondrial products for peptides 4-7 of cytochrome oxidase. Dotted lines indicate higher molecular products present with certain antibodies. Molecular weight standards are shown on right.

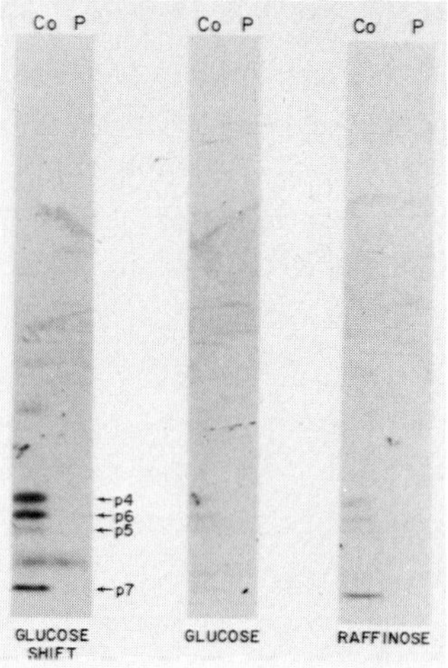

Fig. 11. *In vitro* products directed by RNA isolated from raffinose grown cells, glucose grown cells and cells grown during a high to low glucose shift were immunobound with pre-immune sera (P) and anti-cytochrome oxidase (CO) and fractionated by 15% SDS-PAGE.

Under these conditions the presumptive precursors for subunits VI, V and VII become associated with the mitochondria (data not shown).

Higher molecular forms at 90000, 80000 and 40000 MW immunoreactive with an anti-V antibody do not associate with the mitochondria. Furthermore, glyceraldehyde 3-phosphate dehydrogenase does not bind to mitochondria in substantial levels (data not shown). The specific association of these *in vitro* peptides with mitochondria suggests that the proper 'signal'[19] for specific binding to mitochondria is present in these 'precursors'.

These data further suggest that there are individual precursors for each of the nuclear coded peptides of oxidase. We have found no evidence for the polyprotein precursor described by Poyton[20,21] containing all of the oxidase peptides. Individual precursors have also been found by Schatz in two cases of several nuclear-coded peptides of ATPase[18] and cytochrome bc_1[22].

In order to investigate the regulation of the presumptive precursor, RNAs isolated from derepressive raffinose-grown cells[23], repressive mid-log glucose grown cells[24] and cells grown during a high to low glucose shift (derepressive) were used to program the cell free system. The products were subsequently immunobound with antibodies against holo-cytochrome oxidase (Fig. 11). The precursors are most abundant on raffinose when 'high to low glucose shift' RNA is used to program the system, but are significantly repressed when 'mid-log glucose' RNA is translated.

Thus, all of the cytochrome oxidase 'precursors' are under physiological control in a manner which reflects the derepression of mitochondrial products. Whether these peptides are coordinately regulated under a variety of physiological states is a matter of current investigation.

Future work in our laboratory will focus on the expression of nuclear genes required for derepression of the mitochondrial genome and on the mechanism of coordinate synthesis of proteins such as cytochrome oxidase assembled from both inside and outside the mitochondrion. The role of the mitochondrial RNA polymerase in these processes will be further investigated.

ACKNOWLEDGMENTS

This study was supported in part by National Institutes of Health Grants HL-0442 and HL-09172 to M.R., GM-27795 to J.L. and PHS 5T32 GM-07281 to D.L.; Grant NP-281 from the American Cancer Society, and a grant from the Louis Block Fund of the University of Chicago. J.L. is a Junior Faculty Clinical Fellow of the American Cancer Society.

REFERENCES

1. Locker, J. and Rabinowitz, M. (1979) Methods in Enzymology, 56, (eds. S. Fleischer and L. Packer), Academic Press, New York, New York, pp. 3-16.
2. Borst, P. and Grivell, L.A. (1978) Cell, 15, 705-723.
3. Schatz, G. and Mason, T.L. (1974) Annu. Rev. Biochem., 43, 51-87.
4. Van Ommen, G.J.B. *et al.* (1979) Cell, 18, 511-523.
5. Morimoto, R. *et al.* (1979) J. Biol. Chem., 254, 12461-24170.
6. Jakovcic, S. *et al.* (1979) Biochemistry, 15, 3200-3205.
7. Levens, D. *et al.* (1979) Extrachromosomal DNA. ICN-UCLA Symposium on Molecular and Cellular Biology, 15, (eds. D. Cummings, P. Borst, I. Dawid, S. Weissman, and C.F. Fox), Academic Press, New York, New York, pp. 287-304.
8. Bernardi, G. *et al.* (1972) J. Mol. Biol., 65, 173-189.
9. Slonimski, P.P. *et al.* (1978) Biochemistry and Genetics of Yeast (eds. M. Bacilla, B.L. Horecker, and A.O.M. Stoppani), Academic Press, New York, New York, pp. 391-402.
10. Van Ommen, G.J.B. *et al.* (1980) Cell, 20, 173-183.
11. Merten, S. *et al.* (1980) Proc. Natl. Acad. Sci. USA, 77, 1417-1421.
12. Locker, J. *et al.* (1980) Ann. N. Y. Acad. Sci., in press.
13. Martin, S.A. *et al.* (1975) J. Biol. Chem., 250, 9322-9329.
14. Monroy, G. *et al.* (1977) J. Biol. Chem., 253, 4481-4489.
15. Reeder, R. *et al.* (1977) Proc. Natl. Acad. Sci. USA, 74, 5402-5406.
16. Sollner-Webb, B. and Reeder, R.H. (1979) Cell, 18, 485-499.
17. Blatter, D.P. *et al.* (1972) J. Chromatography, 74, 147-155.

18. Maccecchini, M.-L. *et al.* (1979) Proc. Natl. Acad. Sci. USA, 76, 343-347.

19. Blobel, G. (1980) Proc. Natl. Acad. Sci. USA, 77, 1496-1500.

20. Poyton, R.O. and McKemmie, E. (1979) J. Biol. Chem., 254, 6763-6771.

21. Poyton, R.O. and McKemmie, E. (1979) J. Biol. Chem., 254, 6772-6780.

22. Cote, C. *et al.* (1979) J. Biol. Chem., 254, 1437-1439.

23. Zitomer, R.S. *et al.* (1979) Proc. Natl. Acad. Sci. USA, 76, 3627-3631.

24. Mahler, H.R. *et al.* (1975) Membrane Biogenesis (ed. A. Tzagoloff), Plenum Press, New York, New York, pp. 15-61.

© 1980 Elsevier/North-Holland Biomedical Press
The Organization and Expression of the Mitochondrial Genome
A.M. Kroon and C. Saccone, editors

EXPRESSION OF THE MOUSE AND HUMAN MITOCHONDRIAL DNA GENOME

Jim Battey, Phillip Nagley, Richard A. Van Etten,
Mark W. Walberg and David A. Clayton

Department of Pathology, Stanford University School of Medicine
Stanford, California 94305, USA

INTRODUCTION

It is well known that the ability to investigate biological phenomena is directly dependent on the available technology. In the case of studies on the mechanism of mouse mitochondrial DNA (mtDNA) replication, the ease of isolation and high degree of mtDNA purity obtainable by dye-CsCl gradient centrifugation has facilitated work which has elucidated the basic mechanism of replication of this closed circular DNA[1,2,3]. A major benefit of these results is that animal mtDNA is now well characterized and provides an excellent model system in which to study the expression of a genome known to code for ribosomal RNAs (rRNA), transfer RNAs (tRNA) and messenger RNAs (mRNA) in contrast to the specialized functions of small eukaryotic viral genomes. The advent of DNA and RNA sequencing capabilities, along with the ability to isolate nucleic acid sequences of precise length, has obviated the problems associated with less quantitative protocols for the determination of sequence complementarity. We report here a summary of recent efforts to position mitochondrial RNA (mtRNA) transcripts on the mouse and human mtDNA genomes.

RESULTS AND DISCUSSION

The ribosomal RNA region of mouse mitochondrial DNA. The mouse mitochondrial genome is an approximately 16,000 base pair closed circular DNA which can be conveniently dissected into specific subunits by restriction endonucleases (Figure 1). The positions of the two rRNA genes have been mapped adjacent to each other in the mouse[4], human[5], *Xenopus laevis*[6], *Drosophila melanogaster*[7] and rat[8] systems. In the case of mouse mtDNA, the two rRNA genes are adjacent and the 5'-end of the small (12S) rRNA is within 200 nucleotides of the heavy strand (H-strand) origin of mtDNA replication[4,9] (Figure 1). This general arrangement is also true for human mtDNA[5]. In addition, it is known that 4S RNAs are homologous to DNA sequences adjacent to and between the two rRNA genes in HeLa mtDNA[10]. This phenomenon is demonstrated in Figure 2 which shows mitochondrial rRNA hybridized to human mtDNA linearized by cleavage at a single *Bam*H I restriction endonuclease in human mtDNA. This

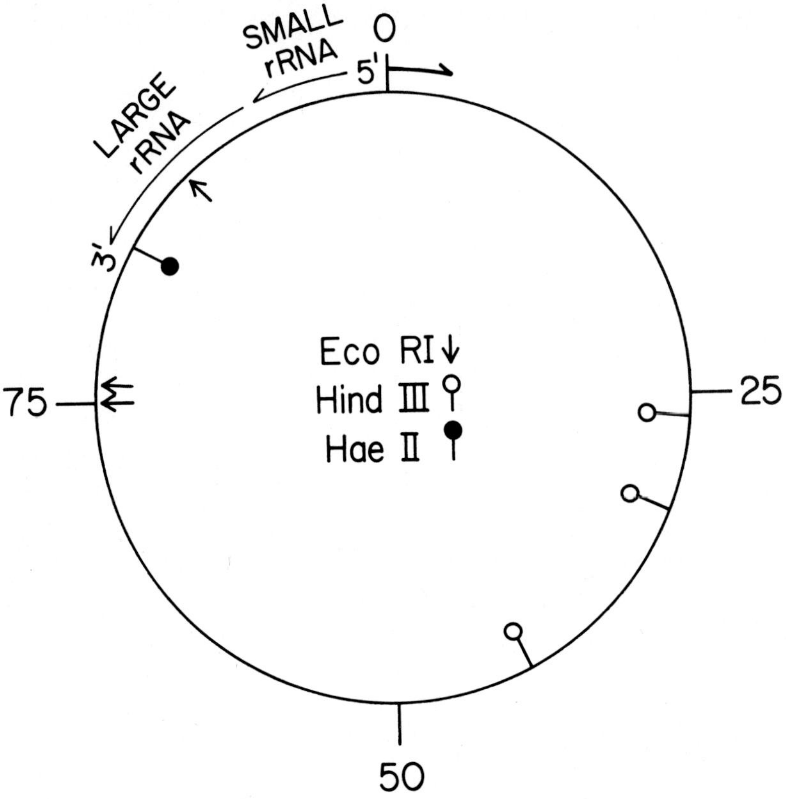

Fig. 1. Restriction endonuclease cleavage map of mouse mtDNA. The cleavage sites are as indicated relative to the D-loop H-strand origin of DNA synthesis at map position zero. The small and large rRNAs are shown and the 5'→3' direction of transcription is indicated.

electron micrograph displays two hybridized regions (R-loops) corresponding to the small and large rRNAs. The intervening region between the two rRNA genes is clearly visible as duplex mtDNA.

We have studied in detail this region of the mouse mtDNA genome. By combining RNA and DNA sequence information we have been able to identify the precise position of the rRNA genes as shown in Figure 3. The small rRNA is 956 nucleotides in size and the large rRNA consists of 1582 nucleotides. Several lines of evidence indicate that these rRNAs are not spliced. In comparing the sequences of the mouse rRNAs with other available rRNA sequences it is clear that significant homologies exist, particularly to *E. coli* rRNAs[11]. The most significant homologies are between regions of 10 to 40 nucleotides in size, and most of these significant homologies share the same order relative to each other in the nucleotide sequence in both RNAs. In contrast, regions of lesser homology do not fall in an ordered pattern and may represent rearrangement of sequence during evolution as opposed to simple deletion or addition of nucleotides. A most interesting region of homology is contained within the terminal 50 nucleotides at the 3'-end of the small rRNA from a number of organisms. In the case of E. coli 16S rRNA this region is known to contain an mRNA binding site[12,13] as well as the potential to form a 10 base pair hairpin loop structure. This 10 base pair hairpin can be constructed for all small subunit rRNAs whose sequence is known, including the mouse mitochondrial 12S rRNA. The actual nucleotide sequence in this region is not highly conserved in the mitochondrial 12S rRNA but the fact that the secondary structure is maintained would argue that it plays a functional role in protein synthesis.

The first evidence that the rRNA region of the mouse mtDNA genome is essentially all coding sequence was obtained by forming DNA-RNA hybrids between total mtRNA and restriction fragments of mtDNA which spanned the rRNA region[9]. Incubation of these hybrids with the single-strand specific nuclease S1, under relatively mild digestion conditions, yielded several classes of RNA-protected DNA regions. The largest of these classes represented several discrete, protected DNA bands slightly greater in length than the sum of the lengths of the 12S and 16S rRNAs. Such a class of protected DNA can be best explained by the presence of several small RNA transcripts interspersed among the rRNAs in the hybrids. If the distances between the ends of the various transcripts in the hybrid were small, then S1 might be expected not to cut the resulting short single-stranded DNA regions under the less stringent conditions. Increasing the stringency of the S1 conditions results in the disappearance of this largest class of protected DNAs and in the appearance of protected DNAs of the same size as mature 12S and 16S rRNAs. These data indicated that four small RNAs mapped in the rRNA region. One of these small RNAs, about 60 nucleotides in length, mapped at the 5'-end of the 12S rRNA gene. A transcript of about 80 nucleotides in length maps in the region between the 12S and 16S rRNA genes, and two small transcripts of about 80 nucleotides each mapped at the 3'-end of the 16S rRNA gene. In addition, to account for this variability in the S1 protection pattern of hybrids with increasing stringency of S1 digestion conditions, these small RNAs had to map within several nucleotides of the ends of the rRNAs (Figure 3).

Direct examination of the DNA sequence of this region has confirmed these conclusions. These tRNA genes were identified by computer analysis of the mtDNA sequence and are located at the 5'-end of the 12S rRNA, between the 12S and 16S rRNAs, and at the 3'-end of the 16S rRNA. All three tRNA genes identified map

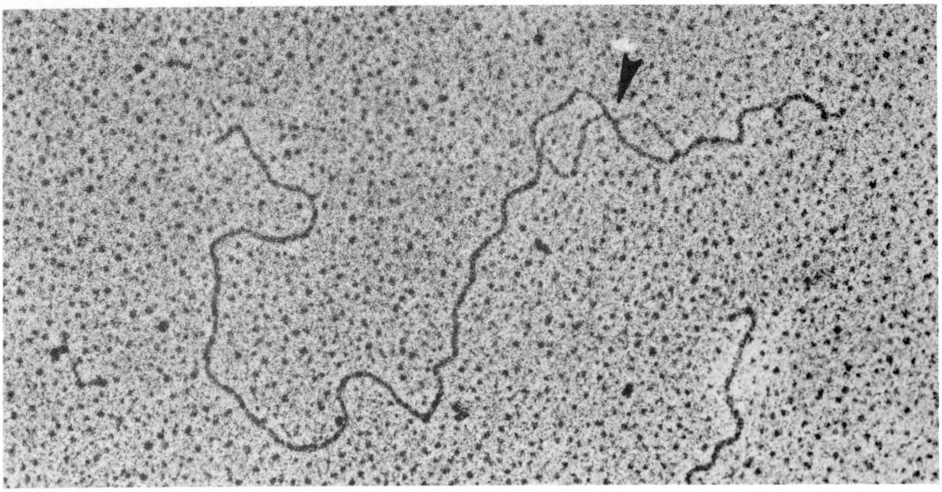

Fig. 2. Electron micrograph of human mtDNA hybridized to both the small and large rRNAs. The mtDNA has been singly-cleaved with BamH I restriction endonuclease. The arrow denotes the short DNA region between the two rRNA genes which codes for a tRNA. The entire DNA molecule is ~16,500 nucleotides in size.

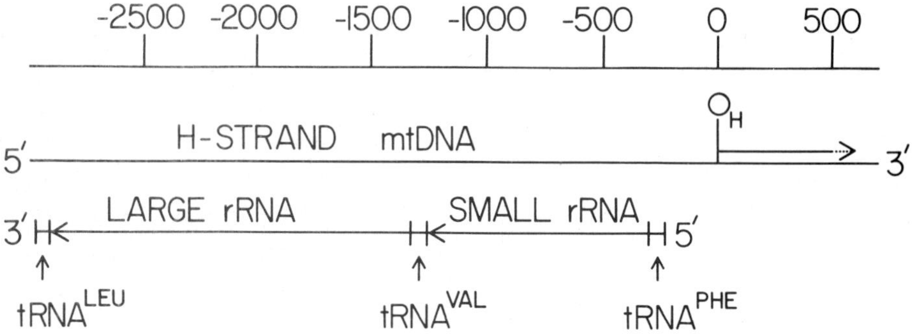

Fig. 3. Line drawing of a portion of the mouse mtDNA genome. The D-loop H-strand origin of DNA synthesis is denoted O_H and the direction of DNA synthesis is indicated by the rightward arrow. The stippled portion of this arrow indicates the variability in the 3'-end position of D-loop H-strands[3]. The two rRNAs and three tRNAs encoded in this portion of the mtDNA H-strand are positioned by their distance from O_H in nucleotides (top line).

within one nucleotide of the ends of the rRNA genes. No second tRNA gene can be identified on either side of the $tRNA^{Leu}$ gene at the 3'-end of the 16S rRNA gene. If this fourth species is a tRNA, then it must be so unusual as to render the gene unrecognizable as such by computer analysis. Alternately, it could be a stable, non-tRNA functional transcript, or a non-functional 3' transcribed segment which is processed from the 3'-end of the $tRNA^{Leu}$. In the E. coli rRNA operons, the 5S RNA gene is located near the 3'-end of the 23S rRNA gene, raising the question of whether this small RNA could be a 5S RNA analog in mitochondria. However, the DNA of the 3' region of the mitochondrial 16S rRNA gene shows no significant homology with any published prokaryotic 5S or eukaryotic 5S or 5.8S sequence[15]. Whatever its identity, it too must map within 1-2 nucleotides of the $tRNA^{Leu}$ gene. The identity of this RNA is currently under investigation.

The 3'-terminus of every prokaryotic and eukaryotic tRNA that has been sequenced has the trinucleotide CCA-OH-3' at the 3'-end of the amino acid stem[16], the 3'-hydroxyl group of the terminal A being the site where the RNA is aminoacylated in the charging reaction. If this is also true for the mouse mitochondrial tRNAs, then these last three nucleotides are not encoded in the DNA sequence, and must be added post-transcriptionally. This is in contrast with some prokaryotic tRNAs, where the CCA is included in the DNA sequence coding for the tRNA[17]. The enzyme tRNA nucleotidyltransferase has been isolated from both E. coli and rabbit liver homogenates, and has the ability to add the 3'-terminal CCA residues in vitro to tRNAs which lack these nucleotides[18]. It seems clear that a similar enzyme must be operating in mouse mitochondria. The gene for such an enzyme would almost certainly be encoded in the nucleus. It is unknown whether the same enzyme processes both cytoplasmic and mitochondrial tRNAs, or whether two distinct enzymes are maintained.

The tRNA genes from mouse mitochondria are very similar to those in human mitochondria[19] and the proposed RNA structures[14] from both mouse and human lack sequences in the D and TΨC loops (and elsewhere) which are known to be conserved in all other non-organelle tRNAs which have been sequenced. It is not apparent from examination of the sequences of mouse mitochondrial $tRNA^{Phe}$ and $tRNA^{Val}$ whether the D and TψC loops can still interact in a manner similar to those in yeast $tRNA^{Phe}$, or whether they assume some different tertiary structure. Although $tRNA^{Phe}$ and $tRNA^{Val}$ contain novel sequences, $tRNA^{Leu}$ is very much a conventional tRNA in terms of its conserved bases and potential tertiary interactions. If one considers any theory in which all mitochondrial tRNAs arose from ancestral prokaryotic-like tRNA genes, then it is not clear why at least one of these genes has changed so little while others have diverged drastically. This unexpected conservation could be accounted for if the $tRNA^{Leu}$ had some other function, in the mitochondrion or elsewhere in the cell, in addition to its role in mitochondrial protein synthesis, which required maintenance of the original tRNA structure.

The organization of the tRNA genes in the rRNA region of mouse mtDNA allows several speculations about the origin, transcription, and processing of the genes to be made. The organization of rRNA cistrons in prokaryotes has been extensively studied, especially in E. coli[20,21]. Sequencing studies of the rrnX locus of E. coli, which is one of 7 rRNA operons, have allowed the delineation of the basic structure of the operon and of its mode of transcription and processing. The order of the genes in the operon is 5'-16S rRNA-23S rRNA-5S rRNA-3'. The spacer region between the 16S and 23S genes is 437 nucleotides long and contains genes for $tRNA^{Ile}$ and $tRNA^{Ala}$. The 5S gene is followed by genes for $tRNA^{Asp}$ and $tRNA^{Trp}$. This general pattern is also followed by several other rrn operons which have been studied[22]. The other tRNA genes of E. coli which are not associated with rrn operons are widely dispersed and appear to be transcribed monocistronically. In contrast, the rrn operons are transcribed polycistronically and subsequently processed by the action of several enzymes to mature rRNAs, 5S RNA, and tRNAs. Comparatively little is known about the organization and transcription of rRNA and tRNA genes in the nucleus of eukaryotic cells. In many eukaryotes, rRNA genes are located on several chromosomes which participate in the organization of the nucleolus. In higher eukaryotes the 18S, 5.8S, and 28S rRNAs are transcribed as a large 40S precursor RNA which is subsequently processed to the mature species[22], but no tRNAs are known to be present in the precursor. The majority of eukaryotic tRNA genes appear to be transcribed as single cistrons[24] or rarely as dimeric precursors[25], with long non-transcribed spacer regions between genes. Also of significance is the fact that rRNA genes and tRNA genes are transcribed by different enzymes in eukaryotes. RNA polymerase I transcribes rRNA genes while a distinct enzyme, RNA polymerase III, transcribes 5S and tRNA genes[26].

In contrast, the rRNA region of the mouse mtDNA genome shows amazing economy in the way the rRNA and tRNA genes are organized. Lack of space between the genes suggests that the entire region is transcribed by a single RNA polymerase into a polycistronic transcript, from which individual rRNAs and tRNAs are generated by a set of single endonucleolytic cleavages. A large molecular weight RNA transcript spanning this region has been observed to be present in low abundance in previous transcript mapping experiments[4], but there is no evidence that a molecule of this sort is a direct precursor to the rRNAs. It is not known whether a single enzyme might be involved in the processing of such a polycistronic precursor or whether several enzymes participate. It is possible that the tRNA sequences could assume a cloverleaf conformation while still in the precursor molecule, and that this structure acts as a guide to the endonuclease, which might function as a mitochondrial analog of E. coli

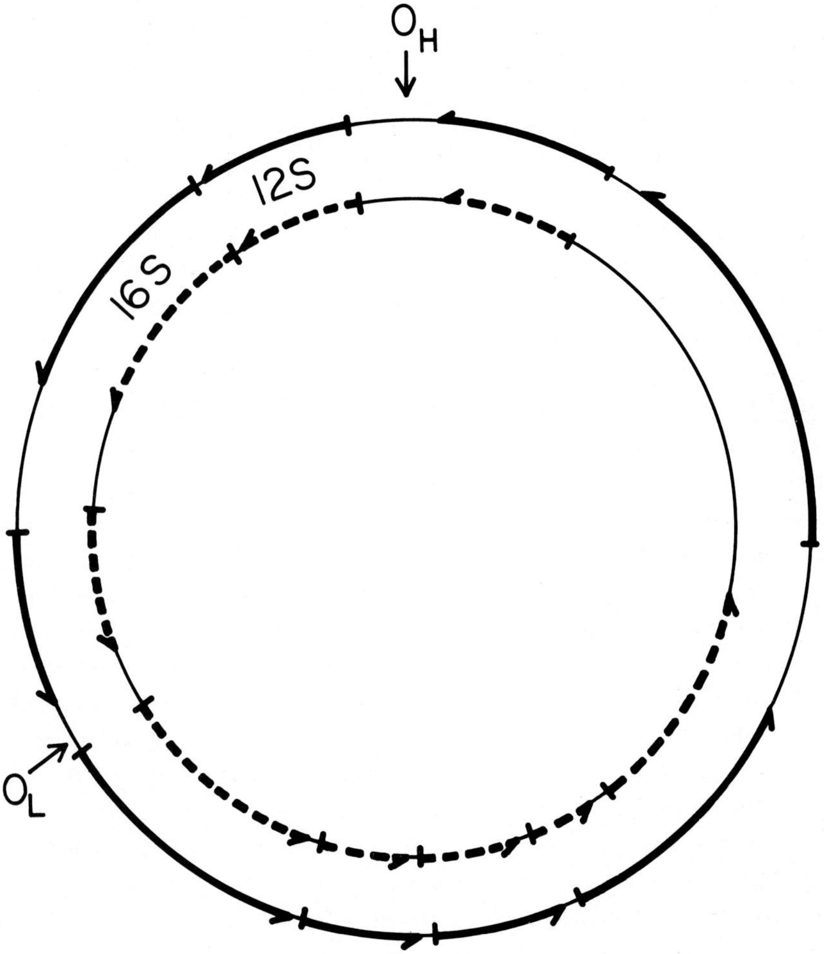

Fig. 4. The transcription map of mouse and human mtDNA heavy strand transcripts. The two mtDNA genomes are aligned by superposition of the heavy strand origin of DNA replication (O_H) and light strand origin of DNA replication (O_L). The outer circle (solid arrows) shows the map positions of seven transcripts from human mtDNA in addition to the 12S and 16S rRNAs. The inner circle (dashed arrows) shows the map positions of seven transcripts from mouse mtDNA in addition to the 12S and 16S rRNAs. The direction of transcription is counterclockwise as indicated.

RNase P. Indeed, it is known in E. coli that RNase P must recognize some aspect of tertiary structure, since primary sequences adjacent to the cleavage site are different in different tRNA precursor species[27].

The structure of the rRNA operon in mouse mtDNA is evidence that the mtDNA genome did not arise as some derivative of the nucleus, for it would be hard to imagine how widely dispersed genes for tRNAs and rRNAs, transcribed by different polymerases from different chromosomes, could be consolidated to form the arrangement shown in Figure 3. It seems more likely that mouse mtDNA arose from a prokaryote-like genome by systematic deletion of non-functional spacer regions as well as of genes whose function could be assumed by the nucleus.

The transcription map of mouse and human mtDNA. We have previously mapped the rRNA and putative mRNA transcripts on the mouse mtDNA genome[4]. The technology employed was to hybridize crude mtRNA to cloned mtDNA restriction fragments, digest non-duplex nucleic acid with S1 nuclease, and then identify the mtDNA sequences which were complementary to mtRNA >300 nucleotides in length. The transcript map of mouse mtDNA is shown in Figure 4. We have pursued the same strategy to map the human KB cell mtDNA genome with the additional approach of hybridizing specific mtDNA sequences to mtRNA which had been covalently coupled to aminothiophenol paper by transfer from a denaturing gel. These two techniques are complementary and permit accurate location and confirmation of the map positions of transcripts.

The human mtDNA transcripts identified account for approximately 80% of the coding capacity of the genome (Figure 4). The positions of the two rRNAs are in agreement with previous data[5] and the relative positions of the putative mRNAs are very similar to those previously determined for mouse mtDNA[4]. As in the case for mouse mtDNA, there is no evidence for intervening sequences in either the rRNA or mRNA coding sequences. It is interesting to note that the superposition of this map on the previously determined map of 4S RNA genes[10] suggests that the 4S RNA genes are interspersed relative to mRNA. Several of the human transcripts can be isolated as "transcript doublets." That is, when analyzed by RNA sizing experiments, several transcripts appear as double bands which are homologous to the same DNA sequence. The map position of the DNA sequence encoding the two components of the doublet is identical for most of their length, and the size difference between the two is of the order of 100 nucleotides[28]. These two observations suggest that tRNAs and mRNAs may be initially co-transcribed and that a major processing event involves precise excision of tRNAs adjacent to mRNAs. In support of this concept, it has recently been shown [19] that the human mtDNA gene coding for cytochrome oxidase subunit II is immediately adjacent to the gene for $tRNA^{Asp}$. The evidence to date suggests tRNA excision may be important in processing both rRNAs and mRNAs from larger precursor RNAs in animal cell mitochondria.

ACKNOWLEDGMENTS

We thank D. Tapper of this laboratory for the electron micrograph of R-looped mtDNA. This work was supported by grant CA-12312 from the National Institutes of Health and grant NP-9 from the American Cancer Society, Inc. J.B., R.A.V. and M.W.W. are Medical Scientist Training Program Trainees (GM-07365) of the National Institutes of Health, P.N. was on leave from Monash University and the recipient of a Fulbright Postdoctoral Award and D.A.C. is a Faculty Research Awardee of the American Cancer Society, Inc. (FRA-136).

REFERENCES

1. Berk, A.J. and Clayton, D.A. (1974) J. Mol. Biol., 86, 801-824.
2. Bogenhagen, D. and Clayton, D.A. (1978) J. Mol. Biol., 119, 69-81.
3. Bogenhagen, D., Gillum, A.M., Martens, P.A. and Clayton, D.A. (1979) Cold Spring Harbor Symp. Quant. Biol., 43, 253-262.
4. Battey, J. and Clayton, D.A. (1978) Cell, 14, 143-156.
5. Wu, M., Davidson, N., Attardi, G. and Aloni, Y. (1972) J. Mol. Biol., 71, 81-93.
6. Ramirez, J.L. and Dawid, I.B. (1978) J. Mol. Biol., 119, 133-146.
7. Klukas, C.K. and Dawid, I.B. (1976) Cell, 9, 615-625.
8. Kroon, A.M., Pepe, G., Bakker, H., Holtrop, M., Bollen, J.E., Van Bruggen, E.F.J., Cantatore, P., Terpstra, P. and Saccone, C. (1977) Biochim. Biophys. Acta, 478, 128-145.
9. Nagley, P. and Clayton, D.A. (1980) Nucleic Acids Res., in press.
10. Angerer, L., Davidson, N., Murphy, W., Lynch, D. and Attardi, G. (1976) Cell, 9, 81-90.
11. Walberg, M.W. and Clayton, D.A., manuscript submitted.
12. Shine, J. and Dalgarno, L. (1974) Proc. Natl. Acad. Sci. USA, 71, 1342-1346.
13. Steitz, J.A. and Jukes, K. (1975) Proc. Natl. Acad. Sci. USA, 72, 4734-4738.
14. Van Etten, R.A. and Clayton, D.A., manuscript submitted.
15. Erdmann, V.A. (1980) Nucleic Acids Res., 8, r3—r48.
16. Sprinzl, M., Grueter, F., Spelzhaus, A. and Gauss, D.H. (1980) Nucleic Acids Res., 8, r1-r22.
17. Young, R.A., Macklis, R. and Steitz, J.A. (1979) J. Biol. Chem., 254, 3264-3271.
18. Deutscher, M.P. and Evans, J.A. (1977) J. Mol. Biol., 109, 593-597.
19. Barrell, B.G., Bankier, A.T. and Drouin, J. (1979) Nature, 282, 189-194.

20. Young, R.A., Bram, R.J. and Steitz, J.A. (1979) rRNA and tRNA Processing Signals in the Ribosomal RNA Operons of E. coli. In Transfer RNA, Abelson, J., Schimmel, P.R., Söll, D., eds. (Cold Spring Harbor Press), in press.

21. Dahlberg, A.E., Tokimatsu, H., Zahalak, M., Reynolds, F., Calvert, P.C., Rabson, A.B., Lund, E. and Dahlberg, J.E. (1978) Proc. Natl. Acad. Sci. USA, 75, 3598-3602.

22. Morgan, E.A., Ikemura, T. and Nomura, M. (1977) Proc. Natl. Acad. Sci. USA, 74, 2710-2714.

23. Hadjiolov, A. and Nikolaev, N. (1976) Prog. Biophys. Mol. Biol., 31, 95-144.

24. DeRobertis, E.M. and Olson, M.V. (1979) Nature, 278, 137-143.

25. Beckman, J.S., Johnson, P.F. and Abelson, J. (1977) Science, 196, 205-208.

26. Roeder, R.G. (1976) Eukaryotic Nuclear RNA Polymerases. In RNA Polymerases, Chamberlain, M. and Losick, R., eds. (Cold Spring Harbor Press) pp. 285-329.

27. Altman, S. (1978) Transfer RNA Biosynthesis. Biochemistry of Nucleic Acids II, 17, 19-44.

28. Battey, J. and Clayton, D.A., manuscript submitted.

The Organization and Expression of the Mitochondrial Genome
A.M. Kroon and C. Saccone, editors

MITOCHONDRIAL DNA POLYMERASE OF EUKARYOTIC CELLS

U. BERTAZZONI and A.I. SCOVASSI
Istituto CNR Genetica, Biochimica, Evoluzionistica, via S. Epifanio 14, Pavia (Italy)

The presence of three distinctive DNA polymerases in animal cells, designated α, β and γ-polymerases, is now well established[1,2]: of these, α and β are related to the nucleus while γ is found both in mitochondrial and nuclear fractions. The mitochondrial DNA polymerase, considered initially as a separate enzyme, appears to be identical to DNA polymerase γ [3-5].

The γ-polymerase (Table 1) is a high molecular weight enzyme, representing a minor constituent of total cellular DNA polymerases. It is characterized by being sensitive to NEM, as opposed to β, and resistant to aphidicolin, as opposed to α. The distinction from α and β is best made by assaying the enzyme activity on the homopolymer system composed of a ribotemplate and a deoxyprimer: α-polymerase is inactive on this system and the response of β-polymerase can be completely inhibited by phosphate.

Concerning its in vivo function, the fact that the only DNA polymerase present in mitochondria is of the γ-type, is a convincing argument for its role in mitochondrial DNA synthesis[4,6].

It has been recently suggested that γ-polymerase is confined only to mitochondria[7]. Nevertheless, DNA polymerase γ appears to be required for adenovirus DNA synthesis in the nucleus[8,9]. Of particular interest is the observation that γ-polymerase synthesizes DNA in vitro in a highly processive fashion[10], suggesting that the enzyme is suitable for the replication of mitochondrial and adenovirus DNAs, which are known to be synthesized continuously without short intermediates.

We have studied the phylogeny of mitochondrial DNA polymerase and we have already shown that, in all the different classes of vertebrates, purified mitochondria contain a unique species of DNA polymerase having the properties of the γ-polymerase[11].

TABLE 1

PROPERTIES OF DNA POLYMERASE γ

Location	mitochondria;nucleus?
Molecular weight, Kdal	150-300
S value	7-9
pH optimum	8.0
Isoelectric point	<7
K_m for dNTPs, μM	0.2-0.6
Preferred cation	Mn^{++}
Associated DNase	uncertain
Type of synthesis	highly processive
Activity with template-primers:	
activated DNA	low
polyA-oligo dT	yes
RNA-primed DNA	no
oligodT-primed mRNA	no
Effects:	
salt (0.15 M NaCl)	stimulation
K phosphate (50 mM)	stimulation
N-Ethyl Maleimide (NEM)	inhibition
dideoxy TTP	inhibition
aphidicolin	no effect
Helix Destabilizing Proteins	no effect
Relative activity (dividing cells)	2-10%
Proposed functions	mt DNA replication adenovirus replication

We are now extending this survey to lower eukaryotes and we are reporting here the results obtained for some invertebrate phyla, yeasts and plants. The operating conditions include the purification of mitochondria, the treatment with digitonin to remove the external membrane of the organelle and contaminating proteins, and the analysis of the DNA polymerase activities in the mitochondrial extracts.

In Table 2 the results obtained by assaying the extracts for α-polymerase and for γ-polymerase are shown.

TABLE 2

MITOCHONDRIAL DNA POLYMERASE ACTIVITY

Organism	Activity(units/mg prot) activated DNA (α-polymerase)		polyA-oligodT (γ-polymerase)	
CV-1 cells	2.0	3%	53.5	97%
L-cells	0.5	6%	8.3	94%
Chick embryo	0.5	4%	13.6	96%
Turtle liver	0.5	8%	5.9	92%
Xenopus oocytes	0.4	0.2%	70.4	100%
Pleurodeles oocytes	0.2	6%	18.4	94%
Trout liver	0.2	25%	0.6	75%
House fly embryos	3.4	79%	0.9	21%
Sea urchin embryos	2.3	89%	0.3	11%
E.fluviatilis (sponge)	3.7	71%	1.5	29%
Yeast: w.t.	1.4	81%	0.2	19%
UV mutant:UVS72	8.2	78%	0.5	22%
petite:UVS72-ρ^-	11.3	95%	0.3	5%
Rice cells	4.7	66%	2.4	34%

It is evident that in all vertebrate organisms the γ-polymerase type of activity is predominant if not unique. On the opposite,the DNA polymerase activity found in lower eukaryotes responds also to the assay of α-polymerase. It is not clear as yet whether this is a distinctive property of mt polymerase of these organisms or if it could arise from contamination by polymerases not belonging to the organelle.

The analysis of the sedimentation profile of mitochondrial extracts from different strains of yeast (Fig. 1) suggests that both α and γ activities are cosedimenting as a single peak. The enzyme is also present in a petite mutant, indicating that the mt DNA polymerase is specified by the nucleus.

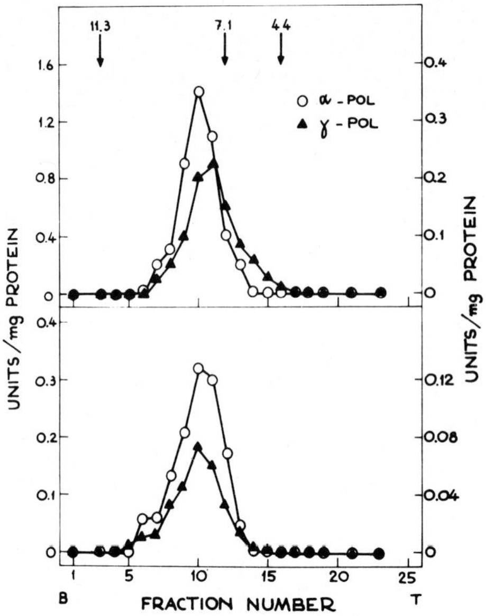

Fig. 1 - Sedimentation analysis of yeast mitochondrial DNA polymerase. Top panel, petite mutant; bottom panel, wild type.

ACKNOWLEDGEMENTS

We are greatly indebted to Drs. A. Goffeau and W. Crosby for the purification of yeast mitochondria and to Dr. M.N. Gadaleta for sea urchin mitochondria. We like to thank Dr. P. Yot in whose laboratory many experiments have been performed. This is contribution n° 1694 of the Radiation Protection Programme of Eur.Commun.Commission.

REFERENCES

1. Falaschi,A. and Spadari,S.(1978) DNA synthesis: present and future, Plenum Publishing Co., pp. 487-515.
2. Stefanini,M.et al.(1980) Lymphocyte Stimulation, Plenum Publish. Co., pp. 33-51.
3. Bolden,A.et al.(1977) J. Biol. Chem. 252, 3351-3356
4. Bertazzoni,U.et al.(1977) Eur. J. Biochem. 81, 237-248
5. Hübscher,U.et al.(1977) Eur. J. Biochem. 81, 249-258.
6. Hübscher,U.et al.(1979) Proc.Natl.Acad.Sci.USA, 76, 2316-2320.
7. Tanaka,S. and Koike,K.(1978) BBRC, 81, 791-797.
8. Brison,O.et al.(1977) J. of Virol., 24, 423-435.
9. Van der Vliet,P.C. and Kwant,M.M.(1978) Nature, 276, 532-534.
10. Yamaguchi,M.et al(1980) Nature, 285, 45-47.
11. Scovassi,A.I.et al.(1979) Eur. J. Biochem.,100, 491-496.

MITOCHONDRIAL RIBOSOME ASSEMBLY AND RNA SPLICING IN NEUROSPORA CRASSA

ALAN M. LAMBOWITZ

The Edward A. Doisy Department of Biochemistry, St. Louis University School of Medicine, St. Louis, Missouri 63104

INTRODUCTION

Our laboratory has focused on mitochondrial (mt) ribosome assembly in Neurospora crassa as a model system, initially to study nuclear-mitochondrial interactions and more recently, to study RNA splicing. The choice of model system is based on two main considerations: First, mt ribosome assembly is subject to regulation by both the nuclear and mitochondrial genetic systems since mt ribosomes consist of rRNAs transcribed from mtDNA and ribosomal proteins, most of which are synthesized in the cytosol[1]. Second, as a model organism, Neurospora offers the important advantage of being an obligate aerobe. Patterns of regulation of mitochondrial biogenesis are likely to differ between obligate and faculative aerobes and it was felt that Neurospora would ultimately prove a more representative organism from this perspective. Neurospora also offers excellent nuclear genetics[2] and the biochemical technology for studying mt ribosomes has been well developed[3]. The new dimension of RNA splicing has been added by recent studies. The gene encoding the large (25 S) mt rRNA has been found to contain an intervening sequence of ca. 2.3 kb[4-7] and nuclear mutants defective in splicing this RNA have been identified[7]. The mutants offer unique opportunities to study the mechanism of RNA splicing and its role in gene expression. The purpose of this article is to review recent work on mt ribosome assembly in Neurospora with particular emphasis on aspects relating to nuclear-mitochondrial interactions and RNA splicing.

MT rRNAS: GENES AND PROCESSING

Fig. 1A shows a detailed physical map of the Neurospora mt rRNA genes based on studies in several laboratories[4-8]. The genes for the two rRNAs (19 S and 25 S) and approximately 90% of the mt tRNA genes have been found to be clustered in a single, 10 to 15 Mdal segment of mtDNA but with the genes for 19 S and 25 S RNA separated by 5 to 6 kb[4,5]. The gene for 25 S RNA contains an intervening sequence of ca. 2.3 kb, located 0.4 to 0.5 kb from one end whereas the gene for 19 S RNA appears to be continuous by all criteria[4-7]. The proximity of the genes for mt rRNAs and tRNAs may be pertinent to the synthesis

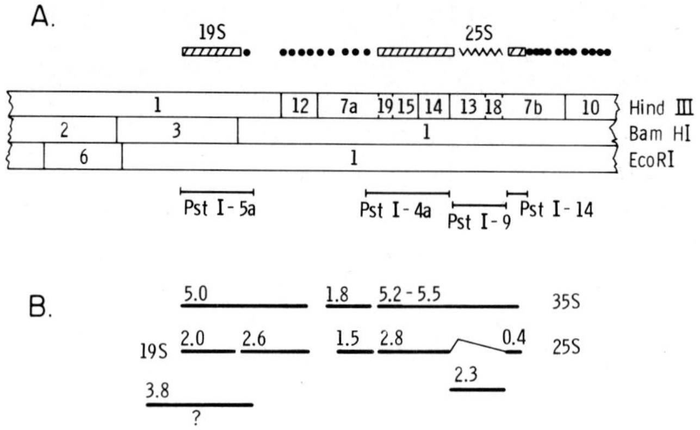

Fig. 1A. Physical map of the mt rRNA region of Neurospora mtDNA. Closed circles represent mt tRNA genes. Data refs. 4-8. B. RNA species detected by RNA gel transfer hybridization experiments using cloned restriction fragment probes. Numbers indicate lengths in kb. Data ref. 9.

of these RNA species. Two clusters of mt tRNA genes have been found flanking the 25 S RNA gene on both sides[5,6,8] and a gene encoding mt $tRNA^{Tyr}$ has been mapped close to the 3' end of the 19 S RNA[5]. Heckman and RajBhandary have shown that the mt rRNAs and most, if not all, of the mt tRNAs are transcribed from the same mtDNA strand[5] and the direction of transcription has been determined to be from the 19 S to the 25 S RNA genes[9].

In general, transcription units have proven difficult to define in mitochondrial systems, in part due to the lack of suitable preparations of mitochondrial RNA polymerase. The number and organization of transcription units in the Neurospora mt rRNA region is not known. Also, it is now clear that the 32 S RNA, which Kuriyama and Luck thought to be a common precursor of 19 S and 25 S RNA[10], was in fact an artifact caused by comigration on gels of true precursor RNAs and aggregates of mature 19 S and 25 S RNA[11]. In collaboration with Drs. Joyce Heckman and U.L. RajBhandary (MIT), we have recently carried out RNA gel transfer hybridization experiments[12] to identify potential precursor RNA species[9]. Fig. 1B summarizes RNA species detected in these experiments using ^{32}P-labeled, cloned restriction fragment probes. Although specific features of processing pathways implied in Fig. 1B must still be considered tentative, two points do emerge from the data: (a) that 19 S and 25 S RNA are synthesized via separate precursors and (b) that high molecular weight RNA species can be identified which span the entire region between and including the 19 S and 25 S RNA genes. The latter finding is consistent with a small number, perhaps even one, transcription unit.

The processing pathway for 25 S RNA is inferred from the RNA gel transfer hybridization experiments combined with information obtained from mutants defective in RNA splicing (see below). The latter accumulate a 5.2 kb RNA which appears to be a colinear transcript of the 25 S RNA gene including the intervening sequence[7]. In wild-type mitochondria, RNA gel transfer hybridizations with probes containing 25 S RNA and/or intron sequences show a 5.2 kb RNA present in low concentrations[9]. The same probes fail to detect precursor RNAs intermediate in size between 35 S and 25 S RNA, but intron specific probes show strong hybridization to a 2.3 kb RNA which is likely to be the excised intron[9]. The results suggest that splicing of 35 S RNA proceeds via a single cleavage-ligation reaction leading to excision of the 2.3 kb intron RNA[9]. The stability of the putative intron RNA could be fortuitous or could reflect some physiological role.

In contrast to the processing pathway for 25 S RNA, the processing pathways for 19 S RNA and mt tRNAs implied in Fig. 1B are based on hybridization data alone and are, therefore, subject to considerable qualification. It is of interest, however, that RNA species detected in the hybridization experiments could correspond to common precursors of 19 S RNA and mt tRNAs or of multiple mt tRNA species[9]. The further characterization of these RNA species is one direction of future research.

ASSEMBLY OF MT SMALL SUBUNITS

A major focus of our research has been the putative mt ribosomal protein S-5 (Mr 52,000) which, in contrast to most mt ribosomal proteins, is synthesized within the mitochondria[13,14]. S-5 has been found to satisfy the following criteria for a ribosomal protein: (a) it is specifically associated with the mt small subunit, (b) it remains associated with mt small subunits after sedimentation through sucrose cushions (1.85 M sucrose, 0.5 M KCl) followed by sucrose gradients containing 0.5 M KCl and (c) it is present in the same stoichiometry as other mt ribosomal proteins as judged by two-dimensional gel electrophoresis[13,15]. The two-dimensional gel analysis shows further that S-5 is a very basic protein, probably with a pI greater than 10 and that S-5 has a very high affinity for RNA[15]. The latter property can be exploited to purify the protein.

The role of S-5 in mt ribosome assembly has been investigated by examining the effect of chloramphenicol on mt ribosome assembly in wild-type *Neurospora*[14,15]. Chloramphenicol was found to rapidly inhibit the assembly of mt small subunits leading to the accumulation of incomplete mt small subunits (CAP-30 S particles) which are enriched in a precursor of 19 S RNA and deficient in S-5 and several other proteins. These effects were observed rapidly (within 15 min) after addition of chloramphenicol, suggesting that they result directly from inhibiting mitochondrial protein synthesis. Also, chloramphenicol was found to have little effect on the assembly of mt large subunits. Considered together, therefore, the results suggest that a mitochondrially-synthesized protein, presumably S-5, is required for the assembly of mt small subunits, to facilitate processing of 19 S RNA and to stabilize the binding of additional mt small subunit proteins[14,15].

If S-5 or some other mitochondrially-synthesized protein is required for mt ribosome assembly, then any mutation which decreases the rate of mitochondrial protein synthesis may secondarily inhibit the assembly of mt small subunits. This prediction has been confirmed for at least four different mutants deficient in mitochondrial protein synthesis, all of which have been found to accumulate incompletely assembled mt small subunits, similar to CAP-30 S particles[16]. In addition, correlation of mutant phenotypes with rates of mitochondrial protein synthesis shows that inhibition of mitochondrial protein synthesis must be greater than 65% before the assembly of mt small subunits is inhibited appreciably[16]. In some mutants (i.e., those with ca. 50% of the rate of wild-type mitochondrial protein synthesis) mitochondrial ribosomes contain essentially normal amounts of S-5 whereas the concentration of cytochrome b is decreased more than 70% and the concentration of cytochrome aa_3 is decreased more than 90%[16]. The latter finding raises the possibility that control mechanisms operate so that S-5 is synthesized preferentially as the rate of mitochondrial protein synthesis decreases.

Two lines of evidence suggest that S-5 binds relatively late in mt ribosome assembly: (a) the CAP-30 S particles which lack S-5 are almost fully assembled mt small subunits[14,15] and (b) S-5 does not appear to be associated with very early mt small subunit precursor particles[13]. The first line of evidence is not definitive since CAP-30 S particles could be abnormally assembled. The second line of evidence seems somewhat stronger. Since the new findings about the 32 S RNA necessitate some reinterpretation of the precursor particle data, it seems worthwhile to review the essential points here. The precursor particles were identified by pulse-labeling cells with ^{3}H-uracil or ^{3}H-leucine (4 min, 25° or 2 hr, 5°) followed by analysis of ribonucleoprotein particles on sucrose gradients containing low salt buffer[13]. Under these conditions, the ^{3}H-uracil and ^{3}H-leucine were found to cosediment as a broad peak centered at *ca*. 30 S. Electrophoretic analysis of pulse-labeled material suggested that the precursor particles contain a specific subset (at least 15) small and large subunit proteins as well as high molecular weight RNA, originally thought to be the 32 S RNA of Kuriyama and Luck[10]. The finding that 32 S RNA is an artifact[11] changes the interpretation in two ways: (a) the RNA species in the particles must now be considered unidentified and (b) it seems likely that the 30 S peak contains separate precursor particles of mt small and large subunits. The precursor particles are present in low concentrations and could not be completely purified. However, they were found to be destroyed by sedimentation through sucrose gradients containing 0.5 M KCl (they are distinguishable from CAP-30 S particles in this respect) so that precursor particle proteins could be defined operationally as pulse-labeled proteins in the 30 S peak under low salt but not high salt conditions. By this criterion, S-5 is not associated with the precursor particles and must be added at a later stage in mt ribosome assembly[13].

In recent studies, we have used equilibrium centrifugation in CsCl gradients to prepare core particles of mature mt small subunits[17]. These studies showed that S-5 remains associated with the core particles even under very stringent gradient conditions (CsCl gradients in the presence of EDTA) where more than half of the proteins are dissociated. The results provide evidence that S-5 is bound directly to 19 S RNA in mt small subunits. The core particle experiments do not permit inferences about the time of addition of S-5 since proteins which interact with RNA could bind either early or late during assembly, so long as RNA binding sites remain open at late stages.

Because S-5 is the only mt ribosomal protein synthesized intramitochondrially, it is pertinent to ask whether it plays some special role in mt ribosome assembly or mitochondrial protein synthesis. Pulse-labeling experiments indicate that S-5 has a small free pool compared to other mt ribosomal proteins[13]. It is possible, therefore, that the rate of synthesis of S-5 controls the rate of mt ribosome assembly, although it is not readily apparent why such control should be exercised at a late stage in the assembly process. Another, perhaps more attractive, possibility is that the intramitochondrial concentration of free S-5 directly regulates other processes. Fallon et al. have proposed that coordinated synthesis of different *E. coli* ribosomal proteins is achieved by a feedback mechanism in which free ribosomal proteins inactivate their mRNAs[18]. Variations of this mechanism could be relevant to the role of S-5. Free S-5 would start to accumulate under conditions in which the supply of cytosolically-synthesized mt ribosomal proteins becomes rate-limiting for mt ribosome assembly. Under such conditions, it would be desirable to decrease the rate of synthesis of S-5, of mt rRNAs and possibly of other mitochondrial translation products so as to coordinate the rates of synthesis of cytosolic and mitochondrial products. In principle, such coordination could be achieved by free S-5 acting as a repressor of transcription or as an inhibitor of translation of mt mRNAs. The high affinity of S-5 for RNA and possibly DNA would be consistent with such a role.

In recent studies on S. cerevisiae, Groot et al.[19] and Terpstra et al.[20] have shown that the mitochondrially-synthesized var-1 protein is associated with mt small subunits and may well be the cognate of Neurospora S-5. Terpstra and Butow[21] presented inhibitor experiments analogous to those for Neurospora which suggest that var-1 might be required for the assembly of mt small subunits. Both laboratories observed that the proportion of mt protein synthesis devoted to var-1 increases as the rate of mt protein synthesis decreases[19-21].

ASSEMBLY OF MT LARGE SUBUNITS

The assembly of Neurospora mt large subunits has recently been of considerable interest because of the role played by intervening sequences and RNA splicing. The key development here has been the identification of two temperature-sensitive nuclear mutants (289-67 and 299-9; isolated by Pittenger and West[22]) which are defective in RNA splicing at the nonpermissive temperature (37°)[7]. When grown at 37°, the mutants show decreased ratios of 25 S to 19 S RNA and accumulate a novel 35 S RNA which appears to be a continuous transcript of the 25 S RNA gene including the intervening sequence[7]. RNA gel transfer hybridization experiments discussed previously suggest that 35 S RNA is related to the normal precursor of 25 S RNA in wild-type mitochondria[9].

Genetic analysis shows that the two mutants are allelic[7]. However, they can be distinguished in temperature shift-up experiments where mt rRNA processing is found to turn off almost immediately in 299-9 and somewhat more slowly in 289-67[7]. The finding that RNA processing turns off rapidly in one mutant suggests that the affected component is one which plays a direct role in RNA splicing, either the enzyme itself or some component which must bind to or modify the RNA for splicing to occur[7].

The structure of 35 S RNA has been studied by Southern and Northern hybridization experiments, S_1-nuclease mapping and R-loop mapping[7,9]. These studies suggest: (a) that 35 S RNA is a continuous transcript of the 25 S RNA gene including the intervening sequence and (b) that 35 S and 25 S RNA have the same 5' and 3' termini within the limits of resolution of electron microscopy[7,9].

The mutants provide attractive possibilities for experiments into the mechanism of RNA splicing and the relationship between RNA splicing and mt ribosome assembly. Ribonucleoprotein particles containing 35 S RNA have been isolated from mutant mitochondria[23]. Surprisingly, these particles contain almost the full complement of large subunit proteins (as judged by two-dimensional gel electrophoresis) despite the presence of the 2.3 kb intron. If most mt ribosomal protein binding sites are located in the 5' 3 kb segment of 25 S RNA, then this segment could be partially assembled even before the intron is transcribed. It should be possible to use the particles to test whether binding of ribosomal proteins is required for 35 S RNA to attain the correct conformation for RNA splicing.

REGULATION OF MT RIBOSOME ASSEMBLY

We anticipate that regulation of mt ribosome assembly will recieve increasing attention because of the availability of novel mutants (see below) and because recombinant DNA technology opens the possibility of direct biochemical analysis of specific nuclear and mitochondrial genes. Studies with yeast and *Neurospora* suggest that mitochondrial biogenesis is subject to regulation by a variety of control mechanisms, but there is presently little information at the molecular level. In general, control mechanisms which regulate mitochondrial biogenesis should have one of two functions: (a) to coordinate the activities of the nuclear and mitochondrial genetic systems or (b) to increase or decrease mitochondrial capacity in response to physiological requirements. The first function seems to be fulfilled by a battery of different control mechanisms which regulate individual processes whereas the second function is often a component of a general cellular response characteristic of an organism or cell type (e.g., glucose repression in yeast[24]; response of mammalian cells to thyroid hormone[25].) Both types of regulation are reflected in the synthesis of mt ribosome constituents in *Neurospora*.

Coordination of the nuclear and mitochondrial genetic systems in *Neurospora* and *S. cerevisiae* is illustrated by the rapid inhibition of mt rRNA synthesis which occurs following inhibition of cytosolic protein synthesis[14,26]. Detailed studies of this phenomenon in *S. cerevisiae* suggest that the mode of inhibition of cytosolic protein synthesis may be significant and specifically that inhibition of elongation on cytosolic ribosomes leads to inhibition of mt rRNA synthesis whereas inhibition of initiation does not[26]. This finding suggests that specific regulatory signals, not just lack of cytosolic proteins, are required for inhibition of mt rRNA synthesis[26].

The more general cellular response in *Neurospora* involves the reaction to impairment of mitochondrial function. Barath and Kuntzel observed that inhibition of mitochondrial protein synthesis results in increased synthesis of mitochondrial constituents (including mt ribosomal proteins and elongation factors) coded for by nuclear genes[27,28]. This response seems likely to be characteristic of an obligately aerobic organism, highly dependent on oxidative phosphorylation for rapid growth and it is noteworthy that the same response may be modulated in *S. cerevisiae* where the predominant mode of regulation is glucose repression[29]. Barath and Kuntzel proposed that the observed increased synthesis of mitochondrial constituents reflected activation of nuclear genes as a result of inhibiting synthesis of a mitochondrially-synthesized repressor protein[27,28]. The evidence for such a repressor remains equivocal and other mechanisms can be envisioned. A detailed discussion has been presented elsewhere[30].

MUTANTS WITH DEFECTS IN MT RIBOSOME ASSEMBLY

The objectives of studying mutants with defects in mt ribosome assembly are (a) to identify nuclear and mitochondrial genes coding for mt ribosomal proteins and other proteins which function in mt ribosome assembly, (b) to discern linkage relationships and interactions among these genes which may be pertinent to their regulation and (c) to elucidate the roles of specific proteins in mt ribosome assembly and protein synthesis. The systematic analysis of *Neurospora* mt ribosome assembly mutants is carried out as part of our long term collaboration with Dr. Helmut Bertrand (University of Regina). We have also been fortunate to have obtained valuable mutants from Dr. Thad Pittenger (Kansas State University). Table 1 lists the mt ribosome assembly mutants which have been analyzed so far. The list includes classical extranuclear mutants (e.g., [*poky*]) which have since been found to have mt ribosome assembly defects[31,32] as well as new mutants which have been isolated recently[16,22,33]. In general, the new mutants have been identified by screening strains deficient in cytochromes *b* and aa_3 for alterations in mt ribosomal subunits or mt rRNAs. More than 75% of such mutants screened in initial experiments were found to have mt ribosome assembly defects; all but one of these showed nuclear inheritance[16,33]. The mutants listed in Table 1, along with additional mutants not yet analyzed, constitute a valuable resource for future studies of mitochondrial biogenesis and mt ribosome assembly in *Neurospora*.

ACKNOWLEDGEMENTS: I gratefully acknowledge my collaborators: Drs. Carmen A. Mannella, Richard A. Collins, Robert R. Goewert; Graduate Students, Michael F. Grimm, Michael R. Green (Washington University), Gian Garriga, Robert J. LaPolla and Lori L. Stohl. Dr. Helmut Bertrand (University of Regina) and Drs. Joyce Heckman and U.L. RajBhandary (MIT). Work from my laboratory was supported by NIH Grants GM 23961 and GM 26836 as well as a Basil O'Connor Grant from the National Foundation--March of Dimes.

TABLE 1: NEUROSPORA MUTANTS WITH DEFECTS IN MITOCHONDRIAL RIBOSOME ASSEMBLY

Mutant	Linkage Group	Phenotype	References
Extranuclear			
[poky]/[mi-1]		small subunit-deficient; decreased ratio 19 S to 25 S RNA	31,13,15
[exn-1]		" "	32,15,16
[exn-2]		" "	32,15,16
[exn-4]		" "	32,15,16
[SG-1]		" "	32,15,16
[SG-3]		" "	32,15,16
[stp B1]		" "	32,15,16
[C93]	complements [poky]	ts small subunit-deficient; normal ratio 19 S to 25 S RNA; probable defect in mitochondrial ATPase	22,33
Nuclear			
289-56	I	ts small subunit-deficient; normal ratio 19 S to 25 S RNA	22,16,34
297-24		small subunit-deficient; decreased ratio 19 S to 25 S RNA	22,16
295-20		cold-sensitive small subunit-deficient	22, +
cni-1		cold-sensitive small subunit-deficient	35
289-4		large subunit-deficient; decreased ratio 25 S to 19 S RNA	22,16
A13		large subunit-deficient	22, +
289-67	I (allelic)	ts defect in splicing 25 S RNA	7,22,16
299-9		ts defect in splicing 25 S RNA	7,22,16
KTS-41	(nonallelic)	ts; deficient in both 19 S and 25 S RNA	36
LAE-5		ts; deficient in both 19 S and 25 S RNA	36
Nuclear Suppressors of [poky]			
su^{I}-1	III		37
su^{I}-3	II		37
su^{I}-4	II		37
su^{I}-5	VII	su^{I}-5 [+] ts defect in large subunit assembly	37,34
su^{I}-10	II		37
su^{I}-14	IV		37
f	V		38

+Stohl and Lambowitz, unpublished data

REFERENCES

1. Boynton, J.E., Gillham, N.W. and Lambowitz, A.M. (1980) In: Ribosomes: Structure, Function and Genetics (G. Chambliss et al., eds.) pp. 903-950 (University Park Press, Baltimore).
2. Davis, R.H. and de Serres, F.J. (1970) Methods Enzymol. 17A, 79-143.
3. Lambowitz, A.M. (1979) Methods Enzymol. 59, 421-433.
4. Hahn, U., Lazarus, C.M., Lunsdorf, H. and Kuntzel, H. (1979) Cell 17, 191-200.
5. Heckman, J.E. and RajBhandary, U.L. (1979) Cell 17, 583-595.
6. de Vries, H., de Jonge, J.C., Bakker, H., Meurs, H. and Kroon, A. (1979) Nucleic Acids Res. 6, 1791-1803.
7. Mannella, C.A., Collins, R.A., Green, M.R. and Lambowitz, A.M. (1979) Proc. Nat. Acad. Sci. U.S.A. 76, 2635-2639.
8. Heckman, J.E., Yin, S., Alzner-De Weerd, B. and RajBhandary, U.L. (1979) J. Biol. Chem. 254, 12694-12700.
9. Green, M.R., Goewert, R.R., Lambowitz, A.M., Heckman, J.E. and RajBhandary, U.L., submitted for publication.
10. Kuriyama, Y. and Luck, D.J.L. (1973) J. Mol. Biol. 73, 425-437.
11. Grimm, M.F. and Lambowitz, A.M. (1979) J. Mol. Biol. 134, 667-672.
12. Alwine, J.C., Kemp, D.J. and Stark, G.R. (1977) Proc. Nat. Acad. Sci. U.S.A. 74, 5350-5354.
13. Lambowitz, A.M., Chua, N.-H. and Luck, D.J.L. (1976) J. Mol. Biol. 107, 223-253.
14. LaPolla, R.J. and Lambowitz, A.M. (1977) J. Mol. Biol. 116, 189-205.
15. Lambowitz, A.M., LaPolla, R.J. and Collins, R.A. (1979) J. Cell Biol. 82, 17-31.
16. Collins, R.A., Bertrand, H., LaPolla, R.J. and Lambowitz, A.M. (1979) Mol. Gen. Genet. 177, 73-84.
17. LaPolla, R.J. and Lambowitz, A.M., manuscript in preparation.
18. Fallon, A.M., Jinks, C.S., Strycharz, G.D. and Nomura, M. (1979) Proc. Nat. Acad. Sci. U.S.A. 76, 3411-3415.
19. Groot, G.S.P., Mason, T.L. and Van Harten-Loosbroek, N. (1979) Mol. Gen. Genet. 174, 339-342.
20. Terpstra, P., Zanders, E. and Butow, R.A. (1979) J. Biol. Chem. 254, 12653-12661.
21. Terpstra, P. and Butow, R.A. (1979) J. Biol. Chem. 254, 12662-12669.
22. Pittenger, T.H. and West, D.J. (1979) Genetics 93, 539-556.
23. LaPolla, R.J. and Lambowitz, A.M. (1979) J. Biol. Chem. 254, 11746-11750.
24. Gillham, N.W. (1978) Organelle Genetics (Raven Press, New York).
25. Jakovcic, S., Swift, H.H., Gross, N.J. and Rabinowitz, M. (1978) J. Cell Biol. 77, 887-901.
26. Ray, D.B. and Butow, R.A. (1979) Mol. Gen. Genet. 173, 227-238, 239-247.
27. Barath, Z. and Kuntzel, H. (1972) Proc. Nat. Acad. Sci. U.S.A. 69, 1371-1374.
28. Barath, Z. and Kuntzel, H. (1972) Nature New Biol. 240, 195-197.
29. Borst, P. and Grivell, L.A. (1978) Cell 15, 705-723.
30. Lambowitz, A.M. and Zannoni, D. (1978) In: Plant Mitochondria (G. Ducet and C. Lance, eds.) pp. 283-291 (Elsevier/North-Holland, Amsterdam).
31. Rifkin, M.R. and Luck, D.J.L. (1971) Proc. Nat. Acad. Sci. U.S.A. 68 287-290.
32. Collins, R.A. and Bertrand, H. (1978) Mol. Gen. Genet. 161, 267-273.
33. Collins, R.A., Bertrand, H., LaPolla, R.J. and Lambowitz, A.M., submitted.
34. Collins, R.A. (1979) Ph.D. Thesis, University of Regina.
35. Kientsch, R. and Werner, S. (1976) Genetics and Biogenesis of Chloroplasts and Mitochondria (Th. Bucher et al., editors) pp. 247-252 (North Holland, Amsterdam).
36. Nargang, F.E., Bertrand, H. and Collins, R.A. (1979) Current Genet. 1, 1-7.
37. Kohout, J. and Bertrand, H. (1976) Can. J. Genet. Cytol. 18, 311-324.
38. Mitchell, M.B. and Mitchell, H.K. (1956) J. Gen. Microbiol. 14, 84-89.

The Organization and Expression of the Mitochondrial Genome
A.M. Kroon and C. Saccone, editors

FUNCTIONAL AND STRUCTURAL ROLES OF PROTEINS IN MAMMALIAN MITOCHONDRIAL RIBOSOMES

T. W. O'BRIEN, N. D. DENSLOW, T. O. HARVILLE, R. A. HESSLER AND D. E. MATTHEWS
Department of Biochemistry and Molecular Biology, J. Hillis Miller Health Center, University of Florida, Gainesville, Florida 32610.

INTRODUCTION

The 55S ribosomes from animal mitochondria resemble bacterial ribosomes and eukaryotic cytoplasmic ribosomes in their fundamental properties, but in terms of their fine structure and physical-chemical properties, they differ unexpectedly from both these kinds of ribosomes, as well as from other kinds of mitochondrial ribosomes[1,2]. Mammalian mitochondrial ribosomes are considered to be members of the prokaryotic class, because they have more homologies with bacterial ribosomes than with eukaryotic cytoplasmic ribosomes[3]. Nevertheless, they contain scarcely half the RNA of bacterial ribosomes, even though they are about the same size[4], the bulk of their mass being contributed by a large number of proteins[5]. The unusual properties of these ribosomes raise questions about their relation to other kinds of ribosomes, and their large number of proteins raises questions about their functional and structural organization, and also about the identity of individual mitoribosomal proteins that are functionally homologous to proteins in other ribosomes. We have chosen the bovine mitochondrial ribosome for these studies to develop a model system for mammalian mitochondrial ribosomes in general, and to form the basis for further studies on the structure, function and evolution of these interesting ribosomes.

RESULTS AND DISCUSSION

Protein content of the bovine mitoribosome

Conditions were used for the preparation of ribosomes that should yield subribosomal particles essentially free of cosedimenting and loosely adsorbed, non-ribosomal proteins. The most rigorous salt-washing condition (KCl/Mg^{2+} = 500/10) under which mitoribosomes retain significant peptidyl transferase activity[5] was used to prepare ribosome subunits for analysis by 2D electrophoresis[5,6]. The schematic diagrams in Figure 1 summarize the analysis of 14 different preparations of mitochondrial small subunits, and 16 different preparations of large subunits, analyzed on more than 150 separate 2D gels. In this manner we have identified a set of 52 different proteins in the large subunit and 39 proteins in the small subunit of bovine mitoribosomes.

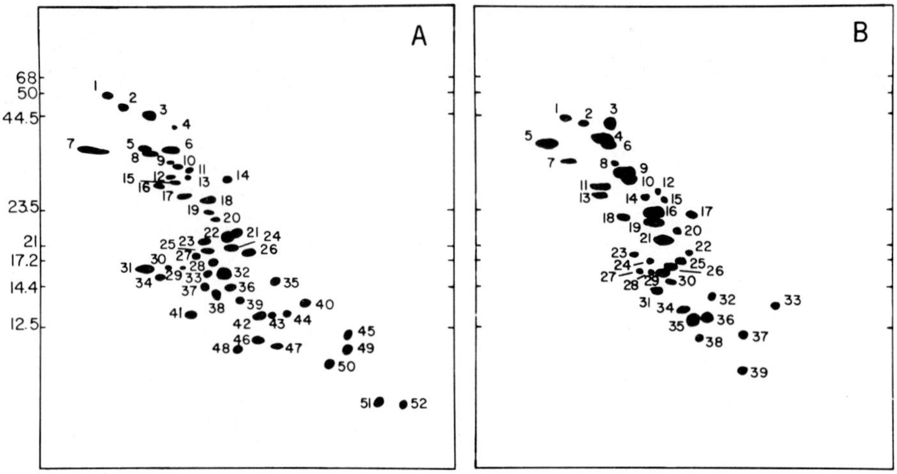

Figure 1. Schematic diagram of the 2D electropherograms of proteins from the large (A) and small (B) subunits of the bovine mitoribosome. Electrophoretic conditions described in reference 6.

Topographic disposition of individual proteins

As a probe for the topographic disposition of individual proteins in bovine mitoribosomes we have used accessibility to Lactoperoxidase-catalyzed radio-iodination[7] of the proteins in monoribosomes (^{125}I) relative to their labelling in subunits (^{131}I) or as free proteins (^{125}I or ^{131}I in urea-LiCl). In this manner, proteins which have tyrosine and histidine residues exposed on the surface of the monoribosome can be distinguished from those having reactive groups in the subunit interface region, obscured by the other subunit, and from those whose tyrosines and histidines are inaccessible in the subunits, essentially buried within the particles. When examined in this manner, the proteins appear to fall within four broad categories, as depicted schematically in Figure 2 and listed in Table 1.

Proteins in the large subunit are distributed among all categories, the most abundant being category 2 proteins, those exposed in the subunit interface region as well as in the monoribosome. Significantly, 8 of the large subunit proteins appear mainly buried within the subunit, where they may serve primarily a structural role. In contrast, all of the small subunit proteins are accessible to iodination, indicating that they are exposed at one point or another in the subunit. These observations suggest a major difference in the structural organization of the two subunits. The small subunit also contains more proteins of strong interfacial character, indicating another difference in the structural organization of the two particles.

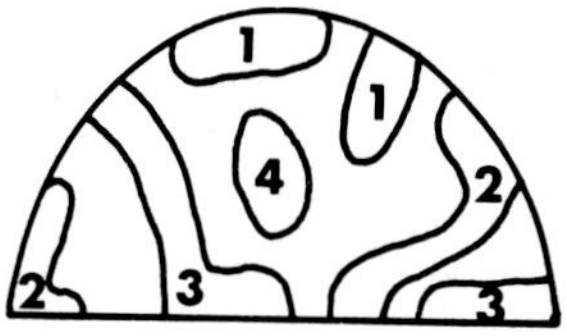

CLASS		
1	M/S $\geq$ 0.75; S/P $\geq$ 0.06	Exposed, No Interfacial Character
2	0.75 $\geq$ M/S $\geq$ 0.4; S/P $\geq$ 0.06	Exposed, Interfacial Character
3	M/S $\leq$ 0.4; S/P $\geq$ 0.06	Exposed, Mainly Interfacial
4	S/P $\leq$ 0.06	Mainly Buried

Figure 2. Assignment of ribosomal proteins to topographic classes based on the extent of their iodination in monoribosomes (M) relative to their labelling in subunits (S), and as free proteins (P) in urea-LiCl.

TABLE 1

TOPOGRAPHIC DISPOSITION OF INDIVIDUAL PROTEINS IN BOVINE MITORIBOSOMES

Topography Class[a]	1	2	3	4
Large Subunit Proteins	16, 29, 30, 31, 34, 40	1, 3, 4, 5, 6, 8, 10, 11, 12, 13, 15, 17, 18, 19, 21, 22, 23, 24, 25, 28, 32, 33, 36, 37, 38, 39, 41, 42, 46	14, 47, 48, 49, 51	2, 7, 26, 27, 35, 43, 44, 50
Small Subunit Proteins	2, 3, 4, 27	1, 5, 6, 7, 10, 11, 12, 13, 16, 19, 22, 24, 25, 26, 28, 29, 33, 37	9, 14, 15, 17 18, 20, 21, 23, 30, 31, 32, 34, 35, 36, 38, 39	

[a]Protein assignments made according to criteria in Figure 2.

Functional centers of the bovine mitoribosome

Bovine mitoribosomes are expected to share certain structural/functional features with bacterial ribosomes, based on their susceptibility to antibacterial antibiotics[8]. Some features of the translocation center are highly conserved in these mitoribosomes, since [^{35}S] thiostrepton binds with high affinity (K_d = 5.6 x 10^{-10}), in unit stoichiometry, to the large subunit of bovine mitoribosomes. Nevertheless, the binding site for the bacterial elongation factor EF-G is not conserved in bovine mitoribosomes, since the mitochondrial and bacterial factors are not functionally interchangeable[9].

As a probe for proteins at the chloramphenicol (CAP) binding site in the peptidyl transferase center of bovine mitoribosomes, we have used a reactive CAP analogue, [^{3}H] iodoamphenicol. One of the large subunit proteins labels specifically with this affinity probe, and this labelling is blocked by CAP (Figure 3). This protein has been identified as L2 by 2D PAGE. L2 has also

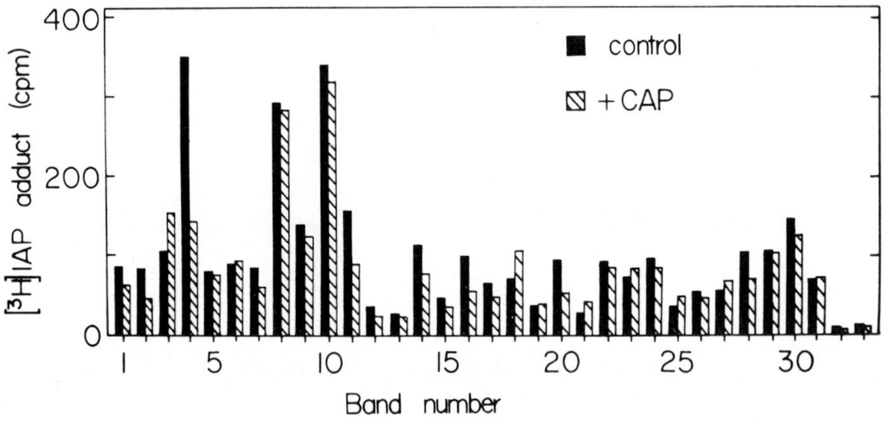

Figure 3. Chloramphenicol blockage of the labelling of mitoribosomal protein(s) by [^{3}H] iodoamphenicol. Mitoribosomal large subunits were incubated for 1 hr with 50 μM [^{3}H] iodoamphenicol in the presence or absence of 4 mM chloramphenicol. Band numbers correspond to protein bands resolved by 1D SDS PAGE.

been implicated in the peptidyl transferase activity of bovine mitoribosomes, since its removal by high salt washes correlates with loss of this activity. It is interesting to note that this protein, having strong interfacial character, is also among the least exposed proteins in the subunit (Table 1). In this regard it is important to note that proteins in the "buried" category (Table 1) need not necessarily be serving merely structural roles.

Despite the fact that they contain scarcely half the RNA of bacterial ribosomes, the gross ultrastructural features of bovine mitoribosomes resemble those of bacterial ribosomes. The "extra" proteins in bovine mitoribosomes therefore probably have structural roles, possibly functional roles, that are served by RNA in bacterial ribosomes.

ACKNOWLEDGEMENTS

We would like to thank Jane Edwards for technical assistance. This work was supported by United States Public Health Service Grants GM 15438 and GM 25888.

REFERENCES

1. O'Brien, T. W. (1977) in International Cell Biology, Brinkley, B. R. and Porter, K. R. (eds.) Rockefeller University Press, New York, pp. 245-255.
2. O'Brien, T. W. and Matthews, D. E. (1976) in Handbook of Genetics, Vol. 5. Robert C. King (ed.), Plenum Publishing Corp., New York, pp. 535-580.
3. O'Brien, T. W. (1976) in Protein Synthesis, Vol. 2, McConkey E. (ed.), Marcel Dekker, New York, pp. 249-307.
4. Hamilton, M. G. and O'Brien, T. W. (1974) Biochemistry 13, 5400-5403.
5. O'Brien, T. W., Matthews, D. E. and Denslow, N. D. (1976) in Genetics and Biogenesis of Chloroplasts and Mitochondria, Th. Bücher et al. (eds.) Elsevier/North-Holland, Amsterdam, pp. 741-748.
6. Matthews, D. E., Hessler, R. A. and O'Brien, T. W. (1978) FEBS Lett. 86, 76-80.
7. Lewis, J. A. and Sabatini, D. D. (1977) J. Biol. Chem. 252, 5547-5555.
8. Denslow, N. D. and O'Brien, T. W. (1978) Eur. J. Biochem. 91, 441-448.
9. Denslow, N. D. and O'Brien, T. W. (1979) Biochem. Biophys. Res. Comm. 90, 1257-1265.

The Organization and Expression of the Mitochondrial Genome
A.M. Kroon and C. Saccone, editors

NEUROSPORA CRASSA MITOCHONDRIAL tRNAs: STRUCTURE, CODON READING PATTERNS, GENE ORGANIZATION, AND UNUSUAL SEQUENCES FLANKING THE tRNA GENES.

SAMUEL YIN, JOYCE HECKMAN, JOSHUA SARNOFF, and UTTAM L. RAJBHANDARY
Department of Biology, Massachusetts Institute of Technology, Cambridge, Massachusetts, 02139 (USA)

INTRODUCTION

Studies in our laboratory on the *Neurospora crassa* mitochondrial system have encompassed sequence analysis of tRNAs,[1,2] mapping and cloning of the tRNA and rRNA genes,[3,4] and DNA sequence analysis of a tRNA gene cluster. Recent results from tRNA sequence analysis have helped to explain how mitochondrial protein synthesis can function with a smaller number of tRNAs than other protein synthetic systems.[5] Mapping efforts have localized fourteen specific tRNA genes on the restriction map of the mitochondrial DNA. DNA sequence analysis has given the direction of transcription of the tRNA and rRNA gene region and has revealed that the genes are often flanked by unusual palindromic repeat sequences containing double recognition sites for the restriction endonuclease Pst I.

tRNA STRUCTURES AND CODON READING PATTERNS

We have determined the nucleotide sequences of the mitochondrial methionine initiator, tyrosine, alanine, leucine$_1$, leucine$_2$, threonine, tryptophan, and valine tRNAs,[1,2,5] and have nearly completed the analyses of the methionine elongator and the glutamine tRNAs. Besides the finding that most mitochondrial tRNAs have some unusual structural features different from prokaryotic or eukaryotic cytoplasmic tRNAs,[1,2] a general finding which has emerged is a correlation between the nature of the nucleotide in the first position of the anticodon and the number of codons to be recognized by a single tRNA. This is exemplified by the two leucine tRNA sequences in Fig. 1. tRNA Leu$_1$ corresponds to the two codons UU^{A}_{G} and as expected contains a modified uridine in the first position of the anticodon. The leucine$_2$ tRNA is the only other detectable leucine acceptor and thus may be the only tRNA corresponding to the four-codon family CUN(N=U,C,A,G). It has an unmodified uridine in the first anticodon, or "wobble", position, an unusual finding since uridines found in this position in tRNAs are nearly always modified. The same pattern holds for five other tRNA sequences. The alanine, valine, and threonine tRNAs, all of which correspond to amino acids with four-codon families and all of which seem to have no other isoacceptors, contain unmodified U in the first anticodon position. Glutamine

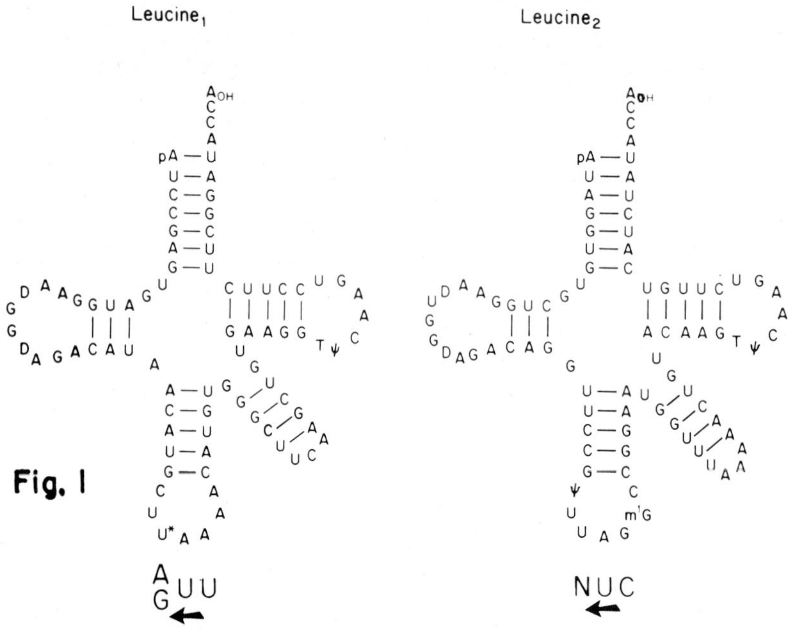

Fig. 1

and tryptophan tRNAs, which have to be restricted to recognizing codons ending in purines only, both contain the same modified U as the leu_1 species in the first anticodon position.

This pattern suggests that the tRNAs containing unmodified U in the anticodon wobble position may be capable of recognizing all four codons of their four-codon families, whereas the tRNAs with the modified U are restricted to the two codons ending in A and G. Such a codon recognition pattern would obviate the need for eight extra isoacceptors required for the eight four-codon families in the genetic code, allowing the mitochondrion to synthesize protein using only about 24 tRNA species. Similar conclusions have been reached for mammalian and yeast mitochondria.[6,7] The presence of the modified U in the wobble position of the mitochondrial tRNA Trp (instead of a C as in normal tryptophan tRNAs) implies that the *N. crassa* mitochondrion, like those of mammals and yeast,[8,9,10] uses the UGA termination codon as a sense codon for tryptophan.

GENE ORGANIZATION AND UNUSUAL SEQUENCES FLANKING THE *t*RNA GENES

The current state of our map of tRNA genes and rRNA genes of the *N. crassa* mitochondrial genome is shown in Fig. 2. The region represents approximately one-third of the total genome. We have previously shown[3] that both rRNAs and all the detectable tRNAs in the two large gene clusters are derived from the

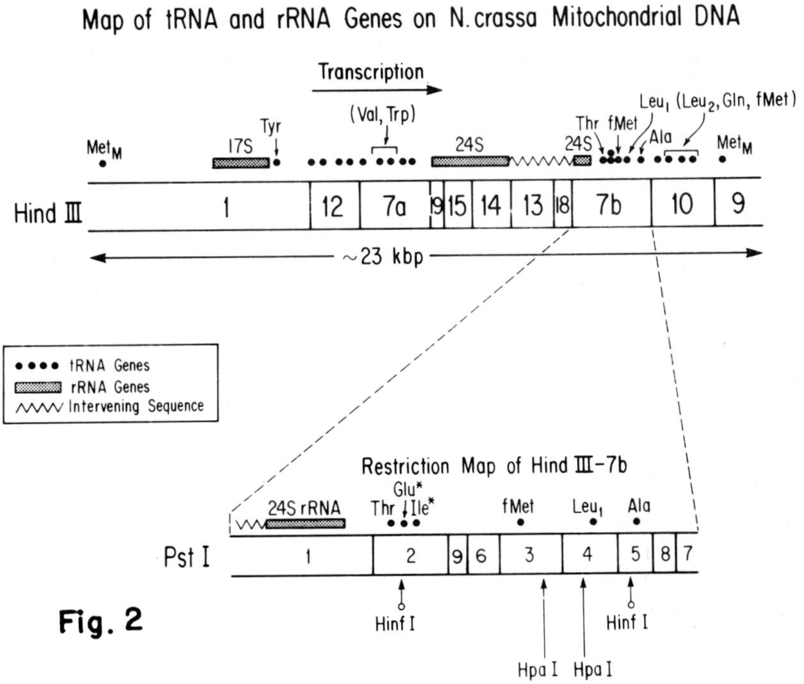

Fig. 2

same DNA strand. DNA sequence analysis in a tRNA gene cluster has now yielded the direction of transcription for the entire region as shown by the arrow. DNA sequencing also allowed identification of the two putative tRNA genes shown with asterisks in Fig. 2, besides those localized by hybridization of purified end-labeled tRNAs. Interestingly, we find two putative genes for the elongator methionine tRNA, at either extremity of this large region. This tRNA and the initiator methionine tRNA are the only two tRNAs which we have found to hybridize at two loci. DNA sequence analysis is under way to determine the nature of the putative double genes.

Because of the tight clustering of tRNA genes within Hind III-7b and 10, we have recently started DNA sequence analysis of this region and have almost completed the sequence of Hind III-7b. We have found that the Pst I sites which occur frequently in Hind III-7b (Fig. 2) and which often seem to flank tRNA sequences all have several unusual features.

The sequences encompassing the Pst I sites all showed an identical 18 base core sequence marked by double Pst I recognition sites, (CTGCAG), and separated by TA. The 18 base sequence is always preceded by a polypyrimidine stretch which generally consists of C's followed by a polypurine stretch which are predominantly G's. This creates a large palindromic sequence with an axis of symmetry between the tandem Pst I sites. A representative sequence at the Pst I-5/8 boundary is shown in Fig. 3 with the common 18 base sequence boxed. The sequence can be paired to generate the structure shown. Another feature common to all the sequences encompassing Pst I sites is their high GC content of approximately 70%. This contrasts with the low overall GC content of the mitochondrial DNA.

Fig. 3

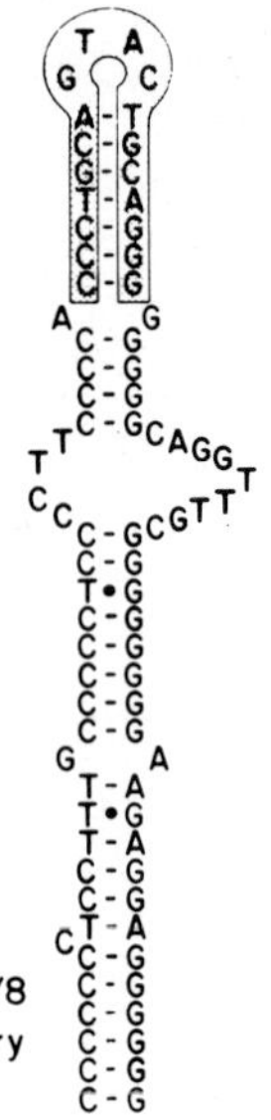

Pst I-5/8 Boundary

Because of the frequency and size of these structures, they occupy a major portion of the tRNA gene cluster. While any comments on the possible function of these sequences must remain speculative at this stage, one possibility is that the stem and loop structures may act as a signal for the processing of transcripts using an enzyme similar to RNase III in *E. coli*.[11]

ACKNOWLEDGMENTS

Supported by grants GM17151 from NIH and NP114 from the American Cancer Society. Work defining the direction of transcription was carried out as part of a collaborative project with Dr. A. Lambowitz and coworkers.

REFERENCES

1. Heckman, J.E., Hecker, L.I., Schwartzbach, S.D., Barnett, W.E., Baumstark, B. and RajBhandary, U.L. (1978). Cell 13: 83-95.
2. Heckman, J.E., Alzner-DeWeerd, B. and RajBhandary, U.L. (1979). Proc. Nat. Acad. Sci. USA 76: 717-721.
3. Heckman, J.E. and RajBhandary, U.L. (1979). Cell 17: 583-595.
4. Heckman, J.E., Yin, S., Alzner-DeWeerd, B. and RajBhandary, U.L. (1979). J. Biol. Chem. 254: 12694-12700.
5. Heckman et al. (1980). Proc. Nat. Acad. Sci. USA 77 in press.
6. Barrell et al. (1980) Proc. Nat. Acad. Sci. USA 77 in press.
7. Bonitz et al. (1980). Proc. Nat. Acad. Sci. USA 77 in press.
8. Barrell, B.G., Bankier, A.T. and Drouin, J. (1979). Nature 282: 189-194.
9. Coruzzi, G. and Tzagoloff, A. (1979). J. Biol. Chem. 254: 9324-9330.
10. Fox, T.D. (1979) Proc. Nat. Acad. Sci. USA 76: 6534-6538.
11. Bram, R.J., Young, R.A. and Steitz, J.A. (1980). Cell 19: 393-401.

The Organization and Expression of the Mitochondrial Genome
A.M. Kroon and C. Saccone, editors

NUCLEOTIDE SEQUENCE AND GENE LOCALIZATION OF YEAST MITOCHONDRIAL INITIATOR $tRNA_f^{Met}$ AND UGA-DECODING $tRNA^{Trp}$

R.P. MARTIN, A.-P. SIBLER, R. BORDONNÉ, J. CANADAY AND G. DIRHEIMER
Laboratoire de Biochimie, Institut de Biologie Moléculaire et Cellulaire du CNRS
15 rue Descartes, 67084 Strasbourg (France)

INTRODUCTION

Saccharomyces cerevisiae mitochondria contain mt tRNAs for the twenty amino acids[1,2]. Although only a few isoacceptors have been identified, it now appears that yeast mt DNA codes for a complete set of tRNAs for use in mitochondrial protein synthesis[1,3]. Most of the mt tRNA genes have been localized on the wild-type genome by deletion mapping with rho^- clones and restriction enzyme analysis[4,5].

Sequences of several yeast mt tRNAs (or their genes) determined in the past two years show that these tRNAs differ from both procaryotic and cytoplasmic tRNAs in overall sequence. Since the mt tRNAs also have original structural features, we found it particularly interesting to investigate the mitochondrial initiator $tRNA^{Met}$ whose sequence we report here. Furthermore, sequence analysis of mitochondrial genes suggest the use of a different genetic code in mitochondria. In particular, several investigators have proposed that the opal terminator codon UGA codes for tryptophan in both human[6] and yeast[7] mitochondria. In view of this, we have sequenced the yeast mt $tRNA^{Trp}$ and tested its UGA-decoding capacity. Finally we have localized both the $tRNA_f^{Met}$ and $tRNA_1^{Trp}$ genes precisely on the mitochondrial genome.

RESULTS AND DISCUSSION

Structural features of mt initiator $tRNA^{Met}$

The yeast mt initiator tRNA, identified by formylation using a mt enzyme extract, has been purified by two-dimensional polyacrylamide gel electrophoresis[1]. The sequence is shown in Fig. 1 where we indicate the features found in the yeast mt $tRNA_f^{Met}$ which are typical of procaryotic initiator tRNAs : no base-pairing at the 5'-end and presence of TψCAA. Although yeast mt $tRNA_f^{Met}$ shows the highest sequence homology with *N. crassa* mt $tRNA_f^{Met}$ [8], they differ in the above-mentioned features as well as in their acceptor arm structures. Interestingly, three organellar tRNAs, including yeast mt $tRNA_f^{Met}$, have an anticodon arm different from either procaryotic or cytoplasmic initiator tRNAs.

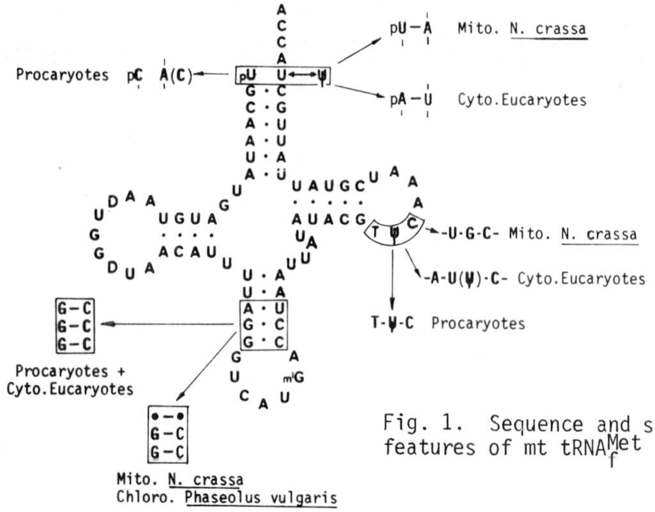

Fig. 1. Sequence and structural features of mt $tRNA^{Met}_{f}$

Nucleotide sequence and UGA-decoding of mt $tRNA^{Trp}$

Two tryptophan isoacceptors have been identified by RPC-5 column chromatography of mt tRNA. The major isoacceptor ($tRNA^{Trp}_{1}$) has been purified and its nucleotide sequence determined. mt $tRNA^{Trp}_{1}$ (Fig. 2) has the anticodon U*CA instead of CCA (or CmCA), usually found in procaryotic or cytoplasmic $tRNAs^{Trp}$. Since the anticodon U*CA could pair with both UGA (opal) and UGG (Trp) using G-U wobble, we therefore tested the coding capacity of mt $tRNA^{Trp}_{1}$ (in collaboration with H. Grosjean and S. de Henau, University of Brussels). mt $tRNA^{Trp}_{1}$ was used in translation of rabbit β-globin mRNA which has a UGA terminator. Either *in vitro* (reticulocyte cell-free system) or *in vivo* (*Xenopus* oocytes micro-injection system[9]), quantitative synthesis of the read-through product of the β-globin chain is observed (Fig.3). It remains to be shown whether mt $tRNA^{Trp}_{1}$ can also read the UGG codon, or whether this codon is translated by the minor

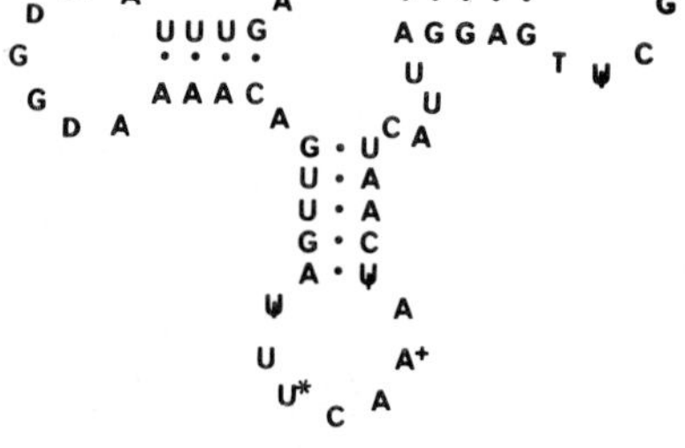

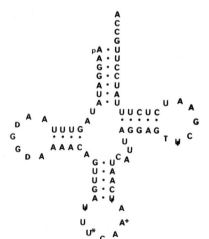

Fig. 2. Structure of yeast mt $tRNA^{Trp}_{1}$. U* is a modified uridine of unknown structure, A^{+} is $i^{6}A$ or $ms^{2}i^{6}A$.

mt $tRNA_2^{Trp}$ isoacceptor which we have detected by column chromatography.

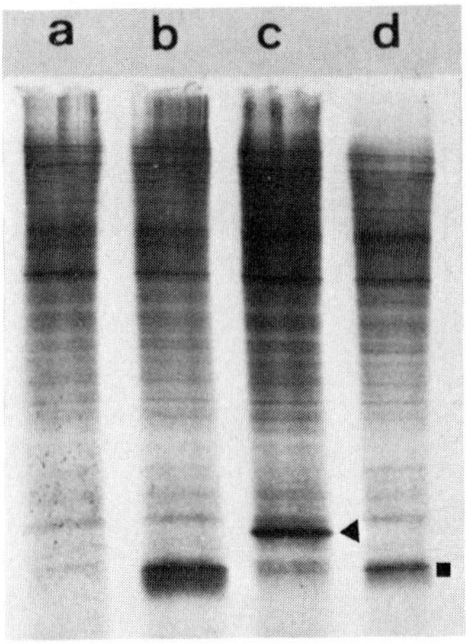

Fig. 3. SDS-polyacrylamide gel electrophoresis of the ^{35}S-labelled polypeptides synthesized in *Xenopus* oocyte cytoplasm after micro-injection of rabbit α + β globin mRNAs + *E.coli* Trp-tRNA synthetase + yeast mt $tRNA^{Trp}$ (lane c) or + yeast cytoplasmic $tRNA^{Trp}$ (lane d).
Controls : no injection (lane a) ; injection of mRNAs alone (lane b).

■ : α + β globin chains
▶ : β-globin read-through product

<u>Localisation of tRNA genes in the oxi 2 - oxi 3 region of mt DNA</u>

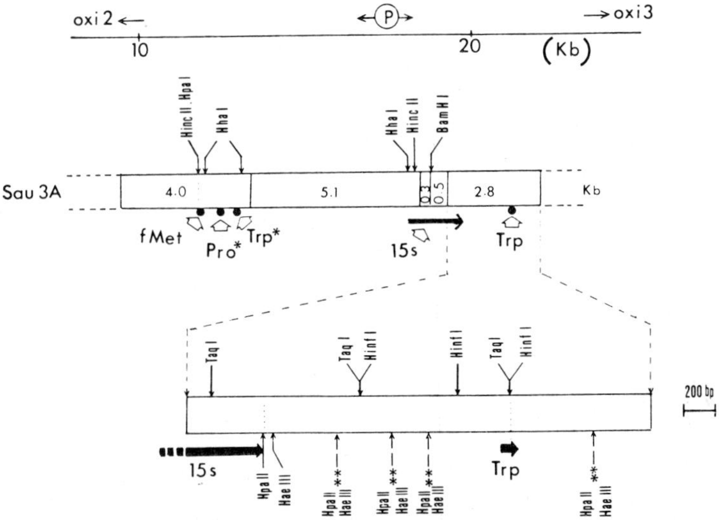

Fig. 4. tRNA gene organization in the oxi 2 - oxi 3 region of mt DNA.
The location and direction of transcription of the 15S rRNA gene is in agreement with the results described in ref. 10. ➡: direction of transcription.
* : the order of these tRNA genes is based on the results in ref. 5.
** : unordered restriction sites.

We have used the Mbo 1 (Sau 3A) cleavage map constructed by Wesolowski *et al.*[5] to localize tRNA genes in the oxi 2 - oxi 3 region of the mt genome. $tRNA_f^{Met}$, $tRNA^{Pro}$ and $tRNA_1^{Trp}$ hybridize to the 4 kb Sau 3A fragment from the oxi 2-par region (Fig. 4). The $tRNA_f^{Met}$ gene is localized very precisely since the DNA corresponding to the D-loop sequence of the tRNA contains the overlapping Hpa I and Hinc II restriction sites of this fragment. Surprisingly, mt $tRNA_1^{Trp}$ also hybridizes to another Sau 3A fragment. Fine mapping of this 2.8 kb fragment cloned in *E.coli* shows that the $tRNA^{Trp}$ gene is located 1.4 kb from the 3'-end of the 15S rRNA gene (Fig. 4). Since the DNA sequence of this tRNA gene (N.C. Martin, personal communication) is colinear with the $tRNA_1^{Trp}$ sequence (Fig. 2) we conclude that two $tRNA^{Trp}$ genes are encoded by the mt DNA. If these two genes have the same sequence, it would be the first case of gene duplication in mt DNA. Alternatively, if the 2 genes have different but similar sequences, it is possible that they code, one for $tRNA_1^{Trp}$ and one for $tRNA_2^{Trp}$.

ACKNOWLEDGEMENTS

This work was supported from grants from the INSERM (CRL 79-1.154.3) and the CNRS (ATP 4256).

REFERENCES

1. Martin, R.P., Schneller, J.M., Stahl, A.J.C. and Dirheimer, G. (1977) Nucl. Acids Res., 4, 3497-3510.
2. Martin, N.C. and Rabinowitz, M. (1978) Biochemistry, 17, 1628-1634.
3. Martin, R.P., Schneller, J.M., Stahl, A.J.C. and Dirheimer, G. (1979) Biochemistry, 18, 4600-4605.
4. Wesolowski, M. and Fukuhara, H. (1979) Mol. Gen. Genet., 1701, 261-275.
5. Wesolowski, M. Monnerot, M. and Fukuhara, H. (1980) Curr. Genet., in press.
6. Barrell, B.G., Bankier, A.T., Drouin, J. (1979) Nature, 282, 189-194.
7. Macino, G., Corruzzi, G., Nobrega, F.G., Li, M. and Tzagoloff, A. (1979) Proc. Natl. Acad. Sci. U.S.A., 76, 3784-3785.
8. Heckman, J.E., Hecker, L.I., Schwartzbach, S.D., Barnett, W.E., Baumstark, B. and RajBhandary, U.L. (1978) Cell, 13, 83-95.
9. Gurdon, J.B., Lane, C.D., Woodland, H.R. and Marbaix, G. (1971) Nature, 233, 177-182.
10. Tabak, H.F., Hecht, N.B., Menke, H.H. and Hollenberg, C.P. (1979) Curr. Genet., 1, 33-43.

The Organization and Expression of the Mitochondrial Genome
A.M. Kroon and C. Saccone, editors

PARTIAL PURIFICATION OF POSTPOLYSOMAL FACTORS ESSENTIAL FOR OPTIMAL RATES OF YEAST MITOCHONDRIAL PROTEIN SYNTHESIS*

ERIC FINZI† AND DIANA S. BEATTIE
Department of Biochemistry, Mount Sinai School of Medicine of the City University of New York, New York, N.Y. 10029 (USA)

INTRODUCTION

The formation of cytochrome oxidase, oligomycin-sensitive ATPase, and the bc_1 complex in yeast mitochondria, requires the coordinated synthesis of proteins in both the mitochondria and cytoplasm[1]. Previous studies have suggested that proteins synthesized in the cytoplasm may control yeast mitochondrial protein synthesis in yeast[2-5]. We have recently reported that protein synthesis by isolated yeast mitochondria can be stimulated 6-8 fold by the addition of yeast postpolysomal supernatant at the start of the incubation[6]. In the present study a partial purification of the factors which stimulate the rate of mitochondrial synthesis *in vitro* is described.

MATERIALS AND METHODS

A modification of the zymolyase digestion method[7] was used to prepare mitochondria for *in vitro* synthesis. Protein synthesis was performed by previously published methods[6] with the modification that 0.02μmol/ml of L-leucine was added. Postpolysomal supernatants were prepared from spheroplasts that were broken by a low speed spin in a Waring blender in 0.6M mannitol/0.1mM EDTA, pH 6.8/1mg/ml of bovine serum albumin/1mM PMSF. The post-mitochondrial supernatants were centrifuged for 60 min at 140,000g. The resulting supernatant was dialyzed for 15 hrs against 20 mM KPi, pH 7.0 containing 1mM EDTA and 10% glycerol (buffer A), before concentration by G-50 beads. Gel electrophoresis followed by autoradiography was done as described[9].

RESULTS AND DISCUSSION

In order to purify the cytoplasmic proteins which regulate mitochondrial protein synthesis, we have used isolated yeast mitochondria to assay for factors which stimulate protein synthesis when added at the start of the

---

*Supported by NIH grant HD-04007 and NSF grant PCM 782435
†Trainee on Medical Scientist Training Grant GM 07280.

incubation. Initially, yeast postpolysomal supernatants were fractionated in buffer A by chromatography on Sephacryl S-200. The void volume and a peak in the 40,000 to 80,000 molecular weight range contained most of the stimulatory activity, while a small peak with activity was observed with a molecular weight less than 10,000. After fractionation of the middle molecular weight peak on Sephacryl S-200, about half of the stimulatory activity was now eluted in the void volume as well as that present in the middle molecular weight range.

Other experiments indicated that concentrating the postpolysomal supernatant to greater than 40 mg protein per ml before fractionation on Sephacryl S-200 resulted in most of the stimulatory activity appearing in the void volume; however, there was a gradual tailing off of activity toward lower molecular weight fractions. These observations suggested that the stimulatory factors present in the postpolysomal supernatants have a tendency to aggregate to higher molecular weight forms. In order to confirm this suggestion, the activity present in the void volume was pooled and dialyzed overnight against buffer A before refractionation on Sephacryl S-200 in an attempt to disaggregate the stimulatory factors. Most of the stimulatory activity now eluted in the middle and low molecular weight peaks with some spreading of activity occurring throughout the fractions. The results suggested that under low salt conditions the stimulatory factor(s) has a strong tendency to aggregate to higher molecular weight forms (>200,000) and that this aggregation can be reversed, in part, by dialysis against 0.5M KPi.

Direct addition of a 1.8M KPi solution to the postpolysomal superantant to a final concentration of 0.5M, caused an even greater disaggregation of the stimulatory activity such that nearly half of the initial stimulatory activity was now eluted in the low molecular weight peak (Fig. 1). The increased activity in the low molecular weight peak combined with the decreased activity in the void volume and middle molecular weight range, leads us to conclude that most of the stimulatory activity eluting in the void volume and some of that eluting in the middle peak activity represent the low molecular weight activator which has aggregated either with itself or other proteins. Consistent with this view was the diffuse elution of the low molecular activator after DEAE-cellulose chromatography possibly because of its aggregation with other proteins with different isoelectric points.

Stimulation of mitochondrial protein synthesis by the low molecular weight activator was sensitive to chloramphenicol, insensitive to cycloheximide and proportional to the protein concentration of activator added. Therefore, the stimulation of amino acid incorporation by the activator represents mitochondrial protein synthesis. In addition, protein synthesis by 1mg of isolated mitochondria was stimulated twenty-fold by 90μg of low molecular weight activator representing a 45-fold purification of the stimulatory activity present in the postpolysomal supernatant. These results confirm our earlier suggestions[8] that freshly isolated mitochondria are deficient in proteins necessary for optimal rates of protein synthesis.

In order to determine whether the low molecular weight activator stimulates the synthesis of some or all mitochondrial translation products, isolated mitochondria were labeled with [^{35}S]methionine in the presence or absence of activator. As seen in Figure 2, the activator stimulated equally the synthesis of all mitochondrial translation products whether added at the beginning of the incubation or after 35 min.

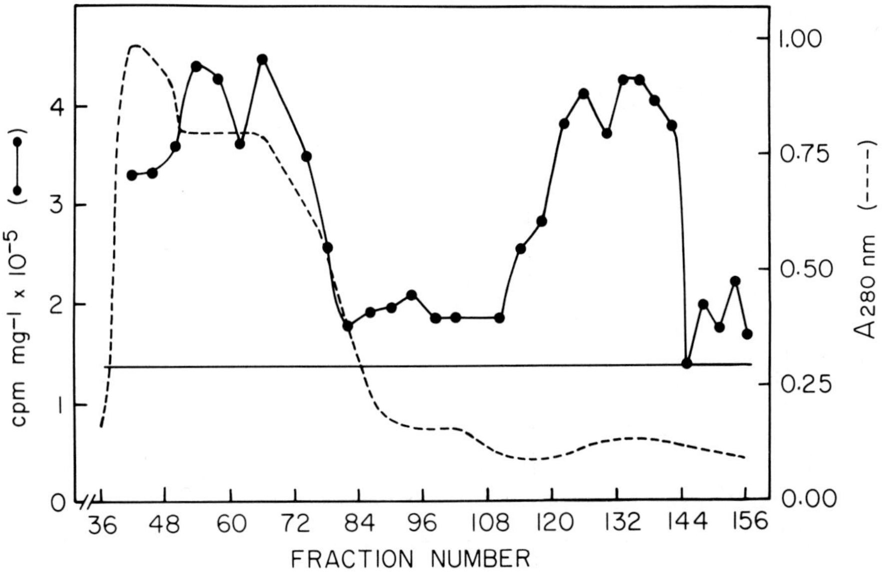

Fig. 1 Elution profile of stimulatory activity after Sephacryl S-200 chromatography. Postpolysomal supernatant (480mg) was loaded onto a 100 x 2.5cm column and eluted in buffer A. Horizontal line represents incorporation by mitochondria alone.

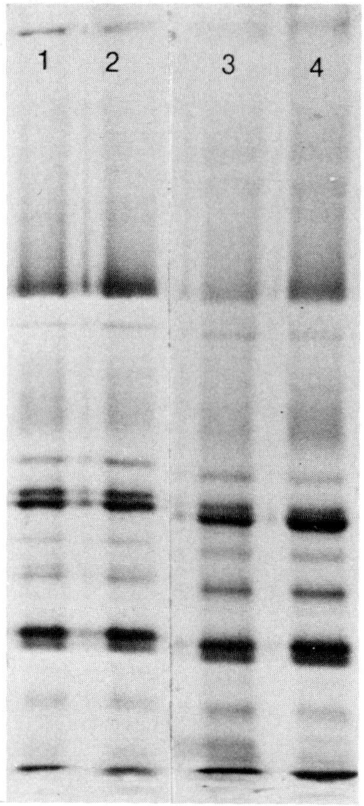

Fig. 2. Electrophoretic analysis of in vitro mitochondrial translation products. Isolated mitochondria were labeled with [^{35}S]methionine. Lane 1-mitochondria alone; Lane 2-activator added at time zero. Lane 3-activator added after 35 min; Lane-4 S-140 added after 35 min. Equal counts were loaded on each lane.

REFERENCES

1. Tzagoloff, A., Macino, G. and Sebald W. (1979) Ann. Rev. Biochem. 48, 419-441.

2. Tzagoloff, A. (1971) J. Biol. Chem. 246, 3050-3056.

3. Ibrahim, N.G., Stuchell, R.N., and Beattie, D.S. (1973) Eur. J. Biochem. 36 519-527.

4. Ibrahim, N.G., and Beattie, D.S. (1976) J. Biol. Chem. 251, 108-115.

5. Poyton, R.O., and Kavanagh, J. (1976) Proc. Nat. Acad. Sci. USA 73, B3947-3951.

6. Everett, T.D., Finzi, E., and Beattie, D.S. (1980) Archives of Biochem. and Biophys. 200, 467-473.

7. Schatz, G., and Kovac, L. (1974) Methods Enzymol. 31, 627-632.

8. Beattie, D.S., Chen, Y.S., Clejan, L., and Lin, L.-F.H. (1979) Biochemistry 18, 2400-2406.

The Organization and Expression of the Mitochondrial Genome
A.M. Kroon and C. Saccone, editors

BIOSYNTHESIS OF MITOCHONDRIAL PROTEINS IN ISOLATED HEPATOCYTES

B. DEAN NELSON, J. KOLAROV[+], V. JOSTE, A. WEILBURSKI and I. MENDEL-HARTVIG
Department of Biochemistry, Arrhenius Laboratory, University of Stockholm, Stockholm, Sweden, and [+]Cancer Research Institute, Slovak Academy of Sciences, Bratislava, Czechoslovakia.

INTRODUCTION

Studies have been recently initiated on the biosynthesis of inner membrane proteins in isolated rat hepatocytes [1-3]. In addition to providing a model for mitochondrial synthesis in well differentiated, hormone-sensitive, cells, hepatocytes offer a number of technical advantages inherent in suspended cell cultures. In the present paper we report studies on the synthesis of cytochrome oxidase in isolated hepatocytes, and on the general influence of thyroid hormones on mitochondrial biosynthesis.

MATERIALS AND METHODS

Hepatocytes were isolated as described previously[4]. Conditions for *in vitro* labeling of hepatocytes[1] and for isolation of mitochondria from labeled cells[4] have been reported. Cytochrome oxidase was purified from rat liver mitochondria by the method of Ades and Cascarano[5], and antisera were raised in rabbits as outlined in[6]. Cytochrome oxidase was immuno-absorbed on Sepharose-protein A[3].

RESULTS AND DISCUSSION

Table 1 shows the effects of inhibitors on labeling of mitochondrial proteins in isolated hepatocytes. The cycloheximide (CHX)-insensitive label present in inner membrane protein has been shown to be distributed primarily between 4 major mitochondrial translation products, with molecular weights of: (I) 45,000, (II) 26,000, (III) 24,000 and (IV) 8-10,000[1,2]. To identify these peptides, we have initiated a series of experiments using specific antibodies against isolated inner membrane components. The results obtained with cytochrome oxidase are given below.

Rat liver cytochrome oxidase isolated as described[5] is resolved into 7 major peptides by electrophoresis in the buffer of Laemmli[7]. The apparent molecular weights of these peptides are: (I) 45,000, (II) 25,800, (III) 16,800, (IV) 12,100, (V) 10,500, (VI) 7,600 and (VII) 5,300. All of these subunits can be precipitated with antibodies to rat liver holo-cytochrome oxidase.

Fig. 1 shows labeling of the 7 subunits immunoabsorbed from mitochondria

TABLE 1

LABELING OF INNER MITOCHONDRIAL MEMBRANE PROTEINS IN ISOLATED HEPATOCYTES.

Labeling conditions	exp. 1		exp. 2	
	cpm/mg protein	%	cpm/mg protein	%
no additions	567,000	100	510,800	100
cycloheximide	63,000	11	46,000	9
cycloheximide+chloramphenicol	28,000	5	26,500	5

Cells were labeled for 4 hours with ^{35}S-methionine, and submitochondrial were prepared as in[4].

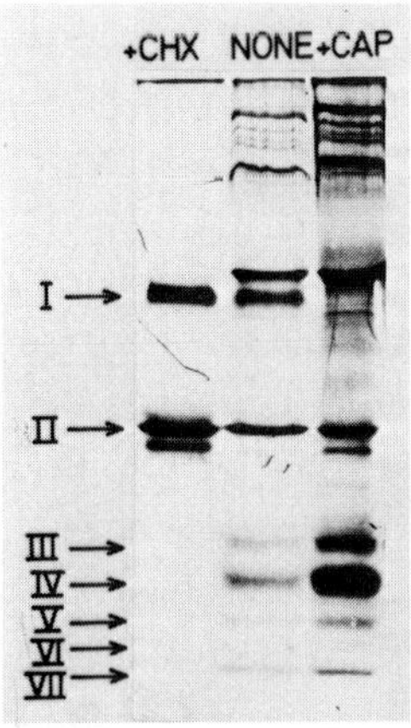

Fig. 1. Biosynthesis of cytochrome oxidase in isolated rat hepatocytes. Hepatocytes were labeled _in vitro_ in the absence of inhibitors or in the presence of either cycloheximide (CHX) or chloramphenicol (CAP). Mitochondria were isolated and cytochrome oxidase was immunoabsorbed on Sepharose-protein A.

TABLE 2

RADIOACTIVITY IMMUNOPRECIPITATED FROM MITOCHONDRIAL MEMBRANES WITH CYTOCHROME OXIDASE ANTISERUM

Labeling conditions	Total cpm in: Intact membranes	immuno-precipitate	cpm in precipitate (%)
no accitions	434,000	19,570	4.6
cycloheximide	48,800	9,600	20.0
chloramphenicol	396,000	18,380	4.6

Cells were labeled for 4 hours with ^{35}S-methionine. Mitochondria were isolated, and cytochrome oxidase was immuno-absorbed with Sepharose-protein A.

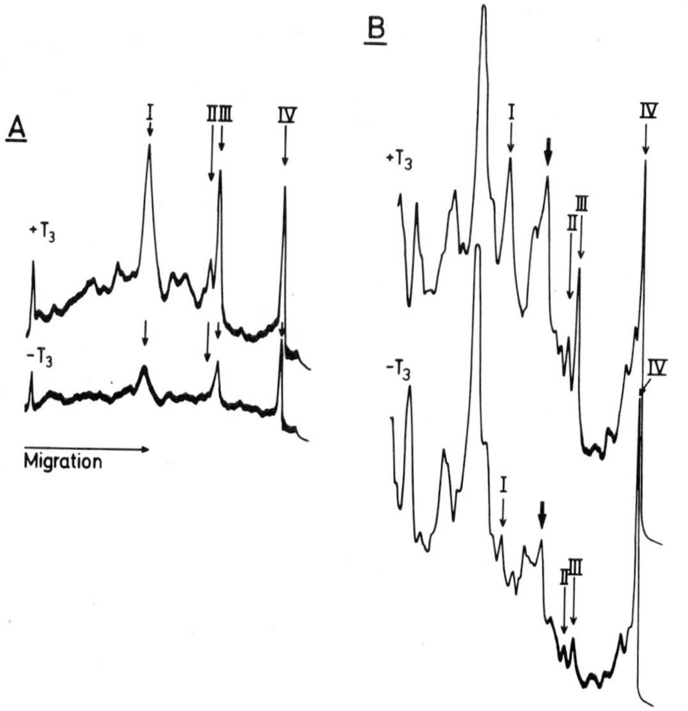

Fig. 2. Effects of thyroid hormone (T_3) on the synthesis of mitochondrial inner membrane proteins. Hepatocytes from hypothyroid rats or from hypothyroid rats treated with T_3, were labeled in vitro in the presence of cycloheximide (A) or in the absence of inhibitors (B). Submitochondrial particles were isolated, and 100 μg protein was electrophoresed in each case. The traces are densiometric Scans of autoradiographs.

isolated from in vitro labeled hepatocytes. Peptides migrating more slowly than subunit I could be absorbed with pre-immune sera, whereas peptides I-VII could not. In the presence of cycloheximide (CHX) only subunits I and II of cytochrome oxidase are labeled in the immunoabsorbed enzyme (Fig. 1) A third, weakly labeled, peptide can also appear in mitochondria from inhibitor-treated cells (Fig. 1). However, since this peptide does not correspond to a coomassie blue staining subunit of the isolated enzyme, and since it is not present in all experiments, we conclude that it is most likely a breakdown product of either subunit I or II.

Chloramphenicol (CAP) inhibits synthesis of subunits I and II (Fig. 1). However, labeling of subunit II appears to be more resistant to CAP than is labeling of subunit I. This could reflect differential turnover in the absence of complete inhibition by CAP. CAP also stimulates labeling of cytoplasmically-translated subunits III and IV (for terminology, see above) (Fig. 1). The effect appears to be greatest with subunit IV. Lack of significant CAP inhibition of total labeling of the immunoabsorbed enzyme (table 2) can be explained by increased labeling of subunits III and IV, and the relative insensitivity of subunit II to CAP (Fig. 2).

Fig. 2 shows the effects of tri-iodothyronine (T_3) on labeling of inner membrane peptides. Hepatocytes were prepared from hypothyroid rats or from hypothyroid rats 24 hrs after a single injection of 20 μg T_3/rat. Cells were labeled *in vitro*, and inner membranes were prepared. Synthesis of the major mitochondrial-translation products is increased after hormone treatment (Fig. 2A), whereas one cytoplasmically-translated mitochondria peptide (Fig. 2B) (heavy arrow) was observed to be influenced by T_3. Thyroid hormone, thus, appears to exert a selective effect on mitochondrially-translated peptides in hepatocytes. Since peptides I and II of the inner membrane preparation (Fig. 2) correspond to immunoabsorbed cytochrome oxidase subunits I and II (not shown), it can be concluded that thyroid hormone controls synthesis of the two largest subunits of cytochrome oxidase.

ACKNOWLEDGEMENTS

This work was supported by grants from the Swedish Natural Science Research Council and the Swedish Cancer Society.

REFERENCES

1. Gellerfors, P., Weilburski, A. and Nelson, B.D. (1979) FEBS Lett. 108, 67-70.
2. Nelson, B.D., Joste, V., Weilburski, A. and Rosenqvist, U. (1980) Biochim. Biophys. Acta, (in press).
3. Kolarov, J., Weilburski, A., Mendel-Hartvig, I. and Nelson, B.D. (submitted for publication).
4. Gellerfors, P. and Nelson, B.D. (1979) Anal. Biochem. 93, 200-203.
5. Ades, I.Z. and Cascarano, J. (1977) J. Bioenerg and Biomembranes 9, 237-253.
6. Nelson, B.D. and Mendel-Hartvig, I. (1977) Eur. J. Biochem 80, 267-274.
7. Laemmli, U.A. (1970) Nature 277, 680-685.

DEVELOPMENTAL AND REGULATORY ASPECTS OF MITOCHONDRIAL BIOGENESIS

© 1980 Elsevier/North-Holland Biomedical Press
The Organization and Expression of the Mitochondrial Genome
A.M. Kroon and C. Saccone, editors

BIOGENESIS OF CYTOCHROME *c* OXIDASE IN *Neurospora crassa*: INTERACTIONS BETWEEN MITOCHONDRIAL AND NUCLEAR REGULATORY AND STRUCTURAL GENES.

HELMUT BERTRAND
Department of Biology, University of Regina, Regina, Sask., CANADA S4S 0A2

INTRODUCTION

The assembly of cytochrome *c* oxidase on the inner mitochondrial membrane involves at least seven different subunit polypeptides, heme *a*, copper and phospholipids [1-6]. The biogenesis of the complex depends on mitochondrial translation for the synthesis of the three largest polypeptides of cytochrome aa_3, subunits 1, 2 and 3, and on cytosolic translation for the synthesis of the remaining four or five subunit polypeptides[7]. It is assumed that the biosynthesis of the different structural polypeptides of cytochrome *c* oxidase by the mitochondrial and nuclear transcription and translation systems, the enzymatic synthesis of heme *a* and phospholipids, and the metabolism of copper all are coordinated by cellular regulatory mechanisms [7-14]. While the recognition of coordinative interactions between these metabolic processes has generated the anticipation for the existence of cellular control over the biogenesis of cytochrome *c* oxidase, the actual confirmation of this concept has been obtained only recently through careful analysis of respiratory mutants in *Neurospora crassa*[15] and yeast[10,14].

This communication presents a summary of genetic and biochemical evidence obtained from *Neurospora crassa* indicating that the biogenesis of cytochrome aa_3 is regulated by several cellular control mechanisms which affect the production and/or processing of specific components of the cytochrome *c* oxidase complex.

MATERIALS AND METHODS

Strains. The *Neurospora crassa* strains used in this study are presented in Table 1 and have been described elsewhere[15,16]. Genetic symbols denoting extranuclear mutations are given in parentheses to facilitate the reading of genotypes.

Analytical procedures. References describing media, growth conditions, the isolation of mitochondria, recording of cytochrome spectra, immunological techniques, and the procedures used to induce cytochrome aa_3 have been published previously[15].

TABLE 1

Mutant	Linkage group	Cytochrome phenotype
(mi-3)	mitochondrial	*aa*$_3$ deficient
(exn-5)	mitochondrial	*aa*$_3$ deficient
(oxi-1)	mitochondrial	*aa*$_3$ deficient
cyt-2-1	VI	*aa*$_3$ and *c* deficient
cya-3-16	VI	*aa*$_3$ deficient
cya-4-23	II	*aa*$_3$ deficient
299-1	unknown (nuclear)	*aa*$_3$ deficient
cyb-1-1	V	*b* deficient
cyb-2-2	VI	*b* deficient, low *aa*$_3$
cyb-3	unknown (nuclear)	*b* deficient, ts

RESULTS AND DISCUSSION

Induction of cytochrome *aa*$_3$ by gene interactions. Normally, the mitochondria from the *(mi-3)* cytoplasmic mutant, and from the *cyt-2-1* and *cya-3-16* nuclear mutants are grossly deficient in cytochrome *aa*$_3$. However, when any one of these three mutant genes is combined genetically with one of two mutations causing a deficiency in cytochrome *b*, namely *cyb-1-1* or *cyb-2-2*, then the double mutants, *(mi-3) cyb-1-1*, *(mi-3) cyb-2-2*, *cyt-2-1 cyb-1-1*, etc., produce cytochrome *aa*$_3$, as illustrated for *(mi-3) cyb-1-1* in Figure 1. However, strains having combinations of *cyb-1-1* or *cyb-2-2* with mutant alleles of other nuclear or extranuclear genes causing cytochrome *aa*$_3$ deficiency phenotypes, such as *cya-4-3*, *cya-5-34*, *(exn-5)*, *(oxi-1)*, etc., do not have cytochrome *aa*$_3$. Clearly, *(mi-3)*, *cyt-2-1* and *cya-3-16* are special mutants, because they have in an intact form all the genetic information necessary for the production of cytochrome *aa*$_3$, but can not express this information unless the *cyb-1-1* or *cyb-2-1* alleles are present. The mere deficiency in cytochrome *b*, however, is not sufficient for the expression of the "silent" genetic information in these mutants: the production of cytochrome *aa*$_3$ can not be induced in *(mi-3)* by introducing the *cyb-3* mutation which causes cytochrome *b* deficiency when double

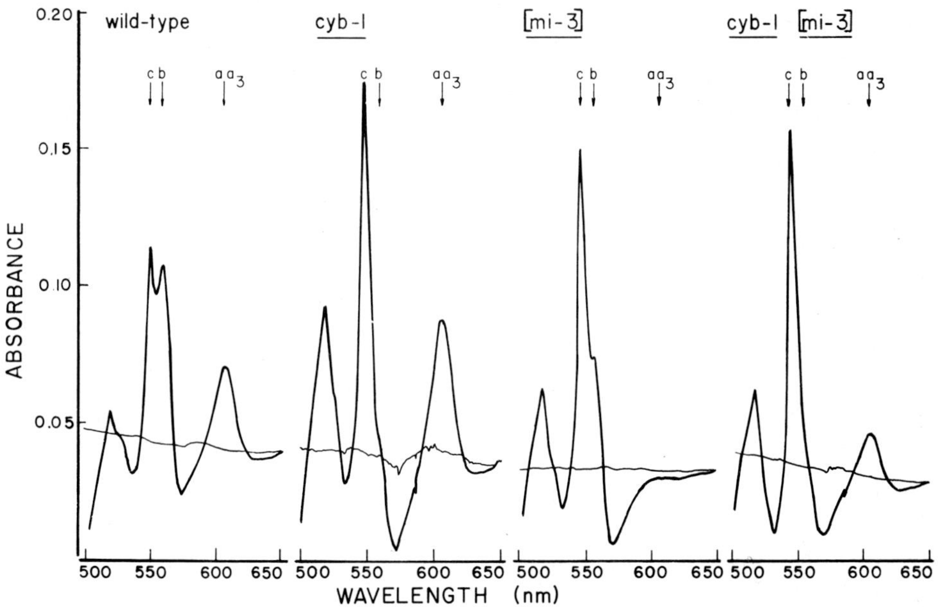

Fig. 1. Absorption spectra depicting the cytochrome *aa*$_3$ content of mitochondria from wild-type, and mutants having the *cyb-1-1*, *(mi-3)* and *cyb-1-1 (mi-3)* genotypes.

mutants are grown at 37°C, but not when the cultures are grown at 25°C (16).

Induction of cytochrome *aa*$_3$ by inhibitors of mitochondrial electron transport. The production of cytochrome *aa*$_3$ in *(mi-3)*, *cyt-2-1* and *cya-3-16* double mutants harbouring *cyb-1-1* or *cyb-2-2* has two possible explanations: either the *cyb* loci are regulatory genes and their products exert positive control on cytochrome *b* synthesis and negative control on the biogenesis of cytochrome *aa*$_3$, or the production of cytochrome *aa*$_3$ is coupled indirectly to a defect in mitochondrial electron transport generated by cytochrome *b* deficiency. The second of these alternatives leads to the prediction that cytochrome *aa*$_3$ can be induced in *(mi-3)*, *cyt-2-1* or *cya-3-16* by treatment of the cells with inhibitors of mitochondrial electron transport, such as antimycin A or 2-heptyl-4-hydroxyquinoline-N-oxide (HQNO). Indeed, when the above three mutants are grown in medium containing 0.1 to 1.0 μg antimycin A/ml of medium, cytochrome *aa*$_3$ is produced, as shown in Figure 2. Antimycin A has no effect on the phenotypes

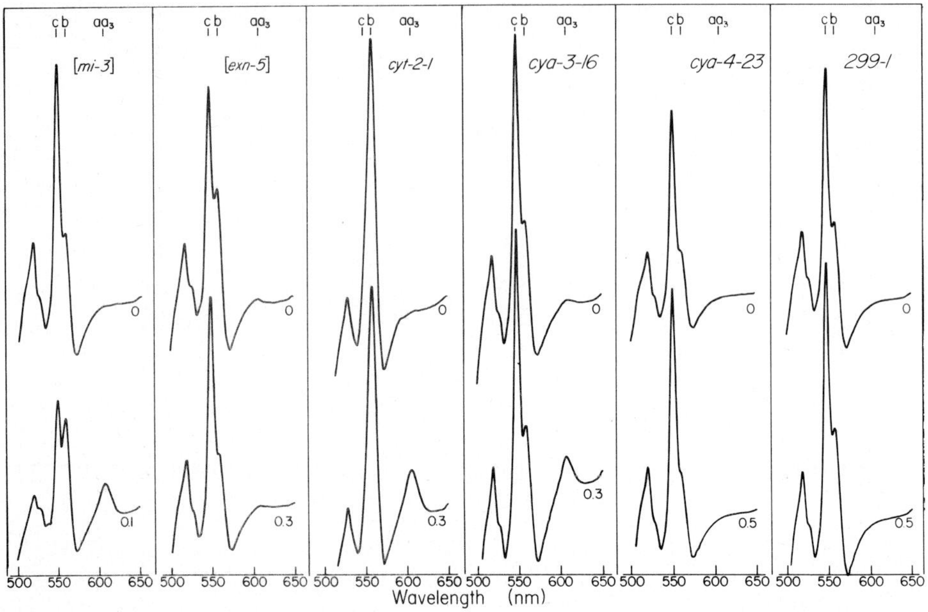

Fig. 2. Absorption spectra reflecting cytochrome $\underline{aa}_3$ contents of mitochondria from untreated (upper row) and antimycin A treated (lower row) cultures of wild type and six cytochrome $\underline{aa}_3$ deficient mutants of *N. crassa*. The numbers on the graph indicate antimycin A concentration in μg/ml.

of other cytochrome $\underline{aa}_3$ deficient extranuclear and nuclear mutants, including *(exn-5)*, *(oxi-1)*, *cya-4-23*, *cya-5-34*, *299-1* and several others that have been isolated recently in this laboratory.

The production of cytochrome $\underline{aa}_3$ also can be induced in *(mi-3)*, *cyt-2-1* and *cya-3-16* mutants by treatments with oligomycin (0.5-2.0 μg/ml) or HQNO (2.0-8.0 μg/ml), but not by the treatment of cells with salicylhydroxamate, an inhibitor of the alternative oxidase. The introduction of genetic alternative oxidase deficiencies into *(mi-3)* also does not result in the production of cytochrome $\underline{aa}_3$.

The stimulation of the production of cytochrome $\underline{aa}_3$ in *(mi-3)*, *cyt-2-1* and *cya-3-16* by antimycin A follows kinetics suggestive of classical enzyme induction, as shown for *(mi-3)* in Figure 3. When the "inducer", in this case an inhibitor of electron-transport, is added to exponentially growing cells, cytochrome *aa*$_3$ begins to accumulate after a lag period and reaches a maximum concentration of about 0.3 nanomoles per nanomole of mitochondrial cytochrome *b* within 8 to 10 hours after the addition of the effector, although very little cell growth occurs during this time.

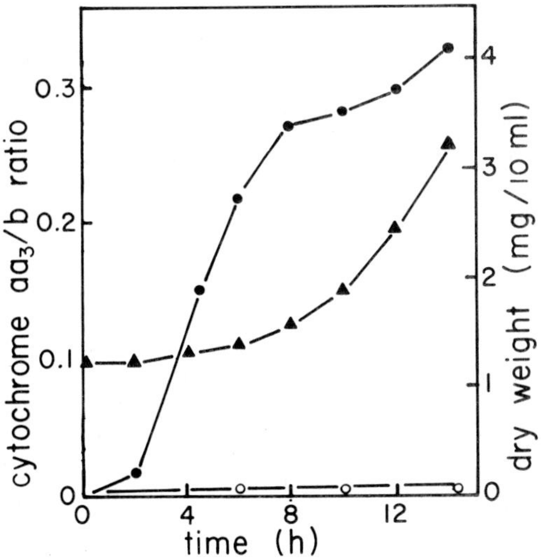

Figure 3. Time course of the induction of cytochrome *aa*$_3$ by antimycin A in the *(mi-3)* cytoplasmic mutant. Antimycin (0.15 μg/ml of culture) was added to exponentially growing cells of a 10 h old culture at zero time. o --- o cytochrome *aa*$_3$/*b* ration for untreated *(mi-3)* control cells, ● --- ● cytochrome *aa*$_3$/*b* ratio for antimycin treated cells, ▲ --- ▲ growth of antimycin treated cells (dry weight).

The cytochrome *aa*$_3$ that is induced in *(mi-3)*, *cyt-2-1* and *cyt-3-16* is a functional enzyme consisting of the seven polypeptide subunits that constitute the cytochrome oxidase complex from wild-type cells[15,17]. This complex is not assembled in mitochondria of uninduced cells of the three mutants[15,17,18], although some of the subunit polypeptides are present[18].

A rudimentary model for the regulation of cytochrome *aa*$_3$ production in *Neurospora crassa*. The observations presented in the previous two sections indicate that some aspect of the synthesis or assembly of cytochrome *aa*$_3$ in *N. crassa* is controlled by two, at least partially independent, regulatory circuits. The first regulatory mechanism controls "constitutive" cytochrome *aa*$_3$ production and is rendered non-functional by mutations in any one of at least three genes, namely the mitochondrial *(mi-3)* locus and the *cyt-2-1* and *cya-3-16* nuclear loci. When the control for "constitutive" cytochrome *aa*$_3$ assembly is inoperative, a second regulatory system is detectable because cytochrome *aa*$_3$ production is inducible. The second regulator appears to function as a "modulator" by which cells increase cytochrome *aa*$_3$ content in response to impairments in the activity of the mitochondrial electron transport chain. It is likely that the connection between the "modulator" and electron transport is an effector molecule which either is generated, or ceases to be generated when mitochondrial electron transport is inhibited. The chemical nature of the effector is unknown, but it is plausible that it is a low molecular weight, diffusible substance which either affects the transcription or translation of genes whose products are required for the assembly of cytochrome *aa*$_3$, or acts as an inducer, activator or inactivator of an enzyme required for the processing of a precursor of one of the components of cytochrome oxidase. It is noteworthy that the mitochondria of the *(mi-3)* mutant contain a 45,000 M_r polypeptide which is processed to the mature 43,000 M_r cytochrome *aa*$_3$ subunit 1 polypeptide and incorporated into the active oxidase complex when cells from the mutant are "induced" with antimycin A[17]. However, since none of the observations indicate that the processing of the subunit 1 polypeptide is directly related to the "induction" process, it is possible that the conversion of the precursor to the mature subunit represents no more than one of the many steps in the assembly process of cytochrome *c* oxidase.

A nuclear suppressor of *(mi-3)*: where does it fit? A mutant allele, *su-1*$^{(mi-3)}$, of a gene located on linkage group I of the nuclear gene complement of *N. crassa* effects a complete suppression of the cytochrome *aa*$_3$ deficiency phenotype of *(mi-3)*, i.e. *(mi-3) su-1*$^{(mi-3)}$ cells have a completely normal cytochrome system[19]. There are two possible mechanisms that could explain the restoration of constitutive cytochrome *aa*$_3$ synthesis in *(mi-3)* by the suppressor. The first possibility assumes that a regulator-operator relationship exists between *(mi-3)* and *su-1*$^{(mi-3)}$. For example, the *(mi-3)* mutation generates a suppressor which binds irreversibly to an operator site, presumably the wild-type allele of the *su-1*$^{(mi-3)}$ locus. In this case, an operator-constitutive mutation yielding the *su-1*$^{(mi-3)}$ allele would restore consitutive cytochrome *aa*$_3$ production. The second possibility is that the *su-1*$^{(mi-3)}$ allele turns the "modulator" from an inducible system into a constitutive system. For example, the modulator system would be converted to constitutive control if an effector, which normally would be generated only when electron transport is impaired, is generated in the *su-1*$^{(mi-3)}$ mutant regardless of the state of electron flow through the cytochrome chain.

Several observations indicate that the *su-1*$^{(mi-3)}$ mutation does not convert the "modulator" to a constitutive regulatory system: 1) the suppressor also restores a full cytochrome aa_3 complement in *(exn-5)* cells, even though the oxidase can not be induced in this extranuclear mutant by antimycin A, oligomycin or HQNO treatments or by genetically caused cytochrome *b* deficiency[15]; and, 2) the *su-1*$^{(mi-3)}$ allele does not suppress the cytochrome aa_3 deficiency phenotype of *cyt-2-1*, as should be the case if the modulator control is changed to a constitutive regulatory system. These findings not only implicate *su-1*$^{(mi-3)}$ as a component of the regulatory mechanism that controls constitutive cytochrome oxidase production, but suggest strongly that the *(exn-5)* locus codes for a cellular component whose function is required for "constitutive" as well "modulatory" cytochrome aa_3 production. The characteristics of the *(exn-5)* mutant provide the only evidence that the regulatory systems are functionally interconnected.

Nuclear control of the expression of mitochondrial structural genes. Nuclear cytochrome aa_3 deficient mutants lacking one of the three intramitochondrially synthesized polypeptides of cytochrome oxidase have been obtained in *Neurospora* as well as yeast. For example, the mitochondria of the *cya-5-34* and *299-1* mutants of *N. crassa* are devoid of polypeptides that crossreact with antibodies specific for subunits 1 and 2, respectively, of the oxidase, or comigrate with the subunit 1 or 2 polypeptides during gel electrophoresis of mitochondrial translation products[18,20]. Similarly, the *pet 494-1* mutant of yeast lacks subunit 3[14]. The phenotype of the *pet 494-1* mutant is suppressible by amber suppressors[14], providing strong reason to assume that a protein which is a nuclear gene product is required for the expression of the mitochondrial locus coding for the subunit 3 polypeptide of cytochrome aa_3. Similarly, in *Neurospora*, the products of the *cya-5-34* and *299-1* nuclear genes are required for the synthesis or integration into the mitochondrial membrane of two mitochondrial gene products, namely the subunit 1 and 2 polypeptides, respectively. While there is no proof that the *cya-4-34* and *299-1* nuclear loci regulate the transcription or translation of the mitochondrial structural genes for subunit 1 and 2, respectively, the absence of mitochondrial gene products in these mutants suggests that the two nuclear loci exert some type of control over the expression of specific mitochondrial genes. It is possible, therefore, that the biogenesis of cytochrome *aa*$_3$ is controlled by a complex regulatory system composed of a series of interacting "circuits", each controlling a specific aspect of the synthesis or assembly of cytochrome *aa*$_3$ including the transcription, translation and processing of each of the subunits of the enzyme.

REFERENCES

1. Weiss, H., Sebald, W. and Bücher, Th. (1971) Europ. J. Biochem. 22, 19-26.
2. Schwab, A.J., Sebald, W. and Weiss, H. (1972) Europ. J. Biochem. 30, 511-516.
3. Sebald, W., Machleidt, W. and Otto, J. (1973) Europ. J. Biochem. 38, 311-324.
4. Schwab, A.J. (1973) FEBS Lett. 35, 63-66.
5. Wohlrab, H. and Jacobs, E.E. (1967) Biophys. Biochem. Res. Commun. 28, 991-997.
6. Werner, S., Schwab, A.J. and Neupert, W. (1974) Europ. J Biochem. 49, 607-617.
7. Schatz, G. and Mason, T.L. (1974) Ann. Rev. Biochem. 43, 51-87.
8. Barath, Z. and Küntzel, H. (1972) Proc. Nat. Acad. Sci. U.S.A. 69, 1371-1374.
9. Sugimura, T., Okabe, K., Nagao, M. and Gunge, M. (1966) Biochim. Biophys. Acta 115, 267-275.
10. Saltzgaber-Müller, J. and Schatz, G. (1978) J. Biol. Chem. 253, 305-310.
11. Charalampous, F.C. (1974) J. Biol. Chem. 249, 1014-1021.
12. Gollub, E.G., Trocha, P., Liu, P.K. and Sprinson, D.B. (1974) Biochem. Biophys. Res. Commun. 56, 471-477.
13. Gordon, P.A., Lowdon, M.J. and Steward, P.R. (1972) J. Bact. 110, 511-515.
14. Ono, B., Fink, G. and Schatz, G. (1975) J. Biol. Chem. 250, 775-782.
15. Bertrand, H. and Collins, R.A. (1978) Mol. Gen. Genet. 166, 1-13.
16. West, D.J. and Pittenger (1977) Molec. Gen. Genet. 152, 77-82.
17. Werner, S. and Bertrand, H. (1979) Europ. J. Biochem. 99, 463-470.
18. Bertrand, H. and Werner, S. (1979) Europ. J. Biochem. 98, 9-18.
19. Bertrand et al. (1976) Can. J. Genet. Cytol. 18, 397-409.
20. Nargang, F.E., Bertrand, H. and Werner, S. (1978) J. Biol. Chem. 253, 6364-6369.

© 1980 Elsevier/North-Holland Biomedical Press
The Organization and Expression of the Mitochondrial Genome
A.M. Kroon and C. Saccone, editors

CHARACTERIZATION OF A MITOCHONDRIAL "STOPPER" MUTANT OF *NEUROSPORA CRASSA*: DELETIONS AND REARRANGEMENTS IN THE MITOCHONDRIAL DNA RESULT IN DISTURBED ASSEMBLY OF RESPIRATORY CHAIN COMPONENTS

HANS DE VRIES, JENNY C. DE JONGE and PETER VAN 'T SANT
Laboratory of Physiological Chemistry, State University, Bloemsingel 10, 9712 KZ GRONINGEN, The Netherlands.

INTRODUCTION

For *Neurospora crassa* a limited number of mitochondrial mutants has been isolated so far. Although it should be possible to isolate simple, *e.g.* antibiotic-resistant mutants, most mitochondrial mutations in *Neurospora* have pleiotropic effects, often resulting in simultaneous deficiency of cytochromes b and aa_3[1]. We have isolated a small number of cyanide-insensitive mutants. One of these mutants, E35, has been characterized with respect to its respiratory chain and ATPase components and to its mitochondrial DNA. E35 has the pleiotropic "stopper" phenotype[1]. In this paper we will describe the characteristics of this interesting mutant.

METHODS AND MATERIALS

Strains: All strains used had type II mtDNA[2] and required inositol (50 μg/ml) for their growth: ANT-1[3] was the source of "wild type" mtDNA; for mutation we used AZS$^-$, a strain which does not contain the azide-sensitive, alternative oxidase[4]. Both strains were kindly provided by Dr. D.L. Edwards, La Jolla, Cal.

Mutation and selection of mutants: conidia of strain AZS$^-$, mating type A, were mutagenized with ethylmethanesulfonate. Selection for cyanide-insensitive (SHAM-sensitive) conidia was performed as described by Rosenberg *et al.*[6].

Genetical techniques were performed as follows[5]: crosses were carried out on corn meal agar slants, ascospores were collected as "random spores" and plated on 4% agar containing Vogel's medium, sucrose and inositol.

Oxygen consumption of whole mycelium was measured polarographically with a Clark electrode, in normal culture medium.

Cytochrome spectra of mitochondria were recorded at room temperature. Cytochrome concentrations were calculated according to Williams[7].

Labeling of mitochondrial translation products. Mycelium was grown overnight

---

* Abbreviations: mt = mitochondrial; SSC = 0.15 M NaCl, 0.015 M Na_3-citrate; SDS = Na-dodecylsulphate; SHAM = salicylhydroxamic acid; kD = kilo Daltons; kbp = kilobasepairs.

at 30 °C and then transferred to sulphate-free medium (20 ml). After 2 h cycloheximide was added to a concentration of 100 μg/ml, followed after 5 min by 1 mCi carrier-free $Na_2[^{35}S]SO_4$. The mycelium was collected after 30 minutes and washed with a cold solution containing 0.1% met, 0.1% cys, 0.1% Na_2SO_4 and 10 μg/ml cycloheximide. All following steps were performed at 0 °C. Mycelium was homogenized with sand in 0.44 M sucrose, 100 mM NH_4Cl, 10 mM $MgCl_2$, 10 mM Tris-HCl, pH 7.4 plus 10 μg/ml cycloheximide and 0.5 mM phenylmethylsulfonylfluoride. Cell debris and nuclei were removed by centrifugation for 5 minutes at 3,000 rpm. Mitochondria were collected by centrifuging the supernatant during 10 minutes at 20,000 xg. Labeling for immunoprecipitation was performed after preincubation in chloramphenicol[8].

Immunoprecipitations were carried out as described by Sebald and Wild[9]. An anti-F_1 antibody was used for precipitation of the whole ATPase complex. For cytochrome oxidase precipitation we used an antiserum against the holoenzyme.

Electrophoretic analysis of mitochondrial translation products and immunoprecipitates was performed as described by Van 't Sant *et al.*[10].

Isolation of mtDNA was carried out according to Terpstra *et al.*[11]. Instead of the final sucrose gradient step, the mtDNA was centrifuged through a Sephadex G-50 column[12] in 0.1 x SSC after two additional phenol-extractions and ethanol-precipitations.

Restriction endonuclease digestion and gel electrophoresis: restriction endonucleases were obtained from Boehringer or Biolabs. mtDNAs were digested using reaction conditions specified by suppliers. Digests were electrophoresed in vertical or horizontal agarose gels as described earlier[11].

Stripfilter hybridization: transfer of gel-resolved digests was performed according to Southern[13]. Filters were preincubated according to Denhardt[14]. DNA fragments to be used as probes were [^{32}P]-labeled by nick translation[15], heat-denatured (5 min at 100 °C) in 0.1 x SSC and diluted to 20 ml with 3 x SSC, 0.1% SDS. Hybridization was for 18 h at 65 °C. Washing and autography were as described earlier[11].

RESULTS

Classification of the mutant E35. Table 1 summarizes the respiratory, growth, spectral and genetical characteristics of the mutants. According to the criteria of Bertrand *et al.*[1] mutant E35 has all features of the mitochondrial "stopper" mutants. These mutants all have a dysfunctional respiratory chain, respiration proceeds via the SHAM-sensitive oxidase.

Mitochondrial protein synthesis and translation products. Table 2 shows the *in vivo* protein-synthetic activity of the mitochondrial and cytosolic systems.

TABLE 1

PHENOTYPIC CHARACTERISTICS OF MUTANT E35

1. Growth and respiration:	resistant to KCN sensitive to SHAM
2. Growth rate:	slow and irregular
3. Cytochrome spectrum:	cytochrome aa_3 absent cytochrome *b* absent cytochrome *c* increased[a)]
4. Cytochrome *c* oxidase activity:	< 3% of wild-type activity
5. Fertility in crosses:	"female": no (no protoperithecia formed) "male": yes
6. Inheritance of mutant phenotype:	non-Mendelian

a) wild-type 0.8, mutant 3.6 nmoles/mg mitochondrial protein

The ratio of wild-type to mutant protein synthesis is about 3.5 both for total cell protein synthesis and for mitochondrial protein synthesis determined in the presence of cycloheximide. Most likely, both this low protein-synthetic activity and the slow growth of the mutant are a reflection of the shortage of energy caused by the low phosphorylating capacity of mitochondria respiring through the alternative oxidase[16]. A defect in the mitochondrial protein-synthetic machinery itself (a possibility favoured by Bertrand *et al.*[1]) is not supported by these data.

Fig. 1A shows the mitochondrial translation products made in cycloheximide-poisoned cells. It is clear that all mitochondrially synthesized subunits of cytochrome *c* oxidase as well as cytochrome *b* apoprotein are made by E35 mitochondria, although much less label is found in E35 oxidase subunit 1.

TABLE 2

SPECIFIC RADIOACTIVITY OF MITOCHONDRIA AND CYTOSOL AFTER [^{35}S]-LABELING *IN VIVO*

Specific activities (dpm/μg protein) in mitochondria were determined after labeling in the presence of cycloheximide and in postmitochondrial supernatant after labeling without inhibitors.

Cell fraction	AZS$^-$	E35	Ratio
Mitochondria	10,600	2,900	3.7
Postmitochondrial supernatant	31,650	9,330	3.4

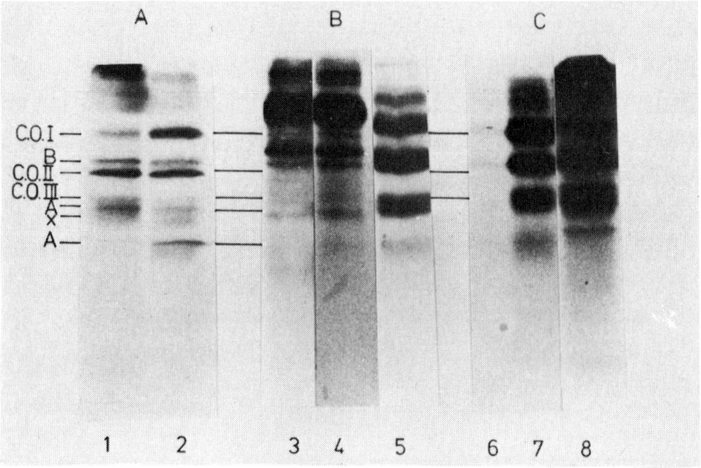

Fig. 1. Mitochondrial translation products and immunoprecipitates, analyzed on SDS gels. 1A: Mitochondrial translation products. 1: E35; 190,000 dpm, 65 μg protein. 2: AZS⁻; 170,000 dpm, 16 μg protein. 1B: Immunoprecipitates obtained with anti-F_1 antibody. No cycloheximide present during labeling. 3: E35; 15,000 dpm, exposed for 400 h. 4: AZS⁻; 70,000 dpm, exposed for 67 h. 5: mitochondrial translation products of AZS⁻. 1C: immunoprecipitates obtained with anti-cytochrome *c* oxidase antibody. No cycloheximide present. 6: E35 5,000 dpm. 7: AZS⁻; 4,500 dpm. 8: mitochondrial translation products of AZS⁻

The molecular weights of these polypeptides are identical for mutant and wild type. The other conspicuous difference between wild type and mutant translation products is the extremely low concentration of the 11 kD translation product which was proposed to be one of the subunits of the membrane (F_O) part of the ATPase[8]. The main argument for this assumption was the fact that this protein is immunoprecipitated by antibodies against F_1. We have used an antibody against the F_1 ATPase to precipitate the total (F_OF_1) ATPase of AZS⁻ and E35 mitochondria. Fig. 1B shows that the 11 kD protein, which is barely visible in Fig. 1A, is present in the immunoprecipitates of both AZS⁻ and E35: in both lanes the 11 kD protein is visible as a very faint band. Table 3 summarizes the ATPase and ATP synthetase activities of AZS⁻ and E35 mitochondria. Both activities per mg protein of the mutant ATPase are about 25% of those of the parent strain. The oligomycin-sensitivity is not significantly different.

Fig. 1C shows that an antiserum against cytochrome *c* oxidase holoenzyme precipitates all subunits from AZS⁻ mitochondria, whereas for E35 only the subunits 1 and 2 are very faintly visible. Hence, we can conclude that, despite

TABLE 3

ATP-ASE AND ATP-SYNTHETASE ACTIVITIES OF MITOCHONDRIA FROM AZS$^-$ AND FROM E35

Mitochondria were isolated from glusulase-treated mycelium as described by Rosenberg *et al.*[6] ATPase activity was measured according to Cleland and Slater[18], ATP-$^{32}P_i$ exchange according to Kagawa *et al.*[19], both in the presence of $MgCl_2$ (2 mM). Activities are given as nmol P_i formed or incorporated per min per mg protein.

Addition	ATPase activity AZS$^-$	ATPase activity E35	ATP-P_i activity AZS$^-$	ATP-P_i activity E35
None	185.9 (100%)	65.8 (100%)	335.5 (100%)	77.3 (100%)
Oligomycin (µg/ml)				
1.3	53.7 (28.9%)	26.5 (40.2%)	-	-
2.0	-	-	0.35 (0.1%)	1.0 (1.3%)
4.0	39.3 (21.1%)	28.0 (42.6%)	0.0 (0%)	1.25 (1.6%)
dinitrophenol (mM)				
0.1	254.1 (136.7%)	69.9 (106.2%)	82.7 (24.8%)	11.0 (14.2%)
DNP (0.1 mM) + oligo (4.0 µg/ml)	47.7 (25.7%)	29.6 (45.0%)	-	-

the presence of the mitochondrially made subunits 1, 2 and 3, no correct assembly to a recognizable complex has occurred in the mutant.

Characterization of the mtDNA of E35: Fig. 2A shows the Eco RI digest of the mutant, 6 months after its isolation. It is evident that both E1 and E2 are absent, and a new fragment EA, estimated at about 27 kbp, is visible. The most likely explanation is that the E1-E2 junction (see Fig. 5) is mutated, probably by a deletion, since E1 and E2 add up to 30 kbp. Again 4 months later Eco RI digestion of a new E35 mtDNA preparation resulted in an even more aberrant pattern (Fig. 2B): besides the EA band a novel band of about the same length as E1 or slightly shorter, EB, showed up. Moreover, the intensities of EA and EB give the impression that these fragments are present in a higher concentration than the other fragments. Also E4 and E6 appear to be amplified now.

Fig. 3 shows that the abnormalities are equally evident in Hind III and Bgl II digests: again the fragments surrounding the E1-E2 junction (Fig. 5) are missing: H9 and H10a; b3 and b6. The Hinc II fragments h9 and h12 are also absent (not shown). Novel fragments are again present: H-A (11.5 kbp); bA (8.8 kbp) and bB(4.6 kbp). Amplified are: H1, HA, b2, bA and the fragments in the 24S region: H7, H12, H13, H14, H15, H18 and H19; b9 and b11. The novel fragment bB is not amplified.

To find out the nature of the novel fragments we hybridized these fragments, after nick translation, to Southern blots of wild type mtDNA digests. Fig. 4

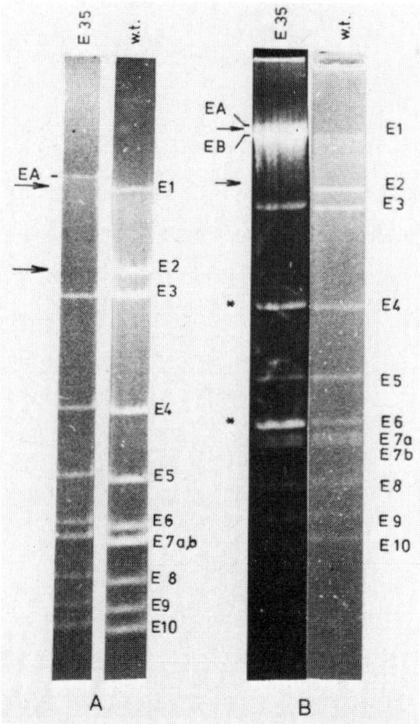

Fig. 2. Comparison of Eco RI digests of E35 and wild type mtDNA. A. E35 mtDNA isolated 6 months after mutagenesis. B. E35 mtDNA isolated 10 months after mutagenesis

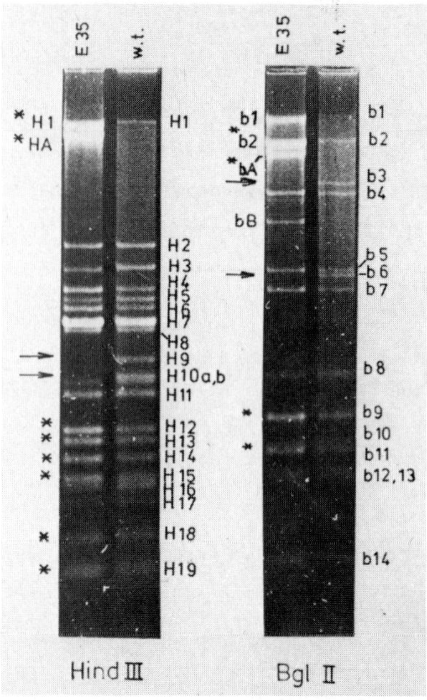

Fig. 3. Comparison of E35 and wild-type tmDNA. The same E35 mtDNA preparation as in Fig. 2B was used. Left: Hind III digests; right: Bgl II digests.

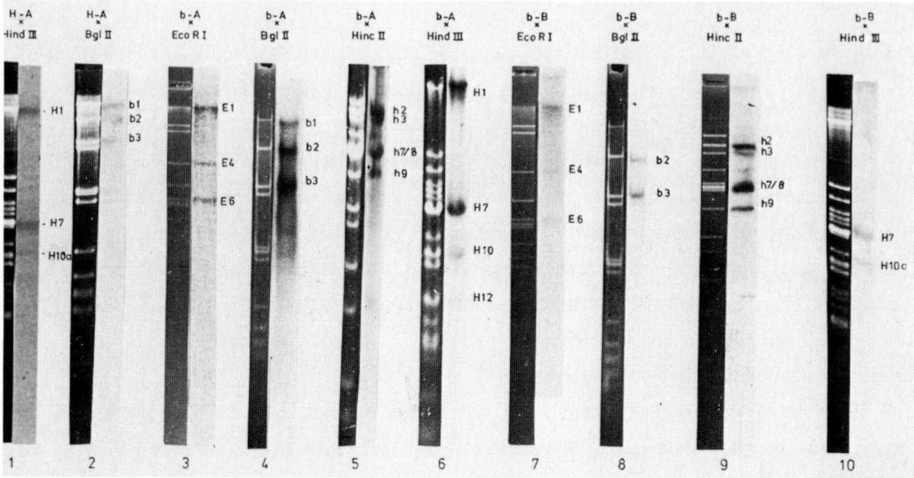

Fig. 4. Hybridization of nick-translated novel fragments of E35 mtDNA to restriction digests of wild-type mtDNA. 1-2: fragment HA; 3-6: fragment bA; 7-10: fragment bB. 1,6,10: Hind III digests; 2,4,8: Bgl II digests; 3,7: Eco RI digests; 5,9: Hinc II digests.

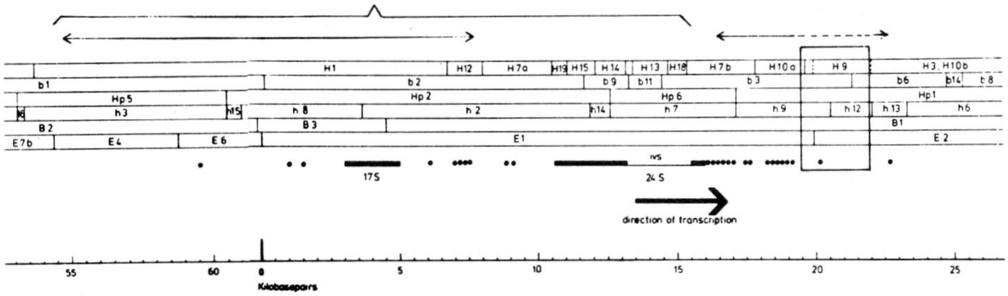

Fig. 5. Cleavage map of part of the type II mtDNA. Fragments are indicated by letters and numbers. H = Hind III, b = Bgl II, Hp = Hpa I, h = Hinc II, B = Bam HI, E = Eco RI. Solid bars denote the rRNA genes, black dots tRNA genes. The differences of E35 mtDNA with wild type mtDNA are indicated as follows: the boxed area is deleted; ↔ : regions of the wild type genome hybridizing with the novel bands HA and bA; ⌒: amplified area.

shows the results of these experiments: H-A hybridizes to H1, H7 and H10a (the latter absent in E35) and to b1, b2 and b3 (again absent in E35). b-A is complementary to E1, 4 and 6; to b1, b2 and b3; to h2, h3, h7, h8 and h9; to H1, H7b, H10, H12. Fragment b-B hybridizes to E1, and weakly to E4 and E6; h2, h3 (weakly), h7, h9, h13 (weakly); b2, b3, b6 (weakly); H1 (weakly), H7, H10. No hybridization of the novel fragments to the disappeared fragments E2, H9 or h12 is found, indicating that the sequences of this part of the wild-type mtDNA (about 2.5 kbp) are no longer present in the E35 mtDNA, whereas those of h9, H10a, b3, which are also absent in E35, are present in all novel fragments. Fragment bB probably has replaced fragment b3 and b6, although a simple deletion of 2.5 kbp is not enough to explain the short length of bB. Moreover, the hybridization of bB to h2 and b2 is difficult to explain. Furthermore, it is evident that the novel fragments HA and bA hybridize to sequences which are far apart on the wild-type genome. In Fig. 5 the area to which these novel fragments are complementary is indicated. It is interesting to notice that no complementarity to the 24S region is found. This is confirmed by hybridization of 24S rRNA to E35 mtDNA digests (not shown): only weak hybridization to HA and bA is found, most likely originating from contaminating 17 S rRNA. Hybridization to bB is stronger, probalby this fragment contains the 3' part of the 24S gene.

DISCUSSION

The complex physiological lesions of the mutant E35, described above, seem to originate from the primary deletion of a segment of the mt genome around the E1-E2 junction. The consequences of this deletion are manifold: first, a mitochondrially synthesized protein of 11 kD is almost absent. Second, the assembly of the respiratory enzymes cytochrome *b* and aa_3 is disturbed: spectrally these are absent, cytochrome *c* oxidase is almost zero, and an antibody against cytochrome aa_3 precipitates only marginal amounts of subunits 1 and 2. On the other hand all three mitochondrially made cytochrome *c* oxidase subunits and cytochrome *b* apoprotein are synthesized by E35 mitochondria, albeit in a lower concentration for subunit 1 of the oxidase. Third, the mutation results in the accumulation and amplification of novel fragments, whereas also some normal fragments are amplified.

The deletion of E35 mtDNA is located next to the tRNA cluster at the 3' side of the 24S rRNA gene. Hence the rRNA and tRNA genes (except perhaps for one) are not affected, a conclusion which can also be drawn from the fact that normal mitochondrial translation products are synthesized. The crucial question about this mutant is, of course, what gene or regulating sequence is present in the deleted part of the mt genome. At the DNA level, the most intriguing phenomenon is the arising of the large amplified fragments (HA, bA, EB), probably products of incorrect recombination. These fragments contain sequences from the parts surrounding the 24S gene, *viz.* from 55 kbp to 8 kbp and from 16.5 to 19 kbp on the map in Fig. 5. The 17S gene, on the contrary, is present in these novel fragments (hybridization data not shown). Moreover, the whole rRNA-tRNA area as well as E4 and E6 have also been amplified. These aberrations are very reminiscent of those described by Mannella *et al.*[17]. Their mutant DNAs contain either an "α" band repeatedly inserted at the E4-E6 junction, or contain the same amplified segment which is present in E35 (H1 to H7b). A major difference between these "poky" mutants and the "stopper" mutant E35 is that in E35 fragments are missing: in none of several E35 mtDNA preparations we could ever find the fragments E2, b3, b6, H9, H10a. Moreover, we could not find any indication for the presence of two types (defective and intact) of mtDNA. The latter possibility was specifically put forward by Mannella *et al.*[17] to explain the "stop-start" growth of stopper mutants. Anyway, a remarkable characteristic of all mutant *Neuropora* mtDNAs described so far is that the fragments E4 and E6 always play a role, either by being amplified or by having an insertion between them. It is possible that in E35 the amplified fragment EB (containing HA and bA) is also inserted there.

The effect of the deletion at the E1-E2 junction not only has its effect on mtDNA replication and/or recombination, but also profoundly disturbs the assembly of the components of oxidative phosphorylation: although the mitochondrial parts of the cytochrome chain are evidently synthesized, no functional, integrated cytochromes *b* and aa_3 are present. It is possible that a product of the deleted area (either an RNA or a protein) plays a role in the integration of the respiratory chain into the membrane. The ATPase of the mutant is normal in its oligomycin-sensitivity and contains the mitochondrial 11 kD subunit as judged from the immunoprecipitation with an anti-F_1 antibody. The low specific activities of the enzyme presumably reflect an overall decrease in the ATPase complex content of the mutant mitochondria. An intriguing point is the difference in the relative intensities of the 19 kD and 11 kD proteins between total mitochondrial translation products and immunoprecipitated ATPase: in Fig. 1A the intensity of the 19kD protein is much higher than that of the 11 kD protein for E35, whereas in Fig. 1B this difference is much less conspicuous. It may well be that the 19 kD protein is overproduced by the mitochondria, and that the 11 kD protein is a key protein in the assembly of the ATPase complex. Its low concentration then results in a low ATPase complex content. The same protein may also play a role in the correct integration and assembly of the respiratory chain components. Present experiments are directed to establish the role of the 11 kD protein in the assembly of the mitochondrial membrane complexes.

ACKNOWLEDGEMENTS

We thank Dr. W. Sebald for sending us an antibody against F_1 and Dr. E. Agsteribbe for antiserum against cytochrome oxidase. We also thank Miss J.F.C. Mak for technical assistance, H. Schokkenbroek for drawing the figures, B. Tebbes for photography, Miss K. van Wijk for preparing the manuscript and Dr. A.M. Kroon for his advice and criticism. This work was supported in part by a grant to A.M. Kroon from the Netherlands Foundation for Chemical Research (SON) with financial support of the Netherlands Organization for the Advancement of Pure Research (ZWO).

REFERENCES

1. Bertrand, H., Szakacs, N.A., Nargang, F.E., Zagozeski, C.A., Collins, R.A. and Harrigan, J.C. (1976) Can. J. Genet. Cytol. 18, 397-409.
2. Mannella, C.A., Pittenger, T.H. and Lambowitz, A.M. (1979) J. Bacteriol. 137, 1449-1451.
3. Edwards, D.L., Chalmers, J.H., Guzik, H.J. and Warden, J.T. (1976) in: Genetics and Biogenesis of Chloroplasts and Mitochondria (Th. Bücher *et al.*, Eds.) North-Holland, Amsterdam, pp. 865-872.
4. Edwards, D.L. and Unger, B.W. (1978) J. Bacteriol. 133, 1130-1134.
5. Davis, R.H. and De Serres, F.J. (1970) Methods in Enzymol. 17A, 79-143.
6. Rosenberg, E., Mora, C. and Edwards, D.L. (1976) Genetics 83, 11-24.

7. Williams, J.N. (1964) Arch. Biochem. Biophys. 104, 537-543.

8. Jackl, G. and Sebald, W. (1975) Eur. J. Biochem. 54, 97-106.

9. Sebald, W. and Wild, G. (1979) Methods in Enzymol. 55, 344-352.

10. Van 't Sant, P., Mak, J.F.C. and Kroon, A.M. (1980) this volume.

11. Terpstra, P., Holtrop, M. and Kroon, A.M. (1977) Biochim. Biophys. Acta 475, 571-588.

12. Penefsky, H.S. (1977) J. Biol. Chem. 252, 2891-2899.

13. Southern, E.M. (1975) J. Mol. Biol. 98, 503-517.

14. Denhardt, D.T. (1966) Biochem. Biophys. Res. Commun. 23, 641-646.

15. Jeffreys, A.J. and Flavell, R.A. (1977) Cell 12, 429-439.

16. Lambowitz, A.M. and Zannoni, D. (1978) in: Plant Mitochondria (Ducat, G. and Lance, C., Eds.) Elsevier/North-Holland, Amsterdam, pp. 283-291.

17. Mannella, C.A., Goewert, R.R. and Lambowitz, A.M. (1979) Cell 18, 1197-1207.

18. Cleland, K.W. and Slater, E.C. (1953) Biochem. J. 53, 547-556.

19. Kagawa, Y., Kandrach, A. and Racker, E. (1973) J. Biol. Chem. 248, 676-684.

The Organization and Expression of the Mitochondrial Genome
A.M. Kroon and C. Saccone, editors

CHARACTERIZATION OF AN UNCOUPLER RESISTANT CHINESE HAMSTER OVARY CELL LINE

KARL B. FREEMAN, RANDALL W. YATSCOFF & JEREMY R. MASON
Department of Biochemistry, McMaster University, Hamilton, Ontario, Canada.

INTRODUCTION

The three suggested hypotheses of oxidative phosphorylation differ mainly in the proposed nature of the energized state used to drive ATP synthesis and other energy dependent processes[1] and give rise to different proposed mechanisms for uncoupler action. The chemical and conformational hypotheses predict that uncouplers act by converting a high energy intermediate into a lower energy form, at a specific site(s) (protein(s)) in the membrane[1,2]. Evidence for such a site(s) has been obtained by the use of photo-active, radio-labelled uncouplers[2] and by equilibrium binding studies[3]. The chemiosmotic hypothesis postulates that the primary energized intermediate is an electrochemical potential gradient[1]. By acting as weak lipophilic acids, uncouplers have been shown to make the lipid bilayer permeable to protons, short circuiting the phosphorylation reaction[2,4].

One approach to understanding the nature of uncoupler action is through the study of uncoupler resistant cells. Uncoupler resistant bacterial[5,6] and yeast cells[7] have been isolated. This paper describes the isolation and characterization of a mammalian cell line resistant to uncouplers.

RESULTS

Isolation and growth characteristics of S-13 resistant cells

The uncoupler resistant cells were isolated from a parental Chinese hamster ovary, thymidine kinase minus (CHO TK^-) cell line after growth of these cells for 96 h in 25 μg bromodeoxyuride/ml followed by plating in medium containing an inhibitory concentration of S-13 (1×10^{-6} M), a potent uncoupler of oxidative phosphorylation[2]. Clones were isolated 3 to 5 weeks later and one clone, UH_5, was characterized.

The differential sensitivity of UH_5 and CHO TK^- cells to uncouplers was examined by observing their growth in the presence of S-13, with UH_5 cells being approximately 5 to 10-fold more resistant to the uncoupler than the parental cells.

Resistance to S-13 could be specific or reflect resistance to uncouplers as a class. Cross resistance of UH_5 cells to other uncouplers was examined by growth and respiratory studies. Growth of UH_5 cells was resistant to SF-6847, an uncoupler having potency similar to S-13[2] and to FCCP, which is approximately 100-fold less potent than S-13 or SF-6847[2]. No marked effect on the growth of wild-type CHO TK^- cells was apparent with the uncouplers DNP, CCCP and TTFB which are at least 100-fold less potent than S-13 or SF-6847[2]. No cross resistance was detected to either oligomycin or venturicidin, or Tevenel, an inhibitor of mitochondrial protein synthesis[8].

Respiration experiments

Cross resistance to uncouplers was also demonstrated by examining cellular respiration. This approach was used with S-13, SF-6847 and with DNP and CCCP. The concentrations of S-13, SF-6847 and DNP required for half maximal stimulation of respiration ($U_{1/2}$) were approximately five-fold higher with UH_5 cells compared to wild-type CHO TK^- cells as shown in Table 1. In contrast to this, less than a 2-fold increase in the concentration of CCCP was required.

TABLE 1

CONCENTRATIONS OF UNCOUPLERS REQUIRED FOR HALF-MAXIMAL STIMULATION OF RESPIRATION ($U_{1/2}$)

	Whole Cells[a]		Isolated Mitochondria[b]	
	CHO	UH_5	CHO	UH_5
Uncoupler	$M \times 10^9$		mol x 10^{11}/mg mit prot	
S-13	9.0	40.0	3.0	30.0
SF-6847	2.7	10.0	2.0	20.0
CCCP	85.0	120.0	ND[c]	ND[c]
DNP	1200.0	5000.0	450.0	450.0
FCCP	ND[c]	ND[c]	6.0	75.0

[a] Endogenous respiration rate was 750 nmol $O_2/h/10^7$ cells.
[b] Mitochondria had a RCR of 2.5 and an ADP:O ratio of 1.4 to 1.8 with succinate.
[c] Not Determined.

Resistance to uncouplers could reflect a change in cell membrane permeability or in mitochondria. These possibilities were resolved by determining the effect of uncouplers on the respiration of mitochondria isolated from UH_5 and CHO TK^- cells. As shown in Fig. 1 and summarized in Table 1 when the $U_{1/2}$ values were measured the respiration of mitochondria from UH_5 cells was about 10-fold more resistant to S-13 (A), SF-6847 (B) and FCCP (C) than in CHO TK^-

mitochondria, while no difference was observed with DNP (D). The results suggest that resistance to S-13, SF-6847 and FCCP is located at the mito-chondrial level and is not the result of a change in cell plasma membrane permeability.

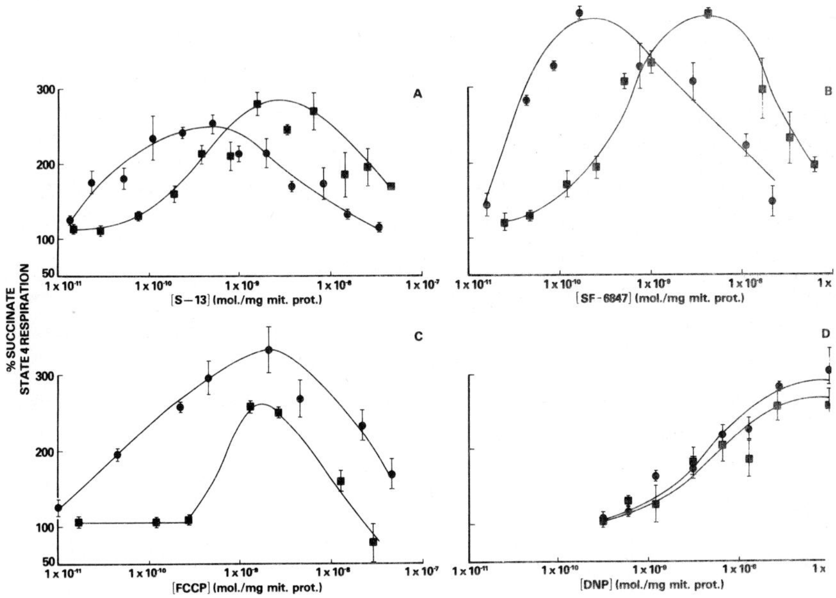

Fig. 1. Respiration of mitochondria from UH_5 and parental cells in the presence of various concentrations of uncouplers.

<u>Mode of inheritance of S-13 resistance</u>

The mode of inheritance (nuclear or mitochondrial) was determined by examining the uncoupler resistance of hybrid cells formed by the fusion of UH_5 cells to cells resistant to the nucleoside analogue 5,6-dichloro-1-B-D-ribofuranosyl-benzimidazole (DRB)[9], but sensitive to uncouplers. The hybrid cells did not grow in the presence of appropriate concentrations of S-13 (1×10^{-6} M) and DRB (6.3×10^{-5}). This result is consistent with S-13 resistance being a recessive trait and, therefore probably nuclear coded. Consistant with this idea, cybrids formed from UH_5 cytoplasts and DRB^R cells[10] were sensitive to S-13.

DISCUSSION

Chinese hamster ovary cells resistant to the uncoupler S-13 were isolated and one clone, UH_5 characterized. UH_5 cells were cross-resistant to some uncouplers but not to a range of compounds which inhibit other aspects of mitochondrial function. Resistance was at the mitochondrial level. These properties are consistent with a specific, rather than non-specific alteration, such as a change in cell membrane permeability. The reason for the absence of cross-resistance to CCCP is not known.

Genetic analysis of UH_5 cells showed that the alteration conferring uncoupler resistance was recessive. Because cells contain many copies of mtDNA, it is highly unlikely that a mitochondrially-inherited recessive character could be selected. This is supported by the fact that all such mutations so far studied have been shown to be dominant in nature [11]. The recessive nature of the alteration is therefore consistent with its being nuclear coded.

The biochemical basis for the uncoupler resistant CHO mutant may reside in protein(s) involved in uncoupler binding or it may be conferred by an alteration in the lipid content of the inner mitochondrial membrane. Further investigations are in progress to resolve the specific biochemical lesions that confer resistance to uncouplers.

ACKNOWLEDGEMENTS

This work was supported by Medical Research Council, grant MT-1940 to KBF. RWY is a recipient of an MRC studentship. We wish to thank Ms. H.V. Patel for excellent technical assistance.

REFERENCES

1. Boyer, P.D. et al. (1977) Annu. Rev. Biochem., 46, 955-1026.
2. Hanstein, W.G. (1976) Biochem. Biophys. Acta, 456, 129-148.
3. Katre, N.V. and Wilson, D.F. (1977) Arch. Biochem. Biophys., 184, 578-585.
4. Bakker, E.P. et al. (1973) Biochim. Biophys. Acta, 292, 78-87.
5. Decker, S.J. and Lang, D.R. (1978) J. Biol. Chem., 253, 6738-6743.
6. Date, T. et al. (1980) Proc. Natl. Acad. Sci. U.S.A., 77, 827-831.
7. Griffiths, D.E. (1974) The Genetics and Biogenesis of Mitochondria and Chloroplasts (Birkey, C.W. et al. eds), Ohio State Univ. Press, Columbus, Ohio, pp. 117-135.
8. Yatscoff, R.W. and Freeman, K.B. (1977) Can. J. Biochem., 55, 1064-1074.
9. Gupta, R. and Siminovitch, L. (1980) Somat. Cell Genet. (in press).
10. Yatscoff, R.W. et al. (1980) Somat. Cell Genet. (submitted).
11. Bunn, C.L. et al. (1974) Proc. Natl. Acad. Sci. U.S.A., 71, 1681-1685.

The Organization and Expression of the Mitochondrial Genome
A.M. Kroon and C. Saccone, editors

RELEASE FROM GLUCOSE REPRESSION AND MITOCHONDRIAL PROTEIN SYNTHESIS IN *S. cerevisiae*.

M.AGOSTINELLI, C.FALCONE and L.FRONTALI
Istituto di Fisiologia Generale, Università di Roma, 00100 Rome (Italy).

INTRODUCTION

Regulation of mitochondrial function by the presence of oxygen and by glucose concentration has been known for a very long time (1-3) and glucose repressed and anaerobically-grown cells have been extensively characterized both from a biochemical and a morphological point of view (1-5). However interactions between nuclear and mitochondrial genomes and biosynthetic apparatus during respiratory induction are still far from clear and even the mitochondrial events underlying this complex phenomenon are not well defined. The latter point, i.e. the assessment of the role of the mitochondrial genome and protein synthesizing machinery in the respiratory induction, is actually strictly bound to the many open problems concerning mitochondrial transcription and processing of transcripts.

We performed, and report here, an analysis of the products of mitochondrial translation in different conditions and in the course of respiratory induction, which might contribute to a better definition of the mitochondrial stages of this process.

The rationale of this work is strictly connected with the problems mentioned above of mitochondrial transcription. In fact, an extremely regular mechanism of complete transcription (as observed in the HeLa cells) followed by rapid splicing of the transcripts would correspond to a simultaneous expression of mitochondrial genes, while a deviation from this pattern would indicate a more complex mechanism involving selective transcription of some genes or differential splicing of transcripts during induction.

In the course of this work we have been using a system in which respiratory induction following release from glucose repression can be obtained in resting cells.

MATERIALS AND METHODS

Strains. Two w.t. yeast strains were used: D273-10B and PS 409 kindly given by Dr. P.P. Slonimski (for the description of the latter strain see ref. 6). Cells were grown aerobically at 28°C on YP medium (1% yeast extract, 1% Bacto peptone) containing 15% glucose or 2% galactose as carbon source.

Induction system. Several parameters have been found to control the respiratory induction, including the stage of the growth curve on glucose and the composition of the induction medium (7). The following conditions allow the optimization of the release from glucose repression. Cells are collected in the late exponential phase of growth on 15% glucose, when the absorbance at 600 mμ has reached 10-12 O.D. units. Cells are then washed three times with water and resuspended at the same O.D. in 0.067 M phosphate buffer, pH 6.5, containing 2% lactate and 0.1% glucose. As a control, part of the cells were resuspended in the same buffer containing 15% glucose. In this system derepression (measured as increase in mitochondrial protein synthesis) can be obtained also for box mutants incubated in a constant 0.1% glucose concentration. At time intervals, 5 ml samples were withdrawn and used for labelling. A typical example of induction in these conditions is reported in Fig. 1.

Labelling procedure.

Labelling with radioactive aminoacids. - Samples of growing or resting cell suspensions were collected and cycloheximide was added to the final concentration of 0.5 mg/ml. After 10 min at 30°C, 7 μCi/ml ^{3}H alanine (specific activity 1.7 Ci/mmole) were added and the incubation continued for 45 min. Labelling was stopped by addition of ice cold trichloroacetic acid (TCA) to the final concentration of 10%. Samples were boiled for 15 min and filtered on GFA Whatman filters. Filters were then washed, dried and counted.

Labelling with $H_2{}^{35}SO_4$. - The procedure reported by Douglas and Butow (8) was essentially followed, except that growing cells were incubated for 1 h with chloramphenicol (4 mg/ml) before cycloheximide treatment and labelling. For resting cells chloramphenicol treatment was omitted; cells were resuspended in 2 ml of the incubation buffer containing 0.5 mg/ml cycloheximide and after 3 min, 1.5mCi $H_2{}^{35}SO_4$ was added. After 45 min incubation, incorporation was stopped by adding casaminoacids and Na_2SO_4. Preparation of mitochondria and solubilization of proteins were as reported by Douglas and Butow (8).

TCA precipitable radioactivity was determined in a Packard scintillation counter and protein concentration by the method of Lowry et al. (9).

Free aminoacid pool was determined by the method reported by Johnston et al. (10), and respiration was measured with an ISI 5331 oxygen monitor.

Gel electrophoresis. - Analysis of mitochondrial proteins was performed by electrophoresis in the presence of SDS on gel slabs containing an exponential gradient of 10-15% polyacrylamide (8). Gels were run overnight at 4°C at 15 mA. After fixing, staining and destaining in 7% acetic acid, gels were soaked for 30 min in water and then treated for 1 h with 1 M sodium salicylate. This treatment allows to substantially reduce (3-5 times) the exposure time of films during autoradiography. Autoradiography was performed by exposing gels for 12 to 24 h to Kodax X-Omatic X-ray films. Autoradiograms were scanned in an Acta III Beckman spectrophotometer equipped with a linear transport scanner.

RESULTS

Under the standardized conditions reported under 'Material and Methods', the induction process shows the time course exemplified in fig. 1 for strain D273-10B. Mitochondrial protein synthesis increases, after a short lag, from undetectable levels to levels approaching those of growing derepressed cells. In these conditions no very important strain-dependent differences were observed in w.t. strains. On the contrary relevant differences were observed in some box mutants in which release from glucose repression can be obtained in a constant low (0.1%) glucose concentration (11). Mitochondrial protein synthesis in cells incubated in buffered 15% glucose is detectable, but not comparable because of the much lower free aminoacid pool in resting repressed cells.

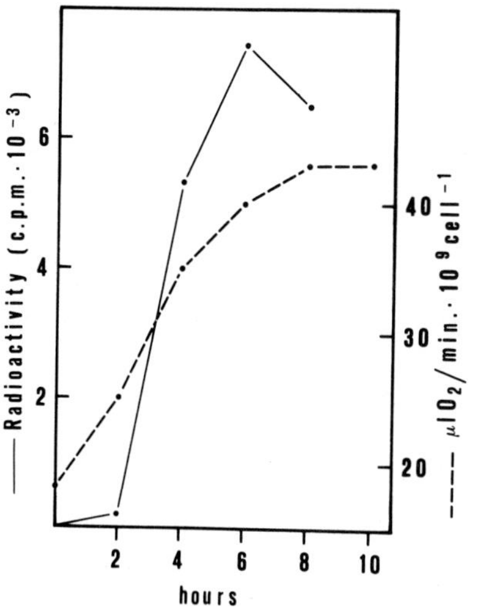

Fig. 1. Oxygen uptake (----) and cycloheximide resistant (———) ^3H alanine incorporation during the derepression process in strain D273-10B in buffered lactate.

At different times during the induction process, aliquots of culture were labelled and the products of mitochondrial protein synthesis analysed by polyacrylamide SDS gel electrophoresis. Results are reported in fig. 2 for strain PS 409 and in fig. 3 for strain D273-10B. Densitometer tracings obtained from autoradiograms are shown in the same figures and the results of quantitative analysis of densitometries are reported in fig.4. Control experiments verified that under the conditions used,the area under each peak was proportional to the amount of radioactivity in the corresponding protein band.

DISCUSSION

Evidence accumulated up to now leaves open two different possibilities concerning transcription of yeast mitochondrial genome.

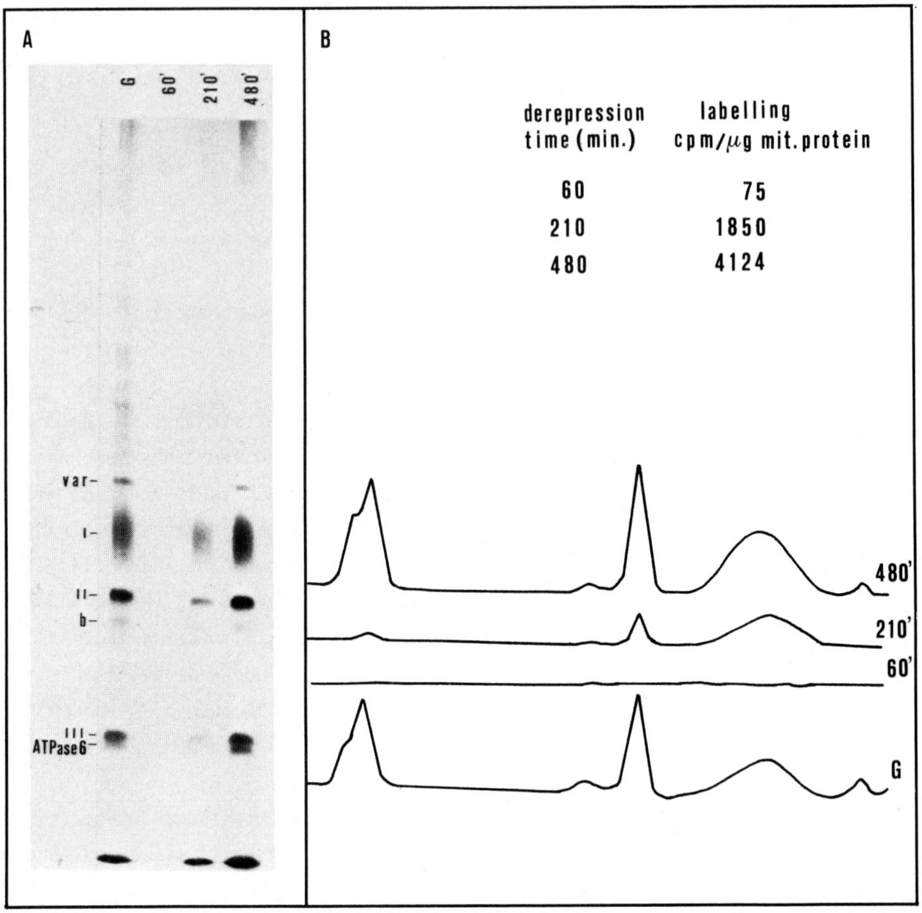

Fig. 2. A. Electrophoretic pattern of mitochondrial translation products in strain PS 409 undergoing release from glucose repression. Aliquots withdrawn at the indicated times were labelled in vivo with $H_2{}^{35}SO_4$ in the presence of cycloheximide and quantities corresponding to 12 μg of mitochondrial proteins were loaded on each slot. The slot marked with G was loaded with 12 μg of mitochondrial protein from cells grown aerobically on galactose. B. Time course of mitochondrial labelling and densitometer tracings of autoradiograms reported in Fig. 2A.

Expression of mitochondrial genes may occur, as it actually does in HeLa cells (12), by a complete full round of transcription of the 25 μ DNA circle, followed by rapid splicing. This hypothesis would account for the presence of

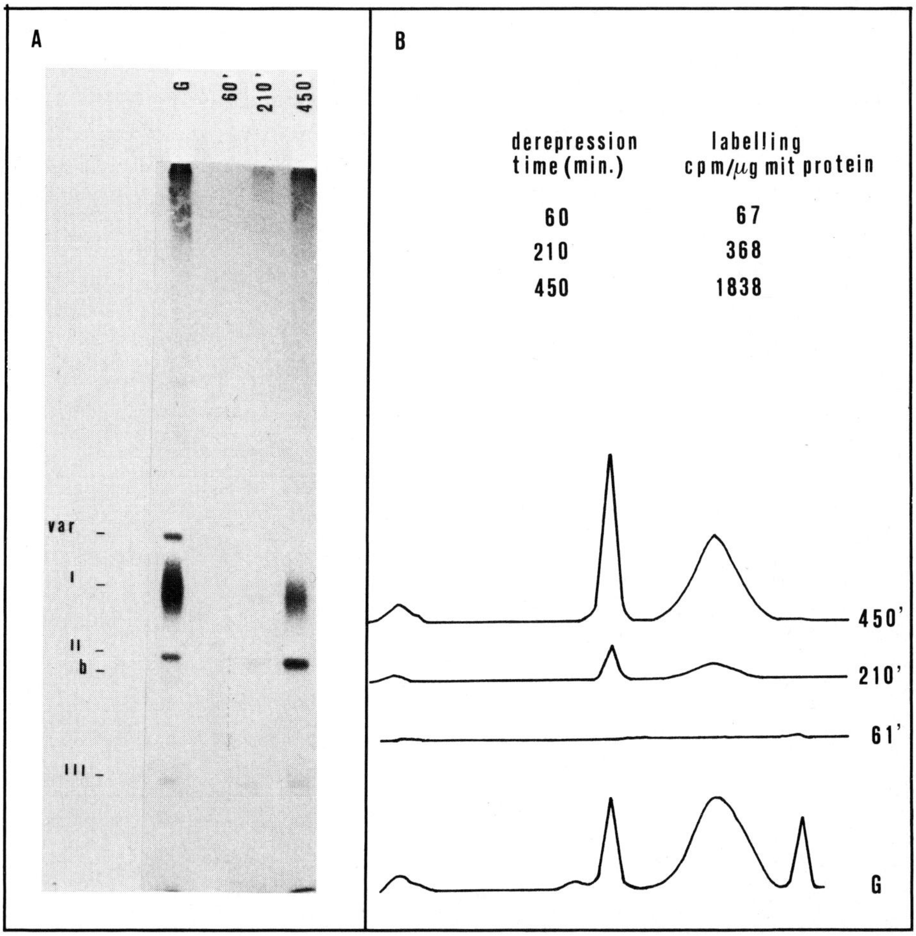

Fig. 3. Electrophoretic pattern of mitochondrial translation products of strain D273-10B undergoing release from glucose repression. 30 µg of mitochondrial protein were loaded on each slot. For all other specifications, see legend to fig. 2.

several precursors of each mRNA and for the absence of obvious promoters and would explain how the genes, whose products are needed in stoichiometric amounts, are coordinately expressed. On the other side, a bacterial-type control of transcription, together with complex and possibly regulated splicing mechanisms could result in differential expression of mitochondrial genes. Some evidence in this sense has been reported (13-15).

Respiratory induction is accompanied by an increase in mitochondrial protein synthesis the extent of which may be inferred from fig. 1 for the case of release from glucose repression. Strong differences in levels of protein synthesis between anaerobic and aerobic cells have been also reported (16). It was therefore of interest to know whether this increase involves all mitochondrial gene products in the same way, i.e. whether respiratory induction is due to a more rapid turning on of the complete transcription and splicing machinery in response to some signal coming from the cytoplasm or, on the contrary, induction is connected with a multifactor control resulting in differential timing of gene expression.

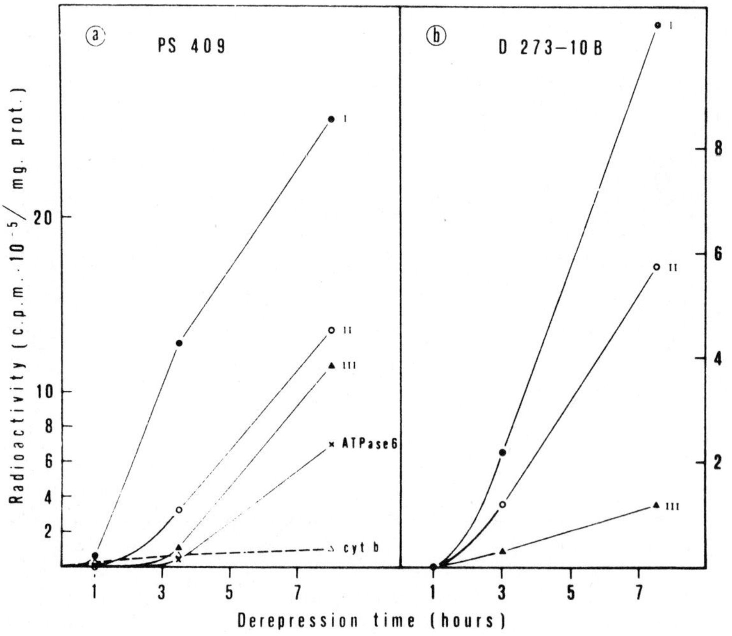

Fig. 4. Quantitative analysis of densitometer tracings from figures 2B and 3B. (●) cytochromoxidase subunit I; (○) subunit II, (▲) subunit III; (X) ATPase subunit VI; (△) cytochrome b.

The electrophoretic patterns of mitochondrial translation products in growing derepressed and glucose repressed cells show that, as in the case of anaerobically grown cells, all mitochondrial proteins are synthesized in the repressed condition. Extensive variability is observed in growing repressed cells with regards to the synthesis of cytochrome oxidase subunits, but on the whole no definite differential repression effect can be stated. It is worthwhile mentioning that in anaerobic cells, mitochondrial protein synthesis is diminished by a factor of 20 as compared to the aerobic ones, while the synthesis of var 1 protein is much less affected (16).

During release from glucose repression in our conditions, the increase of the various subunits of respiratory enzymes has very similar kinetics in strain D273-10B and the percentage of each translation product on the total is remarkably constant. However in strain PS 409 (a 'long strain'), synthesis of subunit 1 of cytochrome oxidase seems to precede that of the other polypeptides, while that of subunit III and ATPase VI is clearly delayed. Correspondingly the percentage of each polypeptide on the total varies with time.

However the most striking effect is the one concerning var 1: while the synthesis of other proteins proceeds in a rather coordinate way during induction, synthesis of var 1 is almost absent in strain D273-10B and very low and late in strain PS 409. Decrease in synthesis of var 1 (more evident for strain D273-10B than for strain PS 409) is also observed in resting repressed cells.

Control of mitochondrial supernatants shows that this effect cannot be explained by differential release of this protein from mitochondria.

These results might be explained either by a mechanism of specific and immediate proteolysis of the var 1 polypeptide in resting conditions, or by a regulated expression of the var 1 gene in connection with growth, e.g. some cytoplasmic factor controlling the expression of var 1 could be lacking in resting conditions.

The difference in the response to respiratory induction between strains D273-10B and PS 409 could be suggestive of a difference in the control of gene expression in 'long' and 'short' strains, but before reaching a conclusion on this point the two mitochondrial genomes must be tested in the same nuclear background.

The existence of box mutants whose sensitivity to glucose repression is increased (6) would indicate a mitochondrial level of regulation: some event in the initiation of transcription, or (more probably) in the complex pathway of sequential splicing, should be influenced by glucose repression and this effect should be enhanced in the mutants. This point is of relevant importance since it is the only evidence indicating a mitochondrial level of control for the respiratory induction process. Experiments are in progress to check mitochondrial protein synthesis during release from glucose repression in the hyper-repressible box mutants.

REFERENCES

1. Slonimski, P. (1953) La Formation des Enzymes Respiratoires chez la Levure, Masson, Paris.

2. Slonimski, P. (1956) Proceedings of the 3rd Congress of Biochemistry, Academic Press, New York, p. 242.

3. Polakis,E.S., Bartley, V. and Meek, G.A. (1964) Biochem. J. 3, 369-374.

4. Perlman, P.S. and Mahler, H.R. (1974) Arch.Biochem.Biophys. 162, 248-271.

5. Stevens, B. (1977) Biol.Cellulaire 28, 37.

6. Pajot, P., Wambler-Kluppel, M.L., Kotylak, Z. and Slonimski, P. (1976) in Genetics and Biogenesis of Chloroplasts and Mitochondria (Th. Bücher et al. eds.),Elsevier/North Holland Biochemical Press, Amsterdam, pp. 443-451.

7. Agostinelli, M., Falcone, C., Frontali, L. and Sacco, M. (1980),in preparation.

8. Douglas, M., Finkelstein, D. and Butow, R.A. (1979) in Methods in Enzymology, vol. 56, part G (S.Fleischer and L.Packer eds.), Academic Press, New York, pp. 58-67.

9. Lowry, O.H., Rosebrough, H.J., Farr, A.L. and Randall, R.J. (1951) J. Biol. Chem. 193, 265-275.

10. Johnston, G.C., Singer, R.A. and McFarlane,E.S. (1977) J. Bacteriol. 132, 723-730.

11. Agostinelli, M., Falcone, C., Frontali, L. and Paoletti, A.M., in preparation.

12. Murphy, W.I., Attardi, B., Tu,C. and Attardi, G. (1975) J. Mol.Biol. 99, 809-814.

13. Saltzgaber-Müller,J. and Schatz,G. (1978) J.Biol.Chem. 253, 305-310.

14. Mahler, H.R. and Johnson, J. (1979) Molec.gen.Genet. 176, 25-31.

15. Baldacci, G., Falcone, C., Francisci, S., Frontali, L. and Palleschi, C. (1979) Eur.J.Biochem. 98, 181-186.

16. Woodrow, G. and Schatz, G. (1979) J.Biol.Chem. 254, 6088-6093.

© 1980 Elsevier/North-Holland Biomedical Press
The Organization and Expression of the Mitochondrial Genome
A.M. Kroon and C. Saccone, editors

INTERACTIONS BETWEEN MITOCHONDRIA AND THEIR CELLULAR ENVIRONMENT IN A CYTOPLASMIC MUTANT OF *TETRAHYMENA PYRIFORMIS* RESISTANT TO CHLORAMPHENICOL

ROLAND PERASSO, JEAN-JACQUES CURGY, FRANCINE IFTODE AND JEAN ANDRE
Laboratoire de Biologie Cellulaire 4, Bâtiment 444, Université de Paris-Sud, 91405 Orsay-Cedex (France)

INTRODUCTION

Tetrahymena pyriformis strain ST, is an amicronucleate ciliate. This strictly aerobic organism is very sensitive to chloramphenicol (CAP). In culture medium containing a relatively low concentration of CAP (100 μg/ml), the growth is blocked and all cells die after 6 days. A spontaneous mutant (STR_1) CAP-resistant has been isolated in our laboratory[1]. This mutant grows at the same rate as the wild strain in normal medium (generation time 4 hrs), but its growth is slowed by 250 or 500 μg CAP/ml (generation time : 6 hrs)[2]. The origin of the mutation has been looked for by way of microinjection of cytoplasm from a resistant cell to a sensitive cell[1]. In 250 μg of CAP/ml, such a cell stays blocked for a few days and then, switches towards resistance and resumes division. This transferability of the resistance character by injection of cytoplasm demonstrates that this character is under cytoplasmic inheritance. Most probably, the mutation affects mitochondrial DNA.

Treatment with 250 μg of CAP/ml during 24 hrs leads to alterations of the structure of mitochondria in sensitive as well as in resistant cells[2]. When the treatment is continued during 48 hours, these alterations are accentuated in the sensitive strain whereas the resistant strain recovers an almost normal morphology[2].

The alteration and then recovery of the morphology of the mitochondria of the resistant cells in selective medium sets an interesting problem to study, in relation with the mechanism of CAP resistance. This paper describes the effects of CAP on mitochondria and mitoribosomes of both strains and proposes an explanation of the mutant behaviour based on nucleo-mitochondrial regulation mechanisms.

I. THE MECHANISM OF RESISTANCE TO CAP

Assuming the fact that CAP blocks the mitochondrial translation by binding with mitoribosomes, we thought it worth to examine if CAP resistance involves changes in mitochondrial ribosomes.

In a first step, we have isolated and purified mitoribosomes and cytoribosomes from both strains, and we have measured their affinity for CAP by incubation in CAP-^{14}C medium. CAP binds with mitoribosomes only. With STR_1 mitoribosomes, it binds two times less than with ST mitoribosomes. This result is in agreement with those obtained with *Paramecium*[3,4], and shows that the resistance character is due to changes in the mitoribosomes.

In a second step, we have tried to disclose differences in the biochemical and morphological characters of STR_1 mitoribosomes compared to those of ST mitoribosomes. Both kinds of particles sediment in coincident 80 S peaks in a 5-20% sucrose gradient and have a similar electrophoretic mobility in polyacrylamide gels. The mitoribosomal RNAs present the same electrophoretic patterns in polyacrylamide gels. Mitoribosomes of both strains, observed by electron microscopy after negative staining, show similar morphology. Only the mitoribosomal proteins are different. The proteins were obtained by incubation of mitoribosomes with RNases A and T_2[5], and were deposited on polyacrylamide gels. A constant difference between electrophoretic patterns is the absence of one peak near the origin, for STR_1 (Fig. 1).

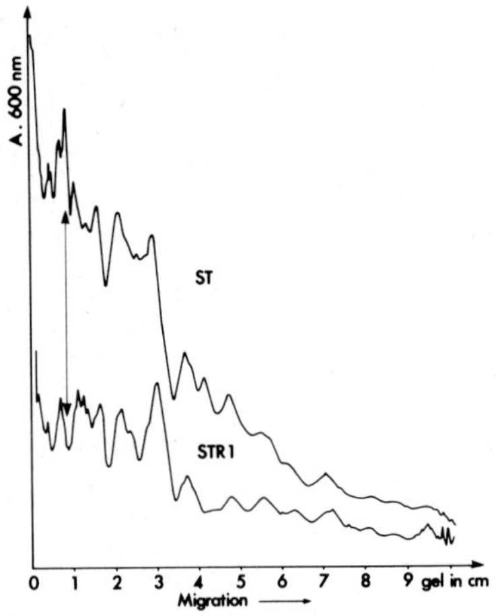

Fig. 1. Electrophoretic analysis of mitoribosomal proteins from ST and STR_1 cells. The proteins were obtained by incubation of mitoribosomes with RNases A and T_2. The gel consisted of 7.5% acrylamide and 30% urea in acetic acid 30%. One peak is absent in the STR_1 pattern (arrow).

This loss of one peak shows that the CAP resistance is due to a modification (or a loss) of a mitoribosomal protein. Most probably this protein interfers with the CAP-binding site. This result suggests that at least one mitoribosomal protein of *Tetrahymena* is coded for by the mitochondrial genome.

II. EFFECTS OF CAP-TREATMENT ON MITOCHONDRIA

Mitochondria of both strains have a similar morphology (Figs. 2 and 3). Exposure to CAP (250 µg/ml) during 24 hrs leads to fragmentation of the mitochondria and profound alterations of the mitochondrial structure in resistant cells as well as in sensitive cells (Figs. 4 and 5) : disappearance of the normal cristae (which are very numerous and tubular), appearance of a small number lamellar cristae, thinning of the matrix. Negative staining of disrupted mitochondria shows that the elementary particles are present in STR_1 strain but rare in ST strain. When the treatment is prolonged until 48 hours, these modifications are accentuated in the sensitive strain (Fig. 6), whereas the resistant one recovers an almost normal morphology (Fig. 7).

Low temperature spectra of purified mitochondrial preparations indicate the presence of a, b and c types cytochromes in both strains with α-bands situated at 618, 559-556 and 551 nm respectively (Figs. 8 and 9). Extracted cytochrome c by hypotonic KCl solution absorbs at 551 nm. After 24 hrs or 48 hrs of exposure to CAP, an important decrease of the concentration in cytochrome b and cytochrome-oxidase is observed in ST mitochondria (Fig. 8). This decrease is less pronounced in STR_1 mitochondria ; it persists even after seven days of treatment (Fig. 9).

Then, the effects of a CAP treatment are 1) perturbations of the mitochondrial inner membrane protein assembly with loss of elementary particles in the wild strain ; 2) a and b cytochromes deficiency. These results suggest that in *T. pyriformis*, as in yeast and *Neurospora*, several proteins of the inner membrane (including at least several polypeptides of cytochrome-oxidase, cytochrome b and ATPase), are produced by the mitochondrial translational system.

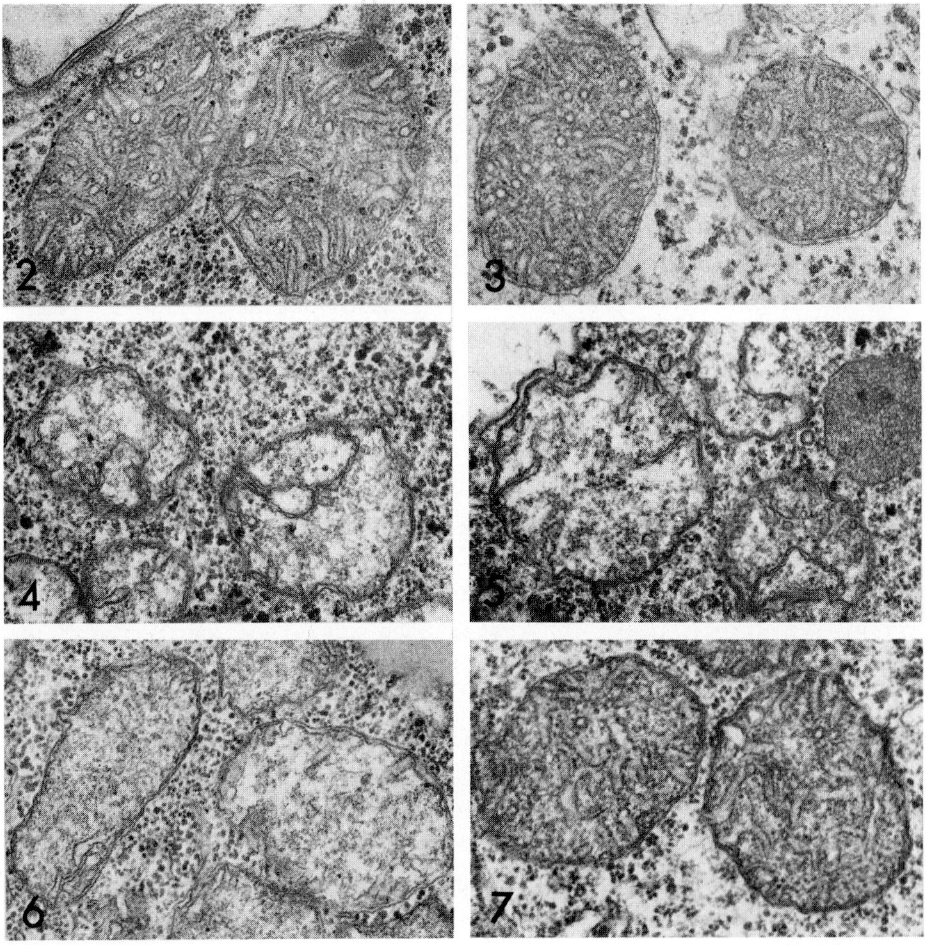

Figs. 2, 4 and 6. Mitochondrial morphology in ST cells grown in normal medium (2) and grown in CAP-containing medium (250 μg/ml) during 24 hrs (4) and during 48 hrs (6). X 30,000.

Figs. 3, 5 and 7. Mitochondrial morphology in STR_1 cells grown in normal medium (3) and grown in CAP-containing medium (250 μg/ml) during 24 hrs (5) and during 48 hrs (7). X 30,000.

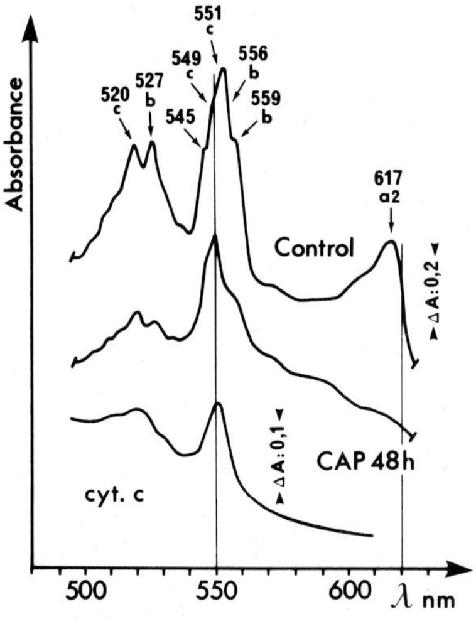

Fig. 8 (strain ST)

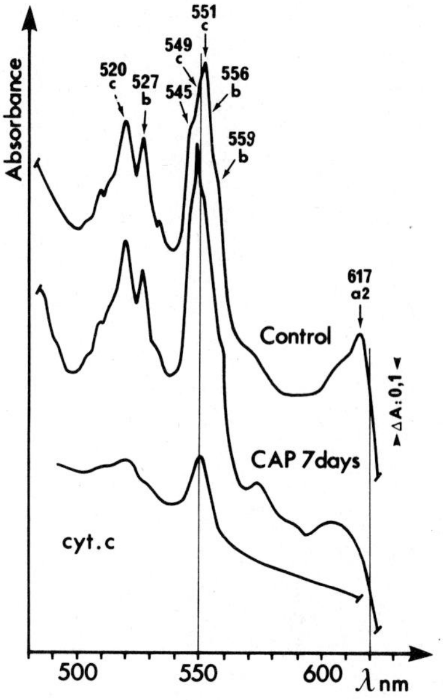

Fig. 9 (strain STR_1)

Figs. 8 and 9. Low temperature spectra of fully dithionite reduced mitochondria isolated from ST (8) and STR_1 (9) cells grown in normal medium (controls), and grown in CAP-containing medium.

III. EFFECTS OF CAP TREATMENT ON MITORIBOSOMES

CAP treatment has a direct effect on the density of mitoribosomes (number of mitoribosomes/surface unit of mitochondrial section). The mitoribosomes have been counted on sections stained using Karnovsky's technique[6] and observed with an electron microscope. The data are summarized in Fig. 10. In absence of CAP

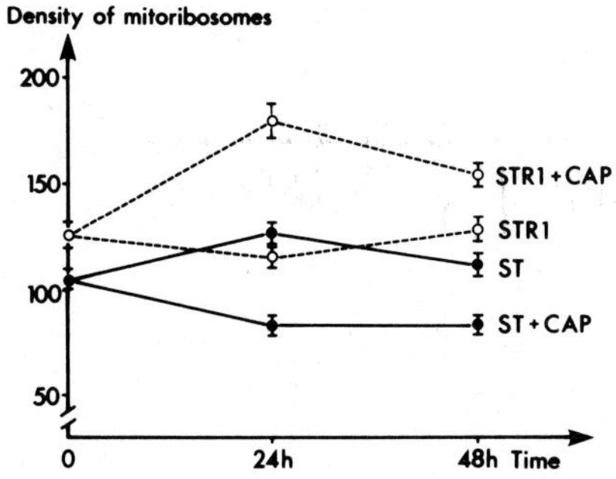

Fig. 10. Number of mitoribosomes per surface unit of section of mitochondria in both strains grown in normal medium and grown in CAP-containing medium.

(controls), the density of mitoribosomes is about the same in both strains. In the sensitive strain, the density of mitoribosomes is decreased by 20% after 24 or 48 hrs of exposure to CAP. At difference, in the resistant strain, the density of mitoribosomes is increased by more 40% after 24 hrs of CAP treatment and by more 20% after 48 hrs.

When the mutant is grown in presence of CAP, its mitoribosomes are not, or very little, affected. They show no change in sucrose gradient sedimentation, nor in electrophoretic migration. We have not detected any modification in their RNAs, in their proteins or in their morphology. At difference, the mitoribosomes from ST cells give an enlargement of the 80S peak in sucrose gradient with an obvious tendency to subunit dissociation (Fig. 11). This tendency is corroborated by electrophoresis that shows clearly two bands corresponding to the mitoribosomes 80S and to the subunits 55S (Fig. 12). Electron microscopy corroborates also this fact :

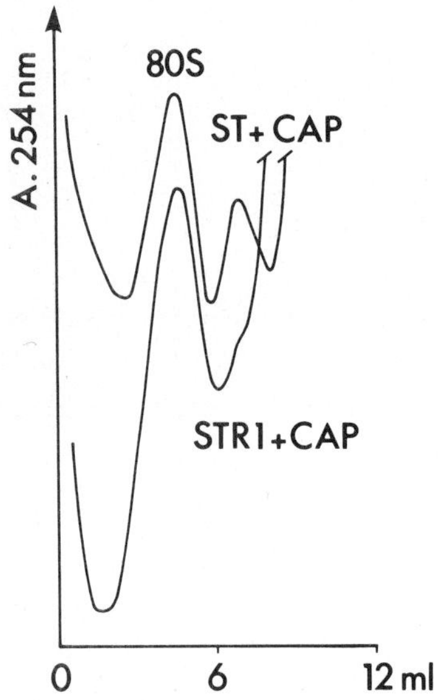

Fig. 11. A_{254} profiles of mitoribosomes from ST and STR_1 CAP-grown cells centrifuged on 5-20% sucrose gradients. The mitoribosomes from ST cells present an obvious tendency to dissociation.

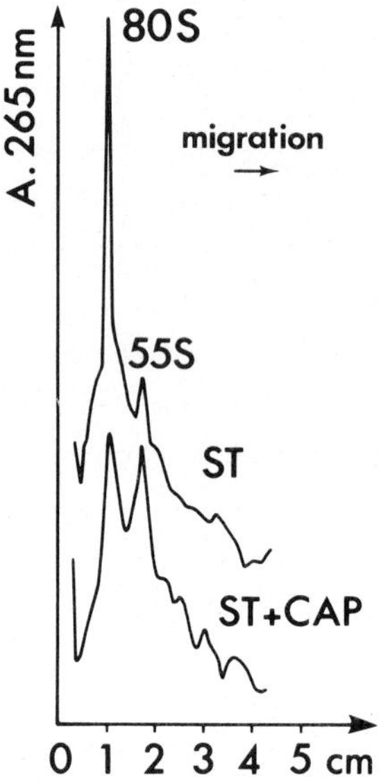

Fig. 12. Electrophoretic analysis of whole mitoribosomes in polyacrylamide gels. Many mitoribosomes from ST CAP-treated cells are dissociated during the migration.

Fig. 13 shows a field of mitoribosomes from untreated ST cells ; all present particles are recognizable mitoribosomes ; at difference, in Fig. 14, the field of mitoribosomes from treated ST cells includes many smaller particles that we interpret as subunits.

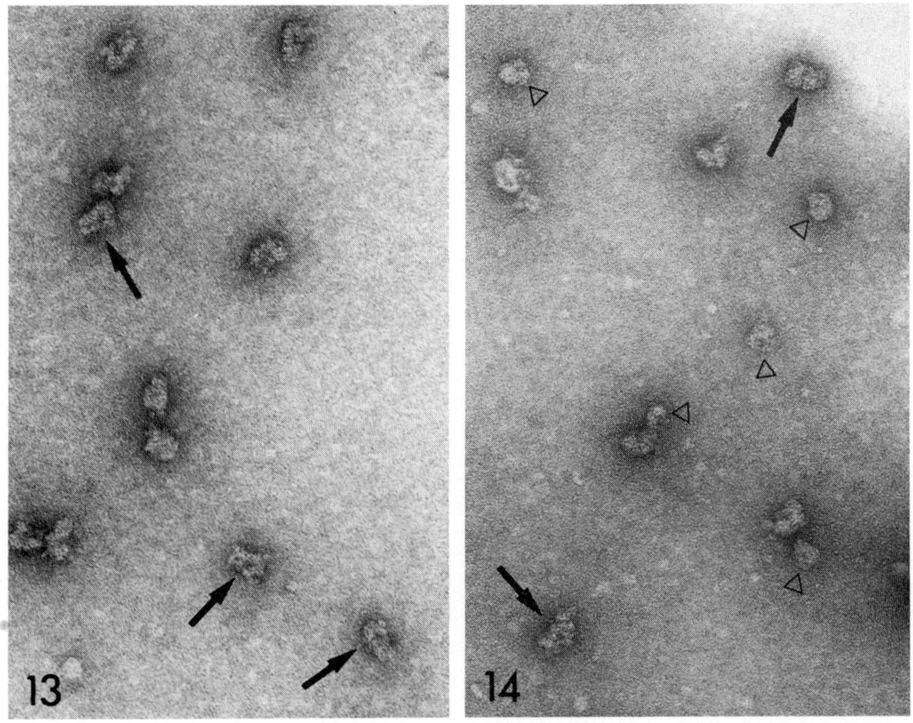

Figs. 13 and 14. Negative staining of mitoribosomes from untreated ST cells (13) and from ST CAP-treated cells (14). Many mitoribosomes are dissociated after a CAP treatment of the ST cells. ➞ : mitoribosomes ; ▷ : mitoribosomal subunits. X 180,000.

The protein composition of ST mitoribosomes is also modified by a CAP treatment of the cells : two bands disappear in the electrophoretic pattern (Fig. 15).

These results show that a CAP treatment modifies differently the density of mitoribosomes according to the strain, and alters mitoribosomes in the sensitive strain only.

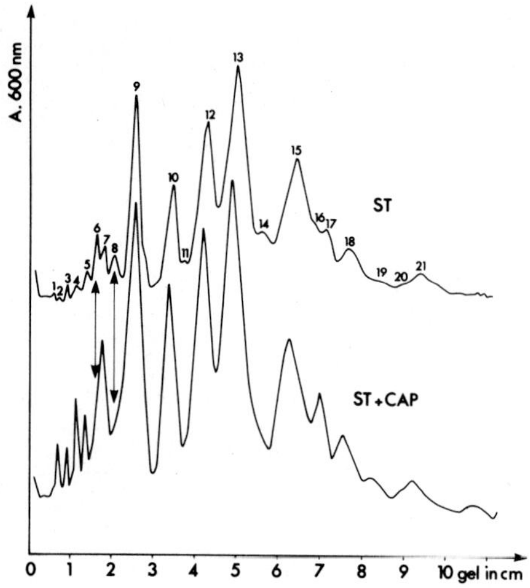

Fig. 15. Electrophoregram of mitoribosomal proteins from ST cells grown in normal medium and grown in CAP-containing medium. The proteins were extracted by acetic acid Mg^{++}. Two bands (6 and 8) are missing in the ST treated cells.

CONCLUSION

These results confirm that the mitochondrion of *T. pyriformis* is a semi-autonomous organelle : its translational system synthesizes several proteins of the inner membrane. Furthermore, the electrophoresis of mitoribosomal proteins show that several proteins of the mitoribosome are translated in the mitochondria and at least one is coded for by the mitochondrial genome. This result is in agreement with precedent works on *T. pyriformis*[7] and *Paramecium*[8]. The tendency to subunit dissociation of the mitoribosomes of ST cells in presence of CAP suggests that the mitoribosomal proteins translated into the mitochondria are necessary to assemble the two subunits of the mitoribosome. These proteins are comparable with that of *Neurospora* which is responsible for mitochondrial ribosome assembly[9,10].

Concerning the behaviour of the mutant exposed to CAP, we propose the following hypothesis : the recovery of a normal mitochondrial morphology and the increase of mitoribosomal density are the results of a nucleo-mitochondrial regulation. The addition of CAP to the growth medium blocks the mitochondrial translation completely in the sensitive cells, but only partially in the resistant ones. This causes in the latter a loss of balance between the decreased translation in the mitochondria, and the unchanged translation in the cytoplasmic background. So, the stoechiometry between the pool of polypeptides of mitochondrial origin and that of cytoplasmic background origin, is no more respected. The result is a perturbation in the inner mitochondrial membrane protein assembly. After a while, the nuclear transcription becomes adjusted to the mitochondrial one : a greater number of mitoribosomes compensates their decreased efficiency. This new balance between the activity of the two translational systems restores the morphology and the functioning of the mitochondria.

ACKNOWLEDGEMENTS

We are grateful to Elisabeth Boissonneau, Cécile Couanon and Annie Charrier for faithful and skillful help in preparation of the manuscript.

REFERENCES

1. Perasso, R., Curgy, J.J., Iftode, F. and André, J. (1980) Biologie Cellulaire, 37, 45-49.
2. Curgy, J.J., Iftode, F., Perasso, R. and André, J. (1980) Biologie Cellulaire, 37, 51-59.
3. Tait, A. (1972) FEBS Letters, 24, 117-120.
4. Spurlock, G., Tait, A. and Beale, G.H. (1975) FEBS Letters, 56, 77-80.
5. Traub, P., Mizushima, S., Lowry, C.V. and Nomura, M. (1971) Methods in Enzymology, vol. XX, Acad. Press, New York, pp. 391-407.
6. Karnovsky, M.J. (1961) J. Biophys. Biochem. Cytol., 11, 729-732.
7. Millis, A.J.T. and Suyama, Y. (1972) J. Biol. Chem., 247, 4063-4073.
8. Beale, G.H., Knowles, J.C.K. and Tait, A. (1972) Nature, 235, 396-397.
9. Lambowitz, A.M., Chua, N.H. and Luck, D.J.L. (1976) J. Mol. Biol., 107, 223-253.
10. Lambowitz, A.M., La Polla, R.J. and Collins, R.A. (1979) J. Cell Biol., 82, 17-31.

The Organization and Expression of the Mitochondrial Genome
A.M. Kroon and C. Saccone, editors

MITOCHONDRIAL BIOGENESIS IN THE COTYLEDONS OF V. FABA DURING GERMINATION

L.K. DIXON, B.G. FORDE, J. FORDE, C.J. LEAVER
Department of Botany, University of Edinburgh, Edinburgh, EH9 3JH, Scotland.

INTRODUCTION

In higher plants a number of developmental changes are linked to dramatic changes in mitochondrial activity. An example is the sequence of events occurring in embryos and cotyledons during seed maturation and germination. We have used cotyledons from germinating seeds of Vicia faba L. to investigate the development of mitochondrial protein synthetic and respiratory activity.

MATERIALS AND METHODS

The methods for isolation of mitochondria, amino acid incorporation into isolated mitochondria, measurement of respiration rate, amino acid analysis and polyacrylamide gel electrophoresis are described by Forde et al[1,2].

Plant material. Vicia faba L. seeds were soaked in 50% H_2SO_4 and surface sterilized with sodium hypochlorite before germinating in darkness at 25°C.

Immunoprecipitation. Mitochondrial pellets were dissociated in 1% SDS and incubated with antisera raised against subunit I of yeast cytochrome c oxidase and formaldehyde fixed Staphylcoccus aureus cells.

Electron microscopy. Pieces of cotyledon tissue were fixed in 3% gluteraldehyde, 2.5% Acrolein, 1.5% paraformaldehyde, 0.1 M Na Cacodylate pH 7.2, 0.3 M sucrose, then postfixed in 1% OsO_4, dehydrated through an alcohol series and embedded in low viscosity resin. Sections containing parenchymal cells were mounted on formavar coated grids and viewed at 75 KV.

RESULTS

a) Respiratory activity during germination

On contact with water cotyledon respiration rate, which is initially negligible, increases rapidly for about 18 hours. This phase coinciding with the period of imbibition is followed after 40 hours by a second phase of increasing respiration rate. Mitochondria isolated from dry seeds possess very little respiratory activity and show no respiratory control. In the first 24 hours of rehydration there is a six-fold increase in respiratory activity of the mitochondrial fraction (per mg protein) and a rapid restoration of both respiratory control and efficiency of oxidative phosphorylation (Fig. 1).

Fig. 1. Respiratory activity of isolated mitochondria during germination was measured as described in methods.

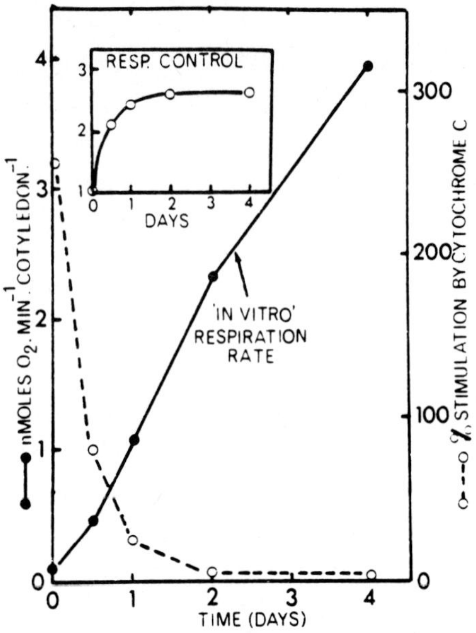

b) <u>Protein synthetic activity of mitochondria isolated during germination</u>

Protein synthetic activity (per mg protein) of mitochondria isolated from cotyledons of dry and cold imbibed seeds is barely detectable and increases only slightly during the first 24 hours even when supported by an energy generating system. During the second day of development protein synthetic activity increases approximately 10 fold (Fig. 2). Amino acid analysis shows this increase is not caused by depletion of the intramitochondrial methionine pool during germination. In dry seed and during the early stages of germination incorporation is higher when ATP is provided by an external energy generating system rather than generated by oxidative phosphorylation. However by the end of the first day (by which time respiratory control and efficiency of oxidative phosphorylation have been restored) endogenously supplied ATP supports higher levels of incorporation (Fig. 2).

Fig. 2. Amino acid incorporation by isolated mitochondria during germination. Incorporation of ^{35}S methionine (expressed per mg mitochondrial protein) into isolated mitochondria was followed during germination using ATP either generated endogenously or provided by an external energy generating system (see Methods).

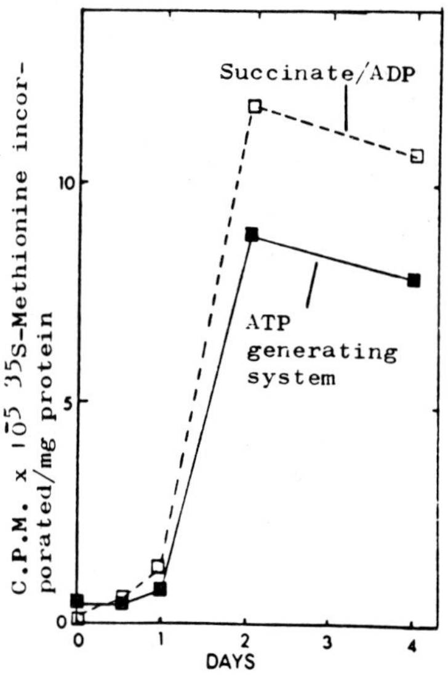

c) Products of *in vitro* protein synthesis by isolated mitochondria

Analysis of translation products shows that mitochondria isolated from 96 hour cotyledons synthesise about 12 major polypeptides. Of these 12 only half are synthesised as major products by mitochondria from dry or 12 hour imbibed seeds. The relative synthesis rate of the remainder increases mainly between 12 and 48 hours.

One of these polypeptides has been tentatively identified by immunoprecipitation as subunit I of cytochrome oxidase. This polypeptide increases in relative synthesis rate over 10 fold between 12 and 96 hours (Fig. 3).

d) Electron microscopy of cotyledons at various stages of germination

Ultrastructural examination of cotyledon parenchymal tissue shows distinct changes in the mitochondrion during germination. Between 12 and 48 hours the surface area of the inner membrane increases as reflected by an increase in cristae number and the matrix becomes more electron dense, containing more ribosome like particles. By 48 hours mitochondria seem fully developed although the number of mitochondrial profiles is still increasing.

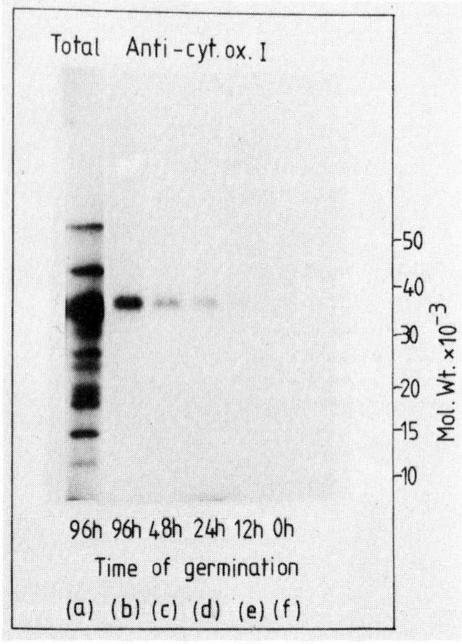

Fig. 3. Immunoprecipitation of cytochrome c oxidase subunit I during germination. An equal number of trichloroacetic precipitable cpm from mitochondria at various stages of germination labelled in vitro were immunoprecipitated with antisera raised against yeast cytochrome c oxidase subunit I. Radioactively labelled polypeptides were detected by fluorography of polyacrylamide gels. a) shows the total products of mitochondrial protein synthesis and (b) to (f) a developmental sequence of immunoprecipitates.

DISCUSSION

We conclude that during the initial 24 hours of germination cotyledon mitochondria are defective in their protein synthesising ability. During this period a number of important changes occur in mitochondria, including an increase in cristae number, a six-fold increase in respiratory activity of isolated mitochondria and a recovery of mitochondrial integrity. Thus it is unlikely that the initial functional and structural development of the organelle is dependent upon protein synthesis by mitochondrial ribosomes.

The lag in development of protein sunthesising ability in mitochondria may be due to a requirement for transcription of new mitochondrial messenger RNA, to deficiencies in the translational apparatus of mitochondria from dry seeds or may depend upon de novo synthesis of cytoplasmic proteins.

REFERENCES

1. Forde, B.G., Oliver, R.J. and Leaver, C.J. (1979) Plant Physiol. 63, 67-73.
2. Forde, B.G., Oliver, R.J. and Leaver, C.J. (1978) Proc. Natl. Acad. Sci. USA, 75, 3841-3845.

© 1980 Elsevier/North-Holland Biomedical Press
The Organization and Expression of the Mitochondrial Genome
A.M. Kroon and C. Saccone, editors

DEFECT IN HEME *A* BIOSYNTHESIS IN *OXI* MUTANTS OF THE YEAST *SACCHAROMYCES CEREVISIAE*.

EZZATOLLAH KEYHANI AND JACQUELINE KEYHANI
Laboratory for Cell Biology and Biochemistry, Institute of Biochemistry and Biophysics, University of Tehran, P.O. Box 314-1700, Tehran, Iran.

Cytochrome *c* oxidase from yeast *Saccharomyces cerevisiae* consists of seven subunits whose apparent molecular weights range from 42 000 to 6 000, two hemes *a*, two copper atoms and 1-10% phospholipids[1, 2, 3]. The three large subunits (I to III) are made on mitochondrial ribosomes, whereas the four smaller subunits (IV to VII) are made on cytoplasmic ribosomes[1, 2]. The structural genes, termed *OXI I*, *II* and *III*[4] coding for subunits II, III and I respectively, are now characterized[5, 6]. In the *OXI* mutants the assembly of functional heme *a* does not occur, and cytochrome *c* oxidase exhibits neither its specific spectral properties nor its enzymatic activity[6].

Since the apoprotein biosynthesis in these mutants is now well characterized, we decided to study the biosynthesis of heme *a*, one of the two prosthetic groups of cytochrome *c* oxidase. In this study, we report that porphyrin *a* was found in the apocytochrome *c* oxidase of *OXI I*, *II*, and *III* mutants, suggesting that mutation by deletion (absence of subunit I in *OXI III*) or point mutation (change in the electrophoretic mobilities of subunits II and III in *OXI I* and *II*) produced a change in the conformation of the apoprotein, so that the final assembly of heme *a* does not occur.

THE BIOSYNTHETIC PATHWAYS FOR LABELING HEME *A* OF CYTOCHROME *C* OXIDASE : USE OF RADIOACTIVE Δ-AMINOLEVULINIC ACID AND MEVALONIC ACID

The biosynthetic pathways for labeling heme *a* of cytochrome *c* oxidase are illustrated in Figure 1. Δ-aminolevulinic acid (AmLev) is a specific precursor of hemes including heme *a* of cytochrome *c* oxidase, while mevalonic acid (MVA) is incorporated into heme *a*, presumably as a precursor of farnesyl pyrophosphate[7]. Since mevalonic acid is also a precursor of lipids[8], the mitochondrion should be delipidated before determination of radioactive mevalonic acid. Under such conditions, the remaining radioactivity would be due to heme *a*.

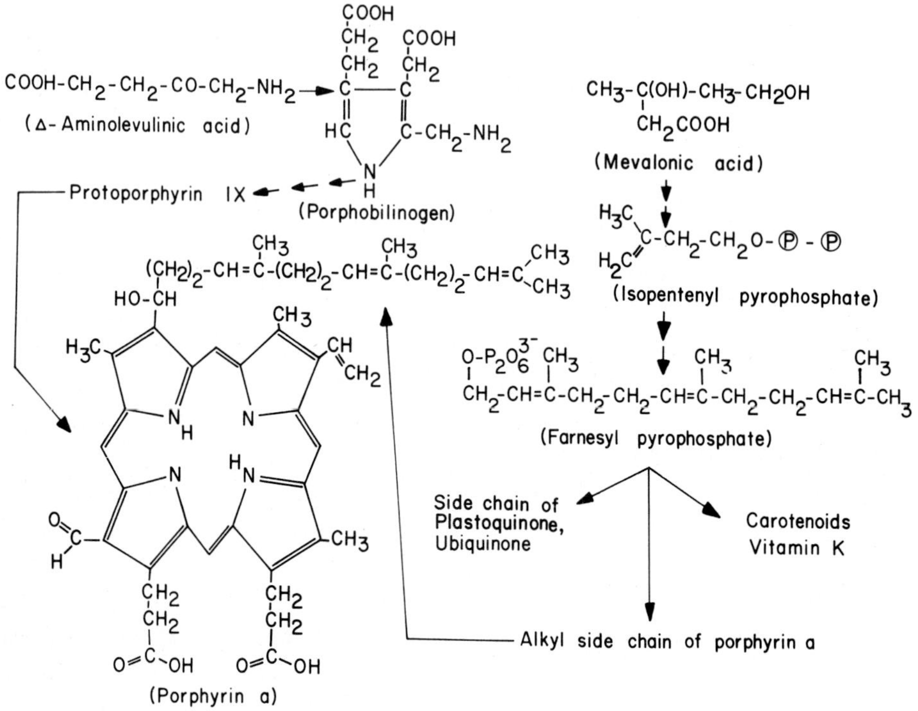

Fig. 1 : Biosynthetic pathways for labeling heme *a* of cytochrome *c* oxidase : use of radioactive δ-aminolevulinic acid and mevalonic acid.

PRESENCE OF RADIOACTIVE [^{3}H]MVA IN DELIPIDATED MITOCHONDRIA AND IMMUNOPRECIPITATED CYTOCHROME *C* OXIDASE.

Figure 2 shows an SDS-polyacrylamide gel of [^{14}C]AmLev, [^{3}H]MVA labeled delipidated mitochondria from wild type and *OXI* mutants. [^{14}C]Amlev incorporation into hemes and porphyrins was resolved into 3 bands of 28 000, 13 500, and 10 000 in all experimental conditions. Hemes covalently bound to proteins, such as that of cytochrome c_1 and cytochrome *c*, would migrate to 28 000, and 13 500 bands, respectively, while hemes non covalently bound, such as hemes of cytochrome *b* and cytochrome *c* oxidase, would

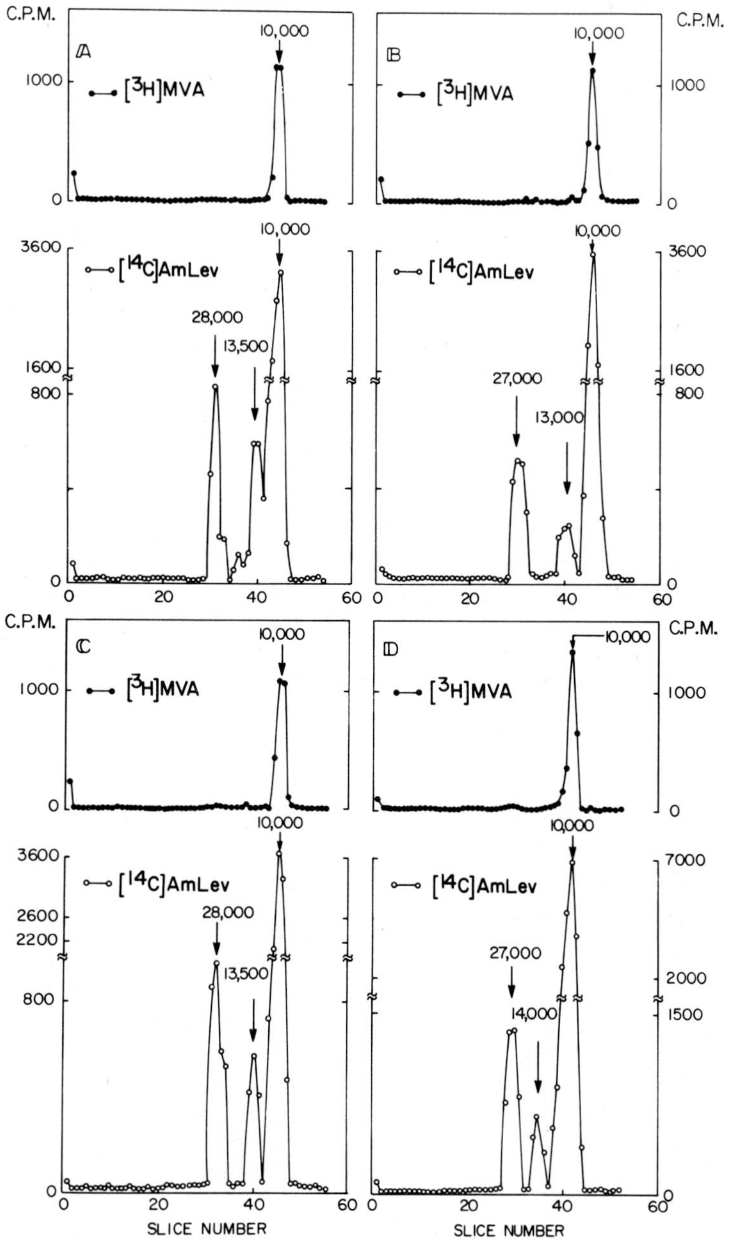

Fig. 2. Distribution of radioactivity among electrophoretic fractions of [^{14}C]AmLev- and [^{3}H]MVA-labeled delipidated mitochondria from wild type (A), *OXI I* (B), *OXI II* (C), *OXI III* (D) cells.

migrate to the low molecular weight region (10 000). [^{3}H]MVA incorporation in delipidated mitochondria was resolved in a single band in the region of 10 000 Mw, corresponding to the region of heme non covalently bound to protein. Compared to the control, *OXI I*, *II* and *III* mutants also showed a single band in the 10 000 Mw region. The presence of radioactivity due to MVA in the delipidated mitochondria of *OXI* mutants as observed in SDS-polyacrylamide gel electrophoresis suggests that a precursor of heme *a* is present in the mitochondria. This point was further investigated in order to see if this precursor of heme *a* is integrated into the apocytochrome *c* oxidase of *OXI* mutants. Table I shows the immunoprecipitated cytochrome *c* oxidase from wild type and *OXI* mutants double labeled with [^{14}C]AmLev and [^{3}H]MVA. The immunoprecipitate after 3 cycles of washing in 0.1 M phosphate buffer, containing 1% Triton X-100, and 3 cycles of delipidation showed a high specific radioactive AmLev and MVA. These observations suggest that, although heme *a* was not present in *OXI* mutants, a porphyrin precursor of heme *a* was synthesized and integrated into the mutated apoprotein.

TABLE I

DISTRIBUTION OF RADIOACTIVE AMLEV AND MVA IN DELIPIDATED IMMUNOPRECIPITATED CYTOCHROME *C* OXIDASE FROM WILD TYPE AND *OXI* MUTANTS IN *SACCHAROMYCES CEREVISIAE*[a]

	[^{14}C]AmLev	[^{3}H]MVA	[^{14}C]/[^{3}H]
Wild type	1,822	254	7.2
OXI I	1,732	211	8.2
OXI II	1,121	137	8.2
OXI III	4,824	733	6.7

[a] Expressed as total radioactivity present in the immunoprecipitate.

THE NATURE OF PORPHYRIN PRESENT IN APOCYTOCHROME *C* OXIDASE OF *OXI* MUTANTS.

The distribution of radioactivity among hemes and porphyrins extracted from mitochondria according to the procedure of Lemberg et al.[9, 10] is shown in Table II. In all three *OXI* mutants, at a concentration of 20% HCl, a solvent which extracts porphyrin *a*, a significant amount of radioactivity was extracted. This result suggests that the porphyrin present in apocytochrome *c* oxidase is porphyrin *a*, the porphyrin precursor of heme *a*[11, 12, 13].

TABLE II
DISTRIBUTION OF [^{14}C]AMLEV AMONG HEMES AND PORPHYRINS EXTRACTED FROM MITOCHONDRIA[a] OF WILD TYPE AND *OXI* MUTANT CELLS

Fraction[b]	[^{14}C]AmLev Distribution (%)[c]			
	Wild Type	*OXI I*	*OXI II*	*OXI III*
HCl 4%	5.9	6.0	2.2	1.9
HCl 8%	1.5	4.8	2.5	11.8
HCl 20%	4.9	14.0	13.2	16.2
Ether	87.6	75.0	82.0	70.0

[a] The extraction procedure of Lemberg et al.[9, 10] was used in these experiments.
[b] HCl 4%, 8%, 20% are solvents which extract, respectively, protoporphyrin IX, cryptoporphyrin and porphyrin *a*[14]. The hemes remain in the ether phase.
[c] The values are expressed as % of the total radioactivity present in the ether phase before HCl extractions.

DISCUSSION

The genetic information encoded in the primary sequence of proteins can be modified by mutation or chemically without destroying the functional capacity of the molecule. Cytochrome *c*, hemoglobin and myoglobin of different species are apparently unaltered by many such substitutions. Such mutations or chemical modifications do not alter the proper folding of the molecule so that the functional capacity of the molecule is preserved. In other cases amino acid substitution can lead to the loss of enzymatic activity, presumably by a change in the active site and improper folding leading to an altered configuration of the molecule. A point mutation in the *E. Coli* gene for tryptophan synthetase leads to the synthesis of tryptophan synthetase molecules in which a single amino acid replacement has produced an altered three-dimensional structure lacking biological activity[14, 15]. Our observations suggest that porphyrin *a*, the precursor of heme *a* is synthesized and is present in the mutated apocytochrome *c* oxidase. Thus the *OXI* mutations by producing an alteration in the configuration of mitochondrially made subunits of cytochrome *c* oxidase, prevent the incorporation of iron into porphyrin *a*, which in turn leads to the formation of inactive cytochrome *c* oxidase.

REFERENCES

1. Tzagoloff, A., Rubin, M.S. and Sierra, M.F. (1973) Biochem. Biophys. Acta, 301, 71-104.
2. Schatz, G. and Mason, T.L. (1974) Ann. Rev. Biochem., 43, 51-87.
3. Malmström, B.G. (1974) Quarterly Rev. of Biophys. 6, 387-431.
4. Slonimski, P.P. and Tzagoloff, A. (1976) Eur. J. Biochem., 61, 27-41.
5. Cabral, F., Saboz, M., Rudin, Y., Schatz, G., Clavillier, L. and Slonimski, P.P. (1978) J. Biol. Chem. 253, 297-304.
6. Keyhani, E. (1979) Biochem. Biophys. Res. Commun. 89, 1212-1216.
7. Keyhani, J. and Keyhani, E. (1978) FEBS Lett., 93, 271-274.
8. Goodwin, T.W. (1965) in "Biosynthetic pathways in higher plants" (Pridham, J.B. and Swain, T., eds), Academic Press, New York, pp. 57-71.
9. Lemberg, R., Bloomfield, B., Caiger, P. and Lockwood, W.H. (1955) Aust. J. Exp. Biol., 33, 435-450.
10. Lemberg, R., Stewart, M. and Bloomfield, B. (1955) Aust. J. Exp. Biol., 33, 491-496.
11. Keyhani, J. and Keyhani, E. (1976) FEBS Lett., 70, 118-122.
12. Keyhani, J. and Keyhani, E. (1978) in "Frontiers of biological energetics : from electrons to tissues" (A. Scarpa, P.L. Dutton and J.S. Leigh, eds),Academic Press, New York Vol. II, pp. 913-922.
13. Keyhani, E. and Keyhani, J. (Submitted for publication).
14. Lemberg, R. and Parker, J. (1955) Aust. J. Exp. Biol., 33, 483-490.
15. Henning, U. and Yanofsky, C. (1962) Proc. Natl. Acad. Sci. U.S.A., 48, 1497-1504.
16. Helinski, D.R. and Yanofsky, C. (1963) J. Biol. Chem. 238, 1043-1048.

© 1980 Elsevier/North-Holland Biomedical Press
The Organization and Expression of the Mitochondrial Genome
A.M. Kroon and C. Saccone, editors

THE ASSEMBLY PATHWAY OF NUCLEAR GENE PRODUCTS IN THE MITOCHONDRIAL ATPase COMPLEX

R. TODD, T. GRIESENBECK, P. McADA, M. BUCK AND M. DOUGLAS
Department of Biochemistry, Univ. of Texas Health Sci. Ctr., San Antonio, Texas 78284

INTRODUCTION

Although it has been known for almost 10 years that approximately 90% of the mitochondrial mass in yeast is synthesized by the nucleocytoplasmic system, it is only through recent studies[1] that an understanding of the molecular events involved in the segregation and assembly of nuclear coded components during mitochondrial biogenesis has begun to emerge. A system ideally suited to elucidate these events is that of the mitochondrial ATPase complex*: approximately 80% of the mass of this major inner membrane complex is synthesized via the nucleocytoplasmic system; in addition it is conveniently obtained in large amounts for biochemical characterization and the production of specific immunological probes[2,3,4].

The present report summarizes studies from this laboratory which attempt to define the structure of the yeast mitochondrial ATPase complex and its assembly pathway in greater detail.

RESULTS AND DISCUSSION

Arrangement of subunits in the ATPase complex

The ATPase complex obtained from yeast by triton extraction[2-5] contains 9-10 non-identical subunits. When examined on a 4-17% gradient gel, subunit 6 is resolved into two bands (Figure 1A) both of which are apparently translated on mitochondrial ribosomes. Studies to determine whether 6a and 6b are unique mitochondrial translation products are currently in progress. Analysis of the enzyme with a specific antiserum indicates that the subunits are assembled into a stable structure. When ATPase complex is immunoprecipitated with subunit 2 antiserum from labeled submitochondrial particles (Figure 2) all of the subunits present in the isolated enzyme are found associated with this large cata-

---

*ATPase complex, oligomycin-sensitive adenosine triphosphate complex; F_1-ATPase, the oligomycin-insensitive complex from yeast containing subunits 1,2,3 and 8; DSP, dithiobis(succinimyidyl propionate); SDS, sodium dodecyl sulfate.

lytic subunit. In addition, these data also allow determination of the number of each subunit present in the immunoprecipitable complex by comparing on slab gels the subunits seen when uniformly labeled ATPase complex is isolated with ATPase complex antiserum and with antisubunit 2 serum (Figure 2). The relative amount of label present for each subunit is calculated by normalizing the areas under the curves of Fig. 2 to subunit 3; the observed stoichiometry is derived by dividing the relative amount of label present in each subunit by its apparent molecular weight. In order of decreasing size, the number of copies of each subunit is 3:3:1:2:1:2:2:1:2:3. Thus the ATPase complex has a molecular weight of 5.8×10^5 which agrees well with the values derived from hydrodynamic data[4,6] and contains a minimum of 20 polypeptide chains. It is noteworthy that the ATPase complex contains per mole multiple copies of both mitochondrially and nucleocytoplasmically synthesized components. This feature of the ATPase, which may be unique among the major mitochondrial complexes, suggests additional complexity in its coordinated assembly.

Analysis with a cleavable crosslinking reagent has been used to establish the arrangement of the subunits within the solubilized ATPase complex. The monodisperse enzyme was treated with the protein-protein crosslinker dithiobis (succinimydyl propionate) and the resulting products analyzed by two-dimensional SDS/acrylamide slab gel electrophoresis[7]. Using the observed major subunit crosslinks and the subunit stoichiometries, a model for the complex can be constructed (Figure 3) which makes several predictions about the assembly pathway of the enzyme. The complex is a closed structure containing a pseudo two-fold axis of symmetry which could be formed by self-assembly of the subunits. The mitochondrially synthesized components (subunits 6 and 9) are clustered in an integral membrane complex and the cytoplasmically translated subunits (subunits 1-5, 7 and 8) are associated with each other in a more peripheral structure. Furthermore, the architecture of the complex suggests that the subunits synthesized in the cytoplasm are associated with each other on the matrix side of the membrane when they are mature.

Assembly pathway of the ATPase complex

Earlier studies from this laboratory (8) have demonstrated that the single nuclear mutation, pet 936, interrupts the assembly pathway of the ATPase complex. The unassembled ATPase subunits associated with the mitochondrial membrane appear to be as stable as those of the assembled parental complex[8]. Even though the mutant contains only 10-15% of the parental level of subunits 1 and 2, freshly prepared mitochondria of the mutant and parental strain bind comparable amounts of subunit specific antiserum (Figure 4). Additional

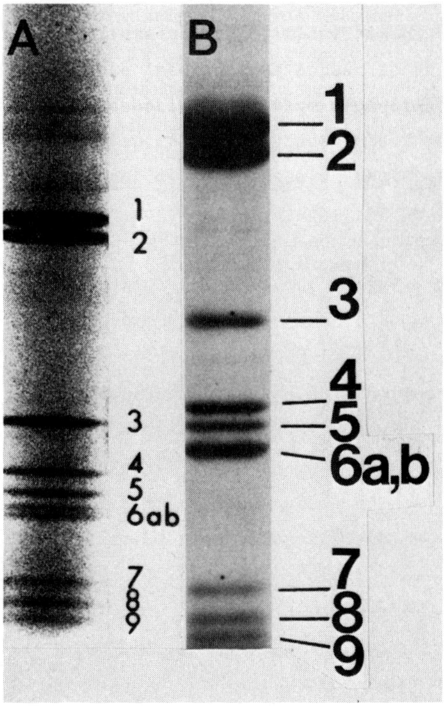

Fig. 1. The yeast mitochondrial ATPase complex. A. Oligomycin sensitive ATPase complex prepared from commerically grown yeast as described[4] was resolved on a 4-17% gradient slab gel in the presence of SDS and stained with coomassie blue. B. A triton extract of mitochondria prepared from cells grown in a ^{14}C-labeled amino acid mixture was treated with subunit 2 and antiserum. The washed immunoprecipitate was resolved and flourographed as described[8]. Only the lower portion of the gradient slab gel is illustrated in each case.

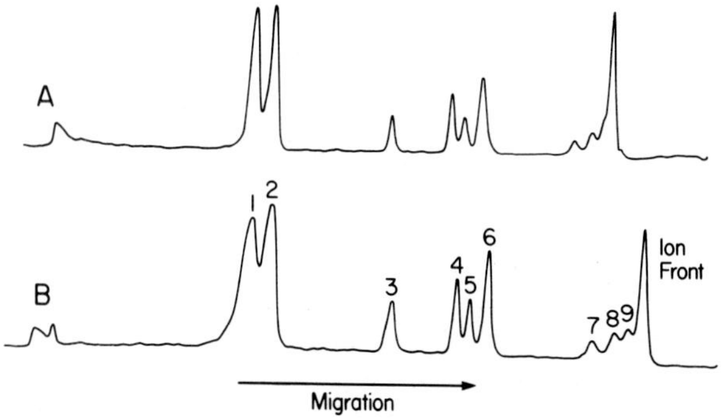

Fig. 2. Subunit quantitation of the ATPase complex. Scans of fluorographs of gels containing labeled ATPase complex immunoprecipitated with antisubunit 2 serum (A) and antiholo enzyme serum (B) were processed as described[4].

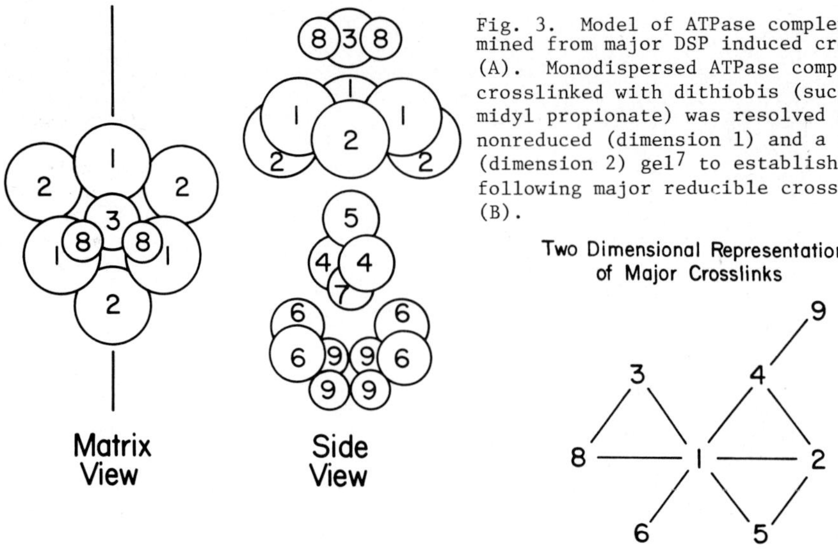

Fig. 3. Model of ATPase complex determined from major DSP induced crosslinks (A). Monodispersed ATPase complex crosslinked with dithiobis (succinimidyl propionate) was resolved on a nonreduced (dimension 1) and a reduced (dimension 2) gel[7] to establish the following major reducible crosslinks (B).

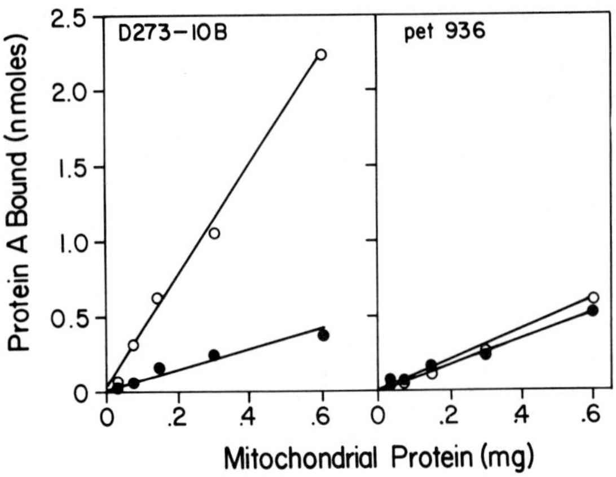

Fig. 4. Localization of F_1-ATPase determinants on the mitochondrial membranes of an assembly defective pet mutant. Mitochondria were prepared from D273-10B and pet 936 yeast spheroplasts and washed 3 times in 0.6M mannitol, 10mM Trisglycylglycine, 1mM EGTA pH 7.0, 0.1% BSA. Freshly prepared untreated (closed circles) and disrupted (open circles) mitochondria in 450 μl of buffer were mixed with 50μl of F_1-ATPase subunit 2 antiserum (diluted 1 to 4 with preimmune serum) or 50μl of preimmune serum alone. After incubation for 1 hour at 4°C, 20 nmole of ^{125}I-protein A (2.3 x 10^5 cpm/nmole) was added and incubated for an additional 60 min. A portion of the reaction mixture was layered over a 1ml cushion of 1M sucrose, 10mM Trisglycylglycine, 100mM NaCl, 1% BSA and centrifuged at 120,000 x g_{max} for 45 min at 4°. The bottom of the tube was sliced off and counted. Values using preimmune serum were subtracted for each mitochondrial point examined.

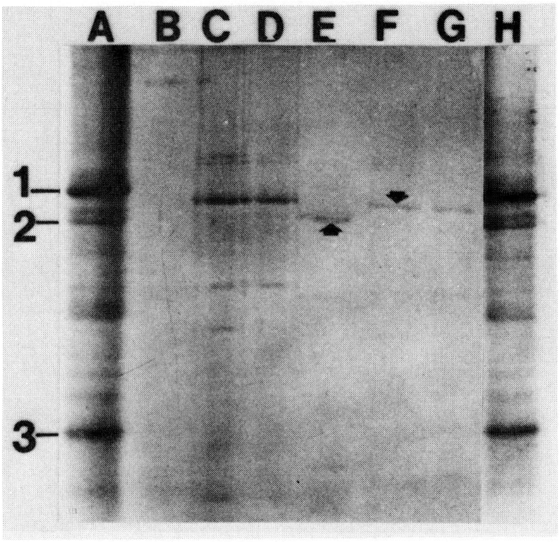

Fig. 5. The binding and processing of F_1-ATPase subunit 2 precursor by mitochondria.

Yeast poly A containing RNA was translated in rabbit reticulocyte lysate[8]. After translation was complete, D273-10B mitochondria (40μg) was added to 50μl of a translation reaction and incubated 20 min at 20°C. Mitochondria were separated from the reaction mixture through a sucrose cushion as described in Fig. 4. Immunoadsorption using antisubunit 2 serum of detergent treated mitochondria and translation reaction mixtures was performed as described[9] and analyzed by slab gel fluorography. A and H. ^{125}I labeled F_1-ATPase standard; B. immunoadsorption of translation using preimmune serum; C and D. using anti-subunit 2 serum; E. immunoadsorption of pelleted mitochondria; F and G. immunoadsorption of reaction supernatant after removing mitochondria.

TABLE 1

END GROUP ANALYSIS OF MATURE ATPase SUBUNITS

The large F_1-ATPase subunits were isolated from enzyme dissociated in 6M guanidine-HCl and 8M urea followed by chromatography on DEAE cellulose. Carboxyterminal sequencing was performed by following the time course of amino acid release upon treatment with carboxypeptidase A and B.

Subunit	Amino terminal	carboxy terminal
1	blocked	•leu•lys•ala
2	blocked	ala•ser•gly•lys

determinants for antibody binding are observed only in the parental membrane after its mechanical disruption. Thus, these data and additional studies indicate that unassembled F_1-ATPase subunits appear to be localized in the parental mitochondria in a manner similar to that of the nuclear mutant.

Since these accessible F_1-ATPase determinants of the parental strain may represent unassembled subunits of the enzyme which are engaged with the membrane in an early stage of the mitochondrial assembly pathway, studies were initiated to characterize the solubilization properties and labeling behavior of these antigenic determinants in relation to those of the assembled complex. This small fraction of the F_1-ATPase determinants (5-8% of the total) is preferentially and quantitatively released by 0.05% Triton X-100 extraction. Glycerol gradient resolution revealed that approximately 40% of the material released by 0.05% triton behaved as free or unassembled subunits while the remainder co-sedimented with assembled ATPase complex. Extraction of membranes at higher triton concentrations released additional assembled ATPase complex. In addition, pulse chase experiments showed that the unassembled fraction of F_1-ATPase determinants from mitochondria were more rapidly labeled and chased than the assembled fraction. When mitochondria were prepared from cells grown in ^{3}H-leucine which received a 20 minute pulse of $^{35}S-SO_4$, the ratio of $^{35}S/^3H$ in the assembled and free fractions was 0.55 and 7.1, respectively. Following a 15 minute chase of the ^{35}S label the ratio of the free fraction had dropped to a value of 0.36 while the ratio of the assembled fraction did not change significantly. These data support the presence of unassembled, transient, cytoplasmically synthesized subunits on mitochondria which may represent a portion of the assembly pathway of the enzyme.

Recent studies[9,10] have shown that the binding and processing of precursor forms of the F_1-ATPase subunits by isolated mitochondria occurs as (a) post translational event(s). As illustrated in Figure 5, the precursor of subunit 2 prepared in an _in vitro_ translation system, when mixed with isolated mitochondria is bound, processed and co-sediments with the organelle. Moreover, this post translational segregation with mitochondria is not observed with labeled mature subunit 2. Thus, the recognition and association of the precursor of a cytoplasmically synthesized ATPase subunit with mitochondria occurs as a post translational event _in vitro_. Similar segregation behavior in either a cotranslational or post translational event is proposed to involve transient peptide sequences present in the F_1-ATPase precursors _in vivo_.

Based on these data and analysis of the mature F_1-ATPase subunits which inidcate that the amino termini are blocked (Table 1), we suggest the following events in the assembly pathway of the cytoplasmically synthesized subunits of the ATPase complex: 1. Recognition and segregation of precursor subunits in a cotranslational or post translational event; 2. Specific endoproteolytic processing of the precursors; 3. Blocking of the amino terminal residue;

4. Import and assembly of all the cytoplasmically synthesized subunits of the enzyme into a stable structure which associates with a complex of mitochondrially synthesized subunits.

ACKNOWLEDGEMENTS

This study was supported by NIH Grant GM 25648 and Biomedical Research Support Grant RR05654.

REFERENCES

1. Schatz, G. (1979) FEBS Letters 103, 203-211.
2. Tzagoloff, A. and Meagher, P. (1971) J. Biol. Chem. 246, 7328-7336.
3. Ryrie, I. and Gallagher, A. (1979) Biochim. Biophys. Acta 545, 1-14.
4. Todd, R., Griesenbeck, T. and Douglas, M. (1980) J. Biol. Chem. 255, 5461-5467.
5. Douglas, M., Koh, Y., Ebner, E., Agsteribbe, E. and Schatz, G. (1979) J. Biol. Chem. 254, 1335-1339.
6. Tzagoloff, A. (1976) in The Enzymes of Biological Membranes (Martonosi, A., ed.) Vol. 4, pp. 103-124, Plenum Press, New York.
7. Todd, R. and Douglas, M., manuscript submitted.
8. Todd, R., McAda, P. and Douglas, M. (1979) J. Biol. Chem. 254, 11134-11141.
9. Maccecchini, M., Rudin, Y., Blobel, G. and Schatz, G. (1979) Proc. Natl. Acad. Sci. 76, 343-347.
10. Nelson, N. and Schatz, G. (1979) Proc. Natl. Acad. Sci. 76, 4365-4369.

The Organization and Expression of the Mitochondrial Genome
A.M. Kroon and C. Saccone, editors

MODIFIED MITOCHONDRIAL TRANSLATION PRODUCTS IN NUCLEAR MUTANT OF THE YEAST *SCHIZOSACCHAROMYCES POMBE* LACKING THE β SUBUNIT OF THE MITOCHONDRIAL F_1 ATPase

MARC BOUTRY and ANDRE GOFFEAU
Laboratoire d'Enzymologie, UCL, Place Croix du Sud, 1, 1348 Louvain-la-Neuve, (Belgium)

INTRODUCTION

Several mutants altered in the mitochondrial ATPase activity have been selected from strains of *Schizosaccharomyces pombe* unable to grow on glycerol medium. Some of them were distinguished by total loss of ATPase activity and specific alteration of one subunit of the F_1 ATPase. Other mutants exhibited deficient assembly of the ATPase complex. Their postmitochondrial supernatant contained a soluble immunoprecipitable F_1 ATPase. All these mutants also exhibited reduced levels of mitochondrial NADH-cytochrome c reductase and cytochrome oxidase activities[1]. We report here, as an example, qualitative modifications in the mitochondrial translation products of a mutant in which the β subunit of the F_1 ATPase is not detected.

MATERIALS AND METHODS

S. pombe 972h$^-$ and the nuclear mutant B59/1 (A. Vassarotti and A.M. Colson, unpublished) were grown aerobically to early stationary phase in a liquid medium containing 9% raffinose, 0.1% glucose and 2% yeast extract.

Uniform radioactive labeling was obtained by incubation of the cells in minimal medium supplemented by 9% raffinose and 200 μCi [^{35}S]-sulphate (carrier-free) per ml. Obtention and analysis of [^{35}S]-labeled products of mitochondrial protein synthesis were carried out as described by Douglas *et al*[2]. Detection of the F_1 ATPase subunits by the immune replica technique was performed as described by Cabral *et al*[3]. Enzymatic activities were detected as described by Tzagoloff *et al*[4].

RESULTS AND DISCUSSION

Figure 1 shows that immunodetection of the F_1 ATPase subunits present in the mitochondrial fraction did not reveal any β subunit in the mutant B59/1. The other subunits α, γ, δ and ε were observed in this mutant as in the parental type. As expected, no ATPase activity was detected in the mutant mitochondrial

TABLE 1

MITOCHONDRIAL ENZYMATIC ACTIVITIES

Strain	Specific activities (μmol x min^{-1} x mg^{-1})		
	ATPase	NADH-cytochrome c reductase	Cytochrome oxidase
Parental strain	3.02	1.480	0.760
B59/1	0.05	0.109	0.067

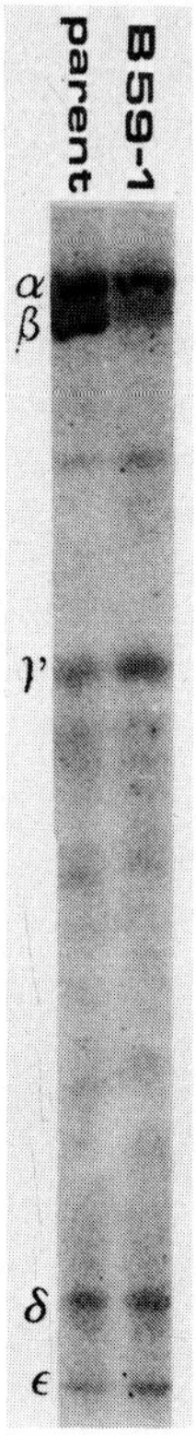

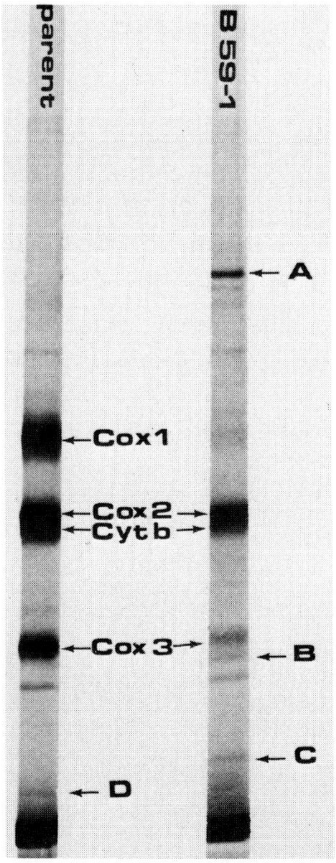

Fig. 1. Immunodetection of the F_1 ATPase subunits. Mitochondria from uniformly labeled cells were isolated and subjected to polyacrylamide gel electrophoresis in the presence of sodium dodecylsulfate and then analysed by the immune replica technique.

Fig. 2. Mitochondrial translated polypeptides. Mitochondria of cells labeled in the presence of cycloheximide were isolated and analysed in polyacrylamide gel electrophoresis in the presence of sodium dodecylsulfate. The gel has been dried and radioautographed.

fraction (Table 1). Moreover the mutant exhibited some pleiotropic deficiencies since the mitochondrial NADH-cytochrome c reductase and cytochrome oxidase activities were reduced to less than 10% of the parental activity (Table 1). However other mitochondrial components such as cytochrome c and cytochrome b_{554} (the synthesis of which are insensitive to chloramphenicol) were present in comparable amount in the mutant and in the parental type (not shown). No differences could be found between B59/1 and the parental type in the polypeptide pattern of a mitochondrial fraction analysed in SDS PAGE and colored by Coomassie blue (not shown). However, several differences were observed between the mutant and the parental strain in the mitochondria-translated polypeptides observed after 40 min of radioactive labeling in the presence of cycloheximide. Figure 2 shows that two mitochondrial polypeptides were not detected in the mutant : subunit 1 of the cytochrome oxidase and polypeptide D of unknown function (apparent molecular weight of 14000). On the other hand, three new labeled polypeptides appeared in the mutant : polypeptides A, B, C of apparent molecular weights of 54000, 20000 and 15000 respectively. Identical modified patterns were obtained for several other mutants characterized by other structural defects of the F_1 ATPase (not shown).

The results demonstrate that the absence of the subunit 1 of cytochrome oxidase which was known to result from mitochondrial box mutations affecting the introns of the split gene of cytochrome b[5] and from other mitochondrial mit^- mutations (Linnane, personal communication) can also result from nuclear mutations in the F_1 ATPase. We conclude that labeling of subunit 1 of cytochrome oxidase in the presence of cycloheximide is more sensitive to non-specific defects of oxidative phosphorylation than that of the other well identified products of mitochondrial protein synthesis. The three new mitochondrial polypeptides detected in the mutant might be either precursors or hydrolytic products of other polypeptides synthesized on mitoribosomes such as cytochrome oxidase subunit 1, component D or non identified components of the F_0 portion of ATPase. Fingerprint studies of these polypeptides are required to verify these possibilities.

REFERENCES

1. Boutry, M. and Goffeau, A., in preparation.
2. Douglas, M., Finkelstein, D., and Butow, R.A. (1979) Methods in Enzymology, 56, 58-66.
3. Cabral, F., Solioz, M., Rubin, Y., Schatz, G., Clavilier, L. and Slonimski, P.P. (1978) J. Biol. Chem. 253, 297-304.
4. Tzagoloff, A., Akai, A. and Needleman, R. (1975) J. Bacteriol. 122, 826-831.
5. Claisse, M., Slonimski, P., Johnson, J. and Mahler, H. (1980) Molec. gen. Genet. 177, 375-387.

© 1980 Elsevier/North-Holland Biomedical Press
The Organization and Expression of the Mitochondrial Genome
A.M. Kroon and C. Saccone, editors

REGULATION OF THE SYNTHESIS OF MITOCHONDRIAL PROTEINS: IS THERE A REPRESSOR ?

PETER VAN 'T SANT, J. FRANCISCA C. MAK and ALBERT M. KROON
Laboratory of Physiological Chemistry, State University, Bloemsingel 10,
9712 KZ GRONINGEN, The Netherlands.

INTRODUCTION

For *Neurospora crassa* a simple mechanism has been proposed for the interaction of the nucleocytoplasmic and mitochondrial genetic systems by Barath and Küntzel[1,2]. They suggested that a mitochondrially coded and synthesized protein is transported to the nucleus and represses coordinately the synthesis of the enzymes involved in mitochondrial replication, transcription and translation. Their hypothesis was based on the observation that the synthesis of nuclearly coded mitochondrial proteins increases when mitochondrial protein synthesis is blocked. Accumulation is reported for mitochondrial (mt)EF-G[1], transformylase[1], mtRNApolymerase[2], mt leu-tRNA synthetase[3], mt phe-tRNA synthetase[3], but also for cytochrome *c*[4] and CN^--insensitive respiration[5]. These observations offer only indirect evidence that a mitochondrial gene product might be involved in keeping the correct balance between the two transcription-translation systems. We have tried to gather direct proof for this repressor model.

METHODS AND MATERIALS

In all experiments the *Neurospora* slime mutant, resolved from the heterokaryon FGSC 327 was used. The culturing conditions, the procedures of labeling the cells with ^{35}S-sulphate and the media and methods used for isolation of nuclei, mitochondria and postmitochondrial supernatant have been described elsewhere[6].

RESULTS AND DISCUSSION

In the first place we studied the alteration of the activity of 8 mt aminoacyl-tRNA synthetases induced by the inhibition of mitochondrial protein synthesis with ethidiumbromide. If coordinate repression is indeed occurring, one would expect comparable accumulation of these mitochondrial enzymes, which are of nucleocytoplasmic origin. The results obtained are given in Table 1. The increases of the phenyl- and leucyl-tRNA synthetases are in good agreement with the results of others[3]. However, only 5 of the 8 synthetases show an increased activity and the degree of this augmentation is quite different. Moreover, the specific activity of the 3 others is decreased. For the arginyl- and seryl-tRNA

TABLE 1

AMINOACYL-tRNA SYNTHETASE ACTIVITIES (pmoles aminoacyl-tRNA/min x mg mt protein)

mitochondrial aminoacyl-tRNA synthetase	control cells	ethidiumbromide treated cells	EtBr / control
Leu	17	70	4
Phe	4,0	19	5
Met	3,0	30	10
Val	0,7	28	40
Tyr	11	120	11
Ser	1120	817	0,7
Arg	330	227	0,7
Lys	57	2	0,04

synthetases this 30% lower specific activity may be largely explained by the relative increase of total mitochondrial protein with about the same percentage. In any case the lysyl-tRNA synthetase activity is strongly reduced. It is clear that the repressor model cannot explain these observations without the assumption of differential inactivation or proteolysis and/or differential inhibition of transport across the mitochondrial membranes.

In a more direct approach we searched for one or more exported mitochondrially synthesized proteins. Cells were labeled with ^{35}S-sulphate in the presence of cycloheximide. Incorporation into proteins of the subcellular fractions was measured by SDS-polyacrylamide gel electrophoresis. Typical fluorograms of the gels are given in fig. 1A-C. The nuclear fraction is slightly contaminated with mitochondrial proteins. In the nuclear and postmitochondrial supernatant fraction a highly labeled band is visible with an apparent molecular weight of about 18 kDaltons. A protein with about the same molecular weight is labeled in the mitochondria. In order to prove that the 18 kDalton protein is a mitochondrial product we tried to block the synthesis with chloramphenicol[9], ethidium bromide[9], and acriflavin[10] which are known as strong inhibitors of the mt translation and/or transcription. The results are shown in fig. 1F-I. Chloramphenicol did not influence the incorporation into the 18 kD protein, whereas all the other proteins labeled in the presence of cycloheximide disappeared. On the contrary ethidium bromide and especially acriflavin clearly diminished the incorporation also into the 18 kD protein.

In view of these unexpected discrepancies we had to consider the possibility that labeling of the 18 kDalton protein is caused by an effect specific for the labeled sulphate. We, therefore, labeled cells with ^{3}H-leucine and with a ^{14}C-aminoacid mixture. In both cases we could not detect a labeled 18 kDalton band in the nuclear and postmitochondrial supernatant fraction. However, in the mito-

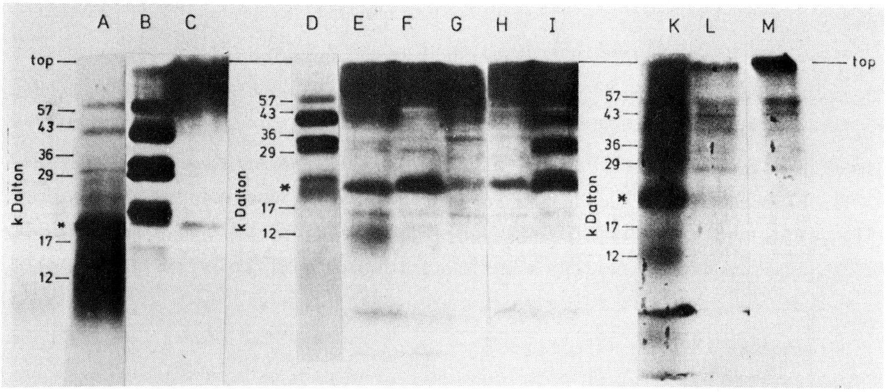

Fig. 1. Fluorograms[7] of 15% SDS-polyacrylamide gels[8] of proteins, ^{35}S-labeled (except of lane D) in the presence of cycloheximide. For further details see text. The asterisk indicates the 18 kDalton site. A. Nuclear fraction. B. mt-fraction. C. Post-mitochondrial supernatant (Pms). D. ^{3}H-labeled mt-fraction. E. Pms. F-I. 50 µg of total protein of cells labeled in the presence of F. Chloramphenicol (2 mg/ml); G. Acriflavin (50 µg/ml); H. Ethidiumbromide (50 µg/ml); I. no extra addition. K. Pms. L. The same prep. after TCA precipitation. M. as L after 4 h in Glycine-NaOH buffer, pH 9.6 at 40 °C.

chondrial fraction we still found the 18 kD protein as can be seen in Fig. 1D. Bertrand[11] has described a protein with a similar molecular weight, the function of which is yet unknown. Another fact that points against a translational labeling of the extramitochondrial 18 kDalton protein is the instability of this product. We noticed that the majority of the label disappears by TCA precipitation (Fig. 1 K+L). The residual label could be removed by mild saponification, a treatment especially suitable for sulphate-ester hydrolysis (Fig. 1M).

Our results, therefore, indicate that the labeling of the extramitochondrial protein is due to some post-translation modification. Also the fact that chloramphenicol did not influence the labeling of this band but clearly blocked the labeling of the other proteins, points against a mitochondrial origin. In yeast similar observations have been made[12], but the labeling of the yeast products was far less pronounced as in our case.

CONCLUSIONS

In spite of our careful and intensive attempts we have been unable to trace a mitochondrial translation product in nuclei and in the post-mitochondrial supernatant. The initial positive results have to be regarded as merely coincidental and are ascribed by us to a sulphatation process. It remains peculiar that the sulphate labeling was so strong and mainly restricted to one band. The further coincidence with the presence of a real mitochondrial translation product of similar molecular weight and unknown function is another pitfall. We, therefore, conclude that there is no indication that mitochondria do really synthesize repressor-type proteins which regulate at the level of nuclear DNA. This is strengthened by the fact that none of the proteins labeled in the presence of cycloheximide showed affinity to DNA under conditions that other proteins did. For this reason we feel that the repressor hypothesis[1,2] in its simple form has to be rejected. Also the various degrees of accumulation and even reduction of mitochondrial enzymes of nuclear origin argue against the model. At best one can imagine regulation controlled by a family of mitochondrial repressors which remain undetectable with the methods used. One should keep in mind, furthermore, that accumulation of at least some mitochondrial proteins of nucleocytoplasmic origin also occurs if mitochondrial proteinsynthesis is not at all impaired, *e.g.* if respiration is blocked[13].

ACKNOWLEDGEMENT

These studies were supported in part by the Dutch Foundation for Chemical Research (SON) with financial aid from the Netherlands Organization for the Advancement of Pure Research (ZWO). The authors wish to thank Harm Schokkenbroek for graphical, Bert Tebbes for photographical, Karin van Wijk and Rinske Kuperus for typographical help.

REFERENCES

1. Barath, Z. and Küntzel, H. (1972) Proc. Natl. Acad. Sci. USA 69, 1371-1374.
2. Barath, Z. and Küntzel, H. (1972) Nature New Biol. 240, 195-197.
3. Beauchamp, P.M. and Gross, S.R. (1976) Nature 261, 338-340.
4. Von Jagow, G., Weiss, H. and Klingenberg, M. (1973) Eur. J. Biochem. 33, 140-157.
5. Edwards, D.L., Rosenberg, E. and Maroney, P.A. (1974) J. Biol. Chem. 11, 3551-3556.
6. Van 't Sant, P., De Jong, L. and Kroon, A.M. (1980) in: Endosymbiosis and Cell Research (W. Schwemmler, and H. Schenk, Eds.) W. de Gruyter, Berlin-New York, in press.
7. Chamberlain, J.P. (1979) Anal. Biochem. 98, 132-135.
8. Laemmli, A.K. (1970) Nature 227, 680-685.
9. Schatz, G. and Mason, T.L. (1974) Ann. Rev. Biochem. 43, 51-87.
10. Woodrow, G. and Schatz, G. (1979) J. Biol. Chem. 254, 6088-6093.
11. Bertrand, H. and Werner, S. (1977) Eur. J. Biochem. 79, 599-606.
12. Terpstra, P., Zanders, E. and Butow, R.A. (1979) J. Biol. Chem. 254, 12653-12661.
13. Lambowitz, A.M. and Zannoni, D. (1978) in: Plant Mitochondria (G. Ducet and C. Lance, Eds.) Elsevier/North-Holland Biomedical Press, 283-291.

The Organization and Expression of the Mitochondrial Genome
A.M. Kroon and C. Saccone, editors

REGULATION OF MITOCHONDRIAL GENOMIC ACTIVITY IN SEA URCHIN EGGS

A.M. RINALDI, I. SALCHER-CILLARI, M. SOLLAZZO and V. MUTOLO
Institute of Comparative Anatomy, University of Palermo, via Archirafi 22
90123 Palermo, Italy

INTRODUCTION

It is generally accepted that nuclei and mitochondria collaborate in making functional the mitochondrial synthetic apparatus. While the extramitochondrial system contributes the majority of mitochondrial proteins, the role of the mitochondria is limited to the production of very few proteins. How can the nuclear and the organellar genomes interplay in regulating each other?

The sea urchin egg represents a very suitable system to ask such a question, because during its maturation and development it operates a peculiar coupling and uncoupling of the synthesis of nuclear and mitochondrial DNAs. It in fact accumulates mitochondria during oogenesis, i.e. in absence of nuclear DNA synthesis. Following fertilization it cleaves very quickly into smaller cells, with a high synthesis of nuclear DNA and no synthesis of new mitochondria, until at least the pluteus stage. Moreover the egg can be divided[1] by centrifugation into a nucleated and a non nucleated halves, both viable and fertilizable, thus offering the unique opportunity to easlily separate the mitochondria from nuclear influence in a viable cell.The non nucleated half, in fact, contains the majority of the mitochondria. By using such technique we studied mit RNA and DNA synthesis and mitochondrial duplication in the presence and in the absence of the nuclear influence, demonstrating that the cell nucleus prevents mitochondrial duplication during early development.

RESULTS AND DISCUSSION

Mitochondria of unfertilized eggs incorporate RNA precursors at a very low rate. The rate is somewhat increased if the egg is fertilized or parthenogenetically activated. If the nucleus has been removed by centrifugation and the non nucleated half parthenogenetically activated, the incorporation increases ten-twenty fold. The permeability to the precursor does not change under these conditions. If the male pronucleus is introduced by fertilizing the non nucleated half instead of parthenogenetically activating it, the RNA synthesis remains at the low level observed in the entire eggs, fertilized or parthenogenetically activated. The RNA synthesized by the mitochondria of the non nucleated half is

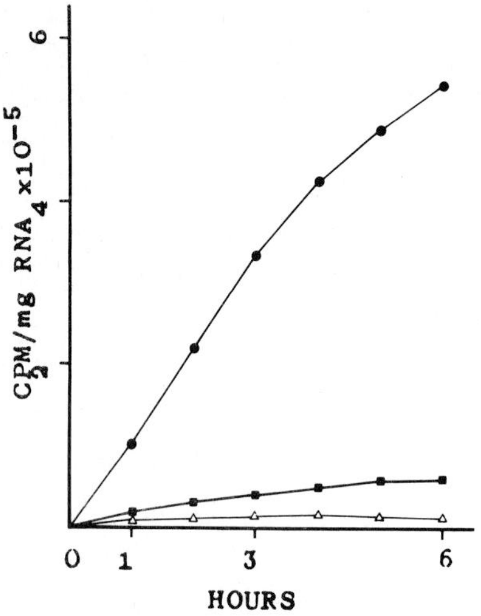

Fig. I. ^{32}P incorporation into mitochondrial RNA. At the times indicated 20 ml samples were collected, washed and mitochondrial RNA purified, as descibed by Rinaldi et al.[2]. ●—● activated non nucleated halves; ■--■ fertilized entire eggs; △—△ fertilized non nucleated halves.

essentially ribosomal and transfer RNA[2]. These results strongly suggest that the removal of the cell nucleus may activate mitochondrial duplication.

In order to investigate that possibility, we studied the incorporation of ^{3}H-thymidine into mit DNA.

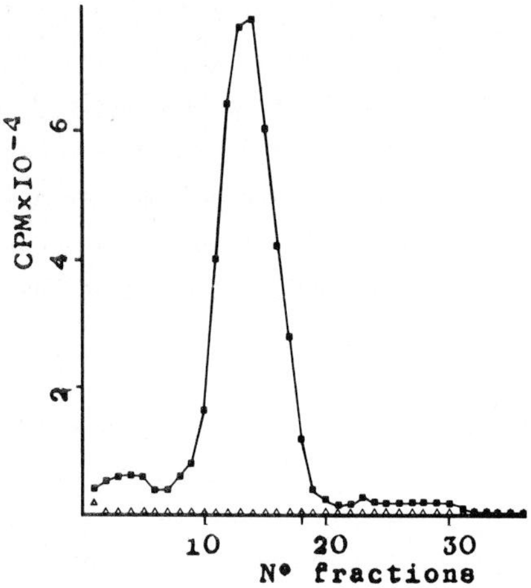

Fig. 2 Cesium chloride-ethidium bromide gradient of mit DNA from enucleated activated halves[3] ●—● ; entire eggs △-△-△

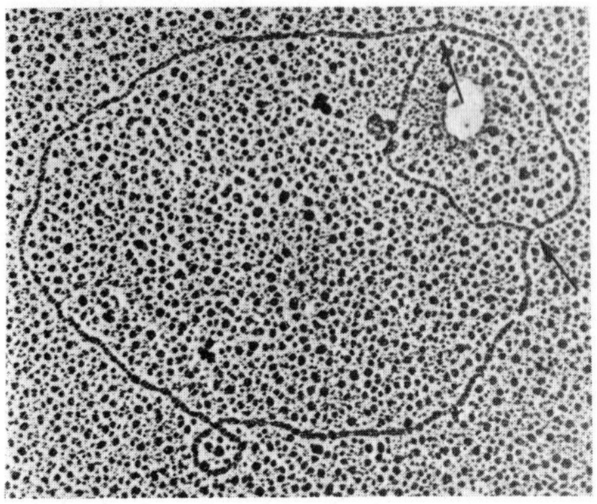

Fig. 3 Mitochondrial DNA replicative form from activated enucleated half with a D-loop.

Fig. 2 shows very little, if any, incorporation in the mit DNA of fertilized or parthenogenetically activated entire egg, in which however nuclear DNA (notshown) exhibits a very active incorporation. However, mit DNA of the non nucleated activated halves very actively incorporates the precursor. If the non nucleated halves are fertilized, no thymidine incorporation at all was observed. An electron microscopic analysis of the DNA purified from the enucleated parthenogenetically activated halves showed many replicative forms whereas no replicative forms were ever observed in the DNA purified from fertilized enucleated halves nor from activated entire eggs. These results show that the removal of the egg nucleus followed by parthenogenetic activation induces mit DNA synthesis together with mit RNA synthesis and therefore suggest that it might activate mitochondria duplication.

In order to verify such an hypothesis the mitochondrial population of the entire or non nucleated eggs was analyzed at the electron microscope. In the enucleated halves we have found many figures (Fig. 4) of mitochondria that can be interpreted as stages of divisio process of the mitochondria. No such figures have been observed in whole eggs and in fertilized enucleated halves.

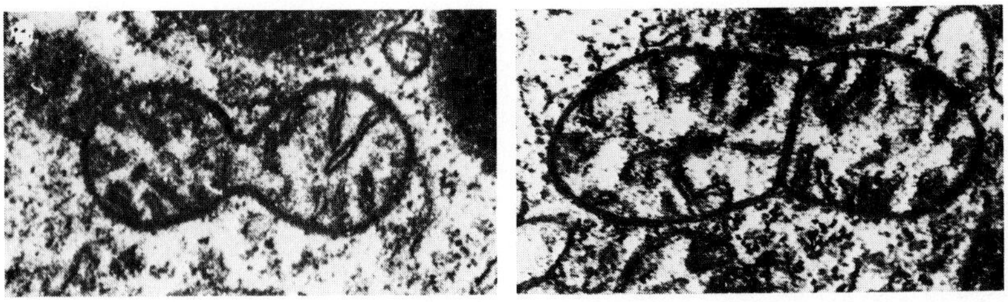

Fig. 4. Mitochondria of parthenogenetically activated enucleated halves.

These results indicate that the cell nucleus somehow exerts a negative control on the mitochondrial genomic activity, i.e. mit DNA replication, RNA transcription and mitochondria division.

How does the nucleus exert such a control? Preliminary results suggest that the electrophoretic pattern of the newly synthesized proteins found in the mitochondria of enucleated parthenogenetically activated halves is different than that of the fertilized or parthenogenetically activated entire eggs.

This negative control might be a general phenomenon of many embryonic systems. Experiments in progress seem to suggest that DNA replication is stimulated in Xenopus laevis eggs if they are enucleated and parthenogenetically activated.

REFERENCES

I) De Leo, G., Rinaldi, A.M., Salcher-Cillari, I. and Mutolo, V.
Acta Embryol. Exptl., 2 (I979) I4I

2) Rinaldi, A.M., Storace, A., Arzone, A., and Mutolo,V.
Cell Biol. Intern. Rep. I (I977) 249

3) Rinaldi, A.M., De Leo, G., Arzone, A., Salcher, I., Storace, A. and Mutolo, V.
Proc. Natl. Acad. Sci. 76 (I979) I9I6

4) Rinaldi, A.M., Salcher-Cillari, I. and Mutolo, V.
Cell Biol. Int. Rep. 3 (I979) I79

The Organization and Expression of the Mitochondrial Genome
A.M. Kroon and C. Saccone, editors

EFFECT OF HYPOTHYROIDISM ON SOME ASPECTS OF MITOCHONDRIAL BIOGENESIS AND DIFFERENTIATION IN THE CEREBELLUM OF DEVELOPING RATS

MARIA N. GADALETA[1], GIULIA R. MINERVINI[1], MARCELLA RENIS[2], G. ZACHEO[3], TERESA BLEVE[3], IDA SERRA[4] AND ANNA M. GIUFFRIDA[4]
[1]Centro di Studio sui Mitocondri e Metabolismo Energetico c/o Istituto di Chimica Biologica, Università di Bari, Italy
[2]Ospedale Psichiatrico "Casa della Divina Provvidenza" Bisceglie, Italy
[3]Istituto di Nematologia Agraria del C.N.R., Bari, Italy
[4]Istituto di Chimica Biologica, Università di Catania, Italy

INTRODUCTION

It has been clearly established, with many different experimental approaches, that thyroid hormones have a marked influence on the functional maturation of the central nervous system in many mammalian species. Particularly studied has been the effect of thyroid hormone deficiency in rat cerebellum, in which 97% of the cells, essentially interneurons, are formed after birth. In the cerebellum of hypothyroid developing rats there is a transient retardation in the rate of cell acquisition which is ultimately compensated by a prolongation of the proliferative phase and, most striking, a persistent retardation in the growth and branching of the dendritic arborization and in the ontogenesis of the dendritic spines of prenatally formed Purkinje cells[1,2]. It has been also reported that in rat brain the number of mitochondria per g tissue increases between 1 and 21 days of age.[3] The changes in mitochondrial concentration has been related to increasing levels of synaptic activity in the neuropil of rat inferior colliculus[4]. Furthermore, in our previous studies, a peak of mitochondrial DNA and protein synthesis was demonstrated in the cerebellum of 10-day old rats[5,6]. Since it is well known that thyroid hormones affect also mitochondrial structure and function[7], we though interesting to study the biogenesis and the differentiation of mitochondria in the cerebellum of developing rat. The biogenesis of mitochondria was followed by measuring either DNA polymerase activity in isolated organelles or the labelling with 5-methyl-^{3}H thymidine of mitochondrial DNA in tissue slices experiments. As an index of mitochondrial differentiation we estimated the content of haemes a a_3 in mitochondria of normal and hypothyroid rats during development. Lastly, electron microscope analysis was conducted on ultrathin sections of cerebellum to visualize the internal organization of mitochondria in 21-day old normal and hypothyroid rats.

MATERIALS AND METHODS

Wistar rats were made hypothyroid by daily treatment of the mother (from the 18th day of gestation) with propylthiouracil (PTU, 50 mg per day) administered by stomach intubation. Free (non synaptosomal) mitochondria were prepared as already reported by Gadaleta et al.[5] γ-DNA polymerase was assayed in sonicated mitochondria according to Hubscher et al.[8]. Tissue slices experiments were according to Giuffrida et al.[6] Haemes a a_3 were estimated with a double beam spectrophotometer following the absorbance variation at 605-630 nm (ε_{mM}=14)[9] after addition of few grains of solid Na-dithionite to the mitochondrial suspension in 50 mM potassium phosphate buffer, pH 7.4. For electron microscope analysis fragments of cerebellum were fixed for 5 h in 3% glutaraldehyde at 4°C, rinsed overnight in 0.05 M cacodylate buffer at pH 7.2, post-fixed for 8 h in 2 % osmium tetroxide in the same buffer with 2.5% sucrose, dehydrated in ethanol and embedded in Spurr[10]. Sections were stained in uranyl acetate and in lead citrate[11] and examined in a JEM 100 B.

RESULTS AND DISCUSSION

Fig.1 and Fig.2 show that mitochondrial DNA synthesis in the cerebellum of hypothyroid developing rats is impaired and probably delayed. Indeed, at 14 days of age, DNA synthesis is higher in hypothyroid rats than in normal animals. At 21 days of age, however, it is quite similar in both groups of animals.

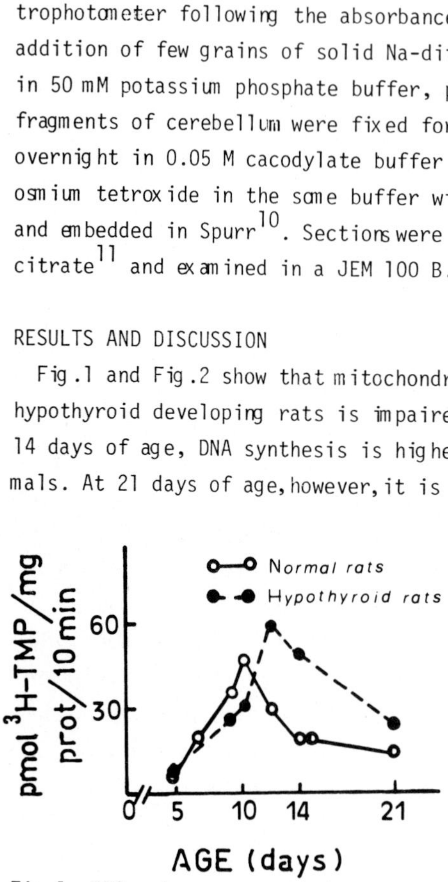

Fig.1 γ-DNA polymerase activity in sonicated mitochondria.

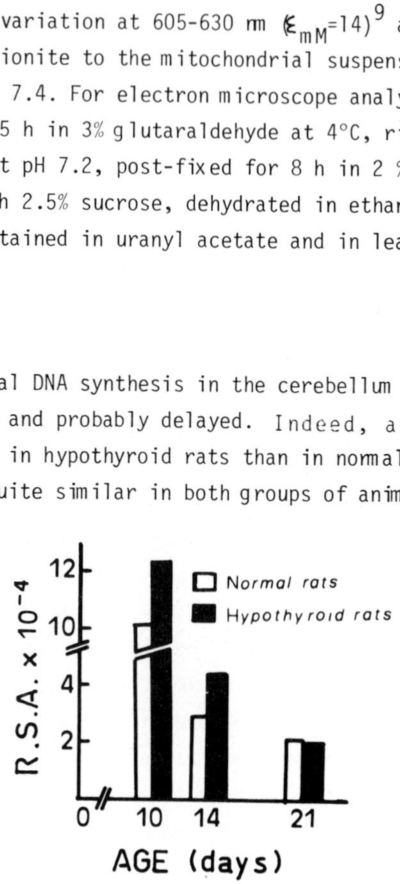

Fig.2 Labelling of mitochondrial DNA by tissue slices experiments.R.S.A.: dpm/ mg DNA/ac.sol. precursor radioactivity

Fig.3 shows the content of cytochrome a a_3 per mg of mitochondrial protein in normal and hypothyroid rats during development. It appears diminished in hypothyroid animals as from the 14th days of age suggesting a minor differentiation

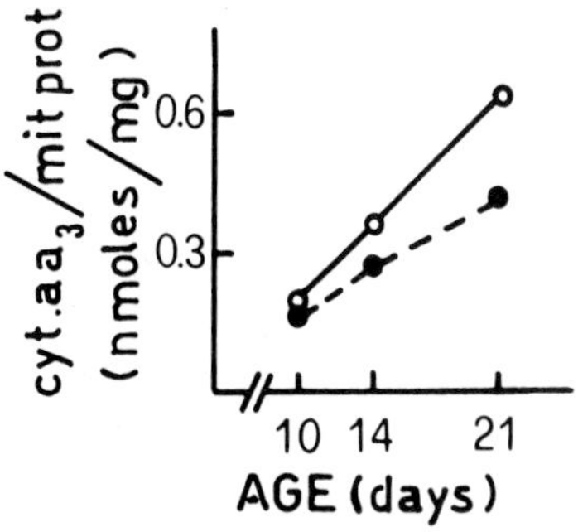

Fig.3 Cytochrome aa_3 in mitochondria isolated from the cerebellum of normal (o——o) and hypothyroid (●– – –●) rats.

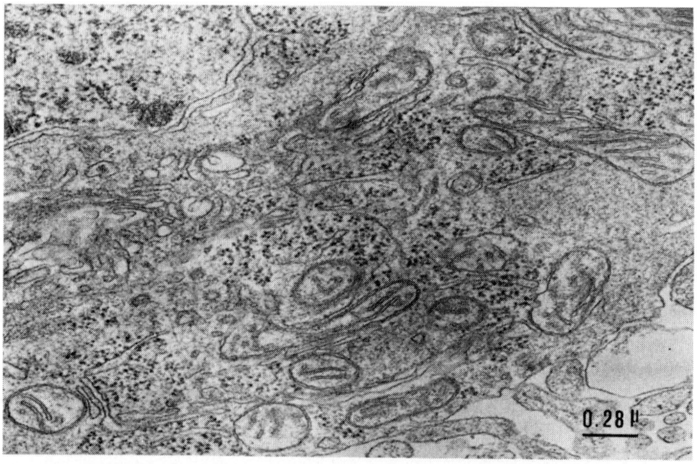

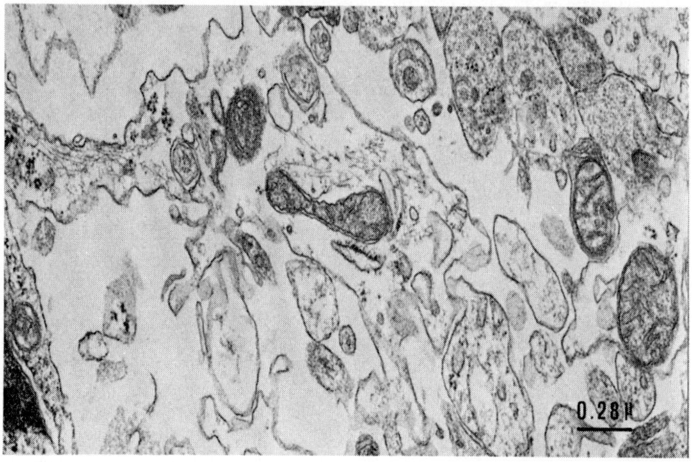

Fig.4 Ultrathin sections of cerebellum of normal (upper) and hypothyroid (lower)

rats at 21 days of age.

of these mitochondria. Electron micrographs of ultrathin sections of cerebellum of normal and hythyroid rats are reported in Fig.4. Mitochondria of hypothyroid rats appear to be internally disorganized with less regular and less extended cristae than mitochondria of normal rats. In conclusion, from the data above reported, it seems that both the biogenesis and differentiation of free mitochondria are impaired in the cerebellum of developing rats by thyroid hormones deficiency.

REFERENCES

1. Patel, A.J., Balazs R., Smith R.M., Kingsbury, A.E. and Hunt, A. (1980) Multidisciplinary approach to brain development (Di Benedetta C., Balazs, R. Gombos, G., Porcellati, G. eds), Elsevier/North-Holland, Amsterdam, pp.261-277.
2. Legrand, J. (1980) Multidisciplinary approach to brain development (Di Benedetta C., Balazs, R., Gombos G., Porcellati, G. eds.), Elsevier/North-Holland, Amsterdam, pp. 279-292.
3. Samson, F.E., Jr. Balfour, W.M. and Jacobs, R.J. (1960) Am. J. Physiol., 199, 693-696.
4. Pysh, J.J. (1970) Brain Research, 18-325-342.
5. Gadaleta, M.N. Giuffrida, A.M. Renis, M., Serra, I., Del Prete, G., Geremia, E., and Saccone C., (1979) Neurochem Res., 4,25-35.
6. Giuffrida, A.M., Gadaleta, M.N., Serra, I., Renis, M., Geremia, E., Del Prete, G., and Saccone, C. (1979) Neurochem Res. 4,37-52.
7. Tata, J.R. (1966) Regulation of metabolic processes in mitochondria. (Tager, J., Papa, S., Quagliariello, E., Slater, E.C., eds), Elsevier, Amsterdam, pp.489--517.
8. Hubscher, U., Kuenzle, C.C. and Spadari, S. (1977) Eur. J. Bioch., 81,249--258.
9. Nicholls, P. and Kimelberg, H.K. (1972) Biochemistry and Biophysics of mitochondrial membranes (Azzone, G.F. Carafoli, E., Lehninger, A.L., Quagliariello, E. and Siliprandi, N., eds), Academic Press, New York, pp. 17-32.
10. Spurr, A.R. (1969) J. Ultrastruct. Res., 26,31-43.
11. Reynolds, E.S. (1963) J. Cell. Biol., 17,208-222.

© 1980 Elsevier/North-Holland Biomedical Press
The Organization and Expression of the Mitochondrial Genome
A.M. Kroon and C. Saccone, editors

ASSEMBLY AND STRUCTURE OF CYTOCHROME OXIDASE IN NEUROSPORA CRASSA

Sigurd Werner*, Werner Machleidt*, Helmut Bertrand** and Gerd Wild*

*Institut für Physiologische Chemie, Physikalische Biochemie und Zellbiologie der Universität, Goethestraße 33, 8 000 München (Federal Republic of Germany) and **Department of Biology, University of Regina, Regina, Saskatchewan (Canada)

Cytochrome oxidase is one of the best structurally and biogenetically characterized components of the mitochondrial membrane system. Because of the special experimental advantages offered by a microorganism, such as *Neurospora crassa*, emphasis is usually placed on the biogenetic aspect in these investigations. Nevertheless, it has become obvious that a clear understanding of the process of enzyme assembly depends on a concommitant supply of structural information. The necessity of such a dual concept will be demonstrated in this communication at the level of precursor polypeptides of the oxidase.

Cytochrome oxidase from *Neurospora crassa* is most probably composed of eight polypeptide subunits[1]. In contrast to the enzyme isolated from various mammalian cells (in which the number of subunits is still under dispute – there may be as many as twelve[2,3]) the polypeptide components within the fungal complex are present in stoichiometric amounts. The homogeneity of the individual polypeptides isolated from the dissociated enzyme was seen by analyzing the material in different electrophoretic systems[4,5] and by the determination of N–terminal amino acid residues or amino acid sequences. A list of the amino termini obtained for subunits 1–8 is given in the table.

UNASSEMBLED SUBUNITS. The three largest polypeptides of the oxidase (1,2 and 3) are coded by the mitochondrial genome and are synthesized inside the organelle (for review see ref. [6] and [7]). Unassembled mitochondrial translation products for the oxidase ('free subunits') were isolated by immunological techniques from wild–type cells and were characterized by pool size and half–life[8].

TABLE

N–TERMINAL AMINO ACID RESIDUES OF CYTOCHROME OXIDASE SUBUNITS FROM NEUROSPORA CRASSA

Subunit	1	2	3	4	5	6	7[a]	8[a]
Apparent molecular weight (X 10^{-3} Daltons)	41	28.5	21	16	14	11.5	10[b]	10[b]
N–terminal amino acid	Ser	Asp	?[c]	Asn	Ala	Ala	Pro	Ser
Number of residues determined	15	40	–	15	40	1	1	1

[a] Previously termed as 7a and 7b.

[b] Partially resolved in SDS–PAGE. Subunit 8 was isolated by column chromatography.

[c] Blocked, but obviously not N–formylated.

These precursor polypeptides have been found to be associated with the inner mitochondrial membrane and display apparent molecular weights identical with those of the genuine enzyme subunits prepared from a functional enzyme complex[9]. Moreover, sequencing of the radioactively labelled 'unassembled subunit 1' revealed that incorporated [^{3}H]–isoleucine and [^{3}H]–tryptophane (see Fig. 1)were recovered in the same degradation steps as the corresponding amino acids of the unlabelled 'authentic subunit' from the holo–enzyme. Similar results were obtained from sequence studies of subunit 2 and its unassembled precursor with the same molecular weight[10]. These experiments supply strong evidence that the N–terminal sequences of these unassembled subunits are identical with those of the corresponding subunits isolated from the cytochrome oxidase complex.

ELONGATED PRECURSORS TO SUBUNITS. A 'subunit 1' polypeptide larger than the authentic subunit 1 of the wild–type enzyme was found in the mitochondria from three cytochrome oxidase–deficient mutants, the nuclear mutants cyt–2–1 and 299–1, and the cytoplasmic mutant mi–3[11]. The cross–reacting material bound to a specific antibody to

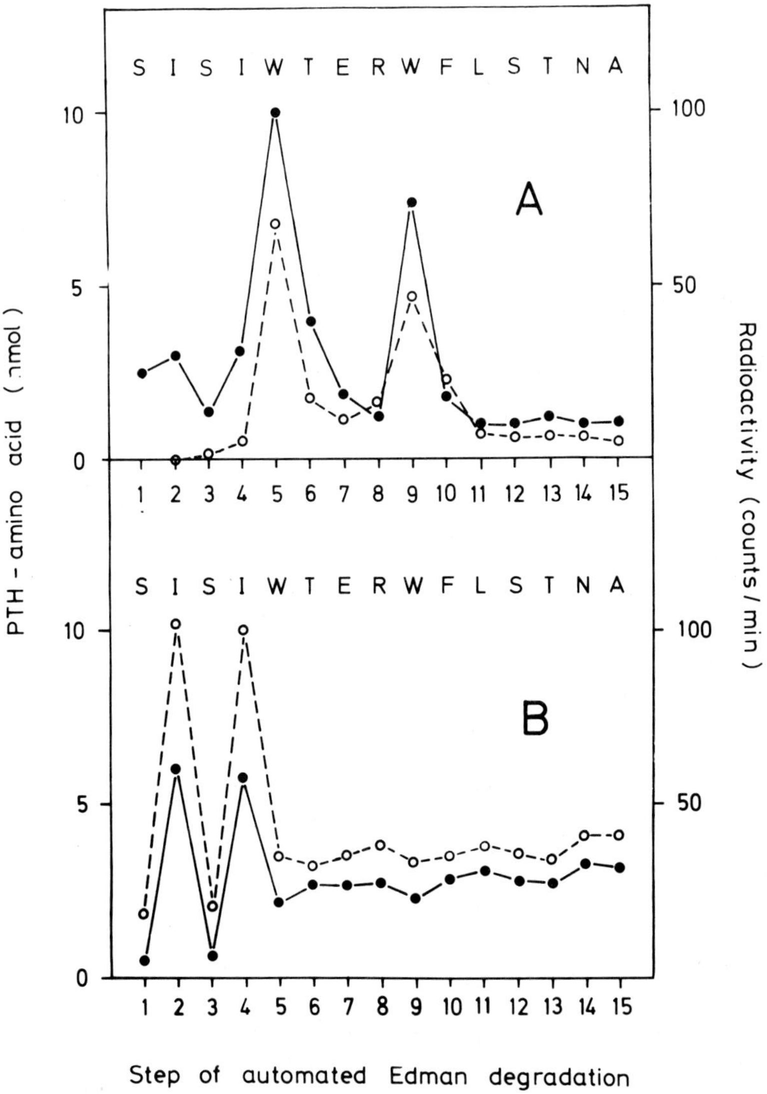

Fig. 1. N—terminal amino acid sequence analysis of the subunit 1 polypeptide from wild type: Yields of PTH—tryptophane (A) and PTH—isoleucine (B) obtained in solid—phase Edman degradation of the protein. PTH—yields (O--O) during the degradation of unlabelled subunit 1 isolated from the holo—enzyme were calculated from the peak areas of HPLC. Radioactivity (●—●) resulting from the degradation of unassembled subunit 1 labelled with [^{3}H]—tryptophane (A) and [^{3}H]—isoleucine (B), respectively, was determined in the HPLC eluates corresponding to these peaks. The amino acid sequence of subunit 1 is given in the upper part of the panels.

subunit 1 migrated more slowly on dodecylsulfate gels than the corresponding enzyme component. An apparent molecular weight of ca. 45 000 was determined for the mi-3 product, in contrast to the wild-type subunit 1 exhibiting a molecular weight of 41 000[12].

What is the biological significance of this larger product? Is it an 'incorrectly' translated polypeptide subunit, or is it a naturally occurring precursor to subunit 1, which is not transformed into the mature polypeptide in these mutants? The latter assumption could be verified by demonstrating that the larger molecular weight polypeptide is processed into a genuine subunit 1 of the oxidase.

This was indeed shown for the 45 000 molecular weight component present in the mi-3 mutant[12]. In this case, the experimental approach is greatly facilitated by the fact that the mi-3 mutant can be induced to synthesize a functional oxidase complex by adding small amounts of antimycin A to the medium[13]. Figure 2 gives an overview of the experimental procedure followed.

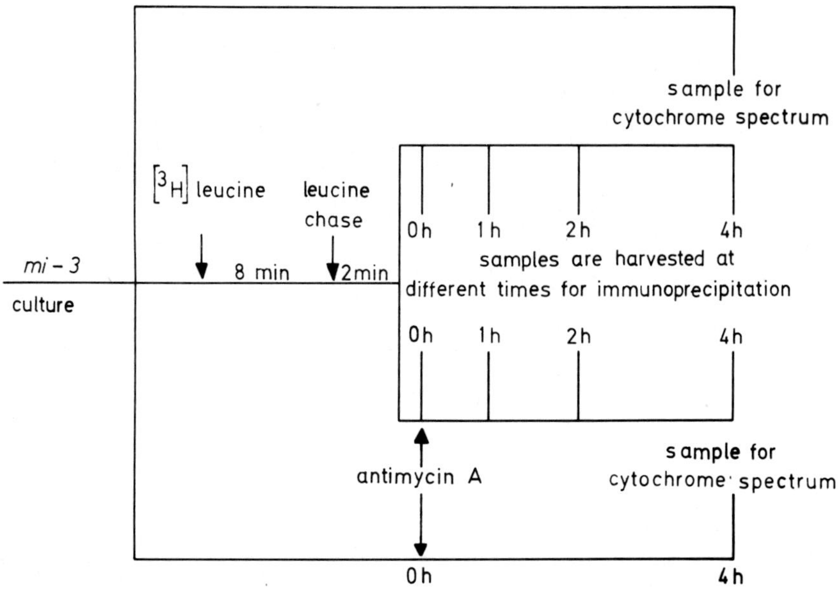

Fig. 2. Flow diagram of the experimental procedure applied to the mi-3 mutant cells to follow the processing of the 45 000 molecular weight mitochondrial product.

An exponentially growing mi-3 culture was initially divided into three portions. Two portions served as unlabelled control samples to demonstrate the induction effect on cytochrome oxidase by the antimycin treatment. The presence of cytochrome aa_3 in the cells was judged from the typical absorption spectra obtained with the samples. Cells of the third portion were pulse chase labelled with [^{3}H]-leucine, as indicated in figure 2. Aliquots were harvested after different times and they were subjected to immunological assays.

Figure 3 shows a gel electrophoretic analysis of the material isolated from the cells by means of a specific antibody to subunit 1. In mitochondria prepared from non-induced cells a polypeptide was found which migrates more slowly on the gel than the marker subunit from wild-type (panel A). Most of this pulse labelled material recognized by the antibody disappeared upon 4 hours of further incubation of the cells in the presence of antimycin. During this period, the formation of cytochrome c oxidase was induced in the mutant cells, as we know from the unlabelled control samples. Figure 4 represents the results of the immunological analysis for assembled cytochrome oxidase. Here, an antibody directed against subunits 4-8 was applied. This antibody is able to precipitate holo-cytochrome oxidase (if present), but does not recognize unassembled subunit 1. In non-induced cells (see panel A), only labelled material was found in the precipitate corresponding to cytochrome oxidase polypeptides 4-8. In induced cells (see panel B), however, radioactively labelled polypeptides comigrating exactly with all marker subunits of the wild-type enzyme were detected – including a large amount of a labelled polypeptide now corresponding to the 41 000 molecular weight subunit 1. Detailed kinetics[12] have shown that the radioactive label previously found in the unassembled larger protein appeared gradually in an assembled subunit 1 polypeptide upon enzyme induction. A typical precursor product relationship was observed.

In summary, the following conclusions have been drawn from the kinetic experiments: (1) The 45 000 molecular weight protein represents a genuine precursor to subunit 1 of the oxidase. (2) This pre-protein, present in the oxidase-deficient mutant, is not associated with other enzyme polypeptides. (3) Upon enzyme induction,

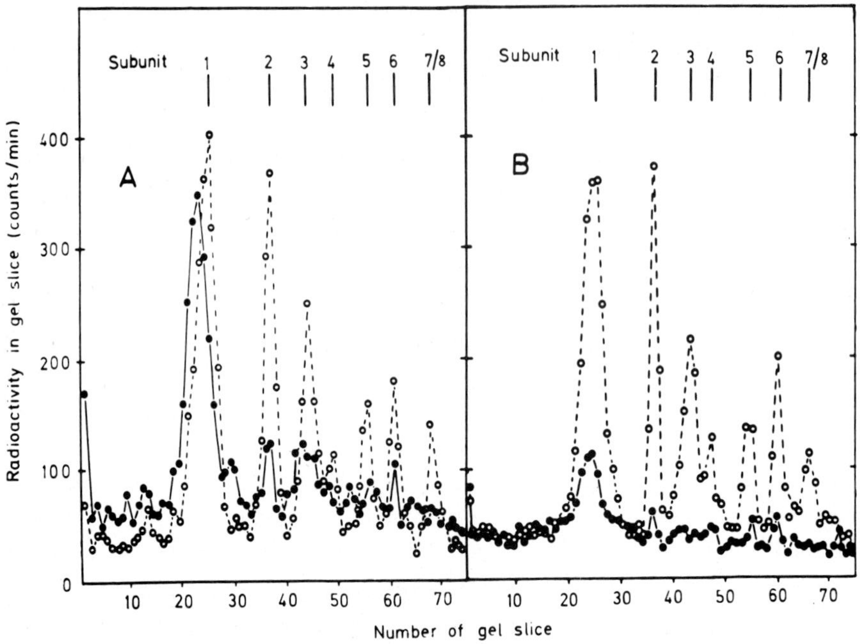
Subunit
1
2
3
4
5
6
7/8
A
B
Radioactivity in gel slice (counts/min)
400
300
200
100
0
10
20
30
40
50
60
70
Number of gel slice

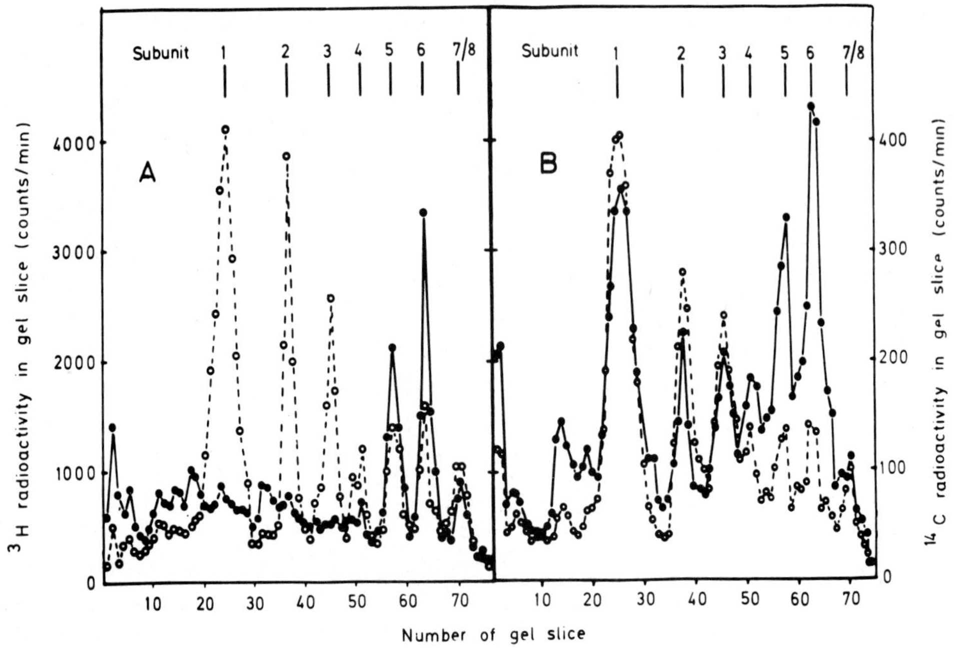
Subunit
1
2
3
4
5
6
7/8
A
B
3 H radioactivity in gel slice (counts/min)
4000
3000
2000
1000
0
14 C radioactivity in gel slice (counts/min)
400
300
200
100
0
10
20
30
40
50
60
70
Number of gel slice

Fig. 3 (upper part of preceding page). Electrophoretic analysis of polypeptides isolated from the mi–3 mutant by antibodies to subunit 1 of cytochrome oxidase. Cells of mi–3 mutant were pulse labelled by addition of [^{3}H]–leucine to the culture. The pulse label was chased with unlabelled leucine. Two min after the application of the chase the culture was divided into two portions. One portion was treated with antimycin A (0.25 µg/ml), the other portion was left untreated and the two cultures were incubated further. After various periods of incubation, samples of both cultures were withdrawn, the cells were harvested and mitochondria were prepared. One portion of each mitochondrial preparation was processed as follows: The protein was solubilized with Triton X–100 and the extract was treated successively with anti–subunit 1 immunoglobulin and protein A/Sepharose. Aliquots of the material released from the Sepharose support were mixed with [^{14}C]–leucine labelled subunits, obtained by immunoprecipitation of cytochrome oxidase from uniformly labelled wild–type cells. The mixtures were subjected to dodecylsulfate gel electrophoresis. The patterns shown are for material isolated from samples harvested (A) after incubation in the absence of antimycin and (B) after 4 h of incubation in the presence of antimycin. (●—●) ^{3}H–radioactivity, (O--O) ^{14}C–radioactivity.

Fig. 4 (lower part of preceding page). Electrophoretic analysis of polypeptides isolated from the mi–3 mutant by antibodies to subunits 5–8 of cytochrome oxidase. The second portion of the mitochondrial preparation, resulting from the procedure described in the legend to figure 3, was examined in the following way: A preparation was first mixed with wild–type mitochondria prepared from cells labelled uniformly with [^{14}C]–leucine. The mixture was solubilized with Triton X–100 and treated with an antiserum against subunits 5–8. The immunoprecipitates were dissolved an subjected to dodecylsulfate gel electrophoresis. The patterns shown are for material isolated from samples harvested (A) after incubation in the absence of antimycin and (B) after 4 h of incubation in the presence of antimycin. (●—●) ^{3}H–radioactivity, (O--O) ^{14}C–radioactivity.

the pre–protein converts into the mature subunit 1 which then participates in the assembly process to a cytochrome oxidase complex.

What are the structural differences between the pre–protein and the mature subunit? As judged by the electrophoretic mobilities of the polypeptides on the dodecylsulfate gel, an extension by about 30 to 40 amino acids should be expected.

The essential feature of the mature polypeptide is its 'open' sequence starting with serine. This is in contrast to the data obtained from the elongated precursor polypeptide. In this case, the N–terminus is 'blocked', and N–formyl–methionine was determined as the 'first' amino acid residue (see Fig. 5). Therefore, an

N–terminal elongation of the precursor polypeptide chain seems to be established. Moreover, structural differences of the processed and unprocessed form of the mitochondrial polypeptide can be shown in proteolytic 'fingerprint' patterns. The 41 000 and the 45 000 molecular weight components were labelled separately *in vivo* with different leucine isotopes, the isolated polypeptides were combined and subjected to digestion using the V 8 protease from *Staphylococcus aureus*. Figure 6 exhibits the separation of the cleavage products by dodecylsulfate gel electrophoresis (panel A) and by column chromatography (panel B). Using both techniques, significant differences in the peptide patterns of the elongated precursor and of the mature subunit have been observed.

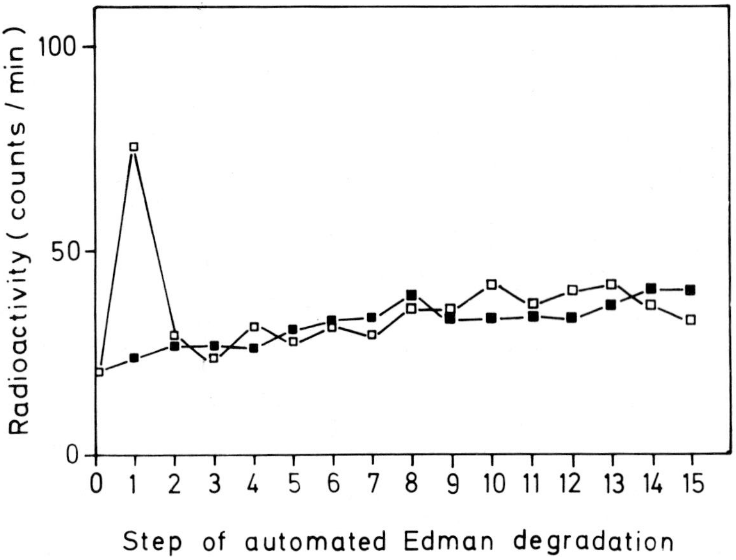

Fig. 5. Solid–phase Edman degradation of the labelled subunit 1 pre–protein isolated from the mutant mi–3. Radioactivity was determined in the HPLC eluates of PTH–methionine resulting from the degradation of the [^{35}S]–methionine labelled protein. Identical samples were degraded before (■—■) and after (□—□) deformylation in 1 N HCl/MeOH. Step 0 contains the radioactivity released by washing the reaction column with trifluoroacetic acid prior to the first sequencer cycle.

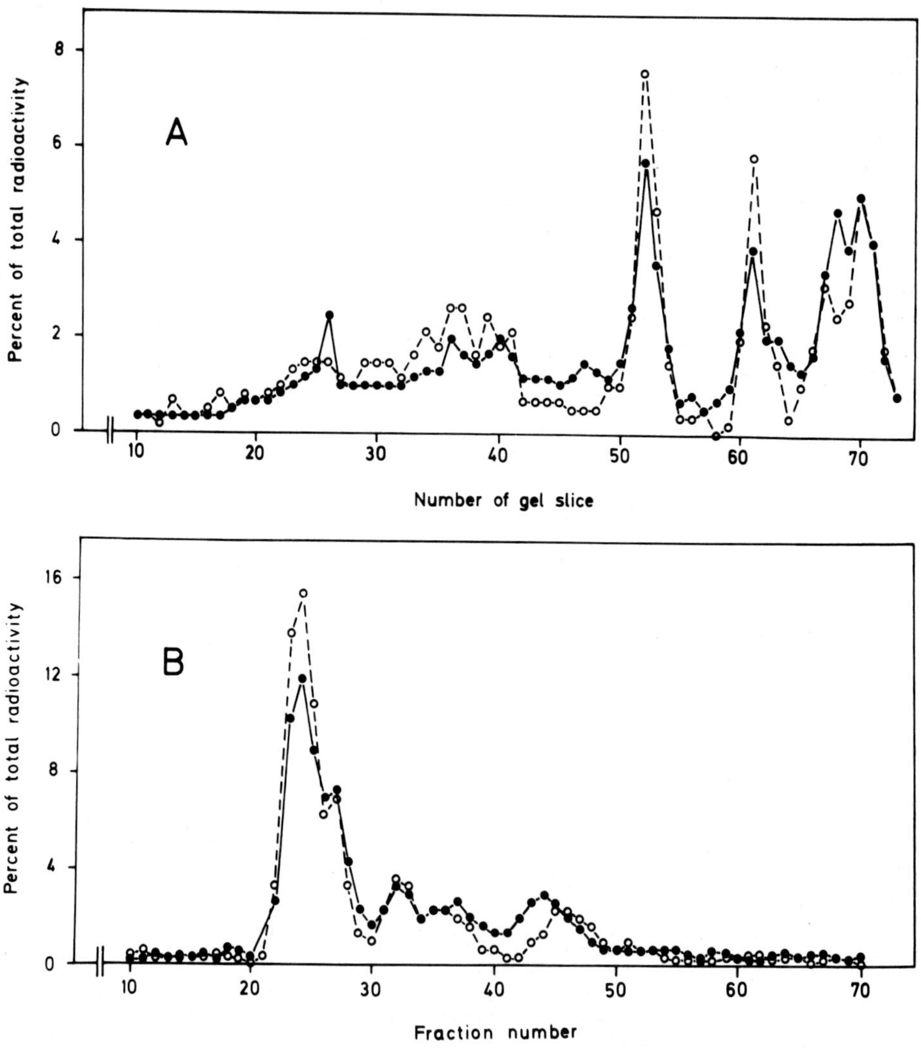

Fig. 6. Peptide patterns resulting from Staphylococcus aureus V 8 protease cleavage of unassembled subunit 1 from wild-type cells labelled with [^{14}C]-leucine (O--O) and of the pre-protein from the mi-3 mutant labelled with [^{3}H]-leucine (●—●). Both proteins were mixed prior to cleavage reaction. The cleavage products were analyzed by dodecylsulfate gel electrophoresis on a 15% polyacrylamide gel (A) and by column chromatography on LH-60 Sephadex eluted with 70% formic acid in n-propanol (B). Radioactivity of the fractions is given as percent of the total radioactivity recovered.

The occurrence of a larger precursor polypeptide seems not to be restricted to subunit 1 of the fungal cytochrome oxidase. There is indirect evidence that an N–terminally extended precursor for another mitochondrially synthesized enzyme subunit – namley subunit 2 – may exist: If one compares the N–terminal amino acid sequence of the beef heart subunit II[14] with that of the corresponding *Neurospora* polypeptide[10], a striking homology is found (12 out of 38 residue positions are identical), and the arrangement of the N–terminus suggests the existence of an elongated precursor protein for this cytochrome oxidase component of the fungus. This prediction has been strongly supported by data obtained from base sequencing of the corresponding mitochondrial DNA fragment from yeast[15]. In figure 7, a convincing 'assembly' of these three lines of information has been attempted.

Yeast DNA   F-Met-Leu-Asp-Leu-Leu-Arg-Leu-Gln-Leu-Thr-Thr-Phe-Ile-Met-Asn-

Yeast DNA   Asp-Val-Pro-Thr-Pro-Tyr-Ala-Cys-Tyr-Phe-Gln-Asp-Ser-Ala-Thr-
1   10
Neurospora   Asp-Ala-Pro-Ser-Pro-Trp-Gly-Ile-Tyr-Phe-Gln-Asp-Ser-Ala-Thr-
Beef heart   F-Met-Ala-Tyr-Pro-Met-Gln-Leu-Gly-Phe-Gln-Asp-Ala-Thr-Ser-

Yeast DNA   Pro-Asn-Gln-Glu-Gly-Ile-Leu-Glu-Leu-His-Asp-Asn-Ile-Met-Phe-
16   25
Neurospora   Pro-Gln-Met-Glu-Gly-Leu-Val-Glu-Leu-His-Asp-Asn-Ile-Met-Asp-
Beef heart   Pro-Ile-Met-Glu-Glu-Leu-Leu-His-Phe-His-Asp-His-Thr-Leu-Met-

Yeast DNA   Tyr-Leu-Leu-Val-Ile-Leu-Gly-Leu-Val-
31
Neurospora   Tyr-Leu-Val-Val-Ile-Leu-Phe-Gly-Val-
Beef heart   -Ile-Val-Phe-Leu-Ile-Ser-Ser-Leu-Val-

Fig. 7. N–terminal amino acid sequences of subunit 2 of cytochrome oxidase. The data is obtained from the base sequence of the corresponding DNA fragment from yeast[15], and from amino acid sequencing of the enzyme subunit isolated from Neurospora[10] and beef heart[14].

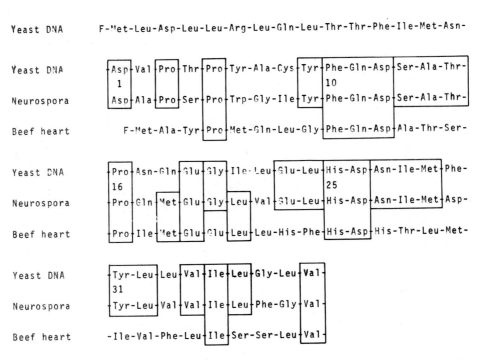

Very recently, the idea of the existence of a larger precursor of subunit 2 of cytochrome oxidase occurring in lower eukaryotic cells has been substantiated in yeast, where a polypeptide exhibiting a molecular weight approximately 1 500 daltons larger than the mature subunit has been immunologically identified[16]. No data is available to date concerning either a possible extended precursor for the third mitochondrially synthesized component of cytochrome oxidase, or for further mitochondrial translation products of other protein complexes.

What could be the biological significance of a pre-piece as observed for the oxidase subunit 1, for example? Since the protein is made inside the mitochondrion and therefore needs no transport across membranes, one is tempted to speculate that the extending portion of the polypeptide chain may be involved in the assembly process and/or in the orientation of the oligomeric protein complex within the membrane.

Is there any evidence for the existence of extended precursor proteins concerning the *cytoplasmically* synthesized subunits of the *Neurospora* cytochrome oxidase? It has been shown in yeast that several of those mitochondrial membrane proteins imported into the organelle – including subunits V and VI of the oxidase – are initially translated in the form of precursors which are larger by about 2 000 to 6 000 daltons compared with authentic components isolated from membrane complexes[17,18]. In order to find out, whether or not elongated precursors to cytosolic oxidase subunits are involved in the formation of the fungal enzyme, polypeptides were synthesized in a wheat germ cell-free system programmed with *Neurospora* poly(A)-containing RNA (details will be presented elsewhere). The proteins synthesized in this heterologous system were analyzed for cytochrome oxidase components with antisera to subunits 6 and 7/8, respectively. Preliminary experiments have shown that in both cases three different protein species can be identified with the antibodies: Two polypeptides migrating upon gel electrophoresis exactly with the marker subunits 6 and 7/8, and in addition one relatively large polypeptide displaying an apparent molecular weight of about 39 000. While the co-precipitation of the two small molecular weight polypeptides could be explained by immunological cross-reactions, the specific isolation of the larger product remains unclear.

Common features of *in vitro* translation products, like those described above, with authentic cytochrome oxidase polypeptides may be demonstrated most convincingly on a structural basis. A prerequisite for this is the determination of amino acid sequences of the oxidase subunits. Examples of N-terminal sequences are given for subunit 4 and 5 (see Fig. 8). It is somewhat surprising that the N-terminal sequence of subunit 4 from *Neurospora* (15 residues determined) is not homologous to any of the cytosolic subunits of the beef heart enzyme. Possibly, the cytoplasmically-made polypeptides of the oxidase complex have been diversified during evolution to a greater degree than the mitochondrial products. On the other hand, the N-terminal amino acid sequence of

```
             Subunit 4
                                                       10
Neurospora   Asn-Ala-Glu-Thr-Lys-Pro-Val-Pro-His-Ile-Lys-Arg-Leu-Ser-Glu-

             Subunit 5
                             5                 10                  15
Neurospora   Ala-Ser-Thr-Met-Pro-Ile-|Ser|-Asn-Pro-|Thr|-Leu-Ala-Asn-Ile-Glu-
Beef heart               Ser-His-Gly-|Ser|-His-Glu-|Thr|-Asp-Glu-Glu-Phe-Asp-

                            20                  25                    30
Neurospora   -Lys-|Arg-Trp|-Glu-Gln-Met-Pro-Met-Gln-Glu-Gln-Ala-Glu-Leu-|Trp|-
Beef heart   -Ala-|Arg-Trp|-Val-Thr-Tyr-Phe-Asn-Lys-Pro-Asp-Ile-Asp-Ala-|Trp|-

                            35                  40
Neurospora   -Met-Ala-|Leu-Arg|-Asp-Arg-|Met|-Lys-Gly-Asn-
Beef heart   -Glu-    |Leu-Arg|-Lys-Gly-|Met|-Asn-Thr-Leu-
```

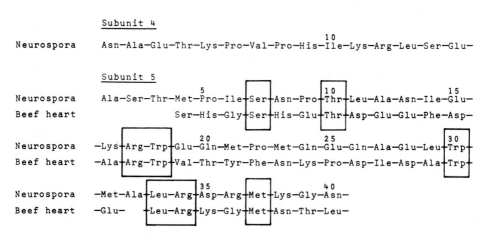

Fig. 8. N-terminal amino acid sequences of the cytoplasmically-made subunits 4 (MW 16 000) and 5 (MW 14 000) as obtained by automated solid-phase Edman degradation of the polypeptides. The N-terminal sequence of subunit V (MW 12 500) from beef heart[19] is included for comparison.

subunit 5 (40 residues determined) shows a faint, but significant homology to the corresponding portion of subunit V from the beef heart enzyme, which has been suggested to be carrying one of the heme groups[19]. Definite conclusions may be drawn, when longer sequences for the *Neurospora* polypeptides are available.

ACKNOWLEDGEMENTS

The authors are grateful to Ms. R. Zauner, Ms. H.M. Rothe, Ms. U. Borchart and Ms. G. Behrens whose technical expertise was essential to this work. The research was supported by grants from the Deutsche Forschungsgemeinschaft (We 706), the Sonderforschungsbereich 51 (B 31), the National Research Council of Canada (A–6351) and the North Atlantic Treaty Organization (NATO).

REFERENCES

1. Werner, S. (1977) Eur. J. Biochem. 79, 103–110
2. Buse, G., Steffens, G.J., Steffens, G.C.M. & Sacher, R. (1978) Frontiers of Biological Energetics (Button, P.L., Leigh, L. & Scarper, A. eds.), Academic Press, New York, Vol. 2, pp. 799–807
3. Merle, P. & Kadenbach, B. (1980) Eur. J. Biochem. 105, 499–507
4. Sebald, W., Machleidt, W. & Otto, J. (1973) Eur. J. Biochem. 38, 311–324
5. Merle, P., Werner, S. & Kadenbach, B. (1980) J. Biol. Chem. submitted
6. Borst, P. & Grivell, L.A. (1978) Cell 15, 705–723
7. Tzagoloff, A., Macino, G. & Sebald, W. (1979) Ann. Rev. Biochem. 48, 419–441
8. Weiss, H., Schwab, A.J. & Werner, S. (1975) Membrane Biogenesis (Tzagoloff, A. ed.) Plenum Press, New York, pp. 125–153
9. Werner, S., Schwab, A.J. & Neupert, W. (1974) Eur. J. Biochem. 49, 607–617
10. Machleidt, W. & Werner, S. (1979) FEBS Lett. 107, 327–330
11. Bertrand, H. & Werner, S. (1979) Eur. J. Biochem. 98, 9–18
12. Werner, S. & Bertrand, H. (1979) Eur. J. Biochem. 99. 463–470
13. Bertrand, H. & Collins, R.A. (1978) Mol. Gen. Genet. 166, 1–13
14. Steffens, G.J. & Buse, G. (1979) Hoppe Seyler's Z. Physiol. Chem. 360, 613–619
15. Coruzzi, G. & Tzagoloff, A. (1979) J. Biol. Chem. 254, 9324–9330
16. Sevarino, K.A. & Poyton, R.O. (1980) Proc. Natl. Acad. Sci. USA 77, 142–146
17. Schatz, G. (1979) FEBS Lett. 103, 203–211
18. Lewin, A.S., Gregor, I., Mason, T.L., Nelson, N. & Schatz, G. (1980) Proc. Natl. Acad. Sci. USA submitted
19. Tanaka, M., Haniu, M., Yasunobu, K.T., Yu, C.A., Yu, L., Wei, Y.H. & King, T.E. (1979) J. Biol. Chem. 254, 3879–3885

© 1980 Elsevier/North-Holland Biomedical Press
The Organization and Expression of the Mitochondrial Genome
A.M. Kroon and C. Saccone, editors

POSTTRANSLATIONAL TRANSPORT OF PROTEINS IN THE ASSEMBLY OF MITOCHONDRIAL MEMBRANES

E.-M. NEHER, M.A. HARMEY, B. HENNIG, R. ZIMMERMANN AND W. NEUPERT
Institut für Physiologische Chemie, Universität Göttingen, Humboldtallee 7, 3400 Göttingen, Germany

INTRODUCTION

Assembly of the mitochondrion involves the transfer of a large number of proteins from the cytosol to the various subcompartments of this organelle. During the last years we have accumulated a body of evidence that this transfer occurs by a posttranslational mechanism [1-9]. Such a mechanism implies the existence of extramitochondrial pools of precursor proteins. In order to understand the transfer on a molecular basis, the overall reaction must be dissected into a number of induvidual steps. Furthermore, the signals on the precursors and the complementary structures on mitochondria which are responsible for the specificity of intracellular traffic must be investigated. A number of representative proteins should be studied to assess whether differences in the assembly pathway exist for the various proteins, e.g. soluble vs. integral membrane proteins, matrix vs. intermembrane proteins. We report here on the transport of cytochrome c, a peripheral membrane protein at the c-side of the inner membrane, on that of the ADP/ATP carrier, an integral transmembrane protein of the inner membrane [10], and on that of "subunit 9" of the ATPase complex, also an integral protein of the inner membrane.

RESULTS AND DISCUSSION

CYTOCHROME C. A peculiar step in the biosynthesis of cytochrome c is the covalent attachement of the haem group to the apoprotein. The question whether the haem group is added before or after completion of the polypeptide chain, was answered in the following way. Neurospora cells were first labelled with ^{3}H leucine and were then pulse labelled with ^{35}S methionine at $8^{o}C$. At various times after the pulse cells were frozen in liquid N_2, disrupted and extracted. Antibodies specific for Neurospora apocytochrome c and holocytochrome c, respectively, were employed to immunoprecipitate the two components from each of the samples. Fig. 1 shows the labelling kinetics of apo- and holocytochrome c. This data demonstrates that apocytochrome c is present in the cells and suggests a precursor-product relationship between apo- and holocytochrome c.

The haem group is linked in a reaction with occurs posttranslationally.

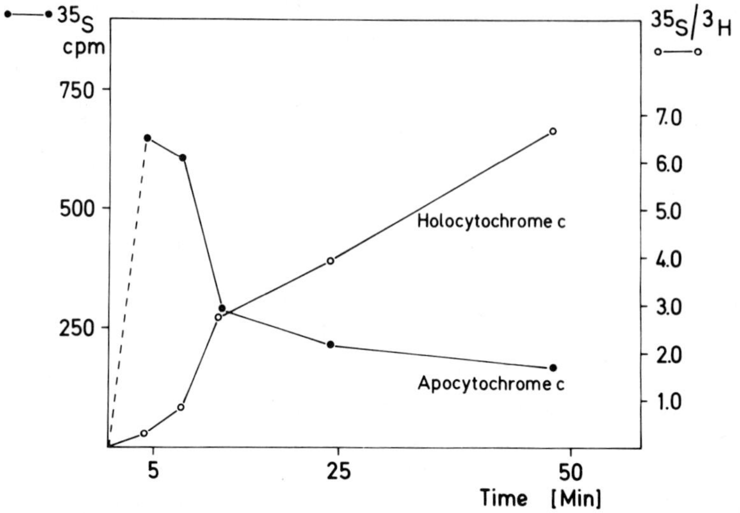

Fig. 1. Kinetics of labelling of apo- and holocytochrome c in Neurospora cells.

Neurospora cells were labelled with ^{3}H leucine after 12h growth. After further 2h cells were cooled to 8°C and after further 1h ^{35}S methionine was added to the culture. At the time points indicated, aliquots were withdrawn, the cells rapidly harvested, frozen in liquid N_2, broken by grinding and extracted with Triton containing buffer. Then from each sample apo- and holocytochrome c were immunoprecipitated with specific antibodies. Immunoprecipitates were analysed by SDS gel electrophoresis and radioactivities in the cytochrome c peaks determined.

In order to decide whether the apocytochrome c found in this experimental system is the primary translation product or a component already processed, apocytochrome c was translated in cell free heterologous systems. The in vitro product had the same size as the isolated apocytochrome c, and the amino terminal sequence was identical to that of holocytochrome c [5]. This is in agreement with recent results on the DNA sequence of the coding region for iso-1-cytochrome c from yeast [11].

To elucidate the intracellular pathway of apocytochrome c, the relation between synthesis and haem incorporation was investigated in reconstituted systems. For this purpose protein synthesis was carried out in homologous and heterologous cell free systems, then postribosomal supernatants were prepared and incubated with isolated mitochondria. Fig. 2 shows that apocytochrome c,

Fig. 2. Protease resistance of cytochrome c transferred into mitochondria in vitro.

Neurospora poly(A)RNA was translated in a reticulocyte lysate in the presence of ^{35}S methionine and the postribosomal supernatant prepared. Two samples of this supernatant were incubated with mitochondria (0.5 and 1 mg protein per ml, respectively) for 60 min. Then mitochondria were reisolated by centrifugation and resuspended in sucrose buffer. One half of each sample was treated with proteinase K for 60 min at 0-4°. Then PMSF was added to all samples and they were lysed with 1% Triton. Immunoprecipitation was carried out with antibody against holocytochrome c. Immunoprecipitates were subjected to SDS gel electrophoresis and autoradiography. Arrow indicates position of stained holocytochrome c. Proteinase K was shown in a seperate experiment to digest holocytochrome c in solution under the conditions applied here.
Lanes 1 and 3: 0.5 mg mitochondrial protein per ml; lanes 2 and 4: 1 mg mitochondrial protein per ml; lanes 1 and 2: control; lanes 3 and 4: treated with proteinase K.

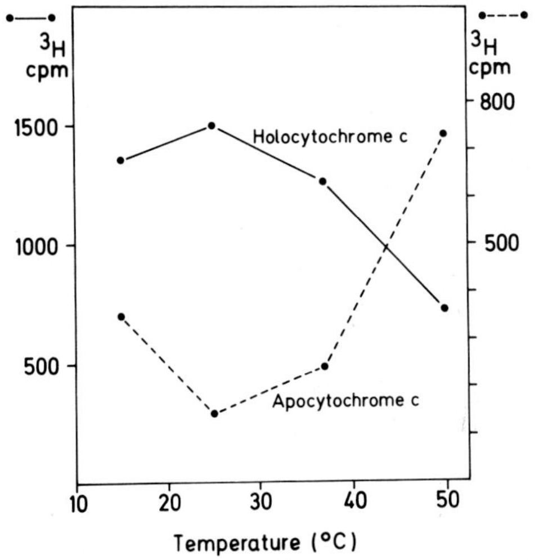

Fig. 3. Temperature dependence of the formation of holocytochrome c from apocytochrome c in a reconstituted system.

A postmitochondrial supernatant of a Neurospora homogenate was incubated with ^{3}H leucine for 10 min. Then a 1h 150 000 x g supernatant was prepared and incubated for 30 min with mitochondria isolated from unlabelled cells at the temperatures indicated. After incubation, the mixtures were lysed with Triton and divided into two equal portions. Immunoprecipitation with antibodies specific for apo- and holocytochrome c was performed, immunoprecipitates were analysed by SDS gel electrophoresis and radioactivities in the cytochrome c peaks determined.

present in the supernatant of a reticulocyte lysate programmed with Neurospora poly(A)RNA is transferred into added isolated Neurospora mitochondria and converted to holocytochrome c. The newly formed holocytochrome c is resistant to added protease. This suggests translocation across the outer mitochondrial membrane. The same observation was made when a supernatant of a homologous cell free system was employed.

Fig. 3 shows the temperature dependence of the apo- to holocytochrome c conversion. There is an optimum at about 25°C. Furthermore, this experiment gives a quantitative evaluation, indicating that the conversion occurs with a high efficiency. More than 90% of the apo form is converted to the holoprotein at optimal temperature.

Intact mitochondria are a prerequisite for linkage (incorporation) of the haem moiety with the apoprotein. Addition to the supernatant of haemin chloride, of detergent-lysed mitochondria, sonicated mitochondria or hypotonically preswollen mitochondria does not lead to the formation of holocytochrome c (Fig. 4). Excess apocytochrome c from Neurospora but not excess holocytochrome c can compete for the transfer and conversion of apocytochrome c synthesized in the cell free system.

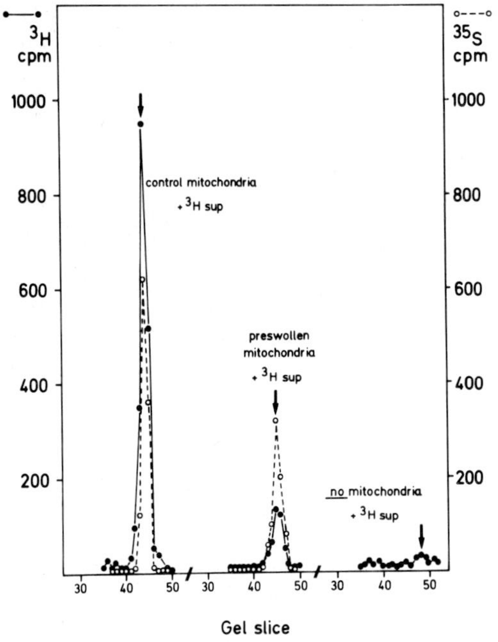

Fig. 4. Dependence of transfer in vitro of cytochrome c on intactness of isolated mitochondria.

A postmitochondrial supernatant of a Neurospora homogenate was incubated with ^3H leucine for 10 min. Then a 1h 150 000 x g supernatant was prepared and incubated a) with mitochondria isolated from cells grown on ^{35}S sulfate, b) with mitochondria as in a, but preswollen with 10 mM Tris HCl, c) without added mitochondria, but with haemin chloride (30 μM). After incubation for 30 min, Triton was added, immunoprecipitation and SDS gel electrophoresis of immunoprecipitates were carried out. Gels were sliced and ^3H and ^{35}S radioactivities determined.
Arrow indicates position of coelectrophoresed holocytochrome c on the gel.

On the basis of this data we propose the following assembly pathway for cytochrome c. Apocytochrome c, synthesized on free polysomes is released into the cytosolic compartment. Its conformation is such that it penetrates the outer membrane with part of the molecule through a pore. An enzyme which catalyses the formation of the thioether bridge between apocytochrome c and haem acts in the intermembrane space. The addition of the prosthetic group triggers the folding of the polypeptide chain in such a way, the molecule is completely translocated across the outer membrane and bound to its functional site on the inner membrane.

ADP/ATP CARRIER

The synthesis of the ADP/ATP carrier protein can be observed in homologous and heterologous cell free systems. The translation product (apparent molecular weight 32 000) in all cases has the same electrophoretic mobility on SDS polyacrylamide gels as the authentic protein in the inner membrane. The in vitro translation product was found in the postribosomal supernatant. Analysis by sucrose density gradient centrifugation and gel filtration showed that the in vitro product occurs in the form of a higher molecular weight complex and that it interacts with detergents.

Reconstitution experiments were carried out to demonstrate the transmembranous transfer in vitro. Reticulocyte lysates were programmed with Neurospora mRNS and the postribosomal supernatant was separated after translation. Mitochondria were isolated from Neurospora spheroplasts. Supernatant and mitochondria were incubated for various time periods, then they were separated again by centrifugation. The mitochondrial samples were divided into two equal portions. One part remained untreated and the other half was treated with proteinase K at $0^{o}C$. The latter treatment leads to digestion of extramitochondrial ADP/ATP carrier but not, or only to a limited degree, of the carrier in the intact mitochondria. ADP/ATP carrier was immunoprecipitated from the supernatants, and from proteinase treated and control mitochondria. The immunoprecipitates were analysed by SDS gel electrophoresis, and yielded one single band (32 K) by autoradiography. The X-ray films were subjected to densitometry and the extinction of the 32 K band was plotted vs. the time of incubation of mitochondria with supernatant. The protein is rapidly bound to the mitochondria (Fig. 5). In studies, in which the radioactivity was determined in sliced gels, 70 - 90% of the in vitro synthesized carrier was found to be bound to the mitochondrial fraction after 60 min incubation. The appearance of protease resistant carrier is also shown in Fig. 5. It makes up ca. 20 - 30% of the total carrier synthesized in vitro.

These in vitro transfer experiments were also made in the presence of carbonylcyanide-m-chlorophenyl-hydrazone (CCCP). As can be seen from Fig. 5, CCCP does not inhibit the binding but it does inhibit transfer of the carrier into a protease resistant position. A similar effect was observed, when the temperature was lowered to 0 - 4^{0}C. Inhibition of transfer of ADP/ATP carrier by CCCP in intact cells has already been reported[7]. Furthermore, with yeast cells it was found that CCCP does not inhibit the synthesis but the proteolytic processing of a number of mitochondrial proteins which are synthesized as larger precursors[12].

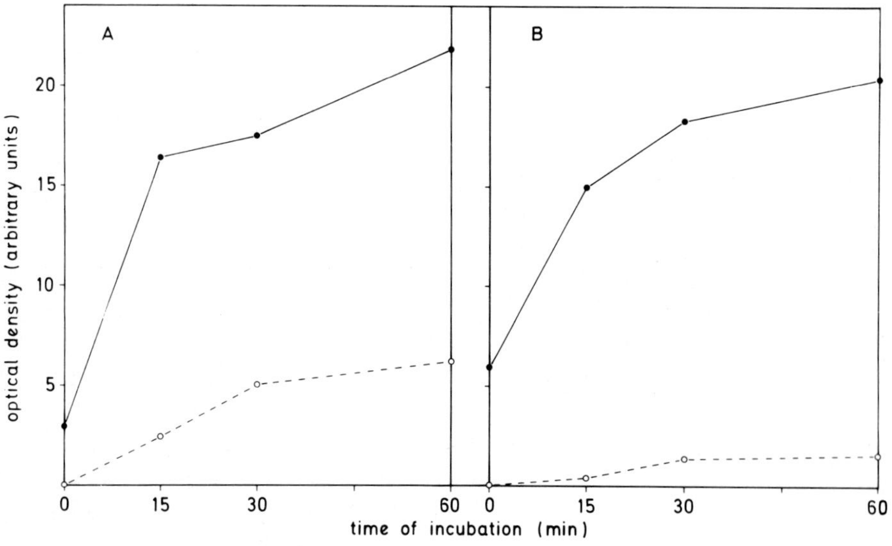

Fig. 5. Transfer of ADP/ATP carrier into mitochondria in vitro.

A reticulocyte lysate was programmed with Neurospora poly(A)RNA and, after incubation of ^{35}S methionine, the postribosomal supernatant was prepared. Mitochondria isolated from Neurospora spheroplasts were resuspended in the supernatant. After incubation for 60 min at 25°C, supernatant and mitochondria were separated again by centrifugation. One half of the mitochondria were treated with proteinase K at 0°C for 60 min. In a parallel experiment, incubation of supernatant with mitochondria was carried out after addition of 12.5 µM CCCP. Then Triton was added, the ADP/ATP carrier was immunoprecipitated from all fractions, the immunoprecipitates were electrophoresed, the gels autoradiographed and the X-ray films subjected to densitometry.
A: without CCCP; B: with CCCP. -●—●- : total ADP/ATP carrier bound to mitochondria; -o--o-:proteinase K resistant ADP/ATP carrier in mitochondria.

Isolation of the ADP/ATP carrier from the membrane includes as the most important step passage of Triton-solubilized mitochondria over hydroxyapatite[10]. Only a few mitochondrial proteins pass through this bed, the major component is the ADP/ATP carrier. When the postribosomal supernatant of a reticulocyte lysate containing newly synthesized carrier and supplemented with Triton was passed over hydroxyapatite, the ADP/ATP carrier protein was found to be completely retained (Fig. 6). However, when mitochondria after transfer in vitro were lysed and subjected to the same procedure, part of the carrier was detected in the eluate. This finding supports the view that part of the carrier after its interaction with the mitochondria, is actually integrated into the membrane.

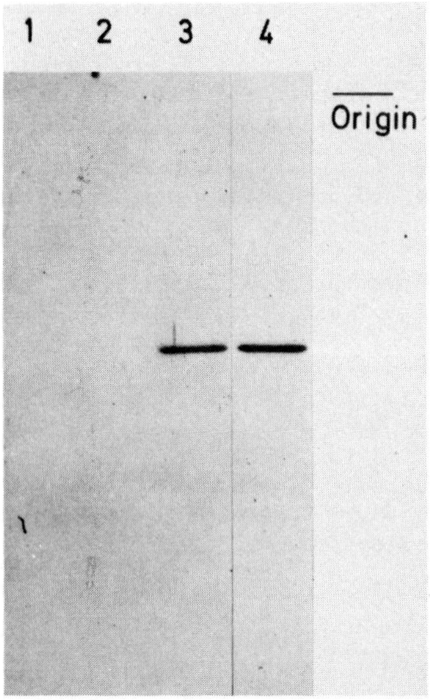

Fig. 6. Chromatography of ADP/ATP carrier on hydroxyapatite before and after transfer in vitro.

After translation of poly(A)RNA in a reticulocyte lysate in the presence of ^{35}S methionine, one aliquot of the postribosomal supernatant was made 1% Triton and passed over hydroxyapatite (lane 2). To a second aliquot, Carboxyatractyloside (CAT) was added and the sample was treated in the same way as in 2 (lane 1). A third aliquot was incubated with isolated mitochondria, mitochondria were reisolated and incubated with CAT; then one half was treated with proteinase K (lane 3) one half remained untreated (lane 4). Mitochondria were solubilized and passed over hydroxyapatite. The eluates were subjected to immunoprecipitation; the immunoprecipitates were analysed by electrophoresis and autoradiography.

These results show that there is an extramitochondrial (cytosolic) precursor form of the ADP/ATP carrier which is transferred into mitochondria in vitro. They clearly confirm earlier conclusions derived from in vivo experiments, that the intracellular translocation of this protein occurs via a posttranslational mechanism[1,7]. They further show that integral membrane proteins can have precursors existing in the cytosol compartment without a "signal extension" in their amino acid sequence which is cleaved upon membrane insertion. The complicated assembly pathway can be divided into at least two steps: binding to the outer membrane and integration into the inner membrane. We envision that the first step entails a "receptor" type molecule at the surface of the outer membrane. The second, rather complex, step is apparently dependent on energy, e.g. in the form of a membrane potential.

"SUBUNIT 9" OF ATPASE COMPLEX

This integral membrane protein is synthesized as a precursor possessing a larger apparent molecular weight, as first shown by translation of Neurospora poly(A)RNA in a wheat germ extract[13]. Fig. 7 shows a comparison of the protein isolated from the membrane with that obtained by translation in a reticulocyte lysate. The apparent molecular weight difference is ca. 6 000. "Subunit 9" shares the property of being synthesized as a larger precursor with a number of other mitochondrial proteins [9, 14-18].

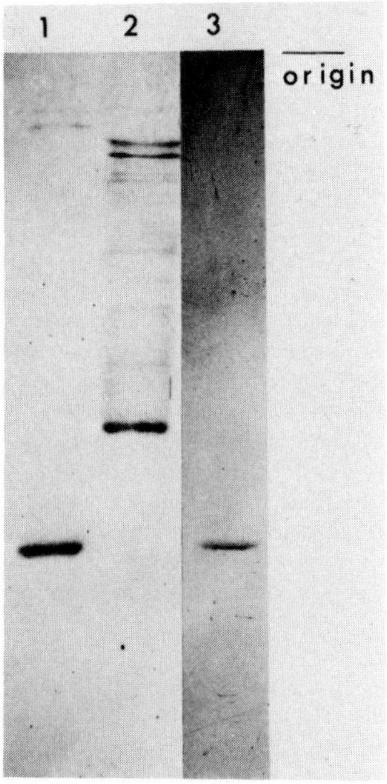

Fig. 7. Synthesis in vitro of "subunit 9" of ATPase complex and transfer in vitro to mitochondria.

A reticulocyte lysate was incubated with poly(A)RNA and ^{35}S methionine for 60 min and the postribosomal supernatant was prepared. From an aliquot "subunit 9" was immunoprecipitated (lane 2). A second aliquot was incubated with isolated mitochondria, then "subunit 9" immunoprecipitated from the reisolated mitochondria (lane 3). "Subunit 9" was also immunoprecipitated from mitochondria isolated from cells grown on ^{35}S sulfate (lane 1). Immunoprecipitates were analysed by SDS gel electrophoresis and autoradiography.

We have also evaluated for this precursor, where and how it is present in the reticulocyte lysate. Similar to the ADP/ATP carrier it was found in the postribosomal supernatant, not as a monomer but with an apparent size larger or comparable to haemoglobin. In vitro transfer was also accomplished by incubating the reticulocyte supernatant with isolated mitochondria. Not only was "subunit 9" then found associated with mitochondria, it was also processed to the size of the authentic membrane protein (Fig. 7).

CONCLUSIONS

The three proteins investigated appear to have some common characteristics with regard to their assembly pathway, but they also show significant differences. A first common feature is that they are translocated by a posttranslational mechanism; a second one is that the extramitochondrial precursors differ in their properties, in particular in their conformation, from the functional products; a third one is that the extramitochondrial precursors pass through the cytosolic compartment.

The type of conformational change which leads to the final product apparently differs among the three proteins. In case of cytochrome c it is the covalent linkage of the haem group which leads to a drastic refolding of the molecule; in case of the ADP/ATP carrier refolding may occur in the step in which the precursor, bound to the outer membrane, is inserted into the inner membrane; and "subunit 9" probably experiences a conformational change when the additional sequence is cleaved. An additional sequence is not necessary for the construction of a precursor molecule; and it can be concluded that the signalling device which directs a precursor protein to its organelle may lie in the tertiary structure ("signal structure"). Those proteins which do have a larger precursor would then contain a "signal structure" which may or may not reside in the additional sequence.

Thus, in several respects posttranslational transfer into mitochondria (and chloroplasts [19]) appears to differ from cotranslational membrane transfer of secretory proteins according to the "signal hypothesis"[20].

ACKNOWLEDGEMENTS

We want to thank Heidi Bliedung, Sabine Krull, Sabine Pitzel and Leopoldine Sander for skilful technical assistance. This work was supported by the Deutsche Forschungsgemeinschaft, Ne 101/17.

REFERENCES

1. Hallermayer, G. et al. (1977) Eur. J. Biochem. 81, 523-532.
2. Harmey, M.A. et al. (1977) Eur. J. Biochem. 81, 533-544.
3. Korb, H. & Neupert, W. (1978) Eur. J. Biochem. 91, 609-620.
4. Zimmermann, R. et al. (1979) Eur. J. Biochem. 99, 247-252.
5. Zimmermann, R. et al. (1979) FEBS Lett. 108, 141-151.
6. Zimmermann, R. & Neupert, W. (1980) Eur. J. Biochem. in press.
7. Hallermayer, G. & Neupert, W. (1976) Genetics and Biogenesis of Chloroplasts and Mitochondria, North-Holland, Amsterdam, pp. 807-812.

8. Harmey, M.A. et al. (1976) Genetics and Biogenesis of Chloroplasts and Mitochondria, North-Holland, Amsterdam, pp. 813-818.

9. Harmey, M.A. & Neupert, W. (1979) FEBS Lett. 108, 385-389.

10. Klingenberg, M. (1976) The Enzymes of Biological Membranes: Membrane Transport, vol. 3., Plenum Publishing Corp., New York, pp. 383-438.

11. Smith, M. et al. (1979) Cell 16, 753-761.

12. Nelson, N. & Schatz, G. (1979) Proc. Natl. Acad. Sci. U.S.A. 76, 4365-4369.

13. Michel, R. et al. (1979) FEBS Lett. 101, 373-376.

14. Maccechini, M.-L. et al. (1979) Proc. Natl. Acad. Sci. U.S.A. 76, 343-347.

15. Schatz, G. (1979) FEBS Lett. 103, 201-211.

16. Raymond, Y. & Shore, G.S. (1979) J. Biol. Chem. 254, 9335-9338.

17. Mori, M. et al. (1979) Proc. Natl. Acad. Sci. U.S.A. 76, 5071-5075.

18. Conboy, J.G. et al. (1979) Proc. Natl. Acad. Sci. U.S.A. 76, 5724-5727.

19. Chua, N.-H. & Schmidt, G.W. (1979) J. Cell Biol. 81, 461-483.

20. Blobel, G. & Dobberstein, B. (1975) J. Cell Biol. 67, 835-851 and 852-862.

The Organization and Expression of the Mitochondrial Genome
A.M. Kroon and C. Saccone, editors

A MATRIX-LOCALIZED MITOCHONDRIAL PROTEASE PROCESSING CYTOPLASMICALLY-MADE PRECURSORS TO MITOCHONDRIAL PROTEINS

PETER BOEHNI, SUSAN GASSER, CHRIS LEAVER AND GOTTFRIED SCHATZ
Biocenter, University of Basel, CH-4056 Basel, Switzerland

INTRODUCTION

Studies in several laboratories[1,2,3] have shown that mitochondrial proteins which are made on cytoplasmic ribosomes under the direction of nuclear genes are imported into mitochondria by a process which is different from "vectorial translation". Mitochondrial polypeptides destined to be imported into mitochondria are initially made on free polysomes[3,4], in many cases as precursors which are between 2000 and 6000 daltons larger than the mature polypeptide[2,3,5]. The completed polypeptides are then released into a cytoplasmic pool and transported into the mitochondria by a mechanism that is dependent on energy-rich phosphate bonds[6] and independent of concomitant protein synthesis[1,2]. If a polypeptide is initially made as a larger precursor, this precursor is cleaved to the mature form immediately prior to, during, or after the import step[2]. Table I lists some of the precursors to mitochondrial proteins in Saccharomyces cerevisiae that have been identified in our laboratory.

As previously shown by Neupert and his colleagues for Neurospora[14,15] no larger precursor form can be detected for cytochrome c and the adenine nucleotide carrier. This is true in yeast as well (Table I). Individual cytoplasmically-made subunits of oligomeric enzymes such as F_1-ATPase or cytochrome c oxidase are not made as polyproteins (contrast 16) but as discrete precursors that are only 2000-6000 daltons larger than the mature subunits[10,12].

The molecular mechanism of mitochondrial protein import is unknown. Here we describe a mitochondrial protease that cleaves cytoplasmically-made precursors to their mature size. Although the available evidence is not yet conclusive, it suggests that this protease is part of the enzymic machinery for transporting polypeptide precursors from the cytosol into the mitochondrial matrix.

TABLE 1

PRECURSORS TO CYTOPLASMICALLY-MADE MITOCHONDRIAL PROTEINS IN SACCHAROMYCES CEREVISIAE

Polypeptide	Location in mitochondria	Approximate molecular weight difference between precursor and mature polypeptide (daltons)	Posttranslational import demonstrated	Reference
F_1-ATPase α-subunit	Matrix	6000	yes	7
β-subunit		2000	yes	7
γ-subunit		6000	yes	7
Mn^{++}-superoxide dismutase		2000	no	8
Cytochrome bc_1-complex subunit V	Inner membrane	2000	no	9
Cytochrome $\underline{c}_1$		4000	no	6
Cytochrome c oxidase subunit V		2000	yes	10,12
subunit VI		6000	no	10,12
Cytochrome c		0	no	11
Adenine nucleotide carrier		0	no	12
Cytochrome c peroxidase	Intermembrane space	6000	yes	13

RESULTS

As reported earlier, a reticulocyte lysate programmed with yeast RNA synthesizes most yeast mitochondrial proteins as larger precursors which can be converted to their mature size by adding unlabeled yeast mitochondria[2]. This processing activity can be almost quantitatively extracted from the mitochondria by hypotonic buffers (Fig. 1, tracks 3, 4); it is sensitive to mM concentrations of the metal chelators o-phenanthroline (Fig. 1, track 5) and EDTA (Fig. 2, tracks 3,4,5). No significant inhibition is observed with phenylmethylsulfonyl fluoride (PMSF), diisopropyl fluorophosphate (DFP), tosyl-L-lysyl-chloromethylketone, tosyl-L-phenylalanyl-chloromethylketone, pepstatin or EGTA. ATP (5 mM) inhibits slightly, perhaps because of its metal chelating properties (Fig. 2, tracks 9 and 10). A similar protease is recovered in hypotonic extracts of mitochondria from rat liver (Fig. 3), rat heart and germinating maize seeds (not shown).

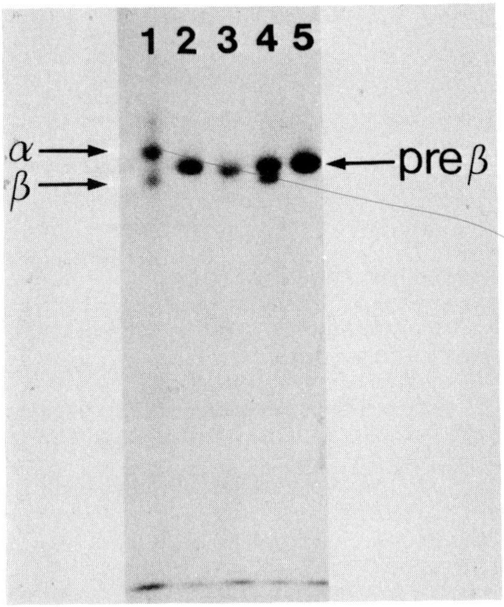

Fig. 1. A hypotonic extract of yeast mitochondria contains an o-phenanthroline-sensitive protease processing cytoplasmically-made precursors to mitochondrial proteins. Yeast mitochondria were incubated for 30 min at 0° with 10 mM Tris-Cl pH 7.4. The mixture was centrifuged and supernatant and pellet were each incubated with a reticulocyte lysate that had been labeled with $\{^{35}S\}$methionine in the presence of yeast mRNA. The mature and the precursor forms of the F_1-ATPase β-subunit were isolated by immunoprecipitation, separated by SDS-polyacrylamide gel electrophoresis and visualized by fluorography. Track 1, mature α- and β-subunit standards[7]; track 2, precursor to ATPase β-subunit; track 3, β-subunit immunoprecipitated from lysate that had been incubated with mitochondrial residue after hypotonic extraction; track 4, same as track 3, but after incubation of hypotonic extract; track 5, same as track 4, but in the presence of 2 mM o-phenanthroline.

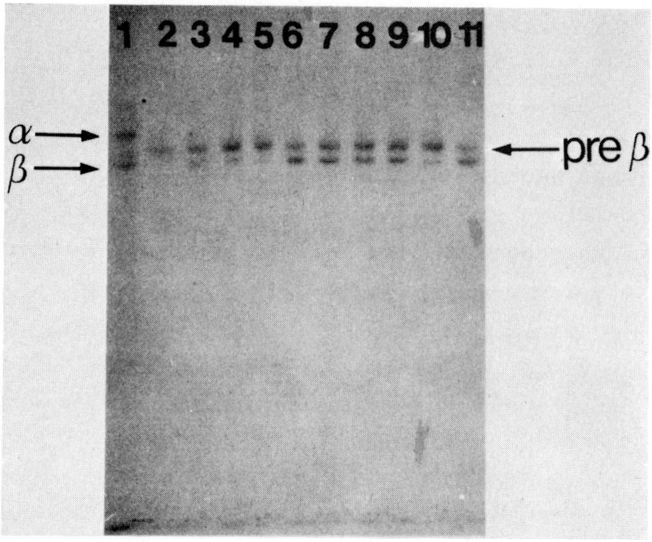

Fig. 2. The mitochondrial processing protease is sensitive to EDTA, but not to EGTA. The experiment was essentially identical to that described in Fig. 1. Track 1, standards of mature α- and β-F_1-ATPase subunits; F_1-ATPase β-subunit precursor; track 11, β-subunit immunoprecipitated from a reticulocyte lysate that had been incubated with a hypotonic extract of yeast mitochondria. The samples run on the remaining tracks shown in the Figure are derived from samples that had been treated identically to that shown in track 11 except that incubation of the lysate with hypotonic extract was carried out with the following additions: Track 3, 0.1 mM EDTA; track 4, 0.5 mM EDTA; track 5, 1 mM EDTA; track 6, 0.1 mM EGTA; track 7, 0.5 mM EGTA; track 8, 1 mM EGTA; track 9, 1 mM ATP; track 10, 5 mM ATP.

The enzyme processed the precursors to the three largest F_1-ATPase subunits and to cytochrome c oxidase subunit V; it was completely inactive against all non-mitochondrial proteins tested so far. These non-mitochondrial proteins (bovine serum albumin, mouse IgG, yeast hexokinase and yeast tryptophan synthetase, either native or denatured by heat or high pH) were iodinated to 10^8-10^{10} cpm/mg, incubated with the crude extract from yeast mitochondria and assayed for proteolytic cleavage by SDS-polyacrylamide gel electrophoresis followed by radioautography. Since yeast tryptophan synthetase is extremely susceptible to proteolysis[17] and because of the high sensitivity of the test system employed, the results strikingly underscore the specificity of the mitochondrial protease described here.

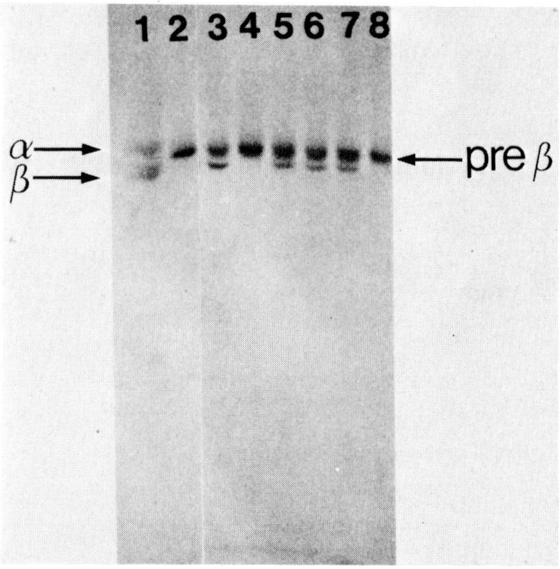

Fig. 3. Solubilization of an o-phenanthroline-sensitive processing protease from rat liver mitochondria. The experiment was analogous to the experiments shown in Figures 1 and 2 except that rat liver mitochondria were hypotonically extracted. The extract was tested for conversion of F_1-ATPase β-subunit precursor to the mature β-subunit. Tracks 1 and 2, same as in Fig. 2; track 3, hypotonic extract (440 μg protein). The samples shown on the remaining tracks were treated identically as the sample of track 3 except that incubation of the lysate with the hypotonic extract was carried out with the following additions: Track 4, 2 mM o-phenanthroline; track 5, 2 mM PMSF; track 6, 2 mM DFP; track 7, 2 mM PMSF and 60 μM pepstatin; track 8, 2 mM o-phenanthroline, 2 mM PMSF and 60 μM pepstatin.

No significant processing of yeast mitochondrial precursors to their mature form was observed with (a) purified samples of the yeast proteinases A and B, (b) a yeast aminopeptidase or (c) with frozen-thawed preparations of purified yeast vacuoles. These preparations degraded the *in vitro* synthesized precursors to small fragments which migrated off the SDS polyacrylamide gels. The proteolytic activity of yeast vacuoles was largely abolished by a combination of phenylmethylsulfonyl fluoride and pepstatin, but not by o-phenanthroline (Fig. 4). In contrast, virtually all of the processing activity of isolated yeast mitochondria was blocked by 2 mM o-phenanthroline (not shown). These data make it highly unlikely that the protease extracted from mitochondrial preparations originates from contaminating vacuoles.

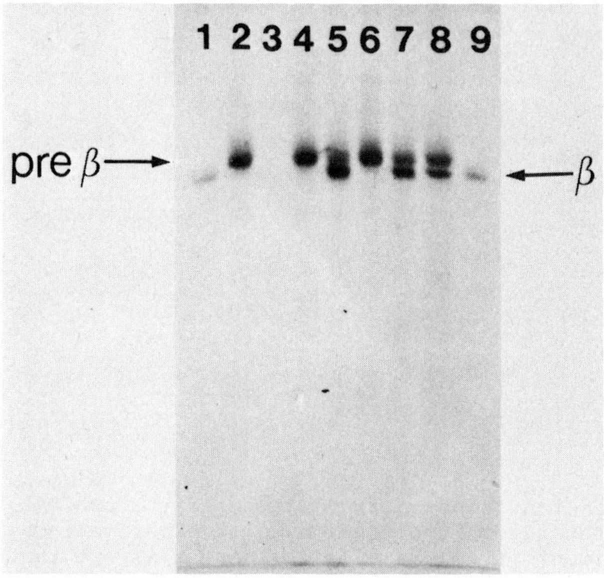

Fig. 4. The o-phenanthroline-sensitive processing activity extracted from yeast mitochondria does not originate from contaminating vacuoles. Isolated yeast vacuoles or a hypotonic extract of yeast mitochondria were tested for the processing of *in vitro* synthesized F_1-ATPase β-subunit precursor as described in the preceding Figures. Track 1, mature F_1-ATPase β-subunit; tracks 2-8, β-subunit immunoprecipitated from a labeled reticulocyte lysate after incubation with the following: Track 2, nothing; track 3, 5 ng yeast vacuoles; track 4, 5 ng vacuoles plus 60 μM pepstatin plus 2 mM PMSF; track 5, 150 μg mitochondrial extract per ml; track 6, 150 μg mitochondrial extract per ml plus 2 mM o-phenanthroline; track 7, 150 μg mitochondrial extract per ml plus 60 μM pepstatin plus 2 mM PMSF; track 8, 150 μg mitochondrial extract per ml plus 120 μM pepstatin plus 4 mM PMSF. Track 9 displays a sample of mature F_1-ATPase β-subunit.

If the o-phenanthroline-sensitive mitochondrial protease is indeed involved in the processing of mitochondrial precursors *in* vivo, then o-phenanthroline should induce the accumulation of these precursors in intact cells. As Fig. 5 shows, this is indeed the case. Unfortunately, o-phenanthroline also strongly inhibits overall protein synthesis so that the results of this experiment must be interpreted with caution.

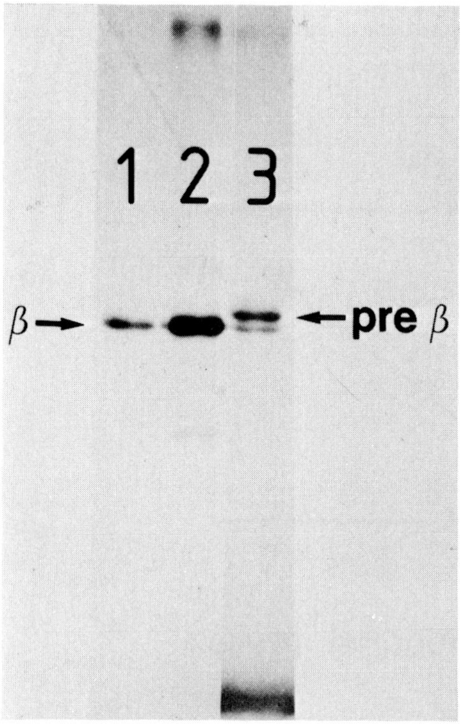

Fig. 5. o-phenanthroline causes the in vivo accumulation of the precursor to the F_1-ATPase β-subunit. Yeast spheroplasts were pulse-labeled with {^{35}S}methionine[9] in the absence or presence of o-phenanthroline, lysed in SDS and subjected to immunoprecipitation with an antiserum against F_1-ATPase β-subunit. Track 1, mature β-subunit standard; track 2, immunoprecipitate from spheroplasts labeled in the absence of o-phenanthroline; track 3, same as track 2, but after labeling in the presence of 2 mM o-phenanthroline. Because of the low amount of radioactivity recovered in the sample labeled in the presence of o-phenanthroline, the sample shown in track 3 was radioautographed five times longer than the samples shown in tracks 1 and 2.

Since the mitochondrial protease failed to cleave all non-mitochondrial proteins tested so far, it must be assayed with in vitro synthesized radioactive mitochondrial precursors as substrates. The assay is as follows. In a first step, a reticulocyte lysate programmed with yeast mRNA is allowed to synthesize yeast proteins with {^{35}S}methionine as labeled precursor. In a second step, the labeled lysate is incubated with the protease. In a third step, an antibody against a cytoplasmically-made mitochondrial polypeptide (such as the F_1-ATPase β-subunit) is added and the in vitro synthesized precursor form as well as any mature form produced by the protease are isolated by immunoprecipitation and separated from each other by SDS-polyacrylamide gel electrophoresis. In a final step, the precursor and mature forms of the polypeptide are visualized by radioautography and the extent of processing is quantitated by scanning the suitably exposed X-ray films. Despite the inherent pitfalls of this assay, it can be used to quantitate the enzyme (Fig. 6). We have defined one unit as the amount of enzyme which processes 50% of the precursor of F_1 β-subunit under our specified assay conditions.

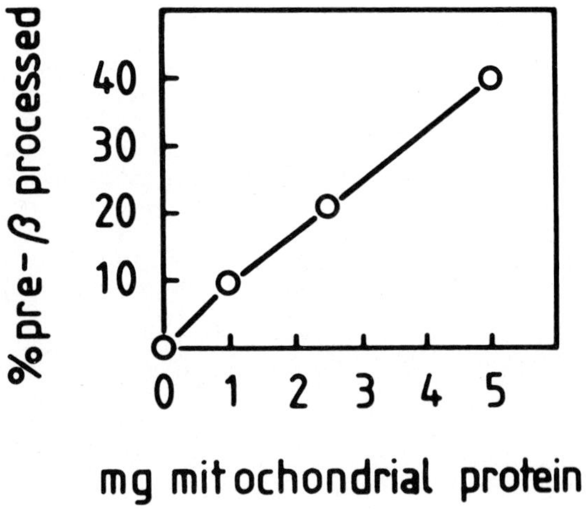

Fig. 6. Relationship between the amount and the observed activity of the mitochondrial processing protease in yeast mitochondria.

Where in the mitochondrion is the protease located? This question is difficult to answer with yeast mitochondria since there exist no procedures for separating these organelles into inner and outer membrane as well as into matrix and intermembrane space compartments. However, such methods have been carefully worked out for rat-liver mitochondria[18]. Since rat liver mitochondria contain a protease closely resembling that extracted from yeast mitochondria (Fig. 3), the intramitochondrial localization of the protease was investigated in rat-liver. The outer membrane and the intermembrane contents were removed by the following three different procedures: exposure to digitonin, passage through a French pressure cell, or mild hypotonic shock. The "mitoplasts" obtained by the first two procedures were further disrupted by freeze-thaw cycles in hypotonic buffers or by mild hypotonic shock. Each of the mitochondrial subfractions was then assayed for mitochondrial marker enzymes as well as for o-phenanthroline-sensitive processing of the precursor to the F_1 β-subunit. The results are plotted in Fig. 7; it is clear that the protease is localized in the mitochondrial matrix. In addition, rat liver mitochondria contain an-

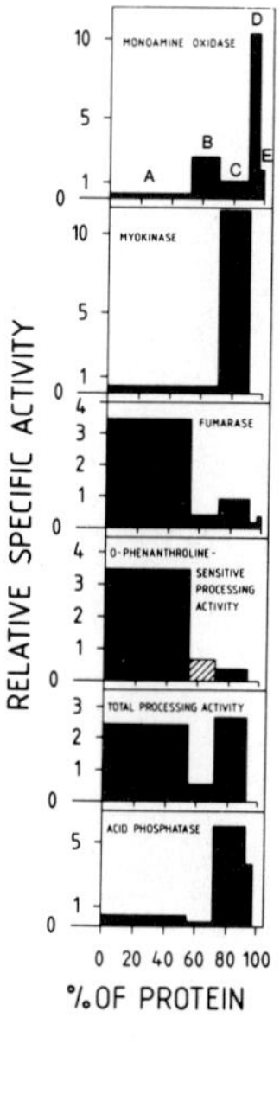

Fig 7. The o-phenanthroline-sensitive processing protease of rat liver mitochondria is located in the matrix space. Rat liver mitochondria were fractionated into their various components by standard procedures[18] and each of the resulting fractions was assayed for appropriate marker enzymes as well as processing activity towards the yeast F_1-ATPase β-subunit precursor in the absence and presence of 2 mM o-phenanthroline. The ordinate denotes the relative specific activity of the respective enzyme in each fraction (specific activity of intact mitochondria = 1). The abscissa denotes the percentage of total homogenate protein contributed by each of the following subfractions: A, matrix; B, inner membrane; C, intermembrane space; D, outer membrane; E, "fluffy layer". The shaded bar represents an upper estimate, the actual value probably being much lower.

other processing activity which is recovered together with the intermembrane contents. This activity, however, does not produce sharp bands of mature polypeptides, is sensitive to phenylmethylsulfonyl fluoride and insensitive to o-phenanthroline. Most likely it originates from contaminating lysosomes whose internal contents cofractionates with the mitochondrial intermembrane space contents. Indeed, the lysosomal marker enzyme acid phosphatase was largely recovered in the "intermembrane space" fraction (Fig. 7). Preparations of rat heart mitochondria (which are essentially free of lysosomes) contained neither the o-phenanthroline-insensitive processing activity nor the lysosomal marker acid phosphatase, but still exhibited the o-phenanthroline-sensitive processing activity (not shown).

So far, all the data described in this preliminary report are consistent with the notion that we have identified a mitochondrial protease that functions in mitochondrial protein import. We are now purifying the protease and are trying to determine whether it processes the mitochondrial precursors at the correct sites within their amino acid sequence.

ACKNOWLEDGEMENTS

This study was supported by grants 3.212.77 and 3.172.77 from the Swiss National Science Foundation and by an EMBO Long-Term Fellowship to Chris Leaver.

REFERENCES

1. Hallermayer, G., Zimmermann, R. and Neupert, W. (1977) Eur. J. Biochem., 81, 523-532.
2. Schatz, G. (1979) FEBS Letters, 103, 203-211.
3. Raymond, Y. and Shore, G.C. (1979) J. Biol. Chem., 254, 9335-9338.
4. Suissa, M. and Schatz, G., in preparation.
5. Michel, R., Wachter, E. and Sebald, W. (1979) FEBS Letters, 101, 373-376.
6. Nelson, N. and Schatz, G. (1979) Proc. Natl. Acad. Sci. USA, 76, 4365-4369.
7. Maccecchini, M.-L., Rudin, Y., Blobel, G. and Schatz, G. (1979) Proc. Natl. Acad. Sci. USA, 76, 343-347.
8. Autor, A. and Schatz, G., in preparation.
9. Côté, C., Solioz, M. and Schatz, G. (1979) J. Biol. Chem., 254, 1437-1439.
10. Lewin, A., Gregor, I., Mason, T.L., Nelson, N. and Schatz, G. (1980) Proc. Natl. Acad. Sci. USA, in press.

11. Gasser, S. and Schatz, G., in preparation.

12. Nelson, N. and Schatz, G. (1979) in: Membrane Bioenergetics (Lee, C.-P., Schatz, G. and Ernster, L., eds.) Addison-Wesley, Reading, pp. 133-152

13. Maccecchini, M.-L., Rudin, Y. and Schatz, G. (1979) J. Biol. Chem., 254, 7468-7471.

14. Korb, H. and Neupert, W. (1978) Eur. J. Biochem., 91, 609-620.

15. Zimmermann, R., Paluch, U., Sprinzl, M. and Neupert, W. (1979) Eur. J. Biochem., 99, 247.

16. Poyton, R.O. and McKemmie, E. (1979) J. Biol. Chem., 254, 6763-6771 and 6772-6780.

17. Dettwiler, M. and Kirschner, K. (1979) Eur. J. Biochem., 102, 159-165.

18. Greenawalt, J.W. (1979) Methods Enzymol., 40, 88-98.

The Organization and Expression of the Mitochondrial Genome
A.M. Kroon and C. Saccone, editors

PROTEASE AND INHIBITOR RESISTANCE OF ASPARTATE AMINOTRANSFERASE SEQUESTERED IN MITOCHONDRIA AND THE FCCP-DEPENDENCE OF ITS UPTAKE

E. MARRA, S. PASSARELLA, S. DOONAN[o], E. QUAGLIARIELLO and C. SACCONE
Centro di Studio sui Mitocondri e Metabolismo Energetico, Istituto di Chimica Biologica, Università di Bari, Italy;
[o]Department of Biochemistry, University College, Cork, Ireland.

INTRODUCTION

We have previously used three different approaches to demonstrate that isolated rat liver mitochondria are permeable to mitochondrial aspartate aminotransferase (mAAT) but not the cytoplasmic isozyme[1,2]. In one approach intramitochondrial AAT activity was measured by a fluorescence method and shown to increase after incubation of the organelles with added mAAT. In our view, results obtained by this approach demonstrate clearly that the enzyme is taken up from suspension and translocated to its site of action in the matrix. However we have considered it worthwhile to confirm this conclusion by direct demonstration that the increase in observable activity is insensitive to an added protease and to a non-permeant inhibitor.

RESULTS AND DISCUSSION

The experimental technique used was essentially as described previously[1] with changes in conditions as noted in the legends to the figures. In the experiment of Fig. 1 mitochondria were incubated in the absence (A) or in the presence (C) of pronase before measurement of intramitochondrial enzyme activity. Incubation with pronase did not affect the integrity of the organelles and it was shown that no internal AAT had leaked into the incubation medium in this time. In B, mitochondrial isozyme (8 μg) was added at the beginning of the 20 min period and resulted in a 30% increase in rate of NADH oxidation as in our previous experiments. In D pronase was added 1 min after addition of mAAT and the rate of change in fluorescence was measured 19 min later. During these 19 mins, the remaining external AAT was inactivate by 30% by action of the protease whereas the amount of enzyme taken up did not decrease (comparison of B and D). These results show that whereas external mAAT is susceptible to proteolysis, it is protected after entry into the organelle.

Fig. 2 shows the results of a similar experiment, but using a chemical inhibitor (p-hydroxymercuriphenylsulphonic acid (PHMPS)) to inactivate the

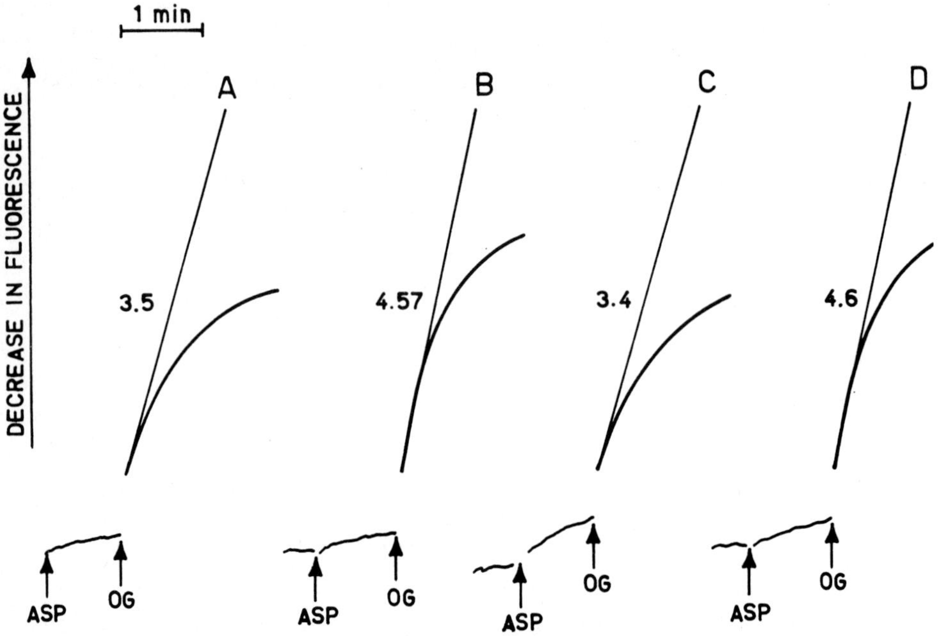

Fig. 1. Insensitivity of incorporated enzyme to pronase. The mitochondria were preincubated for 10 min either with no additions (A), with 270 µg ml^{-1} of pronase (C), with 8 µg of mAAT (B) or with 8 µg of mAAT followed 1 min later by 270 µg ml^{-1} of pronase (D). Substrates were then added in the sequence shown and the resulting fluorescence changes were measured.

remaining external enzyme. It has been shown[3] that this inhibitor at the concentration used here almost completely blocks uptake of the enzyme and decreases its activity by 50%; this observation was confirmed in the present work. Comparison of Fig. 2A and C shows that addition of PHMPS to the mitochondrial suspension did not decrease the intramitochondrial activity (since the inhibitor is non-permeant[4]). Similarly comparison of B and D demonstrates that PHMPS was ineffective in preventing the increased rate of change of fluorescence resulting from externally added enzyme.

In summary the results cited above show that enzyme taken up into mitochondria is inaccessible both to a protease and to a non-permeant inhibitor.

One surprising aspect of our previous results[2] was the apparent failure

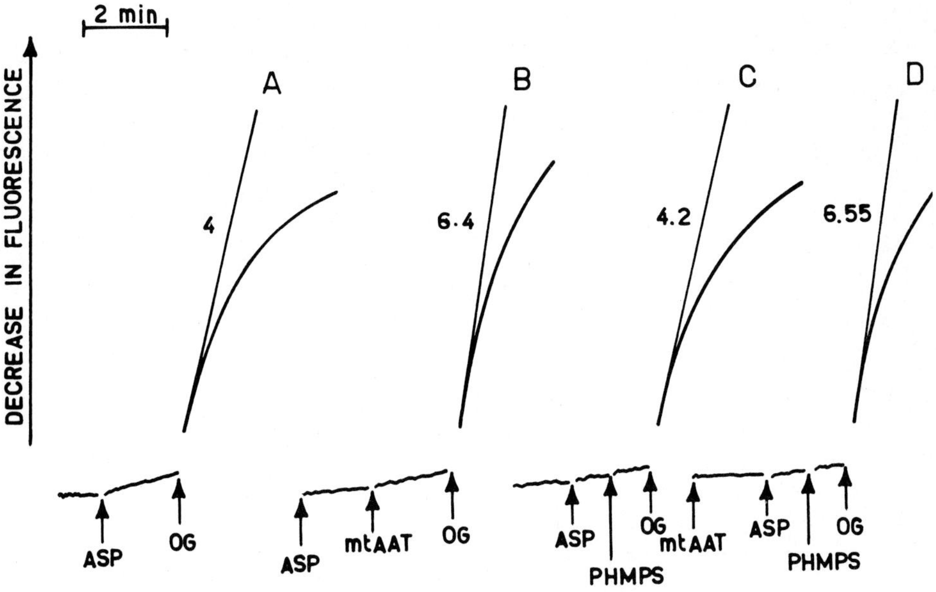

Fig. 2. Insensitivity of incorporated enzyme to PHMPS. Experimental conditions were the same as in Fig. 1 except that the time of preincubation was only 3 min before addition of aspartic acid. The quantity of mAAT was 18 µg and the concentration of PHMPS in C and D was 95 µM.

of the uncoupler FCCP to inhibit uptake of mAAT into mitochondria. In the light of previous results[5,6] we have reexamined this problem and have found that, under appropriate conditions, a complete inhibition of the uptake can be observed. The results are given in table 1. Addition of FCCP 1 min before addition of mAAT completely blocked uptake of the enzyme. On the other hand uncoupler added after addition of the enzyme did not prevent uptake as shown by the 33% increase in the value of V_{FCCP} compared with the control. It is unclear why the addition of FCCP in the absence of external enzyme resulted in a decrease in the rate of change of fluorescence compared with the control; this was not a direct effect of FCCP on the enzyme. What is clear however is that pretreatement of mitochondria with uncoupler abolished their ability

TABLE 1

THE EFFECT OF THE UNCOUPLER FCCP ON THE PERMEATION OF MITOCHONDRIAL ASPARTATE AMINOTRANSFERASE INTO RAT LIVER MITOCHONDRIA IN VITRO.

Mitochondria were incubated at 23°C in 2 ml of a solution of 0.25 M sucrose, 20 mM Tris-HCl pH 7.25, 1 mM EDTA, containing rotenone (2 μg) and 1 mM sodium arsenite. After 3 min the additions indicated were successively made at the following concentrations: aspartate (ASP) 12 mM; oxoglutarate (OG) 2.8 mM; mitochondrial aspartate aminotransferase (mAAT) 10 ug; carbonyl cyanide p-trifluoromethoxyphenylhydrazone (FCCP) 1 μM. The rate of decrease in fluorescence of the intramitochondrial NADH (V) was then recorded and expressed in arbitrary units. V_{FCCP} indicates the rate of decrease in fluorescence in the presence of FCCP. The amount of mitochondrial protein was 2 mg.

ADDITIONS						V	V_{FCCP} (% control)
	T=0	+30"	+30"	+1'	+1'		
MIT	ASP	-	OG	-	-	100	-
MIT	ASP	mAAT	OG	-	-	133	-
MIT	ASP	-	FCCP	-	OG	72	100
MIT	ASP	-	FCCP	mAAT	OG	70	97
MIT	ASP	mAAT	FCCP	-	OG	96	133

to take up externally added enzyme. Since rotenone was present, this was probably an effect of FCCP on ATP hydrolysis or the transmembrane pH gradient.

REFERENCES

1. Marra, E., Doonan, S., Saccone, C. and Quagliariello, E. (1977) Biochem. J., 164, 685-691.
2. Marra, E., Doonan, S., Saccone, C. and Quagliariello, E. (1978) Eur. J. Biochem., 83, 427-435.
3. Marra, E., Passarella, S., Doonan, S., Saccone, C. and Quagliariello, E. (1979) Arch. Biochem. Biophys., 195, 269-279.
4. Fonyo, A. (1978) J. of Bioen. and Biomemb., 10, 171-194.
5. Harmey, M.A., Hallermayer, G. and Neupert, W. (1976) in "Genetics and Biogenesis of Chloroplasts and Mitochondria" (Bücher, Th., Neupert, W., Sebald, W. and Werner, S. eds) Elsevier/North-Holland 813-818.
6. Nelson, N. and Schatz, G. (1979) Proc. Natl. Acad. Sci. USA, 76, 4365-4369.

© 1980 Elsevier/North-Holland Biomedical Press
The Organization and Expression of the Mitochondrial Genome
A.M. Kroon and C. Saccone, editors

THE ORGANIZATION AND EXPRESSION OF THE MITOCHONDRIAL GENOME: CONCLUDING REMARKS AND OUTLOOK

C. SACCONE[1] and A.M. KROON[2]

[1]Istituto de Chimica Biologica, Unviersity of Bari, Italy and

[2]Laboratory of Physiological Chemistry, University of Groningen, The Netherlands

Someone who has read the contents of these Proceedings will realise that it is rather difficult to give a concise summary of all the subjects discussed. Nonetheless we will try to conclude this book with a short chapter indicating where the conference, of which this book is the fruit, has brought us. In general we may conclude that many new and interesting data have been presented. A number of points have been settled; comparisons of similar results with different organisms can be made. It was already predicted in the introduction that much emphasis would be put on the studies of base sequences. It is clear that the powerful techniques of base-sequence analysis and DNA recombination have led us to a point of seemingly unlimited possibilities, but will not enable us to obtain the answers to all questions raised. The first two parts of the book and to a lesser extent the third part show the power of the methods mentioned. The fourth part of the book perfectly illustrates that for the understanding of the mechanisms leading to the biogenesis of mitochondria Maxam ladders and chimeric plasmid DNAs are not the only, not even the most suitable attributes.

MITOCHONDRIAL GENE ORGANIZATION

The kinetoplast DNA is certainly the most peculiar mitochondrial genome from a structural point of view. The network structure containing the maxi- and mini-circles is known for a long time already. It has now been convincingly shown (pp 7-19) for *Trypanosoma brucei* kDNA that the maxi-circle contains genetic information comparable to the information present in the other mtDNAs. The mt-ribosomal RNAs are, however, extremely small: 1080 (12S) and 590 (9S) nucleotides for the "large" and "small" rRNA respectively. This is far shorter than the length of the mammalian mt-rRNAs, which were already considered mini-RNAs. It is an interesting aim for future research to indicate how the main ribosome functions have been conserved in particles with such a low RNA content. For mammalian mitochondrial ribosomes such studies begin to give the first results (pp 301-305). It is especially interesting also in view of the homology that even the 12S and 9S *Trypanosoma* mt-rRNAs show with other mitochondrial and with *E. coli* rRNAs. Homologies between various rRNAs were also observed by others (pp 79-86, 211-219, 231-240, 277-286). The kDNA maxi-circle further contains sequences homologous to mt-poly(A)RNAs. The transcripts localized so far are, however, too small to code for polypeptides like apocytochrome *b* or the largest subunit of cytochrome *c* oxidase.

The organization of the mitochondrial genome of the lower eukaryotes is characterized by the presence of mosaic genes. The split gene structure was first shown for the large rRNA of ω^+ strains of *Saccharomyces cerevisiae*. It appears to be a characteristic of the large rRNA genes of *Neurospora crassa* and *Aspergillus niger* as well. Furthermore, some structural genes have a mosaic structure; the cytochrome *b* gene and the gene for subunit 1 of cytochrome *c* oxidase both in *Saccharomyces cerevisiae*, are the best characterized examples so far (see below). It is noteworthy that the mosaicism of the genes is not obligatory. Some of the structural genes contain the genetic information for the proteins in a continuous base sequence also in *Saccharomyces*.

The petite mutants of *Saccharomyces* have been further characterized (pp 21-31, 33-36). Excision of mtDNA occurs more frequently in the GC clusters than in the AT-rich spacers. Suppressiveness appears to be related to the presence of the origin of replication in the petite genome. The petite mtDNAs, formerly successfully used for the estimation of the gene order in yeast mtDNA via deletion mapping, have now been successfully introduced in studies with other organisms. Due to a degree of homology of mtDNA's of different organisms, well defined petite mtDNAs can be used as probes in the mitochondrial gene localization in other organisms. The method has shown its value for *Neurospora crassa* (pp 51-60) and *Kluyveromyces lactis* (pp 61-64). A striking observation in this context is the apparent homology of probes of the Oli 1 region of yeast with part of the *Neurospora* mtDNA. The DCCD-binding protein, encoded in Oli 1, is a nuclear gene product in *Neurospora*. This raises the question whether the *Neurospora* mtDNA contains a silent gene or remnants of such a gene.

For mammalian mtDNA rapid progress has been made. The data of base sequencing*

---

* Unfortunately the manuscript of the interesting contribution entitled "The organization of the mammalian mitochondrial genome" by J.E. Walker et al. was not received. The editors were notified that the manuscript had been dispatched by mail on July 1, but neither the original nor copies were received. The publisher and editors were unable to obtain the manuscript by other means, and were therefore forced to proceed with publication of the proceedings in mid-August without this contribution. The editors apologise to the other contributors for the unforseen and unwanted delay.

and transcription mapping (pp 103-119) of human mtDNA are fully consistent. It appears that the mammalian mtDNA is a very compact DNA: the bases are highly efficiently used for coding, a number of genes are butt-jointed and it is hypothesized that the stop codons of the messenger RNAs of some genes arise through a posttranscriptional event of cutting adjacent mRNAs and tRNAs with concomitant polyadenylation of the 3'end of the mRNA. The latter activity then gives rise to the formation of a TAA stop signal. The mammalian mtDNA contains a number of as yet unidentified reading frames (URFs), potentially coding for hydrophobic, membrane-type proteins. These reading frames correspond nicely with likewise still unidentified transcripts. The data for bovine (pp 125-130), mouse (pp 277-286), rat (pp 121-124, 211-219, 221-229), hamster (pp231-240) and avian (pp 131-135) mtDNA, although less complete, are all in good agreement with the data for human mtDNA.

MITOCHONDRIAL GENE CHARACTERIZATION

The best characterized mitochondrial gene is at present certainly the gene for apocytochrome *b* in yeast[+]. In a number of papers the characteristics of this gene are discussed at length (pp 37-49, 139-152, 153-156, 157-160, 161-172, 173-177, 179-180). The gene is mosaic. The number of introns (intervening sequences) may vary from strain to strain. In "short" strains there are 3 exons and 2 introns; in "long" strains there are 5 exons and 4 introns. In the latter strains the exons are indicated α, β, γ, δ and ε, the introns $I\alpha\beta$, $I\beta\gamma$, $I\gamma\delta$ and $I\delta\varepsilon$. The exons and introns correspond to the box loci 4/5 (α), 3 ($I\alpha\beta$), 8 (β), 10 ($I\beta\gamma$), 1 (γ), 7 ($I\gamma\delta$), 2 (δ) and 6 (ε) respectively. The data of the various groups show a high degree of consensus. The most interesting feature is without any doubt the presence and expression of reading frames in some of the introns. Especially for box 3 the genetic and biochemical data complement each other nicely. The intron-gene product is a so-called maturase already mentioned in the introduction. It is now postulated that the excision of the $I\alpha\beta$ intron proceeds in a two-step reaction. The first step releases the 10S circular RNA and links the αexon to the reading frame in $I\alpha\beta$. This leads to the production of the maturase responsible for the further processing and the linkage of the α to the β exon. The experiments offer a magnificent example of how posttranscriptional modification may occur and further show how nucleocytoplasmic and mitochondrial activities may be interlinked.

[+]During the conference a summary of the available data was given by Dr. L.A. Grivell

With the characterization of the mitochondrial genes for some of the cytochrome *c* oxidase subunits much progress has been made (pp 37-49, 181-190, 191-194). The Oxi 3 region, coding for subunit 1, is also particularly complex (8 exons and 7 introns). Sequence data show homology between the different introns of Oxi 3, but also between one of the Oxi 3 introns and box 7 of the apocytochrome *b* gene. Studies along these lines may well shed some more light on the interactions of various parts of the mitochondrial genome.

Apart from the genes for the well-known respiratory enzyme components and for the maturase(s) mentioned above, information about mitochondrial genes coding for polypeptides remains scanty. The var 1 gene of yeast is still lost. Nonetheless, some interesting observations have been made in relation to var 1 (pp 195-205). A 16S RNA transcript has been detected that is only partly homologous to mtDNA. To explain this the interesting possibility of coupling of nuclear and mitochondrial transcripts has been raised. This certainly deserves and needs a thorough further investigation. If true it gives another example of the complex interaction of the nuclear and mitochondrial genetic systems.

The URFs and their accompanying transcripts of human mitochondrial DNA have already been mentioned as possible genes. For yeast there are few unidentified reading frames as well, but these would correspond to polypeptides with rather odd aminoacid composition. Moreover, the use of the frames is not substantiated by the presence of transcripts. As soon as the complete base sequence of yeast and other lower eukaryotic mtDNAs will be known, the issue of the presence of other genes can be settled.

MITOCHONDRIAL REPLICATION, TRANSCRIPTION AND TRANSLATION

The mechanism of mtDNA replication is still grossly unknown. Especially for lower eukaryotes there is very little information. For the animal systems the origin of replication is more or less characterized on both strands (pp 277-286). In *Drosophila* replication is initiated at a specific point in the AT-rich region (pp 241-250). The enzymology of mtDNA replication is also poorly understood. The DNA polymerase γ seems the active enzyme in mitochondria, although a role for the DNA polymerase α cannot be fully excluded (pp 287-290). Some data on mitochondrial gyrases and topoisomerases are presented, but the picture is far from complete yet (pp 121-124). The same holds for transcription in fact, although reasonable progress has been made in the characterization of mitochondrial transcripts and their processing and modification (pp 103-119, 253-263, 265-276, 291-300).

The most conspicuous progress has been made with respect to the mechanism of translation. In the first place the divergence of the mitochondrial and the "universal" code appears not to be restricted to the UGA codon, which codes for tryptophan in mitochondria. AUA is supposed to be a start codon as well, whereas also other odd uses of codons are suggested. The small number of tRNAs in mitochondria is now considered compatible with a complete decoding of all codons in mitochondria, mainly on the basis of recognition of groups of 4 codons, differing only in the third base, by one anticodon. With respect to mt-tRNA structure it can be concluded that in the lower eukaryotes the cloverleaf structure is more or less conserved (pp 79-86, 307-310, 311-314). Although minor deviations from the classical tRNA structure occur, the dihydroU- and the TΨC-loops are present in most lower eukaryotic mt-tRNAs sequenced so far. In view of the general belief that mitochondrial ribosomes do not contain a 5S rRNA, one may seriously question the mechanism of interaction between tRNAs and ribosomes as outlined in the most biochemistry textbooks. Anyway such a mechanism will not be operative in mammalian ribosomes, because the mammalian and possibly also other animal mt-tRNAs have a rather deviating structure. Moreover, the 3S rRNA, formerly considered a 5S rRNA analogue is most likely the mt-tRNA for serine (pp 231-240).

Up to now nobody has been successful in setting up a reconstituted system for mitochondrial protein synthesis *in vitro* from mitochondrial ribosomes and factors only. Some mitochondrial messengers have been successfully translated in heterologous systems.

DEVELOPMENTAL AND REGULATORY ASPECTS OF MITOCHONDRIAL BIOGENESIS

Various experimental systems are used to study the regulation of the biogenesis of mitochondria during growth and development but also for the necessary replacement of mitochondria due to loss of mitochondria by turnover under steady-state conditions. The genetic approach remains popular and useful (pp 325-332, 333-342, 343-346, 355-363, 369-374, 383-386). The results of such studies are, however, mainly descriptive and do not self-evidently or easily lead to the understanding of the mechanisms regulating the interactions between the nucleocytoplasmic and mitochondrial genetic systems. The study of special organisms may give additional hints (pp 347-354, 365-368). 391-394, 395-398). The promising hypothesis of a coordinate repression/induction mechanism for the synthesis of mitochondrial proteins in the cytoplasm does not meet with experimental support (pp 387-390). In this respect our knowledge still shows more gaps than facts. Nonetheless, there have been made interesting achievements. Studies on the assembly of mitochondrial enzyme complexes (pp 375-381, 399-411, 413-422, 423-433) have revealed that a number, though not all cytoplasmically synthesized mitochondrial proteins are made as precursors. Furthermore, there is an energy-dependent transfer of these proteins into the mitochondria. Especially the presence and activity of a specific mitochondrial protease is interesting in this context (pp 423-433). The proposition of specific recognition points or receptors on the mitochondrial membranes for the cytoplasmically synthesized mitochondrial proteins is attractive and can be tested (pp 413-422). The observation that also some mitochondrially synthesized polypeptides have to be processed (pp 399-411 and that this processing should partly occur after assembly (pp 325-332) is a complicating factor, but also open for further experimental approach. For the studies of protein transfer through mitochondrial membranes also the aspartate aminotransferase system may be a suitable model (pp 435-438).

The results presented in the fourth part of the book show, that we are definitely but slowly approaching the heart of the matter. It may be expected for the coming years that we finally arrive at the point that we have previously thought to have reached already many times: the intense search for the nucleocytoplasmic-mitochondrial interrelations and the study of the biogenesis of mitochondria as a whole in the strict sense of the word.

SUBJECT INDEX

AUTHOR INDEX